Theory and
Analysis of
Experimental
Designs

Theory and Analysis of Experimental Designs

B. L. Agarwal

M Sc (maths), M Stat, Ph D

Ex-Professor of Statistics and University Head
Department of Statistics and Mathematics
Rajasthan Agricultural University
RCA Campus, Udaipur
Rajasthan

CBS

CBS Publishers & Distributors Pvt Ltd

New Delhi • Bangalore • Pune • Cochin • Chennai

ISBN : 978-81-239-1842-6

First Edition : 2010
Reprint : 2011

Published by Satish Kumar Jain and produced by V.K. Jain for
CBS Publishers & Distributors Pvt. Ltd.,
CBS Plaza, 4819/XI Prahlad Street, 24 Ansari Road, Daryaganj,
New Delhi - 110002, India.   • Website: www.cbspd.com
e-mail: delhi@cbspd.com, cbspubs@vsnl.com, cbspubs@airtelmail.in
Ph.: 23289259, 23266861, 23266867   • Fax: 011-23243014

Branches:
• *Bengaluru:* Seema House, 2975, 17th Cross, K.R. Road,
  Bansankari 2nd Stage, Bengaluru - 560070   Ph.: +91-80-26771678/79
  Fax: +91-80-26771680   • E-mail: cbsbng@gmail.com,
  bangalore@cbspd.com
• *Pune:* Bhuruk Prestige, Sr. No. 52/12/2+1+3/2,
  Narhe, Haveli (Near Katraj-Dehu Road By-pass), Pune - 411051
  Ph.: +91-20-64704058/59, 32342277   • E-mail: pune@cbspd.com
• *Kochi:* 36/14, Kalluvilakam, Lissie Hospital Road,
  Kochi - 682018, Kerala • Ph.: +91-484-4059061-65
  Fax: +91-484-4059065   • E-mail: cochin@cbspd.com
• *Chennai:* 20, West Park Road, Shenoy Nagar, Chennai - 600030
  Ph.: +91-44-26260666, 26208620   • Fax: +91-44-42032115
  E-mail: chennai@cbspd.com

Printed at :
India Binding House, Noida (UP)

to

my mentor

Dr Vijay K. Goyal
Professor of Medicine
R N T Medical College, Udaipur

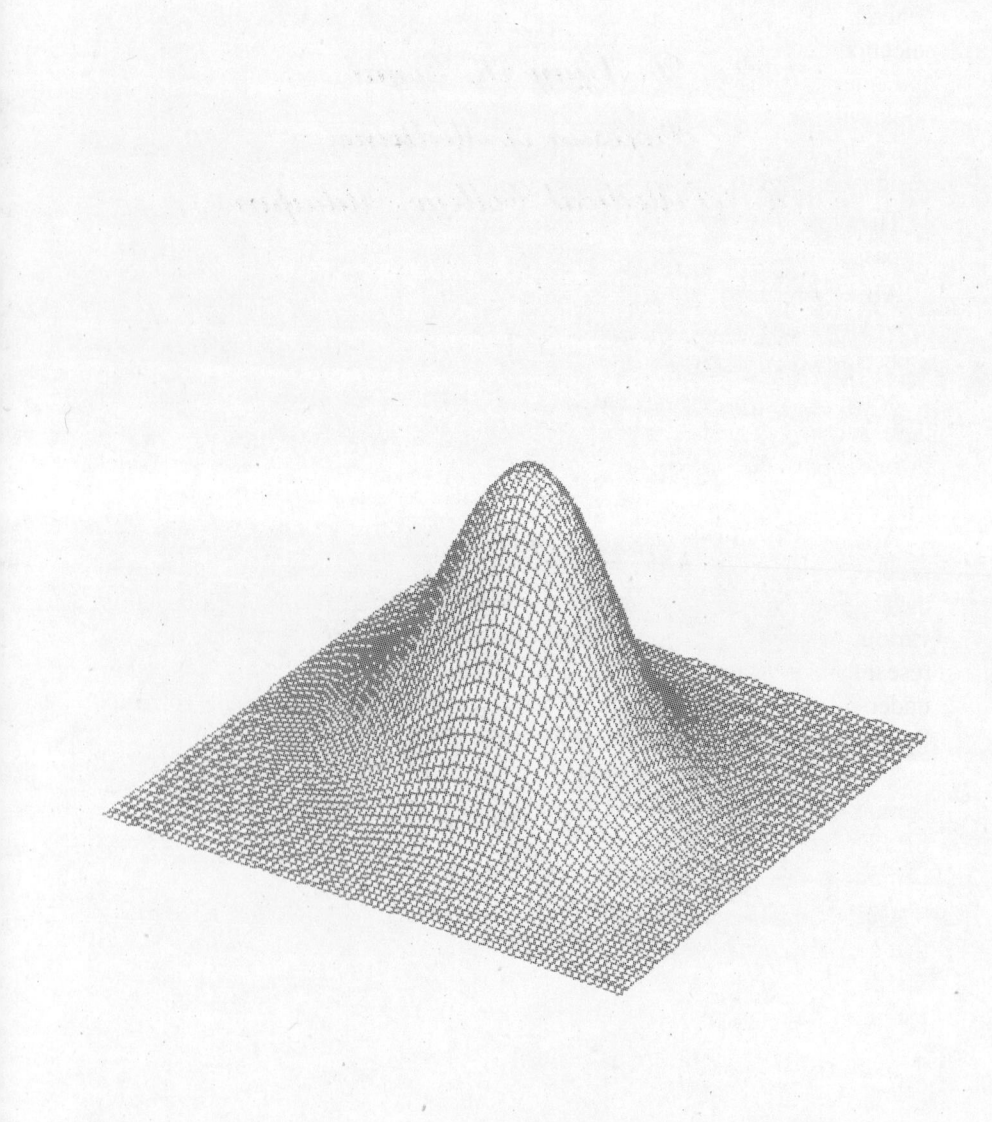

Preface

Statistics is the backbone of all researches and experimental designs are its essential part. This book *Theory and Analysis of Experimental Designs* is meant to fulfill the requirement of all students, teachers and researchers in general. There are five steps in broader sense of experimental designs, namely, selection of design, layout of experiment, collection of data, analysis of data, and interpretation of results. All these aspects are suitably covered in this book without dragging the readers into complex mathematical derivations. The matter in this book on experimental designs equips one for knowing the designs in an exact and clear manner.

There are a large number of books on experimental designs but this book surpasses other books in respect of the following:

All concepts are given without ambiguity in a simple language which provides easily digestible matter. The concepts are substantiated by their applied aspects.

Some old books lack recent advances and they are outdated. At the same time, some recent books contain so much matter which can hardly be covered in one or two academic years. Also, there are a number of topics which neither form the part of syllabi nor have any virtual utility in practice.

Analysis of data is an essential part of the experiments. Therefore, procedures of data analysis are given step-by-step first in general and thereby these procedures are illustrated by analyzing experimental data collected from various sources in cases wherever possible. This will enable the students and researchers to analyze their data independently without difficulty. Unless one understands the analysis part fully, he/she cannot interpret the results correctly. So knowing analysis is a must which one will get aptly in this book.

This is a computer age, hence any recent book should have an element of computer analysis. In view of this, the procedure for analyzing data using SPSS package has been illustrated through Windows exposure for some commonly used experimental designs.

Questions and exercises are adequately given at the end of each chapter so that a reader can check his/her understanding and grasp of the matter covered inside. This will further enable the students to perform well in their examinations and research work.

Bibliography is given at the end of each chapter adequately keeping in view the availability and requirement to study further the topics covered in the chapter.

Analysis of certain examples using SPSS package has been displayed through Windows exposure in appendix of Chapter 2. This also works as verification of results obtained from calculation on a scientific calculator.

The level of presentation of material is intended to make this book useful to all students, teachers and researchers of various science disciplines. Hope, this book will serve better than any other book.

Dr. B. L. Agarwal

Acknowledgements

I am grateful to Dr S.P. Agarwal, Ex-Professor, CCS Agricultural University, Hissar, for going through the entire manuscript and giving his views and suggestions. Also, I thank all my colleagues of Department of Agricultural Statistics and some other departments of Rajasthan College of Agriculture, Udaipur, for their help in succor. I especially thank Dr Yatin Mehta for his substantial help.

My thanks are also due to Dr Bhawna Agrawal, Assistant Professor of Statistics, IILM, New Delhi, for her help in analyzing a few problems on SPSS package and providing window exposures. I thank my grandson Tarun Goyal, a student of class XI, for all time help in computer work. At last I thank my wife.

I gratefully acknowledge and thank the permissions team of Wiley–Blackwell Publishers for granting me permission to reproduce Duncan's table from *Biometrics*, 1955.

Dr. B. L. Agarwal

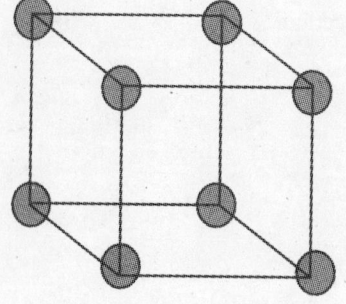

Contents

Appendices

Appendix [A] Introductory Matrix Algebra   443–455

Appendix [B] Statistical Tables   456–489

1

Basics of Experimental Designs

1.1 INTRODUCTION

Scientific researches and investigations are carried through experiments and surveys. This chapter confines to the discussion on experiments or trials and their analysis.

For meaningful, reliable and valid results and conclusions, an investigator comes across two intertwined problems namely, the designing of experiments and analysis of data. Both these aspects are to be studied side by side.

Designing of experiments means planning an experiment for assigning treatments or casual events under investigation to individuals or experimental units in such a way that the nuisance variables (extraneous conditions) are controlled as much as possible.

The procedure of designing an experiment consists of three phases: (*a*) choice of treatments, (*b*) choice of experimental units and (*c*) deciding which treatment to apply to which experimental unit.

Another facet coupled with experimental designs is analysis of data. Analysis part encompasses estimation of parameters and testing of hypotheses about the parameters involved in the statistical model for the experimental design. Each design has a specific statistical model which shows the relationship of the response variable with parameters due to independent factors or variables and an error term. Parameters are mostly estimated using measurement data by the method of least squares whereas testing of hypotheses is carried through *analysis of variance* (ANOVA). In the process of analysis of variance, the total variation present in the response variable for a characteristic or trait under study is split into component variances for the factors included in the model and the hypotheses are tested by F-test. Analysis of variance is always presented in the form of a table, generally abbreviated as ANOVA table. The skeleton of ANOVA table is presented on next page.

Table 1.1.1 Skeleton ANOVA table

Source of variation	Degrees of freedom	Sum of squares	Mean sum of squares	F- value
Source or due to	d.f.	S.S.	M.S.	F - value
A B C | | | Error				
Total				

Factors A, B, C, etc. depend on the model of the experiment. Degree of freedom for any factor in ANOVA is equal to the number of experimental units assigned to a factor minus one. Sum of square is the quantity,

$$\sum_i \frac{X_i^2}{n_i} - \frac{\left(\sum_i X_i\right)^2}{\sum_i n_i} \qquad (1.1.1)$$

where, X_i is the sum of n_i responses for the ith level of a factor which is obtained by dividing the sum of square by its corresponding degrees of freedom. Mean sum of square actually for an effect actually represents the variance due to that factor. F-value is the ratio of the mean sum of square due to a factor and error mean sum of square. F has degrees of freedom (d.f.) υ_1 and υ_2 where, υ_1 and υ_2 are d.f. for mean sum of square in the numerator and denominator, respectively. The decision about the acceptance or rejection of null hypothesis about parameter(s) is taken by comparing the calculated F-value with the tabulated F-value for (υ_1, υ_2) degrees of freedom at a prefixed level of significance α. If calculated F is more than the tabulated value, null hypothesis is rejected otherwise it is acceptable.

Prior to furnishing the details of various experimental designs and their analysis, it seems germane to define certain terms and provide some basics.

Definitions

Experimental unit: This is that entity or subject to which a treatment is assigned randomly which is an independent variable. This entity may be an individual or a group of individuals that may consist of either animates or inanimate. For instance, field plots, animals, patients, machines, etc.

Treatment: It is a substance or item or condition that is assigned to experimental unit to see its effect(s) as a response with regard to some trait(s) or characters. For example, fertilizers, medicines, instruments, methods of application of a substance, dates of sowing, etc.

Nuisance variables: These are unwanted sources of variation in an experiment that affect the response variable(s), e.g. fertility gradient of field plots, temperament of persons, resisting power of insects, etc.

Underlying assumptions for analysis of variance

Analysis of variance involves F-test which is based on certain assumptions. When these assumptions hold good, the results are likely to be valid and reliable. Otherwise the inferences drawn about the parameters remain in jeopardy. Assumptions are given below without describing the methods of testing their validity.

1. Observations have come from normally distributed population.
2. Observations constitute a random sample or in other words, the treatments are assigned to the experimental units randomly.
3. The factors of the model are independent and additive, i.e. the variances in the numerator and denominator of F-statistic are independent.
4. All experimental errors e_{ij} are identically and independently distributed (I.I.D.) normally with mean zero and same variance σ_e^2.
5. The model chosen for the design is appropriate in the sense that it contains all sources of variation that affect the ultimate response.

For in-depth discussion of the above assumptions, the readers are referred to Cochran (1947); Glass, Peckham and Sanders (1972); Rogan and Kaselman (1977); Tomerken and Serlin (1986).

1.2 BASIC REQUIREMENTS FOR A GOOD EXPERIMENTAL DESIGN

Basic requirements of an experimental design are replications, blocking (local control), and randomization. In literature, R.A. Fisher is traditionally credited for emphasizing, if not introducing, the need of replications, randomization and local control. Replication and blocking increase precision in experiments whereas randomization decreases bias and ensures the validity of estimates and tests. This also enables the experimenter to make probability statements.

Replication: Number of times a treatment is repeated on experimental units in an experiment is called its replication. In other words, the number of subjects receiving the same treatment provides the number of replications for that treatment. For example, a dose of nitrogen fertilizer is applied to five plots and a higher dose of same fertilizer to other four plots. Thus, the lower dose has five replications and higher dose has four replications. If a drug is administered to four patients and blood pressure is measured once on each of them, then this drug has four replications. But be wary that repeated observations are not replications. For instance, if on each patient blood pressure is measured five times, then these measurements are called *repeated observations* or *multiple readings* but not replications. In this case, there may be two situations; first the blood pressure is measured by five doctors on each patient at a time, secondly blood pressure is recorded five times a day over an interval of time. First case reflects the variation in measurement process only whereas second situation shows the variation in blood pressure due to response of the drug on the same patient over time. In both the situations, the measurements are not independent and are simply repeated measurements.

Blocking: This is a device to control one or more nuisance variables which are likely to influence the response of treatment(s) on an entity. For instance, the fertility of soil influences the yield. So for comparing the effect of fertilizers, it is necessary that soil fertility should be same for all fertilizers. For this blocking has been done with the help of soil fertility gradient maps prepared through uniformity trials. Blocking is often termed as *local control*. There can be hundreds of basis for blocking in a variety of situations like blocking on the basis of age, breed of animals, region, caste, etc.

Local control: It is a device for dividing the experimental material into homogeneous groups so as to reduce experimental error. In field experiments, a block is formed on a patch of land having same fertility. A block is homogeneous in itself and can differ from any other block. In animals, local control may be on the basis of lactation number, body weight, breed, age, etc. In human population local control or blocking can be done on the basis of income, educational qualification, locality, creed, etc.

Uniformity trial: A crop is grown in a field under similar conditions and the variability is measured among equal size plots for yield or any other trait, and then it is called a uniformity trial. A fertility map is prepared on the basis of fertility gradient. Blocks are formed on the patches that have same fertility.

Randomization: Purpose of randomization is to remove bias from the estimation of treatment effects. Secondly, randomization ensures validity of tests involved in the analysis of experimental data. All tests are based on certain assumptions. A most common assumption is that the observations vis-à-vis the errors therein are independently distributed. This assumption can be ascertained by selecting a random sample from a population or by assigning randomly the treatments to the experimental units. Random sample means that each and every sampling unit has equal chance (same probability) of being included in the sample. Whereas random assignment ensures that all treatments have same chance of being assigned to any experimental unit of an experiment. The process of randomization enables the investigator to proceed as though the independence is attained. Randomization is properly carried on with the help of random number tables.

Quasi-experiments: An experiment in which the treatments are not assigned randomly to experimental units is classified as *quasi-experiments*. The need for such experiments seldom arises in cases where treatments are to be given to pre-existing or naturally occurring units such as persons suffering from a particular disease or subjects who have been mentally retarded, etc.

1.3 ORTHOGONAL DESIGNS

Orthogonality of designs is that property which ensures that classes of different effects that are recorded on experimental material (units) can be estimated and tested separately without any entanglement. An experiment having blocks such that each treatment occurs an equal number of times in each block, the blocks and treatments effects can separately be estimated and tested without influencing or interacting with each other. Such a design is a *balanced design*. So in this type of experiments, blocks and treatments are said to be orthogonal. Randomized block designs hold this property.

An experiment in which each block does not contain all treatments an equal number of times are known as *unbalanced design*. In such a design, comparisons between treatments will be affected by the differences between blocks and vice versa. Thus, the block and treatment effects are not orthogonal. On the contrary, a balanced design holds the orthogonal property.

Often non-orthogonality is deliberately introduced in designs of experiment for reducing the block size so as to maintain the homogeneity of block. In situations where some treatments may be very costly or available in small quantity, fewer replications are taken for such treatments and more on rest of the treatments. Also many a times non-orthogonality is introduced due to failure of experimental units. For example, due to no germination of seeds in certain plot(s), damage in a plot, dying of an animal, failure of a bulb, etc. In this way, missing values convert the balanced design into unbalanced designs. Confounded designs are the ones in which *non-orthogonality* is introduced intentionally.

Balanced versus unbalanced design: Definitely there are some advantages of balanced designs over unbalanced designs.

1. There is conceptual ease of analysis of data in balanced designs.
2. The power of the test is maximum when each treatment has equal number of observations.
3. The test is less sensitive to departure from the assumptions of equality of variances across treatment factors in case of equal number of observations per treatment.

1.4 EXPERIMENTAL ERROR

An experiment is planned in such a way that all the factors known as nuisance variable(s), which are likely to influence the performance of the treatments, are controlled as much as possible and the treatment effects remain totally free from extrinsic influences. In spite of all efforts, some errors creep in due to extraneous factors which are beyond the control of the experimenter and are generally akin to inherent natural variation present in the experimental units. For example, percentage of germination of seeds in the plots, temperament of human beings, animals, etc.

For précised estimates of treatment means and efficient tests for comparison of treatment effects, it is necessary that experimental error should be as small as possible. Experimental error can considerably be reduced by taking homogeneous experimental units, grouping or blocking of units, increasing the number of replications, and so on.

1.5 EFFICIENCY OF A DESIGN

If s^2 is the estimate of error variance based n_e degrees of freedom, then the amount of information available from the experimental data is obtained by the formula,

$$\frac{n_e + 1}{n_e + 3}\left(\frac{1}{s_e^2}\right) \tag{1.5.1}$$

The quantity $(n_e + 1)/(n_e + 3)$ increases rapidly so long as n_e increases from 1 to 11 but shows slower change when $n_e = 12$ or more. Hence, for stabilized information, one should choose the number of replications for treatments which spare at least 12 d.f. for error from an experiment.

Assuming that the cost per experimental unit in two designs is same, the relative efficiency of design 1 over design 2 is,

$$E_{12} = \frac{(n_1 + 1)s_2^2}{(n_2 + 3)} \left/ \frac{(n_1 + 1)s_1^2}{(n_2 + 3)} \right. \tag{1.5.2}$$

It is trivial to verify that if n_1 and n_2 are equal to 12 or more, the ratio $(n_1 + 1)/(n_1 + 3)$ divided by $(n_2 + 1)/(n_2 + 3)$ is almost unity. Hence the formula for relative efficiency reduces to,

$$E_{12} = s_2^2 / s_1^2 \tag{1.5.3}$$

If $s_1^2 < s_2^2$, design 1 is more efficient than design 2 and vice versa.

More simply the efficiency of a design is defined as the inverse of its error variance and the ratio of the efficiencies of two designs results into the relative efficiency of one design over the other.

1.6 STANDARD ERROR

If a treatment mean t is based on r replications and s^2 is the error mean square, then the standard error (S.E.) of t is,

$$\sqrt{\frac{s^2}{r}} \qquad (1.6.1)$$

Standard error of the difference between two treatment means $\bar{t}_1$ and $\bar{t}_2$ of treatments T_1 and T_2 having r_1 and r_2 replications respectively is,

$$\text{S.E.} \left(\bar{t}_1 - \bar{t}_2 \right) = \sqrt{\left[s^2 \left(1/r_1 + 1/r_2 \right) \right]} \qquad (1.6.1)$$

If $r_1 = r_2 = r$, then $\quad \text{S.E.} \left(\bar{t}_1 - \bar{t}_2 \right) = \sqrt{\left(2s^2/r \right)} \qquad (1.6.1.1)$

1.7 NUMBER OF REPLICATIONS

Replications are necessary in an experiment to obtain a valid estimate of experimental error. Also the formula for standard error reveals that greater the number of replications better it is. But there are many limitations which compel the experimenter to take a limited number of replications. Hence, it is worthwhile to consider all those factors which play an important role in determining the number of replications.

1. Material and resources available.
2. Precision required.
3. Degrees of freedom for error.
4. Heterogeneity of experimental units.
5. Size and shape of experimental units particularly in case of field plots.

   - *Material* is a major factor in deciding the number of replications. For instance, the seed of a particular variety are in small quantity or only a limited number of animals of certain breed are available. In such situations, an investigator will have to restrict to a limited number of replications. This number also depends on the availability of personnel and equipment required. Men and material make one to decide about the number of replications.
   - When greater *precision* is required, more replications are to be taken. Numerically, the number of replications can be worked out with the help of the formula for t-statistic,

$$t_\alpha = d \big/ \sqrt{\left(2s^2/r \right)}$$

or $\qquad r = 2t_\alpha^2 \, s^2 \big/ d^2 \qquad (1.7.1)$

where, s^2 is some prior estimate of experimental error variance.

t_α is the tabulated t-value at α level of significance and degrees of freedom of s^2. d is the minimum difference which is aspired to be detected between two treatment means.

- For stable information, the experimenter has to choose the number of replications which provide at least 12 *degrees of freedom* for experimental error.
- An investigator has to take more and more number of replications as the *heterogeneity* among experimental material increases.
- *Size and shape of experimental units* to some extent play their role in deciding the number of replications because of competition between plots in respect of light, water and nutrients, etc. In experiments on animals, there is competition among animals in a pen as some animals are more pugnacious than others and may cause injury to others and eat their food as well. So size and shape are to be determined in such a manner that one experimental unit does not influence the other ones.

1.8 CLASSIFICATION OF STATISTICAL MODELS

A statistical model is specially given for every experimental design which represents the linear relationship of yield (response) with general effect, the treatment effect(s), other factor effects and an error term. The model under consideration is an additive model. These models are classified into three categories namely,

- (i) *Fixed effect model or Model – I or Analysis of variance model,*
- (ii) *Random effect or Model – II or Component of variance model,*
- (iii) *Mixed effect model or Model – III.* The classification is based on the selection of experimental units vis-a-vis the nature of treatments. This concept of model classification came into existence after the paper of Eisenhart (1947).

The classification plays an important role in respect of notations, definitions and the assumptions about the parameters and formulae for expected mean squares. Above all, it makes major difference on the physical interpretation of results. A brief outline of each model is given herewith.

Statistical linear model

Consider an additive model with two treatments and no interaction as given below.

$$y_{ij} = + \mu + \alpha_i + \beta_j + e_{ij} \tag{1.8.1}$$

$$i = 1, 2, \ldots\ldots\ldots, a \text{ and } j = 1, 2, \ldots\ldots\ldots, b$$

where,

y_{ij} – yield or a measurement for any other character taken on the experimental unit receiving the ith level of factor A and jth level of factor B.

μ – general mean which amounts to contribution to y_{ij} even if there were no treatment.

α_i – real effect of the ith treatment A.

β_j – real effect of the jth treatment B.

e_{ij} – error term which subscribes to those extraneous factors which are neither controllable nor directly measurable.

The common assumptions in all the models are:

(*i*) y_{ij} are normally distributed with mean μ + treatment effects, i.e. in the present model, E $(y_{ij}) = \mu + \alpha_i + \beta_j$ and with common variance σ_e^2, i.e. Var $(y_{ij}) = \sigma_e^2$.

(*ii*) e_{ij} are identically and independently distributed (I.I.D) normally with mean 0 and variance σ_e^2.

Fixed effect model

In an experiment, neither the experimental units are considered to have been randomly selected from a population of experimental units nor is the researcher going to draw conclusions with regard to a population of treatments, i.e. the treatments are taken to be of fixed nature, then the model is specified as *fixed effect model*. Hence, the null hypotheses under test are:

$$H_0: \alpha_1 = \alpha_2 = \ldots\ldots\ldots = \alpha_a \text{ and } H_0': \beta_1 = \beta_2 = \ldots\ldots\ldots = \beta_b$$

If statistical model (1.8.1) represents a fixed effect, then following assumptions hold good.

$$\Sigma a_i = \Sigma \beta_j = 0$$

Experiments for assessing the effect of fertilizers, comparing efficacy of certain drugs, etc., are considered appropriate for fixed effect model.

Random effect model

Consider again the model (1.8.1) and **a** levels of factor A and **b** levels of factor B to be a random sample of treatments to represent a population of levels of A and B, then the said model (1.8.1) will be classified as *random effect model*. Mathematically, this model has following assumptions.

$$\alpha_i \text{ are NID } \left(0, \sigma_\alpha^2\right) \text{ and } \beta_j \text{ are NID } \left(0, \sigma_\beta^2\right).$$

In this case, the null hypotheses are: $H_0: \sigma_\alpha^2 = 0$, and $\sigma_\beta^2 = 0$.

In breeding experiments, often the seeds of varieties of cereals are taken to be a random sample from a population of seeds of different varieties. Hence, in this situation a random effect model is more appropriate.

Mixed effect model

When **a** levels of factor A are considered to represent a random sample of the population of levels of factor A and **b** levels of factor B included in the experiment are taken to be of fixed nature, then the model (1.8.1) is to be a *mixed effect model*. In this situation the assumptions are:

$$\alpha_i \text{ are NID}\left(0, \sigma_\alpha^2\right) \text{ and } \Sigma\beta_j = 0.$$

Effects of certain feeds on different breeds of species usually represent such model where feeds are taken to be of fixed nature and breeds of random nature.

1.9 MULTIPLE COMPARISON TESTS (Post-hoc test)

When the null hypothesis, that all treatment means are equal is rejected, i.e. the alternative hypothesis that at least one mean differs significantly from others through F-test in ANOVA is accepted, then the problem arises which of the treatment means are significantly different from each other. One way is to compare all possible pairs of treatment means by student's t-test. But such a procedure is very cumbersome and time consuming. Therefore, some other methods were evolved by different researchers which were found to be appropriate under certain situations. Anyhow, some of them are more frequently used due to their merits and convenience. Hence, a few methods are given in detail and others are referred to as a part of knowledge.

Experimentwise type-I error

As discussed under the head multiple comparison tests, when the null hypothesis of equality of means is rejected, then there is a need to select a test for pair wise comparisons of treatment means or testing the significance of a contrast of treatment means. A number of tests are available in the literature. Thus, one needs a criterion for choosing a test. So, a test procedure will be preferred over the other if it provides smaller experimentwise type-I error. Hence, it becomes obligatory to understand what this error is.

Let there be k treatments in an experiment. Then, there shall be $k(k-1)/2$ pair-wise comparisons among k treatments means. Suppose $k(k-1)/2 = m$. Using α per cent least significant difference (lsd) test (given next), one would expect $100\,\alpha$. m pairs of means significantly different though they are not real differences. So the probability of getting at least one pair of means wrongly significant different is given by the formula,

$$E = [1 - (1 - \alpha)^m] \tag{1.9.1}$$

This probability of wrong decision is known as *experimentwise type–I error rate*. In social science, this error is generally *called familywise type–I error rate*. This error increases as k increases. Let us consider 5% lsd, i.e. $\alpha = 0.05$.

When, $k = 2$, $m = 1$,   E = 0.05   or   5%

$k = 3$, $m = 3$,   E = 0.1426   or   14.26%

$k = 4$, $m = 6$,   E = 0.2649   or   26.49%

$k = 5$, $m = 10$,   E = 0.4013   or   40.13%

$k = 6$, $m = 15$,   E = 0.5367   or   53.67%

$k = 8$, $m = 28$,   E = 0.7622   or   76.22%

Since experiment wise type–I error rate increases rapidly with increase in k, lsd procedure is not preferred when the number of treatments exceeds 3.

1.9.1 Multiple t-test or Fisher's lsd test

It is a commonly used procedure in which the differences among pairs of means out of a set of treatment means are compared with a standard value known as *least significant difference* (lsd) and more often this is called lsd–test. The origin of this test lies in t-statistic meant for testing the equality of two populations means μ_1 and μ_2. The test statistic t is,

$$t = \left(\bar{y}_1 - \bar{y}_2\right) / \left[s_e \sqrt{\left(1/r_1 + 1/r_2\right)}\right] \qquad (1.9.1.1)$$

where, $\bar{y}_1$ and $\bar{y}_2$ are two means based on r_1 and r_2 replications, respectively and s_e is square root of error mean square obtained in ANOVA. If

$$\left(\bar{y}_1 - \bar{y}_2\right) / \left[s_e \sqrt{\left(1/r_1 + 1/r_2\right)}\right] \geq t_\alpha ;$$

where n_e is the error d.f., then μ_1 differs from μ_2 significantly at α level of significance. The inequality reveals that if $\left|\bar{y}_1 - \bar{y}_2\right| \geq s_e \sqrt{\left(1/r_1 + 1/r_2\right)} \cdot t_{\alpha; n_e}$, the difference between means is significant. The right hand quantity of this inequality is known as least significant difference at α level of significance. Thus,

$$\text{lsd} = s_e \sqrt{\left(1/r_1 + 1/r_2\right)} \cdot t_{\alpha; n_e} \qquad (1.9.1.2)$$

If $r_1 = r_2 = r$, then $\text{lsd} = s_e \sqrt{\dfrac{2}{r}} \times t_{\alpha; n_e}$ $(1.9.1.3)$

In routine, α is taken as 0.05. In case α is chosen as 0.01, some workers called this as most significant difference (msd). For application of lsd test, see example 2.1.

Nota bene: Box, Hunter and Hunter (1978) suggested that instead of finding separate lsd for each pair, an approximate single scale factor $s_e \sqrt{2/r'} \cdot t_{\alpha; n_e}$ can be used provided the number of replications do not differ much. Where r' is the average of the number of replications.

R.A. Fisher suggested that lsd test should be applied only when the hypothesis of equality of means is rejected by F-test.

1.9.2 Student-Newman-Keuls test

This test was given by Student in 1927, Newman in 1939 and Keuls in 1952. All the three contributed to the same test independently so, it is named after these three workers jointly and in short it is called S-N-K test. In this test, calculate the standard error for each mean, i.e. $\sqrt{(s_e^2/r)}$, where s_e^2 is the error mean squares presuming that all means are based on the same number of observations r. In this way, all standard errors are equal.

In this method, upper α percent critical studentized ranges Wp are calculated for k, the distance between means in an ordered set of k treatment means. Let $\bar{y}_k$ be the largest mean, $\bar{y}_{k-1}$ the next to largest, and so on, lastly $\bar{y}_1$ be the smallest mean. The distance between $\bar{y}_k$ and $\bar{y}_1$ is k. Similarly, the distance between $\bar{y}_{k-1}$ and $\bar{y}_1$, and also between $\bar{y}_k$ and $\bar{y}_2$ is k-1 and so on. Obviously, k in test statistic is not a constant.

The critical range,

$$W_k = q_{\alpha;\, k,\, n_e} = \sqrt{\left(s_e^2/r\right)} \qquad (1.9.2.1)$$

where, n_e is the d.f. for error mean square, i.e. for s_e^2 in ANOVA.

The values of $q_{\alpha; k,\, n_e}$ for $\alpha = 0.05$ and $\alpha = 0.01$, k and n_e are provided in the table of per cent points of studentized range.

Compare the different $(\bar{y}_k - \bar{y}_1)$ with W_k. If $(\bar{y}_k - \bar{y}_1)$ is less than W_k, then the process stops and it is concluded that k treatment means belong to a homogeneous group. In the reverse situation, it is inferred that $\bar{y}_k$ and $\bar{y}_1$ differ significantly and the process is continued. Now compare the ranges $(\bar{y}_{k-1} - \bar{y}_1)$ and $(\bar{y}_k - \bar{y}_2)$ with W_{k-1}. If either range exceeds W_{k-1}, the corresponding pair of means is considered to be significantly different. Again proceed for the pairs of means which are at a distance of $k - 2$ and so on. The process has to be stopped when the actual range of a subset is greater than the calculated range. If such a subset does not exist, then the process stops at W_2. In this situation, it is concluded that all treatment effects are significantly different from one another. For its application see example 2.8.

1.9.3 Duncan's multiple range test

D.B. Duncan (1955) test is considered better than many other range tests for comparing differences between treatment means. In this test, the chances of committing type II error are less and of type I error more as compared to any other tests. This test is parallel to S-N-K test. To apply Duncan's multiple range test, the treatment means are arranged in ascending/descending order. A table of differences for all possible pairs of means is prepared by writing the means in descending order along the columns leaving the least one and in ascending order along the rows omitting the largest one. The differences among means are written in the cells leaving those pairs which are being repeated.

Duncan's multiple range test uses least significant ranges which take care of the distance between the means similar to S-N-K test. The only difference between S-K-N and Duncan's range test is that it uses Duncan's percent critical values $q^*_{\alpha;\, k, n_e}$

$$D_k = q^*_{\alpha;\, k, n_e} \sqrt{\left(s_e^2/r\right)} \qquad (1.9.3.1)$$

The procedure of comparing actual ranges in an ordered set of k means remains same as in case of S-N-K test except that use D_k in place of W_k. Duncan's critical values of $q^*_{\alpha;\, k, n_e}$ for $\alpha = 0.05$ and $\alpha = 0.01$ are provided in Table B-8. For application see example 2.2.

1.9.4 Dunnet's method

Dunnet (1955) developed a method for comparing $k-1$ treatment means versus control contrasts. Dunnet's procedure is applicable to a set of $k-1$ a priori nonorthogonal (see section 1.10) contrasts for which the correlations between $k-1$ contrasts are equal to 0.5. This situation arises when the number of replications for $k-1$ treatments and control is same, say r.

When the experiment is aimed to find out which of the treatments give a significantly higher response as compared to control treatment, then one sided confidence bounds should be used. Dunnet's method makes use of student's t statistic with equal number of replications for all treatments including control. Dunnet's test statistics tDN is,

$$\text{tDN} = \left(\bar{y}_i - \bar{y}_0\right)\Big/ \sqrt{\left(2s_e^2/r\right)} \qquad (1.9.4.1)$$

$$i = 1, 2, ..., k - 1.$$

where, $\bar{y}_i$ – ith treatment mean.

$\bar{y}_0$ – the control treatment mean.

r – the number of replications for each treatment.

For a one sided test of null hypothesis, absolute calculated value of tDN is compared with Dunnet's tDN tabulated value for α, k, and n_e obtained from Dunnet's table.

If $|\text{tDN}| \geq \text{tDN}_{\alpha;\, k,\, n_e}$, then it affirms that y_i is significantly higher from y_0, otherwise not. For a two-tailed test of null hypothesis, pick-up the value of tDN for $\alpha/2$, k, n_e from table. The decision criteria remains same as for one-tailed test. The only change in interpretation is that y_i is significantly different from y_0, i.e. may it be higher or lower from y_0.

Dunnet's procedure of multiple comparisons has two advantages:

1. Dunnet's procedure controls the probability of falsely rejecting the null hypothesis at α–experimentwise error rate.
2. It is not necessary to perform F-test in ANOVA prior to Dunnet's test for multiple comparisons.

1.10 CONTRAST OR COMPARISON

So far multiple comparisons were confined to paired means. But many times it is considered better to test the significance of combinations of treatment means. Such a combination is called *comparison* or *contrasts* provided they satisfy certain condition(s). Both the terms comparison and contrast are in vogue but in the present text mostly the word contrast will be used. For instance, a researcher may be interested to compare the mean scores secured by the examinees under chalk-duster and charts methods versus slides and computer display methods, dark colour versus light colour effects, alkaline batteries versus dry and heavy duty batteries life, chemical manures versus organic manures effects, etc.

A contrast is a linear combination of treatment (factor) means or totals that have known coefficients in such a way that at least two of them are non-zero and the sum of the coefficients is always zero.

Let $\bar{y}_1, \bar{y}_2, \ldots, \bar{y}_k$ be the k treatment means each having equal number of replications r. The linear combination

$$Z = c_1\bar{y}_1 + c_2\bar{y}_2 + \ldots + c_k\bar{y}_k \qquad (1.10.1)$$

is a contrast if for some i's $c_i \neq 0$ and $\sum_i c_i = 0$, where, $i = 1, 2, \ldots, k$. In case, all the treatments do not have the same number of replications say, treatment T_i has r_i replications, then for Z to be a contrast, the condition is, $r_1 c_1 + r_2 c_2 + \ldots + r_k c_k = 0$ or $\sum_i r_i c_i = 0$.

Linear combinations among three treatment means $\bar{y}_1, \bar{y}_2, \bar{y}_3$ of the type, $\bar{y}_1 - \bar{y}_2, \bar{y}_2 - \bar{y}_3, \bar{y}_1 - 2\bar{y}_2 + \bar{y}_3, \frac{1}{2}\bar{y}_1 + \frac{1}{2}\bar{y}_3 - \bar{y}_2, 2\bar{y}_1 + \bar{y}_2 - 3\bar{y}_3$, etc. are contrasts. Similarly, linear combinations among four treatment means $\bar{y}_1, \bar{y}_2, \bar{y}_3, \bar{y}_4$ of the type, $\bar{y}_1 + 3\bar{y}_2 - 3\bar{y}_3 - \bar{y}_4, \frac{1}{2}\bar{y}_2 + \frac{1}{2}\bar{y}_3 - \bar{y}_4, \bar{y}_3 - \bar{y}_1, 2\bar{y}_4 - \bar{y}_1 - \bar{y}_2, \bar{y}_1 - \bar{y}_2 + \bar{y}_3 - \bar{y}_4$, etc. represent contrasts.

A large number of contrasts can be constructed for $k \geq 3$. But only those contrasts are formed which provide some meaningful effects like linear, quadratic, cubic, quartic effects and so on. In short, a researcher selects only those contrasts which are not redundant.

Note: A contrast may include all the treatment means or only some of them. The means which do not appear in a contrast, their coefficients are taken to be zero.

1.11 PAIRWISE AND NON-PAIRWISE CONTRASTS

A linear combination of treatment means, in which all the coefficients except two of them are zero and the sum of these two non-zero coefficients is zero, is said to be a *pairwise contrast*. In simple words, a pairwise contrast is the difference between two treatment means. Any other contrast having more than two non-zero coefficients of treatment means is a *non-pairwise contrast*.

For instance, $\bar{y}_1 - \bar{y}_2, \bar{y}_2 - \bar{y}_3$, are pairwise contrasts whereas, $\bar{y}_1 - 2\bar{y}_2 + \bar{y}_3$ is a non-pairwise contrast among three treatment means.

1.12 ORTHOGONAL CONTRASTS

Two contrasts for the same set of treatment means, each of them having the same number of replications, are said to be orthogonal if the sum of the product of the corresponding coefficients is zero. The contrasts Z_1 and Z_2,

$$Z_1 = c_1\bar{y}_1 + c_2\bar{y}_2 + \ldots\ldots\ldots + c_k\bar{y}_k \qquad (1.10.2)$$

$$Z_2 = c'_1\bar{y}_1 + c'_2\bar{y}_2 + \ldots\ldots\ldots + c'_k\bar{y}_k \qquad (1.10.3)$$

are orthogonal if and only if, $c_1c'_1 + c_2c'_2 + \ldots\ldots\ldots + c_kc'_k = 0$, i.e. $\sum_i c_i c'_i = 0$ for $i = 1, 2, ..., k$.

For example, the contrasts $Z_1 = \bar{y}_1 - \bar{y}_3$ and $Z_2 = \bar{y}_1 - 2\bar{y}_2 + \bar{y}_3$ are orthogonal contrasts since $1 \times 1 - 0 \times 2 - 1 \times 1 = 0$. When the treatment mean $\bar{y}_i$ is based on r_i replications, then the contrasts Z_1 and Z_2 given by (1.10.2) and (1.10.3) are orthogonal if $\sum_i r_i c_i c'_i = 0$. For k treatments, there cannot be more than $k - 1$ orthogonal contrasts.

1.13 A PRIORI AND POSTERIORI CONTRASTS

A worker mostly has in mind some specific set of hypotheses which he will like to test for his investigation. All such hypotheses about contrasts under test are called *a priori* or *planned* contrasts. But if F-test of the omnibus null hypothesis is significant, then it leads to the conclusion that at least one contrast among treatment means is significant. Now the researcher's interest lies in identifying the contrast which is significant. So he tests different contrasts and tries to extract all information contained in the data. Such contrasts which are constructed following the significant test are called *a posteriori* or *post-hoc* contrasts.

1.14 DATA SNOOPING

Before experimentation, a researcher might have highlighted certain contrasts and treatment means that are of special interest and designed the experiment in such a way which ensures that all of them are estimable. On the availability of data, the researcher examines the data deeply and feels that some contrast(s) which have not been preconceived may reveal some significant conclusions which may not be as anticipated. Thus, allowing the researcher to suggest some new contrast after seeing the data is known as *data snooping*.

1.15 STANDARD ERROR OF A CONTRAST

The standard error of the contrast,

$$Z = c_1 \bar{y}_1 + c_2 \bar{y}_2 + \ldots \ldots + c_k \bar{y}_k$$

among k treatment means, each treatment having same number of replications r, is,

$$\text{S.E.}(Z) = \sqrt{\left[\left(c_1^2 + c_2^2 + \ldots \ldots + c_k^2 \right) s_e^2 / r \right]} \qquad (1.15.1)$$

If ith treatment is replicated r_i times, then the standard error of Z is,

$$\text{S.E.}(Z) = \sqrt{\left[\left(c_1^2 / r_1 + c_2^2 / r_2 + \ldots \ldots + c_k^2 / r_k \right) s_e^2 \right]} \qquad (1.15.2)$$

1.16 SUM OF SQUARE DUE TO A CONTRAST

Consider the contrast $Z = \sum\limits_{i} c_i \bar{y}_i$ among k treatment means, each treatment having an equal number of replication r. The sum of square due to the contrast Z is,

$$\text{SSC} = r Z^2 / \left(c_1^2 + c_2^2 + \ldots \ldots + c_k^2 \right) \qquad (1.16.1)$$

$$= r Z^2 / \Sigma_i c_i^2 \qquad (1.16.1.1)$$

$$i = 1, 2, \ldots \ldots, k$$

If ith treatment is replicated r_i times, then the sum of square for the contrast Z is,

$$\text{SSC} = Z^2 / \left(c_1^2 / r_1 + c_2^2 / r_2 + \ldots \ldots + c_k^2 / r_k \right) \qquad (1.16.2)$$

$$= Z^2 / \Sigma_i c_i^2 / r_i \qquad (1.16.2.1)$$

It is noteworthy that each contrast has one degree of freedom. If one constructs all possible $k - 1$ orthogonal contrasts, then the total of sum of squares of these $k - 1$ orthogonal contrasts equals to treatment sum of squares.

1.17 STEP-DOWN AND STEP-UP PROCEDURES

These procedures are meant for testing hypotheses for all pairwise contrasts among treatment means. In step-down procedure, all the k means are arranged in ascending order. Let the means in increasing order be denoted by $\bar{y}_1, \bar{y}_2, \ldots\ldots, \bar{y}_k$.

Now the procedure starts by testing the hypothesis between smallest and largest means, i.e. H: $\mu_1 - \mu_k = 0$. This is the hypothesis in which the distance between these two means is k steps. If this hypothesis is rejected, then proceed to test the other pairwise hypotheses, H: $\mu_1 - \mu_{k-1} = 0$ and H: $\mu_2 - \mu_k = 0$ in which the means are separated by $k-1$ steps. If these hypotheses are rejected, proceed for the other hypotheses separated by a distance of $k-2$ steps. In this manner continue the process till a pairwise contrast separated by two steps is reached. The process of testing stops at any stage as soon as the null hypothesis for pair of means is accepted. In case, H: $\mu_1 - \mu_k = 0$ is accepted, no further testing is done concluding that all treatment effects do not differ significantly from each other.

In step-up procedure, k means are arranged in descending order. Hypothesis involving first two means are tested. Once this hypothesis is rejected, it results into rejection of all implicit hypotheses. This saves the labour of testing all other hypotheses about pairs of remaining means.

1.18 TUKEY'S HONESTLY SIGNIFICANT DIFFERENCE (HSD) PROCEDURE

Tukey (1953) developed a widely used procedure known as honestly significant difference (HSD) test. It controls type-I error for all pairwise contrasts. Tukey's test is based on the sampling distribution of studentized range, just like t-distribution given by William S. Gosset. Tukey's HSD statistic is,

$$qT = \left(c_1\bar{y}_1 + c_2\bar{y}_2 + \ldots\ldots + c_k\bar{y}_k\right)\Big/\sqrt{\left(s_e^2/r\right)} \qquad (1.18.1)$$

If $qT > q_{\alpha;\,k,\,n_e}$, then the null hypothesis about the contrast is rejected where α, k and n_e are as given earlier. Value of $q_{\alpha;k,\,n_e}$ can be seen in a statistical table's book.

Nota bene: This test can be applied only if all treatments are equi-replicated.

1.19 BONFERRONI'S TEST

Bonferroni propounded a procedure for testing the significance of those contrasts that were selected prior to collection of data. The multiple comparison procedure is based on the following inequality given by an Italian mathematician Carlo Bonferroni. Let B_1, B_2, $\ldots\ldots$, B_p be the p events, in the present case representing the confidence intervals. Then the probability of the intersection of B_i's satisfies the following inequality.

$$P\left(\cap B_i\right) \geq 1 - \Sigma_i \left[1 - P\left(B_i\right)\right] \qquad (1.19.1)$$

$$i = 1, 2, \ldots\ldots, p$$

Making use of the above inequality and taking B_i as an event that a confidence interval for a linear combination of treatment effects contains this contrast such that $P(B_i)$ is the probability of its occurrence. To obtain an over all $100(1 - \alpha)$ % confidence level, Bonferroni established a formula for confidence interval for p preplanned contrasts like $\sum_i c_i \bar{y}_i$, of the treatment response means, i varying over the number of treatments in the experiment. The formula for confidence interval is,

$$\Sigma_i c_i \bar{y}_i \pm t_{n_e, \, \alpha/2p} \sqrt{\text{Var}\left(\Sigma_i c_i \bar{y}_i\right)} \qquad (1.19.2)$$

where, $\text{Var}\left(\sum_i c_i \bar{y}_i\right) = s_e^2 \sum_i c_i^2 / r_i$.

Tabulated value of t for n_e, the error d. f. and $\alpha/2p$ level will not be easily available in t-tables. It has to be obtained by using a computer package or by a approximate interpolation formula.

In case the value of a contrast lies in the interval obtained by the formula (1.19.2), then $H_0 : \sum_i c_i \bar{y}_i = 0$ is accepted, otherwise rejected. The test criteria for a two tailed test of H_0 for a single contrast at α level of significance is,

$$\text{Reject } H_0, \text{ if } \left[\text{SSC}/\text{Var}\left(\sum_i c_i \bar{y}_i\right)\right] > F_{\alpha/p; \, 1, \, n_e}$$

otherwise, accept H_0.

Bonferroni test ensures that the critical level of significance does not exceed during the process of testing preplanned contrasts. Bonferroni suggested the correction that for the test of a contrast, the level of significance be taken as the average value of α, i.e. the prefixed level α divided by the number of contrasts under test.

1.20 SCHEFFE'S TEST

In 1953, Scheffe evolved this test which is used to test the significance of non-pairwise contrasts which are usually constructed by an inspection of data for divulging meaningful information. The test uses F-statistic and is denoted by FS. Let the contrast under test be,

$$Z = c_1 \bar{y}_1 + c_2 \bar{y}_2 + \ldots\ldots\ldots + c_k \bar{y}_k$$

where the mean $\bar{y}_i$ is based on r_i replications.

The statistic,

$$FS = Z^2 / \left[\left(c_1^2/r_1 + c_2^2/r_2 + \ldots\ldots + c_k^2/r_k\right) s_e^2\right] \qquad (1.20.1)$$

In (1.20.1), all notations have their usual meaning. If $FS \geq \upsilon_1 F_{\alpha;\upsilon_1,\upsilon_2}$, $(\upsilon_1 = k - 1)$, and $\upsilon_2 = n_e$, the error d.f. in ANOVA, reject the null hypothesis concluding that the contrast is significant, otherwise accept H_0. As a caution, such contrast be constructed and tested only when F-test rejects H_0.

This test has its own merits and demerits. This test controls type-I error at α or less than α level. It is a robust test. But its power is less than Tukey's HSD test. Scheffe's test should be applied only for those contrasts which were selected after the data were examined.

There are a large number of tests for testing pairwise and non-pairwise contrasts. It is neither possible nor required to provide details of all those tests. Anyhow all these tests are fraught with controversies and problems. Some of them are referred here so that a reader interested in studying them can go through the literature. They are Dunn's (1961) comparison test, Kramer (1956) test, Hayter (1986) and many others not mentioned over here. Each test has one or more virtues and some lacunae too. Virtues comprise of more power, greater control of experimentwise type-I error rate, ease of computation, etc. The tests given in this text have greater power and are quite robust.

1.21 ANALYSIS OF MEANS

Analysis of variance (ANOVA) is the most common procedure for analyzing experimental data. This method is in vogue since the inception of experimental designs. ANOVA has its own limitations. It has been given in section (1.1). A minimal difference between means which an experimenter wants to detect is subjected to the size of the experiment that is to be maintained necessarily. But to meet such a condition is always not possible. Hence some workers, particularly the engineers and industrial personnel preferred to analyze data by the method of analysis of means (ANOM). Ott (1958) was the first to develop the method of ANOM. The main difference between ANOVA and ANOM procedures is that in ANOVA pairwise means are compared to detect which one differs from the others by using a suitable test. But in ANOM procedure, the user finds whether one or more means differ from the overall average of all the values, i.e. general mean. ANOM is a graphical method for comparing treatment means as a substitute to ANOVA. This analysis is mostly carried on MINITAB software, SAS / QC 9.0 and 9.1 from SAS software. In ANOVA, one deals with square of the units of original measurement, whereas ANOM deals in terms of the original unit of measurement. For details of ANOM, the readers are referred to the unique book by Nelson, Wludyka, and Copeland (2005).

The main lacuna of ANOM approach is that this can be applied only for designs with fixed factors, whereas there is no such restriction in case of ANOVA method. ANOVA is applicable to all type of experimental designs with fixed, random, and mixed factors.

As given above, ANOM has been the part of MINITAB. But analysis of means can also be carried out manually. The general approach is to plot the averages on the graph against decision lines which can be obtained by the formula,

$$\overline{\overline{y}} + h_{\alpha,\, k,\, n_e} \text{ S.E.} \left(\overline{y}_i - \overline{\overline{y}} \right) \qquad (1.21.1)$$

where, α – prefixed level of significance

k – number of averages

n_e – error d.f. obtained from ANOVA.

Critical value of $h_{\alpha,\, k,\, n_e}$ can be obtained from the tables given by L.S. Nelson published in the journal of Quality Technology, 15(1), Jan., 1983. These tables are reproduced by Thomas P. Ryan in his book, *Modern Experimental Design*, John Wiley, 2007.

The averages of treatment are plotted on the graph. If all the means lie within the decision lines, then the conclusion is that none of the means differ from overall mean at level α. On the other hand, if any mean lie below the lower decision line or above the upper decision line, that is taken to differ significantly from overall mean at α level of significance.

1.22 CONCEPT OF OUTLIERS

Occasionally an investigator comes across one or more observation in a set of values which are inconsistent with the remaining values. Such value(s) immediately catches the eye and pinches to mind. What can possibly be the reason for the occurrence of such a value? One possibility is that the value arises randomly due to the inherent variability in the population. Secondly, it can occur due to measurement error, faulty recording, biased sampling, improper definition of population, etc. Hence, one should understand what does an outlier mean?

Definition: Broadly an outlier may be defined as an observation that is very large or very small than expectation. This is evinced by a residual which is much larger positive or negative value as compared to other residual values.

Outliers have major effect on the estimators and test statistics which are always the function of sample observations. If outliers are as a consequence of inherent variation in the population, one should be wary of discarding them. Outliers which cannot be attributed to known error are called *discordant* and should not be discarded. Now the problem arises which of the extreme observations can be regarded as outliers. For this, one has to draw line of demarcation for detecting the outliers. Without entangling into the complicacies of the theory of outliers, a simple common procedure is described over here to detect the outliers.

For a normal distribution with mean μ and variance σ^2, the inter-quartile range, Q =1.35σ. If 1.5Q is added to both the sides of the horizontal lines often called *whiskers*, then it provides almost a range of $4Q = 4 \times 1.35\sigma = 5.40\sigma$. This results into the cutoffs as $\mu \pm 2.70\sigma$, for outliers and this range covers 99.3% of the population. Thus, for an ordered sample $y_{(1)}, y_{(2)}, \ldots\ldots, y_{(n)}$ from a normal population, if either of the following inequality holds,

$$Q_1 - y_{(1)} \leq 1.5\sigma \quad \text{or} \quad y_{(n)} - Q_3 \leq 1.5\sigma \qquad (1.22.1)$$

then the sample has no outlier, otherwise one has to look for one or more outliers. In case, the above inequalities do not hold, the simple criteria for a sample value y to be a potential outlier is, either

$$y < Q_1 - 1.5\sigma \quad \text{or} \quad y > Q_3 + 1.5\sigma \qquad ...(1.22.2)$$

When an outlier is detected, a researcher has three options, (*i*) may replace the observation with a new value if justifiable, (*ii*) correct observations if the error is detected and record permits, (*iii*) reject the observation and proceed for the analysis with the remaining values.

QUESTIONS AND EXERCISES

1. What is meant by designing of experiments? Elucidate the concept with the help of real examples.
2. For what purpose, analysis of variance is carried out?
3. State the assumptions underlying the analysis of variance.
4. Define the following terms and give two examples of each.
   (*a*) Experimental unit.
   (*b*) Nuisance variable.
   (*c*) Treatment in experiment.
   (*d*) Blocking.
5. What are the requirements for a good experimental design? Discuss each of them adequately.
6. What is a uniformity trial and what purpose does it serve?
7. Differentiate between balanced and unbalanced design.
8. What do you understand by orthogonality of data of an experiment?
9. In what ways can the experimental error be minimized?
10. How can the efficiency of two experimental designs be compared?
11. Why replications in all experimental designs are so important?
12. What are the factors responsible for choosing the number of replications? Also give the formula which enables an investigator to determine the number of replications and explain its working.

13. Where do we require the standard error of means?
14. Discuss adequately three types of statistical models chosen for the experimental designs.
15. Throw light on the importance of multiple comparison tests.
16. Which test are suitable for pairwise comparisons and which one for non-pairwise comparisons?
17. What are the criteria for using Scheffe's and Bonferroni's procedures and what difference is there between the two procedures?
18. What statistics are used in the following pairwise multicomparison tests?
    (*i*) Least significant test.
    (*ii*) Duncan's test
    (*iii*) Dunnet's test.
    (*iv*) Tukey's test.
    (*v*) Student-Newman-Keul's test.
19. Give an overview of the following.
    (*a*) Orthogonal contrasts.
    (*b*) A priori and posteriori contrasts.
    (*c*) Standard error of a contrast.
    (*d*) Data snooping.
20. What anomaly do outliers create in analysis of data? How can the outliers be detected in a collection of data? How can the data be modified to remove the effect of outliers?
21. Give the concept and method of analysis of means for the treatment means of an experiment.

BIBLIOGRAPHY

1. Bechhofer RE, Dunnet CW. Multiple comparisons for orthogonal contrasts: Examples and Tables. *Technometrics*, 1982: 24, 213–22.
2. Box GEP, Hunter WG, Hunter JS. *Statistics for Experiments: Design, Innovation, and Discovery*, Hoboken, John Wiley, New Jersey, 2005.
3. Cochran WG. Some consequences when the assumptions for the analysis of variance are not satisfied. *Biometrics*, 1947: 3, 22–38.
4. Duncan DB. Multiple ranges and multiple F-test. *Biometrics*, 1955: 2; 1–42.
5. Dunn OJ. Multiple comparisons among means. *Journal of the American Statistical Association*, 1961: 56, 52–64.
6. Dunnet CW. A multiple comparison procedure for comparing several treatments with a control. *Journal of the American Statistical Association*, 1955: 50, 1096–1121.
7. Dunnet CW. New tables for multiple comparisons with a control. *Biometrics*, 1964: 20, 482–91.
8. Eisenhart C. The assumptions underlying the analysis of variance. *Biometrics*, 1947: 3, 1–21.

9. Glass GV, Peckham PD, Sanders JR. Consequences of failure to meet assumptions underlying the analysis of variance and covariance. *Review of Educational Research*, 1972: 42, 237–88.

10. Hunter AJ. The maximum familywise error rate of Fisher's least significant difference test. *Journal of the American Statistical Association*, 1986: 81, 1000–1004.

11. Hicks CR, Turner KV, Jr. *Fundamental Concepts in the Design of Experiments*, 5th ed. Oxford University Press, U.K. 1999.

12. Hinkelman K, Kempthorne O. *Design and Analysis of Experiments*. Vol. 2, *Advanced Experimental Design*, Hoboken, John Wiley, New Jersey, 2005.

13. Hochberg Y, Tamhane AC. *Multiple Comparison Procedures*. John Wiley, New York. Hoboken, John Wiley, New Jersey, 1987.

14. Keuls M. The use of the studentized range in connection with an analysis of variance. *Euphytica*, 1952: 1, 112–22.

16. Kramer CY. Extension of multiple range tests to group means with unequal number of replications. *Biometrics*, 1956: 12, 307–10.

17. Nelson PR, Wludyka PS, Copeland KAF. *The Analysis of Means: A Graphical Method for Comparing Means, Rates and Proportions*. Society for Industrial and applied Mathematics and American Statistical Association. Philadelphia and Alexandria, VA, Respectively, 2005.

18. Newman D. The distribution of range in samples from a normal population expressed in terms of an independent estimates of standard deviation. *Biometrika*, 1939: 31, 20–30.

19. Ott ER. *Analysis of Means*. Technical Report # 1. Department of Statistics, Rutgers University, 1958.

20. Ott ER. *Process Quality Control*. McGraw Hill, New York, 1975.

21. Rogan JC, Keselman HJ. Is the ANOVA F-test robust to variance heterogeneity when sample sizes are equal? An investigation via a coefficient of variation. *American Educational Research Journal*, 1977: 14, 493–98.

22. Scheffe H. A method for judging all contrasts in the analysis of variance. *Biometrika*, 1953: 40, 87–104.

23. Student. Errors of routine analysis. *Biometrika*, 1927: 19, 151–64.

24. Scilling EG. A systematic approach to the analysis of means. Part II. Analysis of contrasts. *Journal of the Quality Technology*, 1973: 5, 147–55.

25. Tomerken AJ, Serlin RC, Comparison of ANOVA alternatives under variance heterogeneity and specific noncentrality structures. *Psychological Bulletin*, 1986: 99, 90–99.

26. Tukey JW. *The Problem of Multiple Comparisons. Unpublished manuscript*, Princeton University, Princeton, 1953.

27. Winer BJ, Brown DR, Michel KM. *Statistical Principles in Experimental Design*. McGraw Hill, New York, 1991.

2

Basic Designs

2.1 COMPLETELY RANDOMIZED DESIGN

This is first in the series of basic experimental designs and is abbreviated as CRD – v, where v is the number of treatments. The prime condition for the use of this design is that all the experimental units should reasonably be homogeneous particularly in respect of the characters which are likely to influence the treatment effects. There is complete flexibility in this design in the sense that any number of treatments can be accommodated and treatments can be replicated any number of times. Therefore, it is called a *no restrictional design*, though it is always preferred to have equal number of replications to ensure equal precision of the estimates of the treatment effects.

All the more, the analysis of data for this design is most simple and is not disturbed even if some units fail to respond for one reason or the other. Even though a treatment fails totally, there is no complication in respect of analysis of data except that the failing units for that treatment are omitted and that treatment is neglected as a part of that experiment.

In CRD, there is no restriction on randomization of experimental units to treatment and thus all experimental units-treatment combinations have equal probability of occurrence. Thus, CRD gives rise to one way classification and provides maximum degrees of freedom for error. But the main drawback of this design is that total variability present among experimental units enters into error. Anyhow, this design is very suitable for laboratory experiment like recording percentage germination in Petri dishes, observing effect of chemicals on insects in test tubes, green house experiments, incubation trials, etc.

Let us consider an experiment with layout in CRD having v treatments and ith treatment be replicated r_i times for $i = 1, 2, \ldots\ldots, v$. Let τ_i be the response of the ith treatment and y_{ij} be the response of the ith treatment allocated to the jth unit. In spite of all efforts to maintain identical conditions during planning and execution, some variability besides treatment effects creeps in which is

24

uncontrollable. Such variability due to extraneous factors is accounted towards experimental error. Hence, another factor e_{ij} is known as error which is a random variable and is distributed with mean zero and variance σ_e^2 enters into the statistical model of the design. e_{ij} are assumed to be mutually independent and normally distributed with mean zero and variance σ_e^2. Thus, the linear model for CRD is,

$$y_{ij} = \mu_i + e_{ij} \tag{2.1.1}$$

$$i = 1, 2, \ldots, v \text{ and } j = 1.2, \ldots, r_i$$

μ_i comprises of the factors μ and τ_i, i.e. $\mu_i = \mu + \tau_i$ where μ is a constant equal to overall mean and τ_i is the true effect of the ith treatment response and represents the positive or negative deviation from μ. Thus,

$$y_{ij} = \mu + \tau_i + e_{ij} \tag{2.1.1.1}$$

For specifically selected treatments, i.e. for fixed effects model (2.1.1.1) is often called *one way analysis of variance model.*

Now consider an experiment in CRD having four treatments A, B, C, and D with replications 3, 5, 2, 6, respectively. The layout of the design will be of the kind given below.

A	C	D	B
D	B	A	D
B	D	C	B
D	A	B	D

2.1.1 Layout plan of CRD

In the above experiment, $v = 4$, $r_1 = 3$, $r_2 = 5$, $r_3 = 2$, $r_4 = 6$. Therefore, total number of observations, $n = 16$. The observations recorded in the experiment can be tabulated as depicted below.

Treatments	Data	Total
A	$y_{11} \ y_{12} \ y_{13}$	y_1
B	$y_{21} \ y_{22} \ y_{23} \ y_{24} \ y_{25}$	y_2
C	$y_{31} \ y_{32}$	y_3
D	$y_{41} \ y_{42} \ y_{43} \ y_{44} \ y_{45} \ y_{46}$	y_4

2.1.1 Data table

Note: In the same way, the layout and data table for responses can be presented for any number of treatments having different or same number of replications.

Least square estimates of the parameters: Two normal equations are obtained as follows by differentiating $\sum_i \sum_j (y_{ij} - \mu - \tau_i)^2$ partially with respect to μ and τ_i respectively and equating to zero.

Table 2.1.1 One way ANOVA

Source	d.f.	S.S.	M.S.	F – value	Expected mean squares Model–I	Expected mean squares Model–II
Among Treats.	$\nu - 1 = \upsilon_1$	T_rSS	$T_rSS/\upsilon_1 = T_rMS$	$T_rMS/EMS = F_{cal}$	$\sigma_e^2 + \left(\sum_i r_i \tau_i\right)^2 / (\nu - 1)$	$\sigma_e^2 + r_0\,\sigma_\tau^2$
Error	$n - \nu = \upsilon_2$	ESS	$ESS/\upsilon_2 = EMS$		σ_e^2	σ_e^2
Total	$n - 1$	TSS				

$$y.. = \sum_i r_i \hat{\mu} + \sum_i r_i \hat{\tau}_i \qquad (2.1.2)$$

$$y_{i.} = r_i \hat{\tau}_i + r_i \hat{\mu} \qquad (2.1.3)$$

From equation (2.1.3), $\qquad \hat{\tau}_i = y_{i.}/r_i - \hat{\mu}$

Taking sum over i, $\qquad \sum_i \hat{\tau}_i = \sum_i y_{i.}/r_i - v \hat{\mu} \qquad (2.1.4)$

Since $\sum_i \hat{\tau}_i = 0$, $\qquad \hat{\mu} = (\sum_i y_{i.}/r_i)/v = \sum_i \bar{y}_i /v \qquad (2.1.5)$

$$= \bar{y} = \text{Average of all treatment means.}$$

Analysis of variance table is presented on page 26 in general for v treatments and r_i replications for treatment i.

Where, $r_0 = (\sum r_i - \sum r_i^2 /\sum r_i)/(v - 1)$. In case, $r_i = r$ for all i, $r_0 = r$. F-statistic in this case has υ_1 and υ_2 d.f.

The hypothesis under test is, $H_0: \mu_i = \mu'_i$ for $i, i' = 1, 2, ..., v$ vs. $H_1 = \mu_i \neq \mu'_i$, for some $i \neq i'$. To decide about H_0, compare F_{cal} with tabulated $F_{\alpha; \upsilon_1, \upsilon_2}$, where α is the prefixed probability of type-I error. Most commonly used value of α is 0.05. If $F_{cal} \geq F_{\alpha; \upsilon_1, \upsilon_2}$, then H_0 is rejected, otherwise accepted. When H_0 is rejected, then the search starts for ascertaining which of the pairwise or non-pairwise contrasts are significant by using any of the tests described in section (1.9).The expected mean squares displayed in ANOVA Table 2.1.1 are expatiated for RBD in the next section. The same explanation holds for CRD. Here the details are avoided for the sake of brevity.

Example 2.1: Four feed stuffs A, B, C, D were tried on 20 alike chicks to see the effect on their body weight gain. All the chicks were treated alike in all respect except the feeding treatments and each feed was given to five chicks randomly. Anyhow, one chick died during experimentation. Data in respect of gain in body weight after fixed period for four feed stuffs were as given below.

Feeds			Weight gain (g)		
A	55	49	42	21	52
B	61	112	30	89	63
C	42	97	81	95	
D	169	137	169	85	154

Obviously the experimental data leads to one way analysis meant for CRD. Here one tests the hypothesis that mean effect of feeds is same, i.e. $H_0: \mu_A = \mu_B = \mu_C = \mu_D$ against H_1: at least two of them differ significantly.

Treatment totals: $y_A = 219$, $y_B = 355$, $y_C = 315$, $y_D = 714$ and $y.. = 1603$.

Also, the means,

Treatment means: $\bar{y}_A = 43.80$, $\bar{y}_B = 71.00$, $\bar{y}_C = 78.75$, $\bar{y}_D = 142.80$.

C.F. $= (1603)^2 / 19 = 135242.58$

Total S.S. $= \sum_i \sum_j y_{ij}^2 = 55^2 + 49^2 + \ldots\ldots + 85^2 + 154^2 - C.F. = 37738.42$

Treat S.S. $= \dfrac{219^2}{5} + \dfrac{355^2}{5} + \dfrac{315^2}{4} + \dfrac{714^2}{5} - C.F.$

$= 161562.65 - C.F. = 26320.07$

Error S.S. $= 37738.42 - 26320.07 = 11418.35$

Table 2.1.2 ANOVA

Source	d.f.	S.S.	M.S.	F-value
Feeds	3	26320.07	8773.36	8773.36/761.22 =11.52
Error	15	11418.35	761.22	
Total	18	37738.42		

The value of $F_{.05;\,3,15}$ from the F-distribution table no. B-7 is 3.29. Since calculated F is greater than the table value, the omnibus hypothesis H_0 is rejected. Now it is further explored which of the feeds differ significantly from one another. Since the number of feeds is only four, lsd test is used to demonstrate how can this test be applied. Since the number of replications for all the four feeds is not the same, so two lsd will be worked out to compare pairwise mean effects of the feeds. Number of replications for feeds A,B, and D is 5 and for C is 4. From Table B-5, $t_{.05,15} = 2.131$.

lsd for A, B, and D, $\mathrm{lsd}_1 = \left(\sqrt{2 s_e^2 / r} \right) \times t_{.05,15}$

$= \left(\sqrt{2 \times 761.22 / 5} \right) \times 2.131$

$= 37.185$

lsd for C vs. A, B, and D, $\mathrm{lsd}_2 = \sqrt{\left[\left(\dfrac{1}{r_1} \right) + \left(\dfrac{1}{r_2} \right) s_e^2 \right] \times t_{.05;15}}$

$= \sqrt{\left[\left(\dfrac{1}{5} + \dfrac{1}{4} \right) \times 761.22 \right] \times 2.131}$

$= 39.441$

While comparing A with B and D or B with D, lsd_1 is used and when C is compared with A, B, and D, lsd_2 is utilized. Similar letters show nonsignificant differences.

Feeds	Treat	Mean
A	43.80	a
B	71.00	a c
C	78.75	a c e
D	142.80	b d f

Above table reveals that feed A has no significant difference with B and C but is significantly inferior than D. B is no better than C but D is superior than B. C is significantly poorer than D.

2.2 RANDOMIZED COMPLETE BLOCK DESIGN

The design is abbreviated as RBD and more specifically it is further written as RBD – v, where v denotes the number of treatments. This is most widely used design in the area of experimental designs. This design controls a nuisance variable which is likely to influence the response, i.e. the dependent variable. Hence, it is known as *one restrictional design*. The ultimate aim of designing is to control factor(s) which contribute towards the variation in response variable. For instance, fertility of land definitely affects the yield, IQ of persons attributes to memorizing and logic power of persons, breed of lactating animals is more akin to milk yield and so on. All such factors are recognizable prior to designing of experiment and are called *blocking factors*. So in RBD, effort is made to control one nuisance variable in such a way that all the units within a block are almost homogeneous whereas block to block differences may be of any level in respect of nuisance variables. In this way, estimates of block differences reduce the effect of nuisance variables on dependent variable which ultimately reduces the experimental error. Finally, a set of homogeneous experimental units is called a *block*.

Another nuisance variable is a covariate or concomitant variable which usually masks or obscures the response. For instance, the number of weeds per plot, infestation of insects on plants, etc. affects the yield. Such variables are measurable in the duration of experimentation and their impacts are adjustable through *analysis of covariance* which shall be discussed in Chapter 10.

The word *complete* ensures that each block consists of as many experimental units as the number of treatments and each treatment occurs once in each block. In this way, each treatment is replicated as many times as the number of blocks and thereby the design is balanced. In common parlance, randomized block design implies randomized complete block design and is abbreviated as RBD. In some recent books, this is abbreviated as RCB design. In RBD, treatments are randomly allocated to experimental units in each block independently or vice-versa. The layout of a randomized block design with six treatments A, B, C, D, E and F in five blocks shall be of the kind given below.

		Blocks		
I	II	III	IV	V
A	C	B	F	D
C	F	D	B	A
D	D	C	E	B
B	A	E	C	F
E	E	A	D	C
F	B	F	A	E

Layout plan of an RBD with $v = 6$ and $r = 5$

Measurement for each treatment in all the blocks can be tabulated in a two-way table as displayed below.

Table 2.2.1 Data table with marginal totals

Treatment	Blocks					Treat. total
	I	II.	III	IV	V	
A	y_{11}	y_{12}	y_{13}	y_{14}	y_{15}	$y_{1\cdot}$
B	y_{21}	y_{22}	y_{23}	y_{24}	y_{25}	$y_{2\cdot}$
C	y_{31}	y_{32}	y_{33}	y_{34}	y_{35}	$y_{3\cdot}$
D	y_{41}	y_{42}	y_{43}	y_{44}	y_{45}	$y_{4\cdot}$
E	y_{51}	y_{52}	y_{53}	y_{54}	y_{55}	$y_{5\cdot}$
F	y_{61}	y_{62}	y_{63}	y_{64}	y_{65}	$y_{6\cdot}$
Block total	$y_{\cdot1}$	$y_{\cdot2}$	$y_{\cdot3}$	$y_{\cdot4}$	$y_{\cdot5}$	$y_{\cdot\cdot}$

The above table provides the treatment totals and block totals.

Statistical model for the design is,

$$y_{ij} = \mu + \tau_i + \beta_j + e_{ij} \qquad (2.2.1)$$

$$i = 1, 2, \ldots\ldots, v \text{ and } j = 1, 2, \ldots\ldots\ldots, k.$$

y_{ij} – response of the experimental unit receiving ith treatment in jth block.

β_j – true effect of the jth block.

All other parameters in the model (2.2.1) are same as in model (2.1.1.1).

where, τ_i is the ith treatment effect for $i = 1, 2, \ldots\ldots, v$.

H_{10}: $\mu_1 = \mu_2 = \mu_3 = \mu_4 = \mu_5 = \mu_6$ versus H'_{10}: At least two of these means are not equal.

Another hypothesis is about block effects.

H_{20}: All the block effects are equal versus H'_{20}: At least two block effects differ from each other significantly.

The model (2.2.1) leads to two-way classification and the analysis of variance will be carried out accordingly. The phrase two-way indicates that there are two primary sources of variation. Partial calculations for ANOVA are as follows:

Grand total $= y..$; $\bar{y} = y.. / k\,v$; C.F. $= y^2.. / kv$

$$\text{Total S.S.} = \sum_i \sum_j y_{ij}^2 - \text{C.F.} = \sum_i \sum_j (y_{ij} - \bar{y})^2 = \text{TSS}$$

$$\text{Treat. S.S.} = \frac{1}{k} \sum_i y^2_{i.} - \text{C.F.} = k\sum_i (\bar{y}_i - \bar{y})^2 = T_r\,\text{SS}$$

$$\text{Block S.S.} = \frac{1}{v}\sum_j y._j^2 - \frac{y^2_{..}}{kv} = \sum_j v\,(\bar{y}_j - \bar{y})^2 = \text{BSS}$$

$$\text{Error S.S.} = \text{TSS} - T_r\,\text{SS} - \text{BSS} = \sum_i \sum_j (y_{ij} - \bar{y}_{i.} - \bar{y}._j + \bar{y})^2 = ESS$$

Table 2.2.2 Two-way ANOVA

Source	d.f.	S.S.	M.S.	F-value	Expected mean squares Model–I	Model–II
Blocks	$k - 1 = \upsilon_1$	BSS	BSS/υ_1 = BMS	BMS/EMS = F_B	$\sigma_e^2 + \dfrac{k}{\upsilon-1}\Sigma\tau_i^2$	$\sigma_e^2 + b\,\sigma_\tau^2$
Treats.	$v - 1 = \upsilon_2$	T_rSS	T_rSS/υ_2 =T_rMS	T_rMS/EMS = F_{tr}	$\sigma_e^2 + \dfrac{v}{k-1}\Sigma_j\beta_j^2$	$\sigma_e^2 + v\,\sigma_\beta^2$
Error	$(k-1) \times (v-1) = \upsilon_3$	ESS	ESS/υ_3 = EMS		σ_e^2	σ_e^2
Total	$kv - 1$	TSS				

In the analysis of RBD, the hypothesis of main interest is H_{10} whereas the test of significance of H_{20} simply reveals whether the blocking has meaningfully contributed to reduce the experimental error or not. Therefore, even though H_{20} is rejected no further test is applied to know which of the blocks differ significantly. To arrive at a decision about H_{20}, F_B has to be compared with the tabulated value $F_{\alpha;\,\upsilon_1,\upsilon_3}$. If $F_B \geq F_{\alpha;\,\upsilon_1,\upsilon_3}$, H_{20} is rejected, otherwise not. Similarly, if $F_{tr} \geq F_{\alpha;\,\upsilon_2,\upsilon_3}$, then reject H_{10}. In this situation, one has to apply some appropriate test to further explore which of the pairwise and/or non-pairwise contrasts are significant and which of them are equally effective.

No design is perfect and hence each design has some uppers and downers. One design may be most appropriate for one experiment and absolutely unsuitable for the other. The suitability of a design depends on the objectives of an experiment and the experimental material and resources available. Some uppers and downers of this design are given over here.

Uppers

1. Randomized block design is more efficient than CRD, if one nuisance variable is to be controlled.
2. Treatment effects are estimated more precisely than CRD.
3. Analysis of data is simple.
4. It controls one nuisance variable at the planning level of the design.
5. If there is one missing value, it can easily be estimated and analysis of data does not create any complication except a few adjustments.

Downers

1. In cases when the number of treatments is large it becomes difficult to form a homogeneous block.
2. If there are two or more known nuisance variables, then this design is no more suitable since RBD cannot control more than one nuisance variable.
3. When two or more missing values occur, they cannot be ignored. The missing value(s) has to be estimated and adjustments in ANOVA have to be made accordingly.

Example 2.2: An experiment was conducted with nine varieties of gram in a randomized block design having four blocks to compare their yield potential. Yields were obtained as tabulated below. Analysis of data has been carried out and the results are interpreted as per points discussed in theory.

Computation:

$$\text{C.F.} = \frac{(935.1)^2}{36} = 24289.22$$

Varieties	Blocks				*Total*	*Mean*
	B_1	B_2	B_3	B_4		
V_1	20.2	41.4	29.0	25.6	116.2	29.050
V_2	18.5	28.8	15.0	24.8	87.1	21.775
V_3	16.5	25.5	11.5	13.5	67.0	16.750
V_4	27.7	30.8	22.5	31.3	112.3	28.075

(Contd.)

Varieties	Blocks				Total	Mean
	B_1	B_2	B_3	B_4		
V_5	32.0	34.7	29.2	30.5	126.4	31.600
V_6	31.3	35.2	16.3	18.7	101.5	25.375
V_7	25.5	43.7	13.2	18.1	100.5	25.125
V_8	39.2	46.3	20.1	20.2	125.8	31.450
V_9	20.6	31.8	19.5	26.4	98.3	24.575
Total	231.5	318.2	176.3	209.1	935.1	

Note: Totals and treatment means are also adjusted along with the data for the sake of brevity.

$$\text{Total S.S.} = 20.2^2 + 41.4^2 + \dots\dots + 19.5^2 + 26.4^2 - \text{C.F.}$$
$$= 26981.03 - \text{C.F.} = 2691.81$$

$$\text{Block S.S.} = \frac{1}{9}(231.5^2 + 318.2^2 + 176.3^2 + 209.1^2) - \text{C.F.}$$
$$= 25516.44 - \text{C.F.} = 1227.22$$

$$\text{Varieties S.S.} = \frac{1}{4}(116.2^2 + 87.1^2 + \dots\dots + 125.8^2 + 98.3^2) - \text{C.F.}$$
$$= 25014.28 - \text{C.F.} = 725.06$$
$$\text{Residual S.S.} = 2691.81 - 1227.22 - 725.06 = 739.53$$

ANOVA Table

Source	d.f.	S.S.	M.S.	F-value	Tabulated F-value
Blocks	3	1227.22	409.07	13.277*	$F_{.05;3,23} = 3.03$
Varieties	8	725.06	90.63	2.941*	$F_{.05;8,23} = 2.80$
Residual	24	739.53	30.81		
Total	35	2691.81			

Note: For F-values see table no. B-7.

* Significant at $\alpha = 0.05$

Calculated value of F for varieties is greater than the tabulated value of F at 5% level of significance, hence it is inferred that varieties differ significantly in respect of their grain yield. Now the question remains to establish which varieties differ significantly from one another and which do not? For this purpose, Duncan's multiple range test has been applied. There are nine varieties and hence for $9C_2$ paired means Duncan's least significant ranges at 5% level of significance and respective distances from 2 to 9 between ordered means are worked out by the formula (1.9.3.1). Critical values of $q^*_{\alpha;\ d,\ n_e}$ where d is taken as the distance between ordered means, are obtained from the Table B-8, provided for Duncan's multiple range test in Statistical tables.

$$S.E. = \sqrt{s_e^2/r} = \sqrt{30.81/4} = 2.775$$

Duncan's least significant ranges are:

$$D_9 = 2.775 \times 3.37 \quad = 9.352$$
$$D_8 = 2.775 \times 3.345 = 9.282$$
$$D_7 = 2.775 \times 3.315 = 9.199$$
$$D_6 = 2.775 \times 3.276 = 9.091$$
$$D_5 = 2.775 \times 3.226 = 8.952$$
$$D_4 = 2.775 \times 3.16 \quad = 8.769$$
$$D_3 = 2.775 \times 3.066 = 8.508$$
$$D_2 = 2.775 \times 2.919 = 8.100$$

Now a table of differences of paired means have been prepared and displayed below.

Difference table

Varieties	Means	V_5 31.600	V_8 31.450	V_1 29.050	V_4 28.075	V_6 25.375	V_7 25.125	V_9 24.575	V_2 21.775
V_3	16.750	14.850	14.700	12.300	11.325	8.625	8.375	7.825	5.025
V_2	21.775	9.825	9.675	7.275	6.300	3.600	3.350	2.800	
V_9	24.575	7.025	6.875	4.475	3.500	0.800	0.550		
V_7	25.125	6.475	6.325	3.925	2.950	0.250			
V_6	25.375	6.225	6.075	3.675	2.700				
V_4	28.075	3.525	3.375	0.975					
V_1	29.050	2.550	2.400						
V_8	31.450	0.150							

The upper triangle elements of the difference table cover all the 36 paired differences of varietal means.

First column differences under V_5 are to be compared with Duncan's critical value from D_9 to D_2 one to one. Only the difference of V_5 from V_3 and V_2 is greater than D_9 and other differences are less than the corresponding critical values. Hence, only V_5 is significantly better than V_3 and V_2 is at par with remaining varieties.

The differences under V_8 are to be compared with Duncan's critical values from D_8 to D_2 one to one. Only the difference of V_8 from V_3 and V_2 is greater than D_8 and other differences are less than corresponding critical values. Hence, V_8 is significantly better than V_3 and V_2.

Similarly comparing differences in other columns, it is apparent that the varieties V_1 and V_4 are superior to V_3 and all other varieties are at par in their performance.

2.3 ANALYSIS OF DATA OF AN RBD WITH ONE MISSING VALUE

Allan and Wishart (1930) presented for the first time the formula for estimating one missing value in RBD and also in Latin square design (discussed ahead). Yates (1933) derived the formula for a missing value by minimizing error mean square. The same is reproduced here.

Consider an experiment in RBD in which an observation is lost due to an unforeseen reason for the ith treatment in jth block. The observations recorded on the experimental units are presented in Table 2.3.1.

Table 2.3.1 Data arranged in two-way table

Blocks

Treats.	1	2	....	j	....	k	Total
1	y_{11}	y_{12}	....	y_{1j}	....	y_{1k}	$y_1.$
2	y_{21}	y_{22}	....	y_{2j}	....	y_{2k}	$y_2.$
.							
.							
i	y_{i1}	y_{i2}	....	X_{ij}	....	y_{ik}	$y'_i.$
.							
.							
v	y_{v1}	y_{v2}	....	y_{vj}	....	y_{vk}	$y_v.$
Total	$y_{.1}$	$y_{.2}$	....	$y'_{.j}$	....	$y_{.k}$	$y'_{..}$

In the above table, $y'_i.$ is the row total of $(k-1)$ available observations in the ith row and $y'_{.j}$ is the column total of $(v-1)$ observations in the jth column. To estimate X_{ij}, find the expressions for sum of squares including the X_{ij} in the usual manner as given below. The estimate of X_{ij} is obtained by the method of least squares in such a way that the sum of square due to error is minimized.

$$\text{Total S.S.} = \sum_{i'} \sum_{j'} y_{i'j'}^2 + X_{ij}^2 - \frac{\left(y'_{..} + X_{ij}\right)^2}{k \times v} = \text{TSS} \tag{2.3.1}$$

Block S.S. $= \frac{1}{v}\sum_{j'} y^2._{.j'} + \frac{1}{v}\left(y'_{.j} + X_{ij}\right)^2 - \frac{\left(y'_{..} + X_{ij}\right)^2}{k \times v} = \text{BSS}$ (2.3.2)

Treat S.S. $= \frac{1}{k}\sum_{i'} y'^2_{i'.} + \frac{1}{k}\left(y'_{i.} + X_{ij}\right)^2 - \frac{\left(y'_{..} + X_{ij}\right)^2}{k \times v} = \text{TrSS}$ (2.3.3)

Error S.S. $= \sum_{i'}\sum_{j'} y^2_{i'j'} + X^2_{ij} - \frac{1}{v}\sum_{j'} y^2._{.j'} - \frac{1}{k}\sum y^2_{i'.} + \frac{\left(y'_{..} + X_{ij}\right)^2}{k \times v}$

$$- \frac{1}{v}\left(y'_{.j} + X_{ij}\right)^2 - \frac{1}{k}\left(y'_{i.} + X_{ij}\right)^2 = \text{ESS}$$ (2.3.4)

To estimate X_{ij}, differentiate partially ESS with respect to X_{ij} and equate it to zero. Let the estimated value of X_{ij} be $\hat{X}_{ij}$. Thus, replace X_{ij} by $\hat{X}_{ij}$.

$$\frac{\partial}{\partial X}(\text{ESS}) = 2\hat{X}_{ij} - \frac{2}{k}\left(y'_{i.} + \hat{X}_{ij}\right) - \frac{2}{v}\left(y'_{.j} + \hat{X}_{ij}\right) + \frac{2}{kv}\left(y'_{..} + \hat{X}_{ij}\right) = 0$$

or $\left(\frac{1}{v} + \frac{1}{k} - \frac{1}{kv} - 1\right)\hat{X}_{ij} = -\frac{1}{v}y'_{.j} - \frac{1}{k}y'_{i.} + \frac{1}{kv}y'_{..}$ Since $2 \neq 0$

$$(k + v - kv - 1)\hat{X}_{ij} = -ky'_{.j} - vy'_{i.} + y'_{..}$$

$$\hat{X}_{ij} = \frac{ky'_{.j} + vy'_{i.} - y'_{..}}{(k-1)(v-1)}$$ (2.3.5)

Replace X_{ij} by $\hat{X}_{ij}$ in the data Table 2.3.1 and prepare ANOVA table with kv observations in the usual manner. Of course, certain adjustments have to made in ANOVA table to avoid biased results. The first and foremost change is to reduce the total degrees of freedom by one. This obviously causes a reduction of one degree of freedom in error d.f. In this way, the error d.f. for error are $[(k - 1) (v - 1) - 1]$. After inserting the estimated value of the missing observation, the treatment sum of square has an upward bias. Hence, this bias has to be adjusted for a valid conclusion. Correction for bias is,

$$C = \frac{\left[y'._{.j} - (v-1)\hat{X}_{ij}\right]^2}{v(v-1)}$$ (2.3.6)

Corrected sum of square = T_r SS $- C = T'_r$SS (2.3.7)

In ANOVA table, T'_rSS will be used for treatment sum of square for all purposes. Total S.S. has also to be adjusted by subtracting the correction factor C from TSS. In this manner, the error sum of square remains same as it was before adjustment.

Therefore, adjusted total S.S. = TSS − C $\qquad$ (2.3.8)

Another rectification that has to be made is in the standard error of means.

S.E. of a treatment mean not having a missing value $= \sqrt{s_e^2/k}$ $\qquad$ (2.3.9)

S.E. of the difference between two such means $= \sqrt{2s_e^2/k}$ $\qquad$ (2.3.10)

Standard error of a paired contrast of treatment means, one having a missing value and any other treatment mean is,

$$= \sqrt{\mathrm{EMS}\left[\frac{2}{k} + \frac{v}{k(k-1)(v-1)}\right]} \qquad (2.3.11)$$

Using the standard errors according to the pair of means their significance can be tested by a suitable test provided the F for treatments in ANOVA comes out to be significant.

Example 2.3: Data of a randomized block design with five sources of sulphur and four blocks pertaining to chlorophyll content of pea leaves in mg/g of fresh weight are presented below. Some how one sample was spoiled due to the negligence of the analyst. Hence, there exists one missing value in the data given below. Partial calculations are also displayed along with the data to save space.

Blocks

Treatments	I	II	III	IV	Total	Mean
Control (S_0)	0.679	0.852	0.513	0.507	2.551	0.6378
Ele. Sulphur (250kg S/ha) S_1	0.952	1.002	X_{13}	0.621	2.575 + X_{13}	0.8250
Gypsum (250 kg S/ha) S_2	0.899	0.919	0.718	0.679	3.215	0.8038
H_2SO_4 (0.1% foliar spray) S_3	0.986	0.949	0.845	0.780	3.560	0.8900
F_e- EDDHA (2% foliar spray) S_4	0.911	0.922	0.668	0.746	3.247	0.8118
Total	4.427	4.644	2.744 + X_{13}	3.333	15.148 + X_{13}	

Analysis of data having one missing value will be carried out as per procedure given in section (2.3).

Estimate the missing value by the formula (2.3.5) is,

$$\hat{X}_{13} = \frac{4 \times 2.744 + 5 \times 2.575 - 15.148}{3 \times 4}$$

$$= 0.725$$

$$C.F. = (15.873)^2/20 = 12.5976$$

$$Total \ S.S. = 0.679^2 + 0.952^2 + \ldots\ldots + 0.725^2 + \ldots\ldots + 0.746^2 - C.F.$$

$$= 13.0363 - 12.5976 = 0.4387$$

$$Block \ S.S. = \frac{1}{5}(4.427^2 + 4.644^2 + 3.469^2 + 3.333^2) - C.F.$$

$$= 12.8616 - 12.5976 = 0.2640$$

$$Treat.S.S. = \frac{1}{4}(2.551^2 + 3.300^2 + 3.215^2 + 3.560^2 + 3.247^2) - C.F.$$

$$= 12.7376 - 12.5976 = 0.1400$$

Correction for treatment S.S. by the formula (2.3.6) is,

$$C = \frac{(2.744 - 4 \times 0.7252)^2}{5 \times 4} = 0.0012$$

Corrected treatment S.S. $= 0.1400 - 0.0012 = 0.1388$

Corrected total S.S. $= 0.4387 - 0.0012 = 0.4375$

ANOVA Table

Source	d.f.	S.S.	M.S.	F-value
Blocks	3	0.2640	0.0880	27.50*
Treats	4	0.1388	0.0347	10.84*
Error	11	0.0347	0.0032	
Total	18	0.4375		

Tabulated values of F for $\alpha = 0.05$ and corresponding d.f. are, $F_{.05; \ 3,11} = 3.59$, $F_{.05; \ 4,11} = 3.36$.

The calculated values of F are significant as they are greater than the corresponding F-values given above. This reveals that there is a significant difference among blocks and among treatment effects. Significant difference among blocks affirms that blocking has meaningfully reduced the experimental error. Since the treatments differ significantly, it is further required that pairwise comparisons be tested to ascertain which of the treatments differ significantly and which are alike.

Standard error of the difference of a pair of treatment means by the formula (2.3.10) is,

$$S.E. = \sqrt{(2 \times 0.0032)/4} = 0.0400$$

Standard error of a contrast of treatment mean having a missing value and any other treatment mean by the formula (2.3.11) is,

$$\text{S.E.}_2 = \sqrt{0.0032 \left(\frac{2}{4} + \frac{5}{4 \times 3 \times 4} \right)} = 0.0440$$

Corresponding lsds will be obtained by multiplying S.E.$_1$ and S.E.$_2$ by the tabulated value of t from table no. B-5, for $\alpha = 0.05$ and error d. f., i.e. $t_{.05; 11} = 2.201$. Thus,

$$\text{lsd}_1 = 0.0400 \times 2.201 = 0.0880$$
$$\text{lsd}_2 = 0.0440 \times 2.201 = 0.0968$$

While comparing treatment S_1 mean with any other treatment mean, use lsd$_1$, otherwise use lsd$_2$.

Treatment means in ascending order are,

S_0	S_2	S_4	S_1	S_3
0.6378	0.8038	0.8118	0.8250	0.8900

In the above sketch, the treatment means which are underlined show that there is no significant difference among them. So, it is concluded that all treatments are superior to control (S_0) but others are equally effective in increasing the chlorophyll.

2.4 ANALYSIS OF AN RBD WITH SEVERAL MISSING VALUES

When there are two or more missing values in a randomized block design, analysis of variance can be carried with the help of analysis of covariance. This will be discussed in Chapter 10. Another method which is often preferred when there are two or three missing values is the *iterative method* though this is not as exact as the method of analysis of covariance. Preference of this method lies in its convenience and direct approach. The same has been discussed over here.

Iterative method: This method is explained step by step.

Step 1: Substitute the average value of all available observations in the data table where the missing values exist except one cell. Estimate this missing value by the formula (2.3.5). Repeat the process of substituting the general mean for another missing and estimate again the missing value by the formula (2.3.5). Continue the process till all missing values are estimated. In this way, the first set of estimates of missing values is obtained. This is called first approximation. After one complete round, insert estimated values at the proper position of all missing values except at one place. Estimate again this missing now at hand as earlier by using the new set of data. Similarly, estimate other missing values one by one. Continue the cycles of estimation unless a set of estimated values is obtained which is almost same as the previous one.

Step 2: In ANOVA, reduce the total degree of freedom by the number of missing values. Thereby this reduces the error degrees of freedom by the same number.

Step 3: After substituting all missing values, analyze the data . . the usual manner.

Step 4: Another amendment in analysis of variance is to reduce the treat ment sum of squares for an upward bias by a correction factor which is introduced due to substitution of missing values. The correction factor is the sum of corrections C calculated by the formula (2.3.6) for each missing value individually. Subtract this total value of correction factor from treatment sum of square to avoid the bias of the test.

Step 5: Also reduce the total sum of square by the correction C so that the error sum of square is not affected because of the adjustment in treatment sum of square.

Step 6: When the treatment effects come out to be significant by F-test, then there arises a need for testing the significance of the pairwise and non-pairwise contrasts. All tests involve the formulae of standard errors of the means and contrasts. The same have been given below in all possible situations.

Taking X_{ij} as the missing value, the estimator of ith treatment effect is given by the formula,

$$\hat{t}_i = \frac{1}{k}\left(y'_{i.} + X_{ij}\right) \tag{2.4.1}$$

The variance of the estimator $\hat{t}_i$ is equal to,

$$\text{Var}\left(\hat{t}_i\right) = \frac{1}{k-1}\left[1 + \frac{1}{k(v-1)}\right] \times \text{EMS} \tag{2.4.2}$$

$$= \frac{1}{k}\left[1 + \frac{v}{(k-1)(v-1)}\right] \times \text{EMS} \tag{2.4.2.1}$$

The variance of an estimator of a treatment mean not having a missing value is,

$$= \text{EMS}/k \tag{2.4.3}$$

The variance of a treatment contrast $\sum_l a_l \, \bar{t}_l = 0$, $(1 = 1, 2, \ldots\ldots, v)$ in which one treatment i has a missing value and others do not, is given by the formula,

$$\text{Var}\left(\sum_l a_1 \bar{t}_1\right) = \frac{\text{EMS}}{k}\left[\sum_{l \neq i} a_1^2 + \sum_i a_i^2 \times \left(1 + \frac{v}{(k-1)(v-1)}\right)\right] \tag{2.4.4}$$

Variance of a pairwise contrast in which both the treatments have one missing value can be obtained by the formula,

$$\text{Var}\left(\hat{t}_i - \hat{t}_u\right) = \frac{2\text{EMS}}{k}\left[1 + \frac{v}{(k-1)(v-1)}\right] \tag{2.4.5}$$

Rest of the analysis procedure remains same as in case of one missing value.

Example 2.4: An experiment was conducted with nine varieties of maize in a randomized block design. The grain yields per plant (in gm) in four replications were as tabulated below. Anyhow two observations, i.e. for variety V_3 in block B_1 and V_7 in B_3, were missed as the plants died during experimentation. Totals and means are also displayed in the table to save space.

Varieties	B_1	B_2	B_3	B_4	Total	Variety means
V_1	20.2	41.4	29.0	25.6	116.2	29.05
V_2	18.5	28.8	15.0	24.8	87.1	21.78
V_3	X_{31}	25.5	11.5	13.5	50.0*	16.67*
V_4	27.7	30.8	22.5	31.3	112.3	28.08
V_5	32.0	34.7	29.2	30.5	126.4	31.60
V_6	31.3	35.2	16.3	18.7	101.5	25.38
V_7	25.5	43.7	X_{73}	18.1	87.3*	27.20*
V_8	39.2	46.3	20.1	20.2	125.8	31.45
V_9	20.6	31.8	19.5	26.4	98.3	24.58
Total	215.0*	318.20	163.1*	209.10	905.4*	26.63*

* The totals and means are based on the available values.

To analyze the data, substitute the value of $X_{73} = 26.63$ and get the first estimate of X_{31} by the formula (2.3.5).

$$X'_{31} = \frac{4 \times 215.0 + 9 \times 50.5 - 932.03}{(4-1)(9-1)} = 15.94$$

Again insert the mean value 26.63 for X_{31} and estimate X_{73}.

$$X'_{73} = \frac{4 \times 163.1 + 9 \times 87.3 - 932.03}{(4-1)(9-1)} = 21.09$$

In this way, the first set of approximate estimates of missing values are available. Now insert the value of $X'_{73} = 21.09$ and estimate X_{31} again.

$$X''_{31} = \frac{4 \times 215.0 + 9 \times 50.5 - 926.49}{24} = 16.17$$

Putting the value of $X_{31} = 15.94$, the second estimate of X_{73} is,

$$X''_{73} = \frac{4 \times 163.1 + 9 \times 87.3 - 921.34}{24} = 21.53$$

Since the values of two sets of estimates are not very close, repeat the estimation process to get a third set of estimates. Now put $X_{73} = 21.53$ and obtain the third estimate of X_{31}.

$$\hat{X}_{31} = \frac{4 \times 215.0 + 9 \times 50.5 - 926.93}{24} = 16.16$$

Similarly on putting the value of $X_{31} = 16.17$, the third estimate of X_{73} is,

$$\hat{X}_{73} = \frac{4 \times 163.1 + 9 \times 87.3 - 921.57}{24} = 21.52$$

Second and third sets of estimates are almost same. Therefore, there is no need to go for fourth set. Finally, take the values, $\hat{X}_{31} = 16.16$ and $\hat{X}_{73} = 21.52$. Now analyze the data as if all values are present making the adjustments as discussed in theory in this section.

Sum of squares due to blocks, treatments and error are calculated in the usual way as displayed below.

C.F. = 24705.55

Total S.S. = 27258.80 − C.F. = 2553.25

Blocks S.S = 25832.61 − C.F. = 1127.06

Treats S.S. = 25438.31 − C.F. = 732.76

There are two missing values in two different blocks; the correction factor for upward bias in treatment sum of square is,

$$C = \frac{(215.0 - 8 \times 16.16)^2 + (163.1 - 8 \times 21.52)^2}{9(9-1)}$$

$$= \frac{7347.92 + 82.08}{72} = 103.19$$

Corrected treat.S.S. = 732.76 − 103.19 = 629.57

Corrected total S.S. = 2553.25 − 103.19 = 2450.06

Error S.S. = 2450.06 − 1127.06 − 629.57 = 693.4

ANOVA Table

Source	d.f.	S.S.	M.S.	F-value	Tabulated F-value
Blocks	3	1127.06	375.69	1127.06/31.52 = 11.92	$F_{.05;\,3,\,22} = 3.05$
Treats.	8	629.57	78.70	78.70/31.52 = 2.50	$F_{.05;\,8,\,22} = 2.40$
Error	22	693.43	31.52		
Total	33	2450.06			

F-values in ANOVA table reveal that block and treatment effects are significantly different. Significance of block effects affirms that blocking has remarkably contributed in reducing the experimental error. The null hypothesis that all varietal effects are equal in magnitude has been rejected. So it becomes necessary to examine which of τ_i's differ significantly from one another and others do not. For this, first calculate the standard errors of the difference of variety means.

Standard error of the difference between two variety means having no missing value $= \sqrt{2 \times \text{EMS}/k} = \sqrt{(2 \times 31.52)/4} = 3.97$.

Standard error of a difference of two variety means in which one treatment has a missing value and other does not,

$$= \text{S.E.}\left(t_i - t_{i'}\right) = \sqrt{\left[\frac{31.52}{4}\left(1 + 1 + \frac{9}{(4-1)(9-1)}\right)\right]} = 4.33$$

Standard error of the difference between two variety means in which each of them has a missing value $= \sqrt{\left[\frac{2 \times 31.52}{4}\left(1 + \frac{9}{(4-1)(9-1)}\right)\right]} = 4.66$

For pairwise comparisons of varietal means, the author has preferred to use Fisher's lsd test rather than S-N-K test or Duncan's multiple range test. These tests will be complicated in the sense that for each critical difference one has to use a different standard error that too should be divided by $\sqrt{2}$ as it is implies in $q_{\alpha; k, \upsilon}$ values. For the same distance among varietal means, different critical values have to be worked out. Those in very efficient test should use either of these tests.

(i) lsd at 5 % level of significance for the pairs of varieties having no missing value $= 3.97 \times 2.074 = 8.23$ since, $t_{.05; 22} = 2.074$, from table no. B-5.

(ii) lsd at 5 % probability for a pair of varieties in which a variety has a missing value and other with no missing value $= 4.33 \times 2.074 = 8.98$

(iii) lsd at 5 % significance level for a pair of varieties in which both are having a missing value $= 4.66 \times 2.074 = 9.66$

Now arrange the varietals means in descending order and compare the differences between means with corresponding lsd given above as per the situation. Start from the highest mean and find the difference from each of the lower order mean value one by one. If this difference is greater than or equal to the corresponding lsd, then it is significant otherwise they are at par.

Repeat the process of comparing second highest variety mean and decide about the significance of differences. Continue the process till the second last variety V_2 is compared with V_3^*. Star over variety indicated that this particular variety has a missing value. Further similar letters in a column show non-significant differences.

Varieties	Varietal means	Significant or non-significant differences
V_5	31.60	a
V_8	31.45	a c
V_1	29.05	a c e
V_4	28.08	a c e g
V_7^*	27.20	a c e g m
V_6	25.38	a c e g m p
V_9	24.58	a c e g m p u
V_2	21.78	b d e g m p u z
V_3^*	16.67	b d f h o p u z

From the above diagram, it is amply clear that varieties V_5 and V_8 are significantly better than varieties V_2 and V_3. Varieties V_1, V_7, V_4 are significantly superior than V_3. Rest all pairs of varieties are at par in their performance with regard to grain yield.

2.5 DERIVATION OF EXPECTED MEAN SQUARES

It has been clearly stated that each experimental design is specified by a statistical model. The model may be classified as fixed, random or mixed effect model. Each type of model is subjected to various restrictions. These restrictions are bound to have a bearing on analysis of experimental data. The consequences of model type give rise to different expected mean squares of factors and these expected mean squares determine the manner in which the null hypothesis about factor effects are to be tested in ANOVA.

As a simple rule, in a fixed effect replicated experiment, all hypotheses about fixed effects be tested against experimental error, i.e. as a ratio of effect M.S. to error M.S. providing the value of F-statistic. On the other hand, in random effect model, main effects are tested against proper interaction term and the significance of interaction is tested against experimenter error.

The users of SPSS, MINITAB, or any other computer package need not to know the expected mean squares. Of course, they will have to instruct the computer which model is being assumed for analysis. Further computer automatically takes care of the model specification. From the discussion so far, it is evinced that expected mean squares are implicit in analysis of variance. Also, an investigator is never required to derive the expressions for expected mean squares for analysis of data. But one would always be curious to know about the origin and form in which they occur.

As already discussed, an additive model is considered for all designs. Each model is specified by certain assumptions about the nature of factors involved in the model. Once the foregoing assumptions have been made, the analysis of experimental data consists of two parts, (i) the estimation of parameters, (ii) testing of hypotheses about the parameters comprising the model. Once the expected mean squares are determined, it becomes simple to adapt a proper test statistic for testing a null hypothesis. In case exact tests are available for the hypotheses of a model, then this is known as *completely specified model*, otherwise an *incompletely specified model*. In case of incompletely specified models, theory of *preliminary test of significance* (PTS) has to be applied or some synthesized variance ratio tests are used as suggested by Cochran (1951), Satterthwaite (1941, 1946), Davenport and Webster (1973) and some other workers.

Note: The author has some what digressed from the theme of expected mean squares. Anyhow, the author returns back to the estimation of parameters and derivation of mean squares for fixed effect model for a randomized block design. The estimates of parameters and expected mean squares will directly be given for all other designs henceforth.

Derivation: Consider the fixed effect model for RBD.

$$y_{ij} = \mu + \tau_i + \beta_j + e_{ij} \qquad (2.5.1)$$

$$i = 1, 2, \ldots\ldots, v; \quad j = 1, 2, \ldots\ldots, k.$$

Under model-I, the null hypothesis H_{10} against alternative hypothesis H'_{10} under test is,

$$H_{10} : \mu_i = \mu_{i'} \text{ versus } H'_{10} : \mu_i \neq \mu_{i'} \text{ and also } \sum_i \tau_i = 0$$

Similarly, the hypotheses about blocks are,

$H_{20} : \beta_j = \beta_{j'}$ versus $H'_{20} : \beta_j \neq \beta_{j'}$ where, $j \neq j'$ for some j and $\sum_j \beta_j = 0$.

Also e_{ij} are identically and independently distributed (i.i.d.) as $N(0, \sigma_e^2)$.

Here the estimation of parameters is presented stepwise by the method of least squares.

$$e_{ij} = \left(y_{ij} - \mu - \tau_i - \beta_j \right) \qquad (2.5.2)$$

Squaring both sides of (2.5.2) and taking sum over all $v\,k$ observations, the equation is,

$$\sum_i \sum_j e_{ij}^2 = \sum_i \sum_j \left(y_{ij} - \mu - \tau_i - \beta_j \right)^2 \qquad (2.5.3)$$

Let us take the quantity, $\sum_i \sum_j e_{ij}^2 = Q$. To estimate μ, τ_i, β_j, differentiate both sides of (2.5.3) partially with respect to μ, τ_i, β_j respectively and equate them to zero. In this way, three normal equations are obtained. Solving the three equations under the assumptions of fixed effect model, one gets the estimators of these three parameters as follows.

$$\frac{\partial Q}{\partial \mu} = -2\sum_i \sum_j \left(y_{ij} - \hat{\mu} - \tau_i - \beta_j \right) = 0$$

or

$$\sum_i \sum_j \left(y_{ij} - \hat{\mu} - \tau_i - \beta_j \right) = 0$$

$$\sum_i \sum_j y_{ij} - kv\hat{\mu} - k\sum_i \tau_i - v\sum_j \beta_j = 0$$

$$kv\hat{\mu} = \sum_i \sum_j y_{ij} = y.. \quad \text{Since } \sum_j \tau_i = 0, \ \sum_j \beta_j = 0$$

$$\hat{\mu} = y../k \, v = \bar{y} \tag{2.5.4}$$

$$\frac{\partial Q}{\partial \tau_i} = -2\sum_j \left(y_{ij} - \hat{\mu} - \hat{\tau}_i - \beta_j \right) = 0$$

or

$$\sum_j y_{ij} - k\bar{y} - k\hat{\tau}_i - \sum_j \beta_j = 0$$

$$y_{i.} - k\bar{y} - k\hat{\tau}_i = 0 \qquad \text{Since } \sum_j \beta_j = 0$$

$$\hat{\tau}_i = \frac{1}{k} y_{i.} - \bar{y} = \bar{y}_i - \bar{y} \tag{2.5.5}$$

$$\frac{\partial Q}{\partial \beta_j} = -2\sum_i \left(y_{ij} - \bar{y} - \tau_i - \hat{\beta}_j \right) = 0$$

or

$$\sum_i y_{ij} - v\bar{y} - \sum_i \tau_i - v\hat{\beta}_j = 0$$

$$y_{.j} - v\bar{y} - v\hat{\beta}_j = 0 \qquad \text{Since } \sum_i \tau_i = 0$$

$$\hat{\beta}_j = \frac{1}{v} y_{.j} - \bar{y} = \bar{y}_j - \bar{y} \tag{2.5.6}$$

To obtain the expected mean squares for various factors, consider total variance of an experimental data. This variance can be partitioned into factor variances namely, variance due to blocks, treatments, error, etc.

Firstly write corrected sum of squares and then on dividing them by their corresponding degrees of freedom, one gets mean sum of squares which are in no way different from their respective variances. The derivation of expected mean squares is as follows.

$$\sum_i \sum_j \left(y_{ij} - \bar{y} \right)^2 = \sum_i \sum_j \left[\left(\bar{y}_i - \bar{y} \right) + \left(\bar{y}_j - \bar{y} \right) + \left(y_{ij} - \bar{y}_i - \bar{y}_j + \bar{y} \right) \right]^2$$

$$= k\sum_i \left(\bar{y}_i - \bar{y} \right)^2 + v\sum_j \left(\bar{y}_j - \bar{y} \right)^2 + \sum_i \sum_j \left(y_{ij} - \bar{y}_i - \bar{y}_j + \bar{y} \right)^2 \tag{2.5.7}$$

Because all cross-product terms vanish as all observations are taken to be independent and hence sum of deviations from respective means are zero.

$$\sum_i \sum_j \left(y_{ij} - \bar{y} \right)^2 = \left(\frac{\sum_i y_{i.}^2}{k} - \frac{y..^2}{kv} \right) + \left(\frac{\sum_j y_{.j}^2}{v} - \frac{y..^2}{kv} \right) + \text{Error S.S.} \tag{2.5.8}$$

$$= \text{Treat. S.S.} + \text{Block S.S.} + \text{Error S.S.} \tag{2.5.8.1}$$

Again taking sum over j in the model (2.5.1) and dividing by k, the equation is,

$$\frac{1}{k}\Sigma_j\, y_{ij} = \mu + \tau_i + \frac{1}{k}\Sigma_j\, \beta_j + \frac{1}{k}\Sigma_j\, e_{ij}$$

$$\bar{y}_i = \mu + \tau_i + \bar{e}_i \qquad (2.5.9)$$

Similarly taking sum over i in the model (2.5.1) and dividing by v, we get,

$$\bar{y}_j = \mu + \beta_j + \bar{e}_j \qquad (2.5.10)$$

Now summing over i and j and then dividing by kv, we obtain,

$$\bar{y} = \mu + \bar{e} \qquad (2.5.11)$$

Expected value of error sum of square

Substituting the value of y_{ij}, $y_{i\cdot}$, $y_{\cdot j}$ and $\bar{y}$ in the expression for error S.S., i.e.

$$\text{Error S.S.} = \Sigma_i\,\Sigma_j\left(y_{ij} - \bar{y}_i - \bar{y}_j + \bar{y}\right)^2$$

$$= \Sigma_i\,\Sigma_j\left(\mu + \tau_i + \beta_j + e_{ij} - \mu - \tau_i - \bar{e}_i - \mu - \beta_j - \bar{e}_j + \mu + \bar{e}\right)^2$$

$$= \Sigma_i\,\Sigma_j\left(e_{ij} - \bar{e}_i - \bar{e}_j + \bar{e}\right)^2$$

$$= \Sigma_i\,\Sigma_j\left(e_{ij}^2 + \bar{e}_i^2 + \bar{e}_j^2 + \bar{e}^2 - 2e_{ij}\bar{e}_i - 2e_{ij}\bar{e}_j + 2e_{ij}\bar{e} + 2\bar{e}_i\bar{e}_j - 2\bar{e}_i\bar{e} - 2\bar{e}_j\bar{e}\right)$$

$$= \Sigma_i\,\Sigma_j\, e_{ij}^2 + k\Sigma_i\,\bar{e}_i^2 + v\Sigma_j\,\bar{e}_j^2 + kv\bar{e}^2 - 2\Sigma_i\,\bar{e}_i\,\Sigma_j\, e_{ij} - 2\Sigma_j\,\bar{e}_j\,\Sigma_i\, e_{ij}$$

$$\qquad + 2\bar{e}\,\Sigma_i\,\Sigma_j\, e_{ij} + 2\Sigma_i\,\bar{e}_i\,\Sigma_j\, e_j - 2k\bar{e}\,\Sigma_i\,\bar{e}_i - 2v\bar{e}\,\Sigma_j\,\bar{e}_j$$

$$= \Sigma_i\,\Sigma_j\, e_{ij}^2 + k\Sigma_i\,\bar{e}_i^2 + v\Sigma_j\,\bar{e}_j^2 + kv\bar{e}^2 - 2k\Sigma_i\,\bar{e}_i^2 - 2v\Sigma_j\,\bar{e}_j^2$$

$$\qquad + 2kv\bar{e}^2 + 2kv\bar{e}^2 - 2kv\bar{e}^2 - 2kv\bar{e}^2$$

$$= \Sigma_i\,\Sigma_j\, e_{ij}^2 - k\Sigma_i\,\bar{e}_i^2 - v\Sigma_j\,\bar{e}_j^2 + kv\bar{e}^2$$

Thus, the value of expected error sum of square is,

$$\text{Exp(ESS)} = \Sigma_i\,\Sigma_j\, \text{E}\left(e_{ij}^2\right) - k\Sigma_i\,\text{E}\left(\bar{e}_i^2\right) - v\Sigma_j\,\text{E}\left(\bar{e}_j^2\right) + kv\text{E}\left(\bar{e}^2\right)$$

$$= kv\sigma_e^2 - k\sigma_e^2 - v\sigma_e^2 + \sigma_e^2$$

$$= (kv - k - v + 1)\sigma_e^2$$

$$= (k-1)(v-1)\sigma_e^2$$

Expected value of error mean sum of square is,

$$\sigma_e^2 = \frac{\text{ESS}}{(k-1)(v-1)} \qquad (2.5.12)$$

Expected value of treatment mean sum of squares

As given in (2.5.7),

$$\text{TrSS} = k \sum_i (\bar{y}_i - \bar{y})^2 \tag{2.5.13}$$

Substituting the value of $\bar{y}_i$ and $\bar{y}$ in terms of errors from the expressions (2.5.9) and (2.5.11) in (2.5.13) the equation is,

$$\text{TrSS} = k \sum_i (\mu + \tau_i + \bar{e}_i - \mu - \bar{e})^2$$

$$= k \sum_i (\tau_i + \bar{e}_i - \bar{e})^2$$

$$= k \sum_i (\tau_i^2 + \bar{e}_i^2 + \bar{e}^2 + 2\tau_i \bar{e}_i - 2\tau_i \bar{e} - 2\bar{e}_i \bar{e})$$

$$= k \sum_i \tau_i^2 + k \sum_i \bar{e}_i^2 + kv\bar{e}^2 + 2k \sum_i \tau_i \bar{e}_i - 2k\bar{e} \sum_i \tau_i - 2k\bar{e} \sum_i \bar{e}_i$$

Under the assumptions of the model, $\sum_i \tau_i = 0$ and $\sum_i \bar{e}_i = v\bar{e}$

Thus, $\quad \text{TrSS} = k \sum_i \tau_i^2 + k \sum_i \bar{e}_i^2 - kv\bar{e}^2$

Expectation of treatment sum of square is,

$$\text{E}(\text{TrSS}) = k \sum_i \text{E}(\tau_i^2) + k \sum_i \text{E}(\bar{e}_i^2) - kv\text{E}(\bar{e}^2) \tag{2.5.14}$$

Obviously under fixed effect model, $\text{E}(\tau_i) = \tau_i$ and $\text{E}(\bar{e}) = \bar{e}$ because $\bar{e}$ is a constant. Now the values of $\text{E}(\bar{e}_i^2)$ and $\text{E}(\bar{e}^2)$ are worked out separately in (2.5.14).

$$\text{E}(\bar{e}_i^2) = \text{E}\left(\frac{1}{k} \sum_j e_{ij}\right)^2$$

$$= \text{E}\left(\frac{1}{k^2} \sum_j e_{ij}^2\right) + \frac{1}{k^2} \text{E}\left(\sum_{j \neq j'} e_{ij} e_{ij'}\right)$$

Since e_{ij} and $e_{ij'}$ are independently $\text{N}(0, \sigma_e^2)$, $\text{E}(\sum e_{ij} e_{ij'}) = 0$, for $j \neq j'$

$$\text{E}(\bar{e}_i^2) = \frac{1}{k^2} \text{E}(\sum_j e_{ij}^2)$$

$$= \frac{1}{k^2} \sum_j \text{E}(e_{ij}^2)$$

$$= \frac{1}{k^2} k \, \text{V}(e_{ij}) = \frac{1}{k} \sigma_e^2$$

Similarly, $E(\bar{e}^2) = E\left(\frac{1}{kv}\Sigma_i \Sigma_j e_{ij}\right)^2$

$$= \frac{1}{k^2 v^2} E\left(\Sigma_i \Sigma_j e_{ij}\right)^2$$

$$= \frac{1}{k^2 v^2}\Sigma_i \Sigma_j E\left(e_{ij}\right)^2 + \frac{1}{k^2 v^2}\underset{i \neq i'}{\Sigma}\underset{j \neq j'}{\Sigma} e_{ij} e_{i'j'} \text{ for } i \neq i' \text{ and } j \neq j'$$

The second term on the right hand side becomes zero as e_{ij} are i.i.d. $N(0, \sigma_e^2)$. Therefore,

$$E(\bar{e}^2) = \frac{1}{k^2 v^2}k v \sigma_e^2$$

$$= \frac{1}{kv}\sigma_e^2$$

$$E(\text{TrSS}) = k\Sigma_i \tau_i^2 + kv\frac{1}{k}\sigma_e^2 - kv\frac{1}{kv}\sigma_e^2$$

$$= k\Sigma_i \tau_i^2 + (v-1)\sigma_e^2$$

As already given,

$$\text{Tr M.S.} = \frac{1}{v-1}\text{TrSS}$$

$$E(\text{TrM.S.}) = \frac{1}{v-1}E(\text{TrSS})$$

$$= \frac{1}{v-1}\left\{k\Sigma_i \tau_i^2 + (v-1)\sigma_e^2\right\}$$

$$= \frac{k}{v-1}\Sigma_i \tau_i^2 + \sigma_e^2 \qquad (2.5.15)$$

For model–II, in the same manner it can easily be proved that,

$$E(\text{TrM.S.}) = k\sigma_\tau^2 + \sigma_e^2 \qquad (2.5.16)$$

Proceding in the like manner it is tenuous to derive the expectation of block sum of squares. For model–I, it will be come out to be

$$E\ (\text{Bl.S.S.}) = \frac{v}{k-1}\Sigma_j \beta_j^2 + \sigma_e^2$$

And for model–II,

$$E(Bl.S.S.) = v\sigma_\beta^2 + \sigma_e^2$$

The derivation of expected mean squares makes one to understand its concepts clearly and can decide about the F-test in ANOVA without ambiguity. The expected mean squares for treatments and blocks reveal that F-test for treatments and blocks against experimenter error result into exact F-test. Hence, the models for randomized block complete designs are completely specified models.

Nota bene: The derivation of expected mean squares will not be given henceforth. But they will be presented according to the statistical model of the design directly and used as per requirement of the analysis. Those interested in the derivation of expected mean squares for other designs may do so at their own level.

2.6 SUBSAMPLING IN A RANDOMIZED BLOCK DESIGN

There are many situations especially in biological experiments where the observations are not recorded on the experimental units as such but a sample of individuals is selected randomly and independently from each experimental unit. Observations are made on each individual on the characteristic in which the interest of the researcher lies. For instance, one is interested to know the intensity of infection of plants in experimental plots treated by different pesticides. In such a case, it is usually assiduous to record the infection on each plant of a plot. Hence, a sample of some plants is randomly selected from each plot and infection is recorded on every selected plant only. In feeding experiments on insects, a fixed number of insects are taken in Petri dishes and then observations on growth, mortality or any other trait of interest are recorded under each feed. Such experiments lead to subsampling and the analysis of variance involves two error terms, namely the experimental error and sampling error. Practically the experimental error may be construed as treatment × block interaction. Let us consider a randomized block design with subsampling having v treatments, k blocks and m sampling individuals from each experimental unit. The data so collected can always be tabulated as follows:

Data table

| Treats. | Blocks | Subsampling units | | | | | | Units Total | Treats. Total |

Treats.	Blocks	1	2	-----	u	-----	m	Units Total	Treats. Total
	B_1	y_{111}	y_{112}	-----	y_{11u}	-----	y_{11m}	$y_{11\cdot}$	
	B_2	y_{121}	y_{122}	-----	y_{12u}	-----	y_{12m}	$y_{12\cdot}$	
T_1									$y_{1\cdot\cdot}$
	B_k	y_{1k1}	y_{1k2}	-----	y_{1ku}	-----	y_{1km}	$y_{1k\cdot}$	
	B_1	y_{211}	y_{212}	-----	y_{21u}	-----	y_{21m}	$y_{21\cdot}$	
	B_2	y_{221}	y_{222}	-----	y_{22u}	-----	y_{22m}	$y_{22\cdot}$	
T_2									$y_{2\cdot\cdot}$
	B_k	y_{2k1}	y_{2k2}	-----	y_{2ku}	-----	y_{2km}	$y_{2k\cdot}$	
	B_1	y_{i11}	y_{i12}	-----	y_{i1u}	-----	y_{i1m}	$y_{i1\cdot}$	
	B_2	y_{i21}	y_{i22}	-----	y_{i2u}	-----	y_{i2m}	$y_{i2\cdot}$	
T_i									$y_{i\cdot\cdot}$
	B_k	y_{ik1}	y_{ik2}	-----	y_{iku}	-----	y_{ikm}	$y_{ik\cdot}$	
	B_1	y_{v11}	y_{v12}	-----	y_{v1u}	-----	y_{v1m}	$y_{v1\cdot}$	
	B_2	y_{v21}	y_{v22}	-----	y_{v2u}	-----	y_{v2m}	$y_{v2\cdot}$	
									$y_{v\cdot\cdot}$
T_v									
	B_k	y_{vk1}	y_{vk2}	-----	y_{vku}	-----	y_{vkm}	$y_{vk\cdot}$	

Sum of squares for various sources of variations are calculated in the following manner.

Total of block $B_1 = y_{11\cdot} + y_{21\cdot} + \ldots + y_{i1\cdot} + \ldots + y_{v1\cdot} = y_{\cdot1\cdot}$

Similarly the totals of other blocks are calculated and displayed below.

Total of $B_2 = y_{\cdot2\cdot}$, $B_3 = y_{\cdot3\cdot}$, $\ldots$, $B_k = y_{\cdot k\cdot}$.

Now the sum of squares,

$$C.F. = \frac{y_{\cdots}^2}{vkm} \tag{2.6.1}$$

$$\text{Block S.S.} = \frac{1}{vm}\left(y_{\cdot1\cdot}^2 + y_{\cdot2\cdot}^2 + \ldots + y_{\cdot k\cdot}^2\right) - C.F. = BSS \tag{2.6.2}$$

$$\text{Treat S.S.} = \frac{1}{km}\left(y_{1\cdots}^2 + y_{2\cdots}^2 + \ldots + y_{v\cdots}^2\right) - C.F. = TrSS \tag{2.6.3}$$

$$\text{Units S.S.} = \frac{1}{m}\left(y_{11\cdot}^2 + y_{12\cdot}^2 + \ldots + y_{i1\cdot}^2 + y_{i2\cdot}^2 + \ldots + y_{v1\cdot}^2 + y_{v2\cdot}^2 + \right.$$

$$\left. \ldots + y_{vk\cdot}^2\right) - C.F.$$

$$= USS \tag{2.6.4}$$

$$\text{Experimental error S.S.} = USS - BSS - TrSS = ESS \tag{2.6.5}$$

$$\text{Total S.S.} = y_{111}^2 + y_{112}^2 + \ldots + y_{vkm}^2 - C.F. = TSS \tag{2.6.6}$$

$$\text{Sampling error} = TSS - USS = SeSS \tag{2.6.7}$$

Table 2.6.2 ANOVA table

Source	d.f.	S.S.	M.S.	F-value
Blocks	$K-1$	BSS	$\dfrac{BSS}{k-1} = BMS$	$\dfrac{BMS}{EMS} = FB$
Treats.	$v-1$	TrSS	$\dfrac{TrSS}{v-1} = TrMS$	$\dfrac{TrMS}{EMS} = Ftr$
Exptl. Error	$(k-1)(v-1)$	ESS	$\dfrac{ESS}{(v-1)(k-1)} = EMS$	$\dfrac{EMS}{SeMS} = Fe$
Sampling error	$kv(m-1)$	SeSS	$\dfrac{SeSS}{kv(m-1)} = SeMS$	
Total	$Kv(m-1)$	TSS		

For the test of null hypothesis about treatments, the question arises whether to pool the experimental and sampling errors or not. For this preliminary test of significance (PTS) provides the answer. Under PTS, first the significance of experimental error is tested against sampling error by F-test. Such a test is called *preliminary test of significance*. If F comes out to be non-significant, pool the error sum of squares. To obtain the pooled error sum of square, divide the sum of ESS and SeSS by their combined degrees of freedom. This quantity is the pooled error mean square for testing treatment mean square by F-test. In case the statistic F for experimental error comes out to be significant, treatments mean square is tested against experimental error. Use of PTS increases the power of the final test as compared to the power under never pool test. Since PTS are not in vogue, Paul (1950) investigated the impact of always pooling and never pooling of experimental error and sampling error sum of squares. Paul found that in case of always pooling, the final F-test tends to produce too many significant results when actually the null hypothesis is true. In such cases, the size of the test is much higher than the predecided level. Anyhow, if the researcher does not apply PTS and follows the principle of never pooling, it is safer relative to always pooling. Now it is left to the discretion of the analyst whether to apply PTS or to resort to never pool test. The author, Agarwal (1990) on the basis of his own research work on PTS and experience, suggests that the use of higher level of significance for preliminary test than that of = 0.05 or 0.10 is preferable. Level of significance α = 0.25 for preliminary test is considered appropriate.

Example 2.6: A study was conducted to compare the effect of six interventions namely, I_1– Synthetic iron group (SIG); I_2 – Food-based iron group (FBIG); I_3 – Dictary improvement group (DIG); I_4 – Synthetic iron and folic acid group (IFG); I_5 – Synthetic iron vitamin E group (IVEG); I_6 – Control group (CG) administered with placebo. Five groups of six persons were formed whereas each group was homogeneous in itself. Six interventions were given to persons in each group randomly and independently for 15 days. Thereafter, they were given the Masters test (MT) and their pulse rates were determined by the instrument at four stages, i.e. at rest, immediately after MT, 2 minutes after MT, 4 minutes after MT. These data are suitable for the analysis of an experiment in RBD involving subsampling.

This study pertains to testing the hypotheses whether there is a significant difference amongst interventions also whether each person shows the same amount of variability in pulse rate with lapse of time under six interventions?

Data table

Inter-ventions	Patients	At rest	Just after MT	2 mins. after MT	4 mins. after MT	Patients total
I_1	G_1	76	120	104	90	390
	G_2	74	114	102	84	374
	G_3	77	126	110	92	405
	G_4	75	122	108	92	397
	G_5	74	124	108	90	396
I_2	G_1	78	136	120	102	436
	G_2	76	124	108	86	394
	G_3	76	122	110	84	392
	G_4	77	134	112	100	423
	G_5	75	122	108	88	393
I_3	G_1	76	124	108	84	392
	G_2	76	126	112	94	408
	G_3	77	126	110	92	405
	G_4	74	122	106	92	394
	G_5	−75	124	108	88	395
I_4	G_1	76	126	110	94	406
	G_2	75	120	106	92	393
	G_3	77	138	122	86	423
	G_4	75	112	100	82	369
	G_5	76	114	112	90	392
I_5	G_1	76	126	116	90	408
	G_2	77	136	116	92	421
	G_3	77	122	110	102	411
	G_4	76	126	112	92	406
	G_5	75	124	114	94	407
I_6	G_1	75	126	110	88	399
	G_2	77	132	114	92	415
	G_3	76	124	112	100	412
	G_4	82	140	130	86	438
	G_5	75	112	100	86	373

Intervention totals,

$I_1 = 1962$, $I_2 = 2038$, $I_3 = 1994$, $I_4 = 1983$, $I_5 = 2053$, $I_6 = 2037$

Grand total, $G = 12067$, $\text{C.F.} = \dfrac{(12067)^2}{120} = 1213437.41$

Group totals,

$G_1 = 2431$, $G_2 = 2405$, $G_3 = 2448$, $G_4 = 2427$, $G_5 = 2356$

Total S.S. $= 76^2 + 120^2 + \ldots\ldots + 100^2 + 86^2 -$ C.F.

$$= 1258423.00 - \text{C.F.} = 44985.59$$

Interventions S.S. $= \dfrac{1}{20}(1962^2 + 2038^2 + \text{-------} + 2053^2 + 2037^2) -$ C.F.

$$= 1213769.55 - \text{C.F.} = 332.14$$

Patients S.S. $= \dfrac{1}{4}(390^2 + 374^2 + \text{-------} + 438^2 + 373^2) -$ C.F.

$$= 1215376.75 - \text{C.F.} = 1939.34$$

Groups S.S. $= \dfrac{1}{24}(2431^2 + 2405^2 + 2448^2 + 2427^2 + 2356^2) -$ C.F.

$$= 1213648.12 - \text{C.F.} = 210.71$$

Experimental error S.S. $= 1939.34 - 210.71 - 332.14 = 1396.49$

Sampling error S.S. $= 44985.59 - 1939.34 = 43046.25$

Source	d.f.	S.S.	M.S	F-value
Groups	4	210.71	52.68	< 1
Interventions	5	332.14	66.43	< 1
Exptl. error	20	1396.49	69.82	0.146
Sampling error	90	43046.25	478.29	
Total	119	44985.59		

F-value for experimental error is less than one and hence it is non-significant. In this situation, it is appropriate to pool the experimental and sampling errors. Groups and interventions be tested against pooled error mean square.

Pooled error mean square $= \dfrac{1}{110}(1396.49 + 43046.25)$

$$= 403.39$$

On dividing the mean squares for groups and interventions by the pooled error mean square respectively, the corresponding F-values are still less than unity. Hence, both factors are non-significant. No further investigations is required. These results reveal that interventions are equally effective and also there is no remarkable variation among groups.

2.7 ANALYSIS OF NON-ORTHOGONAL DATA

Non-orthogonality has already been discussed in section (1.3). There it has been made amply clear that an unbalanced design results into non-orthogonal data. So the author presents in this section analysis of non-orthogonal data without entangling the readers into its mathematical derivation. The methodology is supported by a numerical example. Consider an experiment RBD-v with unequal number of observation per cell. The statistical model for the experimental design can be given as follows.

$$y_{iju} = \mu + \tau_i + \beta_j + e_{iju} \tag{2.7.1}$$

$$i = 1, 2, \ldots, v; \quad j = 1, 2, \ldots, k; \quad u = 1, 2, \ldots, n_{ij}$$

The number of observations in the (i, j)-th cell is n_{ij}. All parameters involved in the model are same as in the model (2.2.1) and underlying assumptions are also the same.

Blocks

Treats.	1	2	....	j	....	k	No. of obs.	Treat. totals
1	y_{111} y_{112} | | | y_{11n11}	y_{121} y_{122} y_{12n12}	....	y_{1j1} y_{1j2} y_{1jnj}	....	y_{1k1} y_{1k1} y_{1knk}	$n_{1.}$	
Cell totals	$y_{11.}$	$y_{12.}$	....	$y_{1j.}$	....	$y_{1k.}$		$y_{1..}$
2	y_{211} y_{212} | | | y_{21n21}	y_{221} y_{222} y_{22n22}	....	y_{2j1} y_{2j2} y_{2jn2j}	....	y_{2k1} y_{2k2} y_{2kn2k}	$n_{2.}$	
Cell totals	$y_{21.}$	$y_{22.}$	....	$y_{2j.}$	....	$y_{2k.}$		$y_{2..}$
	| | |							| | |
i	y_{i11} y_{i12} | | | y_{i1ni1}	y_{i21} y_{i22} y_{i2ni2}	....	y_{ij1} y_{ij2} y_{ijnij}	....	y_{ik1} y_{ik2} y_{iknik}	$n_{i.}$	

(Contd.)

Treats.	1	2	....	j	....	k	No. of obs.	Treat. totals
Cell totals	$y_{i1.}$	$y_{i2.}$	——	$y_{ij.}$	——	$y_{ik.}$		$y_{i..}$
v	y_{v11} y_{v12} y_{v1nv1}	y_{v21} y_{v22} y_{v2nv2}	....	y_{vj1} y_{vj2} y_{2jnvj}	....	y_{vk1} y_{vk2} y_{vknvk}	$n_{v.}$	
Cell totals	$y_{v1.}$	$y_{v2.}$	....	$y_{vj.}$	....	$y_{vk.}$		$y_{v..}$
Block totals	$y_{.1.}$	$y_{.2.}$	....	$y_{.j.}$	....	$y_{.k.}$		G.T.= $y_{...}$
No. of obs.	$n_{.1}$	$n_{.2}$	....	$n_{.j}$	....	$n_{.k}$	$n_{..}$	

Total number of observations = $\sum_i \sum_j n_{ij} = n_{..}$

By the method of least squares, it is easy to obtain the following normal equations.

$$n_{..}\,\hat{\mu} + \sum_i \hat{\tau}_i + \sum_j n_{.j}\hat{\beta}_j = y_{...} \qquad (2.7.2)$$

$$n_{i.}\,\hat{\mu} + n_{i.}\hat{\tau}_i + \sum_j n_{ij}\hat{\beta}_j = y_{i..} \qquad (2.7.3)$$

$$n_{.j}\,\hat{\mu} + \sum_i n_{ij}\hat{\tau}_i + n_{.j}\hat{\beta}_j = y_{.j.} \qquad (2.7.4)$$

To obtain the estimates, one has to solve these equations in terms of observational values.

From equation (2.7.4),

$$\hat{\mu} + \hat{\beta}_j = \frac{1}{n_{.j}}\left(y_{.j.} - \sum_i n_{i.}\hat{\tau}_i\right) \qquad (2.7.5)$$

Substituting for $\hat{\mu} + \hat{\beta}_j$ in the normal equation, a set of equations in $\hat{\tau}_i$ is obtained as,

$$\left(n_{i.} - \sum_j \frac{n_{ij}^2}{n_{.j}}\right)\hat{\tau}_i - \sum_{i \neq 1}\left(\sum_j \frac{n_{ij}}{n_{.j}}n_{lj}\right)\hat{\tau}_1 = Q_i \qquad (2.7.6)$$

Where, $Q_i = y_{i..} - \sum_j \dfrac{n_{ij}}{n_{.j}}y_{.j.}$ $\qquad (2.7.7)$

These equations are not independent. So to solve these equations, impose the condition $\sum_i \hat{\tau}_1 = 0$. $\hat{\mu}$ and $\hat{\beta}_j$ can be estimated from the equations (2.7.2) through (2.7.4) by imposing the condition $\sum_j \hat{\beta}_j = 0$. The estimates of $\hat{\tau}$'s can be obtained from the equation (2.6.6).

The reduction in sum of squares due to (μ, τ, β) under model (2.7.1) say, $R(\mu, \tau, \beta)$ is calculable by the relation,

$$R(\mu, \tau, \beta) = y_{..}\hat{\mu} + \sum_i \hat{\tau}_i y_{i..} + \sum_j \hat{\beta}_j y_{.j.} \qquad (2.7.8)$$

The residual (error) mean square is,

$$= \frac{1}{(n_{..} - v - k + 1)}\left[\sum_{i,j,k} y_{ijk}^2 - R(\mu, \tau, \beta)\right] \qquad (2.7.9)$$

Again the reduction in sum of squares omitting the effect of τ, say $R(\mu, \beta)$ is,

$$= \sum_j \frac{y_{.j.}^2}{n_{.j}} \qquad (2.7.10)$$

Further, with a little algebraic manipulation, the relation between $R(\mu, \tau, \beta)$ and $R(\mu, \beta)$ can be established as given below.

$$R(\mu, \tau, \beta) = R(\mu, \beta) + \sum_i \hat{\tau}_i Q_i \qquad (2.7.11)$$

or $\qquad \sum_i \hat{\tau}_i Q_i = R(\mu, \tau, \beta) - R(\mu, \beta) \qquad (2.7.11.1)$

$\qquad\qquad$ = S.S. due to treatments eliminating block effects.

The analysis of variance can be carried out by preparing the following ANOVA Table 2.7.1.

Table 2.7.1 ANOVA table for non-orthogonal data

Source	d.f.	S.S.	M.S.	F-value
Blocks ignoring Treatments	$k - 1$	$\sum_j \dfrac{y_{.j.}^2}{n_{.j}} - \dfrac{y_{...}^2}{n_{..}} = BSS$	$\dfrac{BSS}{k-1} = BMS$	$\dfrac{BMS}{EMS} = F_b$
Treats. eliminating Block effects	$v - 1$	$\sum_i \hat{\tau}_i Q_i = TrSS$	$\dfrac{TrSS}{v-1} = TrMS$	$\dfrac{TrMS}{EMS} = F_t$
Residual	$n_{..} - k - v + 1 = n_e$	TSS – BSS – TrSS = ESS		
Total	$n_{..} - 1$	$\sum_{i,j,u} y_{iju}^2 - \dfrac{y_{...}^2}{n_{..}} = TSS$		

Test of equality of block effects, i.e. $\beta_j = \beta$ becomes redundant as M.S. due to blocks implicitly contain treatment effects. Any conclusion in respect of blocks is not going to result into a substantive conclusion. Therefore, the test of hypothesis H_0: $\tau_i = \tau$ by F-test for $(v - 1)$ and n_e degrees of freedom provides a reliable inference.

Example 2.7: An experiment was conducted with layout in randomized block design having 4 blocks to compare three varieties of wheat. Instead of plot yield, the yields were recorded through crop cutting experiment in each plot. Some how the number of areas cut could not be maintained uniformly under crop cutting. Grain yield data (in kg) were as given below. Grain yield data (kg) for the plots of size 3×3 m^2 were as given below.

Blocks

Varieties	1	2	3	4	Total
V$_1$	10.6 8.6 9.4 6.8	7.6 8.1 9.0	7.4 7.5 8.7	9.5 7.3	
Sub-total	35.4	24.7	23.6	16.8	100.5
V$_2$	6.4 5.9 7.6	7.4 7.1	6.1 7.0 5.7	6.7 7.3 6.7	
Sub-total	19.9	14.5	18.8	20.7	73.9
V$_3$	8.7 9.5 8.0	7.7 7.1 7.3	7.2 8.9	5.4 5.0 6.5	
Sub-total	26.2	22.1	16.1	16.9	81.3

Note: Yield totals and sub-totals are also displayed along with the data.

Abridged data have been presented in the example for the sake of brevity and better understanding of calculations. Since the number of observations is not same, the data are non-orthogonal and analysis will be carried by the procedure given in section (2.7).

For the given data, the values of various notations are as follows.

$n_{11} = 4$, $n_{12} = 3$, $n_{13} = 3$, $n_{14} = 2$ and $n_{1.} = 12$

$n_{21} = 3$, $n_{22} = 2$, $n_{23} = 3$, $n_{24} = 3$ and $n_{2.} = 11$

$n_{31} = 3$, $n_{32} = 3$, $n_{33} = 2$, $n_{34} = 3$ and $n_{3.} = 11$

$n._1 = 10$, $n._2 = 8$, $n._3 = 8$, $n._4 = 8$ and $n.. = 34$

$y_{11}. = 35.4$, $y_{12}. = 24.7$, $y_{13}. = 23.6$, $y_{14}. = 16.8$ and $y_1.. = 100.5$

$y_{21}. = 19.9$, $y_{22}. = 14.5$, $y_{23}. = 18.8$, $y_{24}. = 20.7$ and $y_2.. = 73.9$

$y_{31}. = 26.2$, $y_{32}. = 22.1$, $y_{33}. = 16.1$, $y_{34}. = 16.9$ and $y_3.. = 81.3$

$y._1. = 81.5$, $y._2. = 61.3$. $y._3. = 58.5$, $y._4. = 54.4$ and $y... = 255.7$

Using the above values, the normal equations from (2.7.2) through (2.7.4) come out to be as follows.

$$34\hat{\mu} + 12\hat{V_1} + 11\hat{V_2} + 11\hat{V_3} + 10\hat{\beta_1} + 8\hat{\beta_2} + 8\hat{\beta_3} + 8\hat{\beta_4} = 255.7 \qquad \text{(i)}$$

$$12\hat{\mu} + 12\hat{V_1} + 4\hat{\beta_1} + 3\hat{\beta_2} + 3\hat{\beta_3} + 2\hat{\beta_4} = 100.5 \qquad \text{(ii)}$$

$$11\hat{\mu} + 11\hat{V_2} + 3\hat{\beta_1} + 2\hat{\beta_2} + 3\hat{\beta_3} + 3\hat{\beta_4} = 73.9 \qquad \text{(iii)}$$

$$11\hat{\mu} + 11\hat{V_3} + 3\hat{\beta_1} + 3\hat{\beta_2} + 2\hat{\beta_3} + 3\hat{\beta_4} = 81.3 \qquad \text{(iv)}$$

$$10\hat{\mu} + 4\hat{V_1} + 3\hat{V_2} + 3\hat{V_3} + 10\hat{\beta_1} = 81.5 \qquad \text{(v)}$$

$$8\hat{\mu} + 3\hat{V_1} + 2\hat{V_2} + 3\hat{V_3} + 8\hat{\beta_2} = 61.3 \qquad \text{(vi)}$$

$$8\hat{\mu} + 3\hat{V_1} + 3\hat{V_2} + 2\hat{V_3} + 8\hat{\beta_3} = 58.5 \qquad \text{(vii)}$$

$$8\hat{\mu} + 2\hat{V_1} + 3\hat{V_2} + 3\hat{V_3} + 8\hat{\beta_4} = 54.4 \qquad \text{(viii)}$$

[*Nota bene:* In all the normal equations (i) to (viii), the sum of the coefficients of V's and of β's is same as that of $\hat{\mu}$.]

The above equations involve eight unknown parameters, $\hat{\mu}$, $\hat{V_1}$, $\hat{V_2}$, $\hat{V_3}$, $\hat{\beta_1}$, $\hat{\beta_2}$, $\hat{\beta_3}$, $\hat{\beta_4}$. The solution of these equations will provide the required estimates. The above equations in the matrix notation can be presented as given below.

$\hat{\mu}$	$\hat{V_1}$	$\hat{V_2}$	$\hat{V_3}$	$\hat{\beta_1}$	$\hat{\beta_2}$	$\hat{\beta_3}$	$\hat{\beta_4}$				
34	12	11	11	10	8	8	8	$\hat{\mu}$		255.7
12	12	0	0	4	3	3	2	$\hat{V_1}$		100.5
11	0	11	0	3	2	3	3	$\hat{V_2}$		73.9
11	0	0	11	3	3	2	3	$\hat{V_3}$		81.3
10	4	3	3	10	0	0	0	$\hat{\beta_1}$	=	81.5
8	3	2	3	0	8	0	0	$\hat{\beta_2}$		61.3
8	3	3	2	0	0	8	0	$\hat{\beta_3}$		58.5
8	2	3	3	0	0	0	8	$\hat{\beta_4}$		54.4

((A))   ((B))   ((Y))

8 × 8   8 × 1   8 × 1

In matrix notation, the equation is,

$$AB = Y$$

or

$$B = A^{-1}Y$$

Matrix A can be inverted by Pivotal Condensation method manually on a scientific desk calculator or more easily on a computer provided all equations be independent but such is not the case here. The packages for inversion of matrix are easily available in most of the packages for science and technology like SPSS, SAS and many others.

Now calculate Q_i for $i = 1, 2, 3$ by the relation (2.7.7).

$$Q_1 = 100.5 - \left(\frac{4 \times 81.5}{10} + \frac{3 \times 61.3}{8} + \frac{3 \times 58.5}{8} + \frac{2 \times 54.4}{8} \right)$$

$$= 100.5 - 91.125 = 9.3750$$

$$Q_2 = 73.9 - \left(\frac{3 \times 81.5}{10} + \frac{2 \times 61.3}{8} + \frac{3 \times 58.5}{8} + \frac{3 \times 54.4}{8} \right)$$

$$= 73.9 - 82.1125 = -8.2125$$

$$Q_3 = 81.3 - \left(\frac{3 \times 81.5}{10} + \frac{3 \times 61.3}{8} + \frac{2 \times 58.5}{8} + \frac{3 \times 54.4}{8} \right)$$

$$= 81.3 - 82.4625 = -1.1625$$

[Check: $\Sigma_i \, Q_i = 0$]

Varietal effects are estimated by (2.7.6). Three equations are obtained as follows.

For the first equation, $i = 1, j = 1, 2, 3, 4$ and $1 = 2, 3$.

$$\left\{ n_{1.} - \left(\frac{n_{11}^2}{n_{.1}} + \frac{n_{12}^2}{n_{.2}} + \frac{n_{13}^2}{n_{.3}} + \frac{n_{14}^2}{n_{.4}} \right) \right\} \hat{V}_1 - \left(\frac{n_{11}n_{21}}{n_{.1}} + \frac{n_{12}n_{22}}{n_{.2}} + \frac{n_{13}n_{23}}{n_{.3}} + \frac{n_{14}n_{24}}{n_{.4}} \right) \hat{V}_2$$

$$- \left(\frac{n_{11}n_{31}}{n_{.1}} + \frac{n_{12}n_{32}}{n_{.2}} + \frac{n_{13}n_{33}}{n_{.3}} + \frac{n_{14}n_{34}}{n_{.4}} \right) \hat{V}_3 = Q_1$$

Now putting the values for n's, the first equation is,

$$\left\{ 12 - \left(\frac{4^2}{10} + \frac{3^2}{8} + \frac{3^2}{8} + \frac{2^2}{8} \right) \right\} \hat{V}_1 - \left(\frac{4 \times 3}{10} + \frac{3 \times 2}{8} + \frac{3 \times 3}{8} + \frac{2 \times 3}{8} \right) \hat{V}_2$$

$$- \left(\frac{4 \times 3}{10} + \frac{3 \times 3}{8} + \frac{3 \times 2}{8} + \frac{2 \times 3}{8} \right) \hat{V}_3 = 9.3750$$

$$7.65 \hat{V}_1 - 3.825 \hat{V}_2 - 3.825 \hat{V}_3 = 9.3750 \qquad (1)$$

Similarly for the second equation, $i = 2, j = 1, 2, 3, 4$ and $1 = 1, 3$.

The second equation is,

$$\left\{11-\left(\frac{3^2}{10}+\frac{2^2}{8}+\frac{3^2}{8}+\frac{3^2}{8}\right)\right\}\hat{V}_2-\left(\frac{3\times4}{10}+\frac{3\times2}{8}+\frac{3\times3}{8}+\frac{3\times2}{8}\right)\hat{V}_1$$

$$-\left(\frac{3\times3}{10}+\frac{2\times3}{8}+\frac{3\times2}{8}+\frac{3\times3}{8}\right)\hat{V}_3=-8.2125$$

$$7.35\hat{V}_2-3.825\hat{V}_1-3.525\hat{V}_3=-8.2125 \qquad (2)$$

For the third equation, $i=3$, $j=1, 2, 3, 4$ and $1=2, 1$.
The third equation is,

$$\left\{11-\left(\frac{3^2}{10}+\frac{3^2}{8}+\frac{2^2}{8}+\frac{3^2}{8}\right)\right\}\hat{V}_3-\left(\frac{3\times3}{10}+\frac{3\times2}{8}+\frac{2\times3}{8}+\frac{3\times3}{8}\right)\hat{V}_2$$

$$-\left(\frac{3\times4}{10}+\frac{3\times3}{8}+\frac{2\times3}{8}+\frac{3\times2}{8}\right)\hat{V}_1=-1.1625$$

$$7.35\hat{V}_3-3.525\hat{V}_2-3.825\hat{V}_1=-1.1625 \qquad (3)$$

To solve the equations (1), (2) and (3), rearrange the equations as follows:

$$7.65\hat{V}_1-3.825\hat{V}_2-3.825\hat{V}_3=9.3750 \qquad (4)$$

$$-3.825\hat{V}_1+7.35\hat{V}_2-3.525\hat{V}_3=-8.2125 \qquad (5)$$

$$-3.825\hat{V}_1-3.525\hat{V}_2+7.35\hat{V}_3=-1.1625 \qquad (6)$$

It can easily be verified that the determinant of the coefficient matrix is zero. It reveals that the coefficient matrix is singular. Further on eliminating $\hat{V}_3$ from the equations (4) and (5), we get the same equation as we get after eliminating $\hat{V}_3$ from equations (5) and (6), i.e.

$$\hat{V}_1-\hat{V}_2=1.549628 \qquad (7)$$

That is why the solution of these equations is not imminent. Therefore, to obtain a solution make use of the restriction, $\hat{V}_1+\hat{V}_2+\hat{V}_3=0$. On dividing equation (4) by 3.825, the equation (4) comes out to be,

$$2\hat{V}_1-\hat{V}_2-\hat{V}_3=2.4510 \qquad (8)$$

Under the restriction, $\hat{V}_2+\hat{V}_3=-\hat{V}_1$

∴   Equation (8) is, $3\hat{V}_1=2.4510$

or $\hat{V}_1=0.8170 \qquad (9)$

Substituting the value of $\hat{V}_1$ in (7), estimated value of $\hat{V}_2$ is available, i.e.,

$$\hat{V}_2 = -0.732638 \tag{10}$$

Again substituting the values of $\hat{V}_1$ and $\hat{V}_2$ in (8), we get,

$$\hat{V}_3 = -0.084382 \tag{11}$$

Substituting the values of $\hat{V}_1$, $\hat{V}_2$ and $\hat{V}_3$ back in the normal equations (i) to (viii), a new set of eight reduced equations (12) to (19) in $\hat{\mu}, \hat{\beta}_1, \hat{\beta}_2, \hat{\beta}_3, \hat{\beta}_4$ is obtained. Making use of the condition $\hat{\beta}_1 + \hat{\beta}_2 + \hat{\beta}_3 + \hat{\beta}_4 = 0$ and solving these equations, the block effects and $\hat{\mu}$ are easily estimated as presented below.

$$34\hat{\mu} + 2\hat{\beta}_1 = 254.883 \tag{12}$$

$$12\hat{\mu} + \hat{\beta}_1 - \hat{\beta}_4 = 90.696 \tag{13}$$

$$11\hat{\mu} - \hat{\beta}_2 = 81.959 \tag{14}$$

$$11\hat{\mu} - \hat{\beta}_3 = 82.228 \tag{15}$$

$$10\hat{\mu} + 10\hat{\beta}_1 = 80.683 \tag{16}$$

$$8\hat{\mu} + 8\hat{\beta}_2 = 60.567 \tag{17}$$

$$8\hat{\mu} + 8\hat{\beta}_3 = 58.416 \tag{18}$$

$$8\hat{\mu} + 8\hat{\beta}_4 = 55.217 \tag{19}$$

[Check: It is interesting to note that the sum of equations (13), (14), (15) and also the sum of equations (16) to (19) reduces to equation (12).

Eliminating $\hat{\beta}_1$ from equations (12) and (16),

$$\hat{\mu} = 7.460825$$

Again substituting the value of $\hat{\mu}$ in the equation (16) to (19), the estimates of $\hat{\beta}$'s are obtained.

$$\hat{\beta}_1 = 0.607475$$

$$\hat{\beta}_2 = 0.11005$$

$$\hat{\beta}_3 = -0.158825$$

$$\hat{\beta}_4 = -0.5587$$

$$\left[\text{Check: } \Sigma_j \hat{\beta}_j = 0 \right]$$

Now calculate R (μ, τ, β) from (2.7.8).

$$R (\mu, \tau, \beta) = 255.7 \times 7.460825 + (0.8170 \times 100.5$$
$$- 0.732638 \times 73.9 - 0.084382 \times 81.3)$$
$$+ (0.607475 \times 81.5 + 0.11005 \times 61.3$$
$$- 0.158825 \times 58.5 - 0.5587 \times 54.4)$$
$$= 1945.4099$$

Reduction in sum of squares omitting the effect of τ, R (μ, τ) by the formula (2.7.10) is,

$$= \frac{81.5^2}{10} + \frac{61.3^2}{8} + \frac{58.5^2}{8} + \frac{54.4^2}{8}$$
$$= 1931.6374$$

Sum of squares due to treatments eliminating block effects by the formula (2.7.11.1) is,

$$\sum_i \hat{\tau}_i Q_i = 1945.4099 - 1931.6374$$
$$= 13.7725$$

Aliter: $\sum_i \hat{\tau}_i Q_i$ can also be worked out directly as,

$$\sum_i \hat{\tau}_i Q_i = 0.817 \times 9.375 + 0.732838 \times 8.2125 + 0.08438 \times 1.1625$$
$$= 13.7742$$

Sum of squares due to blocks ignoring treatments,

$$= \frac{81.5^2}{10} + \frac{61.3^2}{8} + \frac{58.5^2}{8} + \frac{54.5^2}{8} - \frac{255.7^2}{34}$$
$$= 1931.6374 - 1923.0144$$
$$= 8.6230$$

Total S.S. $= 10.6^2 + 8.6^2 + \ldots\ldots + 5.0^2 + 6.5^2 - \text{C.F.}$
$$= 1976.39 - 1923.0144$$
$$= 53.3756$$

ANOVA table for non-orthogonal data

Source	d.f.	S.S	M.S	F-value
Blocks ignoring varieties	3	8.6230	2.8743	2.598 NS
Varieties eliminating blocks	2	13.7742	6.8871	6.2249*
Error	28	30.9784	1.1064	
Total	33	53.3756		

F-value for varieties in the above ANOVA reveals that there is a significant difference between varieties as the calculated F is greater than the tabulated

F from table no. B-7, at $\alpha = 0.05$ and (2, 28) d.f. is $F_{.05;2,28} = 3.34$. For paired comparisons of varieties, Fisher's lsd test is used.

$$\text{lsd for } V_1 \text{ versus } V_2 \text{ or } V_3 = \left[\sqrt{s_e^2 \left(\frac{1}{12} + \frac{1}{11} \right)} \right] \times t_{.05,28}$$

$$= \left[\sqrt{1.1064 \left(\frac{1}{12} + \frac{1}{11} \right)} \right] \times 2.048$$

$$= 0.8992$$

$$\text{lsd for } V_2 \text{ versus } V_3 = \sqrt{\left(2s_e^2 / 11 \right)} \times 2.048$$

$$= \sqrt{2 \times 1.1064 / 11} \times 2.048$$

$$= 0.9185$$

Variety means in ascending order are,

$\overline{V}_1$	$\overline{V}_2$	$\overline{V}_3$
8.375	7.391	6.718

A comparison between variety means with the respective lsd's leads to the conclusion that variety V_1 has significantly better yield on average than varieties V_2 and V_3. Further varieties V_2 and V_3 are at par in respect of grain yield.

2.8 LATIN SQUARE DESIGN (LSD)

Randomized block design utilizes one blocking factor to control one nuisance variable. But in a variety of situations that is not enough. At many occasions, an experimenter has to put control on two nuisance variables. For example, to see the effect of various feeds on lactating animals of different species, it seems logical to maintain homogeneity in respect of their lactation and breed. In field experiments, the patch of land shows fertility variation in horizontal as well as vertical directions. Hence, an experimenter should form the blocks which are homogeneous in both the directions. In such trials, a scientist uses two blocking factors. In quest of such experimental designs, Latin square (LS) designs are ones which provide the facility of using two blocking factors. Therefore, Latin square designs are called *two restrictional designs.* Latin square designs are frequently used in agriculture, animal science and industrial research. As the name imply, Latin square is an array of say, v rows and v columns arranged in such a manner that each of the v Latin letters occurs once in a row and once in a column. In this way, each row is a complete homogeneous block according to one criterion and each column is a complete homogeneous block in respect of other criterion. Such a Latin square is denoted as LSD-v and is known as a Latin square of order-v. Latin squares are classified in conformation of position of letters. Fisher and Yates (1948) gave the following definitions of classified Latin squares.

Definitions

Standard Latin square is a configuration in which the Latin letters in the first row and first column are kept in alphabetical order. Such Latin squares are also called *reduced Latin squares*. By permuting all v rows and all v columns in all possible ways, $v! \, (v - 1)!$ squares can be generated out of a standard Latin square. There is only one standard Latin square of order-3. Illustratively,

$$
\begin{array}{ccc}
A & B & C \\
B & C & A \\
C & A & B
\end{array}
$$

(2.8.1) Standard Latin square of order-3

But there can possibly be four standard Latin squares of order-4. The set of four standard Latin squares is displayed below.

A B C D	A B C D	A B C D	A B C D
B C D A	B A D C	B D A C	B A D C
C D A B	C D B A	C A D B	C D A B
D A B C	D C A B	D C B A	D C B A

(2.8.2) Standard Latin squares of order-4

The set of standard Latin squares can be formed by interchanging the rows and columns barring the first ones.

Self-conjugate Latin square is one in which the sequence of letters in corresponding rows and columns is same. Further two Latin squares are said to be *conjugate* if the columns of one are identical to the rows of the other. This is tenuous to verify that all standard Latin squares are self-conjugate.

Adjugate Latin squares are those formed by permuting rows, columns and letters of a standard Latin square. But all of them are not necessarily different.

Orthogonal Latin squares

Two Latin squares are said to be orthogonal Latin squares when one is superimposed over the other, then each letter of one square matches with each letter of the other square once and only once in the composite Latin square. Such composite designs facilitate to isolate the third nuisance variable. If one Latin square out of two orthogonal Latin squares consists of Greek letters and the other of Latin letters, the composite design is called a *Graeco-Latin square*. A Latin square of order-v can have at the most $(v - 1)$ mutually orthogonal Latin squares (MOLS). Fisher and Yates (1963) have presented complete sets of MOLS for all odd prime numbers or powers of prime numbers and multiples of four. But in 1934, they showed that not a single pair of orthogonal Latin squares of order 6 exists. Later it was propounded by Euler that the

Latin squares of order $(4m + 2)$, where m is an integer, have no pair of orthogonal Lain squares. But Bose, Shrikhande and Parker (1960) proved that orthogonal Latin squares of order $(4m + 2)$ exist for $m > 1$ and they constructed MOLS for the Latin squares of order 10, 14, etc. Orthogonal Latin squares are also useful in the construction of balance incomplete block designs (BIBD), particularly lattice square designs. As an illustration, MOLS of order 3×3, 4×4, 5×5 are presented below.

```
A B C          A B C
B C A          C A B
C A B          B C A
```

MOLS of order 3 × 3

```
A B C D        A B C D        A B C D
B A D C        C D A B        D C B A
C D A B        D C B A        B A C D
D C B A        B A D C        C D A B
```

MOLS of order 4 × 4

```
A B C D E      A B C D E      A B C D E      A B C D E
B C D E A      D E A B C      C D E A B      E A B C D
C D E A B      B C D E A      E A B C D      D E A B C
D E A B C      E A B C D      B C D E A      C D E A B
E A B C D      C D E A B      D E A B C      B C D E A
```

MOLS of order 5 × 5

For complete set of MOLS of order higher than 5×5, the readers are referred to Fisher and Yates (1963), pp. 88 – 89.

Selection of a Latin square design

Latin square design implicitly meets two requirements of a good design due to its structure namely, the local control and replications. So to avoid all sort of proclivity and for the validity of estimation and testing, process of randomization has to be adopted.

Consider first LSD-3. There can be in all $3! \, (3 - 1)! = 12$ Latin squares. Out of these 12 Latin squares, only one is a standard Latin square and rest eleven are other arrangements. There are two alternatives at the disposal of the experimenter. One option is to select one of the twelve arrangements (LS) randomly. Second choice is to take the standard Latin square as given below.

Columns

		1	2	3
	1	A	B	C
Rows	2	B	C	A
	3	C	A	B

Now select six random numbers from 1 to 3. Say, they are (3, 2, 1, 2, 1, 3). Supposing that the first three digits indicate the order of rows. Thus, the rows of the above Latin square are ordered according to the first three digits (3, 2, 1) resulting into the following square.

Columns

		1	2	3
	1	C	A	B
Rows	2	B	C	A
	3	A	B	C

Finally the columns are ordered according to the last three digits (2, 1, 3). This provides the final Latin square which has to be used for the experimental layout.

Columns

		1	2	3
	1	A	C	B
Rows	2	C	B	A
	3	B	A	C

For randomization in LSD-4, following procedure can be adopted. There can be two approaches. One way is to construct all possible Latin squares of order 4. It is already given there can be in 4! (4 – 1)! = 144 Latin squares. Out of these 144 squares, there are 4 standard Latin squares of order-4. Hence, the total number of arrangements = 4 × 144 = 576. Select one square randomly out of 576 squares and use this for experimentation. But to construct 576 Latin squares is very cumbersome and practically infeasible. Hence, one has to adopt a more pragmatic approach.

The second approach is very handy. At the outset select nine digits randomly from 1 to 4. Let these numbers from one digit random number table come out to be (4, 2, 1, 4, 3, 3, 2, 4, 1). The first digit provides the number for selection standard Latin given in set (2.8.2). Here standard Latin square no. 4 is selected, i.e.,

Columns

		1	2	3	4
	1	A	B	C	D
Rows	2	B	A	D	C
	3	C	D	A	B
	4	D	C	B	A

Suppose the next four numbers (2, 1, 4, 3) lead to the random order of the rows. Then the rearranged Latin square is,

		Columns			
		1	2	3	4
	1	B	A	D	C
Rows	2	A	B	C	D
	3	D	C	B	A
	4	C	D	A	B

The last four digits (3, 2, 4, 1) provide the random basis for rearranging the columns of the preceding Latin square resulting into the final Latin square design to be adopted for experimentation.

		Columns			
		1	2	3	4
	1	D	A	C	B
Rows	2	C	B	D	A
	3	B	C	A	D
	4	A	D	B	C

Above two examples affirm that the second approach of random selection of a Latin square design is quite convenient and feasible. All the more this approach can be applied for the random selection of a Latin square from the set of Latin squares of any higher order.

Analysis of data of a Latin square. The analysis of data is governed by the model of the experimental design and the assumptions made about it. The additive model for LS design is,

$$y_{ijk} = \mu + \tau_i + \beta_j + \rho_k + e_{ijk} \qquad (2.8.1)$$

$$\text{For, } i, j, k = 1, 2, \ldots\ldots, v$$

ρ_k is the kth row effect and all parameters represent the same effects as in model (2.2.1). Considering it a fixed effect model, the assumptions are,

(i) $\sum_i \tau_i = \sum_j \beta_j = \sum_k \rho_k = 0$.

(ii) e_{ijk} are i.i.d. N $(0, \sigma_e^2)$.

(iii) The interactions of rows x treatments and columns x treatments have been assumed to be absent. Even if they exist, they are accounted towards experimenter error.

Least square estimates: To estimate the parameters of the model (2.8.1) differentiate partially the residual sum of square with respect to μ, τ_i, β_j and ρ_k respectively and equate them to zero. Also replace the parameters by their estimates by putting crow (^) over them. In this way, a set of four normal equations is obtained. Solving these equations, the required estimators are obtained.

The residual sum of square say, Q is,

$$Q = \sum_{i,j,k} (y_{ijk} - \mu - \tau_i - \beta_j - \rho_k)^2 \qquad (2.8.2)$$

$$y_{...} = v^2\hat{\mu} + v\sum_i \hat{\tau}_i + v\sum_j \hat{\beta}_j + v\sum_k \hat{\rho}_k \qquad (2.8.3)$$

$$y_{i..} = v\hat{\mu} + v\hat{\tau}_i + \sum_j \hat{\beta}_j + \sum_k \hat{\rho}_k \qquad (2.8.4)$$

$$y_{.j.} = v\hat{\mu} + \sum_i \hat{\tau}_i + v\hat{\beta}_j + \sum_k \hat{\rho}_k \qquad (2.8.5)$$

$$y_{..k} = v\hat{\mu} + \sum_i \hat{\tau}_i + \sum_j \hat{\beta}_j + v\hat{\rho}_k \qquad (2.8.6)$$

Under restriction (i) imposed on the parameters of the model (2.8.1), the estimates are,

$$\hat{\mu} = \frac{y_{...}}{v^2} = \overline{y} \quad \text{(Over all mean)} \qquad (2.8.7)$$

$$\hat{\tau}_i = \frac{y_{i..}}{v} - \hat{\mu} = \overline{y}_{1..} - \overline{y} \qquad (2.8.8)$$

$$\hat{\beta}_j = \frac{y_{.j.}}{v} - \hat{\mu} = \overline{y}_{.j.} - \overline{y} \qquad (2.8.9)$$

$$\hat{\rho}_k = \frac{y_{..k}}{v} - \hat{\mu} = \overline{y}_{..k} - \overline{y} \qquad (2.8.10)$$

In this process of estimation, one thing is to be kept in mind that there are v^2 units in total. So while summing up in each equation of (2.8.2), a subscript is dropped as necessitated.

Analysis of variance: Consider an LSD-v. Let the kth row total be denoted. $R_{..k}$, and jth column total by $C_{.j}$, ith treatment total by $T_{i..}$ and grand total by $y_{...}$ (where $i, j, k = 1, 2, ..., v$). Then, the correction factor and sum of squares can be obtained by the following expressions.

$$\text{C.F.} = \frac{y_{...}^2}{v^2}$$

$$\text{Total S.S.} = \sum_{i,j,k} y_{ijk}^2 - \text{C.F.} = \text{TSS}$$

$$\text{Row S.S.} = \frac{1}{v}\left(R_{..1}^2 + R_{..2}^2 + \cdots\cdots\cdots + R_{..v}^2\right) - \text{C.F.} = \text{RSS}$$

Table 2.8.1 ANOVA for LSD-v

Source	d.f.	S.S.	M.S.	F-value	Expected Mean Squares Model-I	Model-II
Rows	$v-1=v$	RSS	$\dfrac{RSS}{v-1}=RMS$	$\dfrac{RMS}{EMS}=F_R$	$\sigma_e^2+\dfrac{v}{v-1}\sum_k\rho_k^2$	$\sigma_e^2+v\sigma_R^2$
Columns	$v-1=v$	CSS	$\dfrac{CSS}{v-1}=CMS$	$\dfrac{CMS}{EMS}=F_C$	$\sigma_e^2+\dfrac{v}{v-1}\sum_j\beta_j^2$	$\sigma_e^2+v\sigma_C^2$
Treats.	$v-1=v$	TrSS	$\dfrac{TrSS}{v-1}=TrMS$	$\dfrac{TrMS}{EMS}=F_{Tr}$	$\sigma_e^2+\dfrac{v}{v-1}\sum_i\tau_i^2$	$\sigma_e^2+v\sigma_T^2$
Error	$\begin{array}{c}(v-1)(v-2)\\=v(v-1)\end{array}$	ESS	$\dfrac{ESS}{v(v-1)}=EMS$	σ_e^2	σ_e^2	σ_e^2
Total	v^2-1	TSS				

$$\text{Column S.S.} = \frac{1}{v}\left(C_{.1.}^2 + C_{.2.}^2 + - - - - - - + C_{.v.}^2\right) - \text{C.F.} = \text{CSS}$$

$$\text{Treat S.S.} = \frac{1}{v}\left(T_{1..}^2 + T_{2..}^2 + \ldots\ldots\ldots + T_{v..}^2\right) - \text{C.F.} = \text{TrSS}$$

$$\text{Error S.S.} = \text{TSS} - \text{RSS} - \text{CSS} - \text{TrSS} = \text{ESS}$$

The expected mean squares are displayed in the ANOVA (2.8.1) without derivation and decoding of terms as they are self-explanatory.

Expected mean squares under both the models reveal that the row, column and treatments mean squares can be tested against error mean square without any ambiguity. The same has been applied in the above ANOVA table to obtain the exact F-test. F for each factor has the same degrees of freedom $\{\upsilon, \upsilon(\upsilon - 1)\}$. The inferences about the null hypotheses for the row, column and treatment effects can be drawn by comparing the F-value with the respective tabulated F-value for α level of significance and $\{\upsilon, \upsilon(\upsilon - 1)\}$ degrees of freedom. If for an individual factor, $F_{cal} \geq F_{\alpha; \{\upsilon, \upsilon(\upsilon - 1)\}}$, reject H_0, otherwise not. Interpretation of results is given in accordance of the problem and objectives of the investigation. In case the null hypothesis about the treatments is rejected, the analyst should further apply a proper test for testing the significance of the pairwise contrasts and any other non-pairwise contrasts of interest.

ANOVA table for a Latin square design of order $v \times v$ having single observation per experimental unit contains $(v - 1)(v - 2)$ d.f. for error. Hence for LSD-2, the d.f. for error is zero. This prohibits the use of any statistical test of hypothesis. Also for LSD-3, error d.f. is 2 and for LSD-4, error d.f. is 6. In both these designs, error d.f. is much less than the requirement for reliable tests of hypotheses. Evidently error d.f. for LSD-5 is 12, which is sufficient for valid conclusions. Hence, if one has to conduct an experiment in a single Latin square design, it should at be of order 5×5. This is only possible when there are five treatments at least in hand to be tested, which is not the case in many situations. Thus, to experiment with 2, 3, 4 treatments in Latin square design, the experimenter has two options.

(i) More than one observation per unit is taken as per need and facility.

(ii) A group of Latin squares be used for experimentation.

In both these situations, a proper degrees of freedom for error is available which enables an investigator to have a valid testing of hypothesis. Experiments in both of these situations are expatiated in the next section.

Example 2.8: An experiment was conducted to compare the efficiency of five soil conditioners namely, A – Control, B – Phospo. + Gyp., C – Gypsum, D – Pyrite, E – FYM in respect of uptake in plants. The experiment was conducted in a Latin square design so as to control two way variation in soil fertility. The layout of the experiment along with the measurements as calcium in plants (mg/gm) is given below. Row and column totals are also displayed alongside.

		Columns				
		1	2	3	4	5	Row totals
	1	D (24.80)	B (25.07)	A (24.27)	E (26.13)	C (23.73)	124.00
	2	B (31.20)	A (24.80)	C (21.87)	D (28.53)	E (27.20)	133.60
Rows	3	C (23.33)	D (28.00)	E (22.93)	A (24.53)	B (24.27)	123.06
	4	E (25.87)	C (23.70)	D (25.06)	B (29.33)	A (27.73)	131.69
	5	A (26.13)	E (25.87)	B (26.40)	C (25.87)	D (26.93)	131.20
Col. totals		131.33	127.44	120.53	134.39	129.86	643.55

$$\text{C.F.} = (643.55)^2/25 = 16566.26$$

$$\text{Total S.S.} = 24.80^2 + 25.07^2 + \ldots\ldots + 25.87^2 + 26.93^2 - \text{C.F.}$$

$$= 16675.50 - \text{C.F.} = 109.24$$

$$\text{Row S.S.} = \frac{1}{5}(124.00^2 + 133.60^2 + 123.06^2 + 131.69^2 + 131.20^2) - \text{C.F.}$$

$$= 16584.88 - \text{C.F.} = 18.62$$

$$\text{Col. S.S.} = \frac{1}{5}(131.33^2 + 127.44^2 + 120.53^2 + 134.39^2 + 129.86^2) - \text{C.F.}$$

$$= 16588.06 - \text{C.F.} = 21.80$$

Conditioners total,

$$A = 24.27 + 24.80 + 24.53 + 27.73 + 26.13 = 127.46$$

Similarly, B =136.27, C = 118.50, D = 133.32, E = 128.00

ANOVA table

Source	d.f.	S.S.	M.S.	F-value
Rows	4	18.62	4.66	4.66/2.65 = 1.76
Cols.	4	21.80	5.45	5.45/2.65 = 2.06
Conditioners	4	37.01	9.28	9.28/2.65 = 3.50*
Error	12	31.81	2.65	
Total	24	109.24		

* Significant at = 0.05
Tabulated value of $F_{.05; 4, 12} = 3.26$

Conditioners S.S. $= \frac{1}{5}(127.46^2 + 136.27^2 + 118.50^2 + 133.32^2 + 128.00^2) - \text{C.F.}$

$$= 16603.21 - \text{C.F.} = 37.01$$

Error S.S. $= 109.24 - 18.62 - 21.80 - 37.01 = 31.81$

F-test reveals that there is a significant difference between soil conditioners. Therefore, this necessitates comparison of all possible pairs of conditioner means applying multiple range tests. Here the pairwise contrasts are tested by Student-Newman-Keul multiple range test. To do so it will be convenient to prepare a table of differences of paired means. The estimates of the five treatment means are:

$$\bar{A} = 25.49, \ \bar{B} = 27.25, \ \bar{C} = 23.70, \ \bar{D} = 26.66, \ \bar{E} = 25.60$$

Writing the means in ascending order along rows and in descending order along columns, the difference table is as displayed below.

Difference table

Treat.		C	A	E	D
	Means	**23.70**	**25.49**	**25.60**	**26.66**
B	**27.25**	3.55	1.76	1.65	0.59
D	**26.66**	2.96	1.17	1.06	–
E	**25.60**	1.90	0.11	–	–
A	**25.49**	1.79	–·	–	–

The upper triangle differences provide the differences for all pairs of means. Therefore, the lower triangle differences are omitted.

Critical ranges are calculated by the formula (1.9.2.1). From the ANOVA, it is evident that $r = 5$, $s_e^2 = 2.65$ and the values of $q_{.05;k,12}$ are obtained from the table of studentized range test.

$$W_5 = 4.51 \times \sqrt{\frac{2.65}{5}} = 3.28$$

$$W_4 = 4.20 \times \sqrt{\frac{2.65}{5}} = 3.06$$

$$W_3 = 3.77 \times \sqrt{\frac{2.65}{5}} = 2.74$$

$$W_2 = 3.08 \times \sqrt{\frac{2.65}{5}} = 2.24$$

Comparing first row differences with W_5, W_4, W_3, and W_2, respectively, it is inferred that treatments B and C differ significantly and others do not. All differences in second, third and fourth rows are smaller than corresponding critical values. So they affirm a non-significant difference among other paired means. Finally, it leads to conclude that the availability of calcium for plants under the conditioner Phospho + Gyp is significantly higher than Gypsum alone whereas all other soil conditioners are at par.

2.9 ESTIMATING MISSING VALUE IN LATIN SQUARE DESIGN

An experimenter conducts a trial with all care and caution. In spite of all precautions, the experiments do not always terminate without accidents. Such accident(s) result into one or more missing values in an experiment. The plants in a plot are eaten by animal(s), seeds do not germinate in a plot, an animal dies during the period of experimentation, a value is mutilated, and so on. In all such situations, the data are incomplete and affect the assumptions of its model, especially the orthogonality of data is disturbed. Due to some mishap, an experimenter cannot discard the trial so easily, because the time, money, his efforts matter a lot to him. To overcome such situations, statisticians came forward and provided sound solutions to their problems for analyzing data which could produce reliable results. The author has no hitch in saying that once an actual value is lost, there is no way to get back that value. An error is committed which definitely amounts to loss of information. But the remedies are invented by different workers, keeping in view the kind and manner in which the value(s) happened to miss. If one or two values are missing, they are estimated by a suitable method, more commonly by the method of least squares.

Allen and Wishart (1930) were the first to present formula for estimating a missing value in Latin square design. Later Yates (1933) proved that the formula given by Allen and Wishart was same as he obtained by minimizing the error sum of square. If there are two or more missing values, an iterative method of estimation proposed by Yates (1933) is generally acceptable. Some times it happens that the yields of two or more plots are mixed up and a combined value for all such plots is at the disposal of the experimenter. Bose and Mahalanobis (1938) came out with a solution for estimating the individual plot values and finally the analysis of data. Grundy (1951) discussed a situation in which a treatment is incorrectly applied to an experimental unit. He suggested a solution for such an anomaly. Yates (1936) considered situations in which an experimenter comes across a variety of calamities in a Latin square design like failure of a treatment in all plots, in a row or a column as whole. Yates put forth the methods for analyzing data in all such abnormal situations. This kind of work was extended further by Yates and Hale (1939) giving the procedures of analysis of data of LS design in which two or more treatments, rows or columns are missing. Another approach to deal with the missing values is analysis of covariance (ANCOVA). Bartlett (1937) used this technique for estimating one or more missing values and analyzing data of an experiment in Latin square design. Nair (1940) made use of ANCOVA for analyzing data of field experiments with several missing values or mixed-up values. DeLury (1946) summarized the results for estimating missing values and analyzing data in Latin squares or sets of Latin squares. Anyhow, estimation of missing values and analyzing data with the help of concomitant variables are dealt within Chapter 10 on Analysis of Covariance. So this procedure is deferred over here.

There are hundred of eventualities which can lead to missing or mixing of values. It is not possible to enumerate all of them here. A general discussion of missing values has been given here which is applicable to other designs as well. For further reading on Latin square designs, the readers are referred to Giesbrecht and Gumpertz (2004), Dean and Boss (1999) and Montgomery (2005).

One missing value in LS design

Consider an experiment with layout in Latin square design having v treatments. Unfortunately, a value is lost for ith treatment occurring in kth row and jth column. Without entangling into derivation for estimating the missing value by minimizing error sum of square, the formula for estimation of missing value is presented directly. Supposing that the totals of the kth row, jth column and ith treatment having a missing value based on $(v - 1)$ observations are $Y'_{..k}, Y'_{.j.}$, and $Y'_{i..}$ respectively. Also let the sum of all $(v^2 - 1)$ available observations be $Y'_{...}$. The estimate of the missing value is,

$$\hat{y}_{ijk} = \frac{v(Y'_{i..} + Y'_{.j.} + Y'_{..k}) - 2Y'_{...}}{(v-1)(v-2)} \qquad (2.9.1)$$

Immediately after estimating the missing value, one can proceed for the analysis of data. Substitute the estimated value in place of missing value. In this way, the data are complete for analysis purpose. Analysis of variance is carried out in the usual manner with certain amendments to compensate for the loss of information as delineated below.

(*i*) First reduce the total degrees of freedom by one in this case (by the number of missing values in general). This obviously results in reduction of error degrees of freedom by one.

(*ii*) It was experienced that there is an upward bias in treatment sum of square due to the substitution of estimated value for the missing value. Hence, this bias is adjusted by subtracting a correction factor C from treatment sum of square for which the formula is,

$$C = \frac{\left[Y'_{...} - Y'_{i..} - Y'_{.j.} - (v-1)\hat{y}_{ijk}\right]^2}{(v-1)^2(v-2)^2} \qquad (2.9.2)$$

Thus an improved treatment sum of square,

$$\text{Tr'SS} = \text{TrSS} - C \qquad (2.9.3)$$

This value of treatment sum of square is used in ANOVA table. Rest of the procedure for analysis of data remains same.

Example 2.9: Consider example (2.8) supposing that the value for treatment D in fourth row and third column is missing. The same has been marked by enclosing it in a box.

For analysis of data, the missing value is estimated first and then complete analysis is carried out. From the given data in example (2.8), it is simple to calculate the following values.

$$Y'_4 = 106.63, \ Y'_{.3.} = 95.47, \ Y'_{..D} = 108.26, \ Y'_{...} = 618.49$$

$$\hat{y}_{43D} = \frac{5(106.63 + 95.47 + 108.26) - 2 \times 618.49}{(5-1)(5-2)}$$

Say $\qquad D^* = \dfrac{314.82}{12} = 26.23$

On substituting the estimated value for the missing plot, the new row, column, treatment and grand totals are calculated and displayed below.

Row totals	Column totals	Treat. totals	
$Y_{1..} = 124.00$	$Y_{.1.} = 131.33$	A = 127.46	
$Y_{2..} = 133.60$	$Y_{.2.} = 127.44$	B = 136.27	
$Y_{3..} = 123.06$	$Y_{.3.} = 121.70$	C = 118.50	and $Y_{...} = 644.72$
$Y_{4..} = 132.86$	$Y_{.4.} = 134.39$	D = 134.49	
$Y_{5..} = 131.20$	$Y_{.5.} = 129.86$	E = 128.00	

$$\text{C.F.} = \frac{(644.72)^2}{25} = 16626.56$$

Sum of squares for different components are:

Row S.S. = $124.00^2 + 133.60^2 + \ldots\ldots\ldots + 131.20^2 - \text{C.F.}$
   $= 16646.79 - \text{C.F.} = 20.23$

Col. S.S. = $131.33^2 + 127.44^2 + \ldots\ldots\ldots + 129.86^2 - \text{C.F.}$
   $= 16644.74 - \text{C.F.} = 18.18$

Treat. S.S. = $127.46^2 + 136.27^2 + \ldots\ldots\ldots + 128.00^2 - \text{C.F.}$
   $= 16665.87 - \text{C.F.} = 39.31$

Total S.S. = $24.80^2 + 25.07^2 + \ldots\ldots\ldots + 25.87^2 + 26.93^2 - \text{C.F.}$
   $= 16735.51 - \text{C.F.} = 108.95$

Correction factor for treatment sum of square,

$$C = \frac{(618.49 - 106.63 - 95.47 - 4 \times 108.26)^2}{(5-1)^2 (5-2)^2}$$

$$= 1.92$$

Corrected Tr.S.S. $= 39.31 - 1.92 = 37.39$

ANOVA Table

Source	d.f.	S.S.	M.S.	F-value
Rows	4	20.23	20.23/4 = 5.06	5.06/3.01 = 1.68
Cols.	4	18.18	18.18/4 = 4.54	4.54/3.01 = 1.50
Treats.	4	37.39	37.39/4 = 9.35	9.35/3.01 = 3.10
Error	11	33.15	33.15/11 = 3.01	
Total	23	108.95		

Table value of $F_{.05;\ 4,11} = 3.36$

Comparing the calculated F-values for rows, columns and treatments with the tabulated F-value for = 0.05 and (4,11) d.f., it is concluded that all effects are non-significant. One point to be noticed here is that a missing value has totally changed the result. The treatments which differed significantly has now come out to be at par. This shows how critical is a missing value.

Qualities of a Latin square design

Every experimental design has some qualities and also some weaknesses. Hence, main points in favor and against a Latin square design are delineated.

Uppers

1. Two ways blocking in a Latin square design controls two nuisance variables resulting into more précised estimates and tests as compared to CRD and RBD.
2. There is always a smaller error mean square in LSD with the same data as compared to CRD and RBD.
3. The analysis of data is simple as that of any other basic design.
4. The missing value does not complicate the analysis more than that of randomized block design.
5. Analysis of data for one or more missing treatments, rows or columns is available in the literature. For details, readers are referred to Yates (1933), Yates and Hale (1939), Searle (1987).
6. A 4 × 4 Latin square design is as efficient as a randomized block design with five replications and four treatments.

Downers

1. The design requires as many levels of blocking factors as the number of treatments. Such a restriction is practically not appealing. Therefore, Latin square designs of order 8 × 8 or more are seldom used.

2. Latin square designs of order 4 × 4 or less do not provide adequate error degrees of freedom. So either one has to use more than one square or more than one observation per experimental unit. So Latin squares of order 2, 3 or 4 may not be as efficient as CRD or RBD.

3. The assumption that there is no interaction between rows, columns and treatments is too restrictive. Such an assumption may not hold good in many situations.

4. Randomization process of layout of a Latin square is not direct, rather needs greater understanding and skill.

2.10 GRAECO-LATIN SQUARE DESIGN

As discussed, a Latin square design controls two way variations, i.e. it isolates the effects of two nuisance variables. But in actual research process, there are situations which may call for isolating three nuisance variables. For instance, in case of experiments on lactating animals to compare the effect of various feeds on their milk yield, it looks logical to eliminate the effects of lactation number, breed and age. Three-way variations due to such factors can be controlled through a Graeco-Latin square (GLS) design. The name of this design has so emerged because it consists of Greek and Latin letters. Atkinson and Bailey (2001) paper recalls on Tippett's use of a Graeco-Latin square to investigate defective cotton spindles. A GLS can simply be obtained by superimposing a Latin square of order-v having v Latin letters A, B, C, ..., over an orthogonal Latin square having v-levels of a third nuisance variable denoted by the Greek letters α, β, γ, ... Summarily, in a GLS one nuisance variable is taken along rows, the second along columns and the third by Greek letters. For illustration, the layout of a GLS of order-4 can be obtained as given below.

$$
\begin{array}{cccc}
A & B & C & D \\
B & A & D & C \\
C & D & A & B \\
D & C & B & A
\end{array}
\qquad
\begin{array}{cccc}
\alpha & \beta & \gamma & \delta \\
\delta & \gamma & \beta & \alpha \\
\beta & \alpha & \delta & \gamma \\
\gamma & \delta & \alpha & \beta
\end{array}
$$

(*a*) LS-4 with Latin letters   (*b*) LS-4 with Greek letters

On superimposing LS (*b*) on (*a*), the compound Latin square so called a Graeco-Latin square is,

A α	B β	C γ	D δ
B δ	A γ	D β	C α
C β	D α	A δ	B γ
D γ	C δ	B α	A β

(*c*) GLS – 4

Further, a Graeco-Latin square design can be analyzed for two factors which have same number of levels and are non-interacting. In such a situation, only two nuisance variables are isolated. Analysis procedure remains same as in case of LS design except the changes in interpretation of results. Analysis of a GLS design can be carried out in the same manner as it is done for an LS design with little modifications.

Statistical model for a GLS design with one observation per cell is,

$$y_{ijklu} = \mu + \tau_i + \xi_j + \phi_k + \theta_l + e_{iklu} \qquad (2.10.1)$$

where $i, j, k, l = 1, 2, 3, \ldots\ldots, v.$

and ξ_j, ϕ_k and θ_l represent row, column and Greek letter effects respectively.

Considering model (2.10.1) as fixed effect model, ANOVA with expected mean squares will be as given below.

Table 2.10.1 ANOVA table for GLS design

Source	d.f.	S.S.	M.S.	F- value	Expected mean square
Rows	$v-1$	RSS	$\dfrac{RSS}{v-1} = RMS$	$\dfrac{RMS}{EMS} = F_R$	$\sigma_e^2 + \dfrac{v}{v-1}\sum_j \xi_j^2$
Columns	$v-1$	CSS	$\dfrac{CSS}{v-1} = CMS$	$\dfrac{CSS}{EMS} = F_C$	$\sigma_e^2 + \dfrac{v}{v-1}\sum_k \phi_k^2$
Latin letters (Treatments)	$v-1$	TrSS	$\dfrac{TrSS}{v-1} = TrMS$	$\dfrac{TrMS}{EMS} = F_{Tr}$	$\sigma_e^2 + \dfrac{v}{v-1}\sum_i \tau_i^2$
Greek letters	$v-1$	GSS	$\dfrac{Gss}{v-1} = GMS$	$\dfrac{GMS}{EMS} = F_G$	$\sigma_e^2 + \dfrac{v}{v-1}\sum_l \theta_l^2$
Residual	$(v-1)$ $\times (v-3)$	ESS	$\dfrac{ESS}{(v-1)(v-3)} = EMS$		
Total	$v^2 - 1$	TSS			

Expected mean squares in the above ANOVA table clearly reveal that all effects are to be tested against residual (error) mean square. The calculation of sum of squares is similar to Latin square design and hence expressions for the same have been omitted.

Like other designs, Graeco-Latin square designs too have some merits and demerits. The same have been discussed here.

Merits

1. In this design, an experimenter is able to control three nuisance variables at a time. This increases the efficiency of the design.
2. Graeco-Latin squares exist for all odd numbers.
3. Graeco-Latin squares are available for all numbers from 3 to 13 except for 6 and 10.
4. Graeco-Latin squares have the capacity to test the significance of two non-interacting treatment factors easily.

Demerits

1. Graeco-Latin squares are not suitable if any of the nuisance variables interact with the treatment or among themselves.
2. The assumption that all interaction effects are absent is always dubious.
3. The restriction that all factor variables should have the same number of levels is arduous.
4. A Graeco-Latin square of the order 5 × 5 or less does not provide adequate degrees of freedom for error.
5. These designs are seldom used because of too many limitations.

2.11 HYPER-GRAECO-LATIN SQUARE

This design has the facility of controlling four nuisance variables. Its construction is also simple as it can be obtained by superimposing an orthogonal Latin square over a Graeco-Latin square. It holds all the properties of a Graeco-Latin square. Thereby, the details are omitted. The layout of a hyper-Graeco-Latin square (HGLS) will be of type given below.

A α 1	B β 2	C γ 3	D δ 4
B γ 4	A δ 3	D α 2	C β 1
C δ 2	D γ 1	A β 4	B α 3
D β 3	C α 4	B δ 1	A γ 2

(*d*) Layout of an HGLS

In the above design, it is easy to verify that every Greek letter occurs with every Latin square only once and every numeral occurs with every Greek and Latin letter once and only once. There can be in all 165888 hyper-Graeco-Latin squares of order 4 × 4. Out of such a large number HGLS, only one is presented over here. Skeleton ANOVA table for a 4 × 4 HGLS is displayed over here.

Table 2.11.1 Skeleton ANOVA table

Source	d.f.
Rows	3
Columns	3
Latin letters	3
Greek letters	3
Numerals	3
Error	zero
Total	15

ANOVA Table 2.11.1 reveals that there is no degree of freedom for error and as such no test of hypothesis is feasible. Hence, an HGLS of order 4×4 has no relevance. To avail error degrees of freedom, one must use more than one HGLS. In this situation, the interaction Treatment × Square provides the error d.f. vis-a-vis the error variance. Hyper-Graeco-Latin squares of order more than 4×4 can also be constructed and analyzed in the like manner.

2.12 CROSS-OVER DESIGN

The designs considered so far necessarily required as many experimental units as the number of treatments added up to their replications. But availability of certain type of experimental units as per our requirement is not always possible particularly in medical sciences, industry, animal sciences, psychological research and so on. For instance, patients of a rare disease, animal of uncommon species, high cost machines, etc. are generally not available in sufficient number for experimentation. Hence, a design is needed in which all treatment effects can be measured accurately having used a fewer experimental units. Also many treatments have a long-term effects. This has to be paid attention and be taken care of.

This does not mean that cross-over designs cannot be used when complete set of subjects are at the disposal of the investigator. In that situation, cross-over design compensates the effects of superior and inferior, individual temperament, tendency and the like. When treatments are expected to have *carry over effect* or *residual effect*, cross-over designs are very appropriate.

Definition: Carry over effect means that a treatment effect which may be observed for certain period even if the treatment has been discontinued. In cross-over designs, the treatments are applied or administered to experimental units in cycles, i.e. one treatment after the other in a sequence. To surmount the problem of carry over effect(s), one way is to nullify this effect by maintaining some reasonable time gap between two consecutive treatment applications. The gap period is termed as *rest period*. In rest period, no observation is recorded. But many times it is not feasible to ascertain whether the treatment effect has completely vanished. Hence, in such situations cross-over design seems to be more appropriate than others.

Cross-over designs are adapted in a variety of research areas and thus different names were given by different workers. To name a few are; *change-over design, switch-over design, switch back design, reversal design, double change-over design, rotation design, etc.* Main features of a cross- or change-over design are as follows:

1. Experimental units receive one treatment level, a second treatment level, a third one and so on in a sequence. Generally, the experimental units remain the same like machines, animals, plots or any other subjects. Over all experimental units receive treatments in such a manner that each treatment level is followed by the other treatment levels an equal number of times. Therefore, cross-over or change-over design is a balanced design.

2. The layout of this design is usually a randomized block design or a set of Latin squares. For balancing a design, it is necessary that number of blocks should be an integral multiple (≥ 2) of number of treatments. Also the number of blocks or Latin squares should be such that they provide minimum required degrees of freedom for error. In case of Latin squares of order 3 or 5, the number of Latin squares should be multiple of 2.

3. Cross-over design in Latin squares controls two nuisance variables, one along rows and other along columns. The rows represent successive periods of application of treatments whereas columns stand for subjects.

4. Number of units is usually less than the total number of treatment levels and their repetition.

5. The treatment effects last longer even after recording of observations.

6. An experimenter has to decide cautiously the time of taking observations so that optimum performance of the treatments is being recorded.

7. The researcher should carefully decide the length of rest period.

8. Cross-over design is generally exploited for small number of treatments, i.e. two to four.

9. It is easy to contemplate that there shall be six sequences in a cross-over design with three treatments and twelve sequences with four treatments.

10. The analysis of data of this design is based on the assumption that there are no interactions between blocks, time periods and treatment levels. If this assumption does not hold good, the results shall be vitiated.

Layout and analysis of cross-over designs

The simple cross-over design is one which has only two treatments leveled as a_1 and a_2. In this design, half of the subjects will receive treatment a_1 followed by a_2 and the other half a_2 followed by a_1. The experimental layout with eight blocks (subjects) in RBD is of the type given below.

Subjects

Rows	I	II	III	IV	V	VI	VII	VIII
Time period 1	a_2	a_1	a_1	a_2	a_1	a_2	a_2	a_1
Time period 2	a_1	a_2	a_2	a_1	a_2	a_1	a_1	a_2

(*a*) Cross-over design in RBD

The same experiment can be conducted as set of four Latin squares of order 2×2.

Rows	Square I		Square II		Square III		Square IV	
Time period 1	a_2	a_1	a_1	a_2	a_1	a_2	a_2	a_1
Time period 2	a_1	a_2	a_2	a_1	a_2	a_1	a_1	a_2

(*b*) Cross-over design in a set of four Latin squares

Skeleton ANOVA for the above two types of layouts shall be as given below.

(2.12.1) Skeleton ANOVA for cross-over design in RBD

Source	d.f.
Blocks	7
Periods	1
Treatments	1
Error	6
Total	15

(2.12.2) Skeleton ANOVA for cross-over design in four LS

Source	d.f.
Squares	3
Cols. within squares	4
Rows within squares	4
Treatments	1
Error (Treat. X squares)	3
Total	15

Comparison of degrees of freedom in the above two ANOVA tables reveals that cross-over design in blocks results more error degrees of freedom than the set of Latin square. In this respect, cross-over block design is better than the set of Latin squares. But this advantage can be hailed only if the differences between time periods over all subjects are uniform and thus are removed by single degree of freedom. Sum of squares can be calculated in the usual manner. The same will further be comprehended though the numerical example given ahead. On the other hand, error degree of freedom is only 3 in case of four Latin squares of order 2×2. But the differences between periods and subjects are better taken care of.

The layouts of a change-over design with three treatments in six blocks and a set of two Latin squares of order 3×3 are displayed below without any explanation as they are self-explanatory.

Blocks

Rows	I	II	III	IV	V	VI
Time period 1	a_1	a_1	a_3	a_2	a_2	a_3
Time period 2	a_2	a_3	a_1	a_3	a_1	a_2
Time period 3	a_3	a_2	a_2	a_1	a_3	a_1

(c) Change-over design with six blocks

	Square I			Square II		
Time period 1	a_1	a_2	a_3	a_1	a_2	a_3
Time period 2	a_3	a_1	a_2	a_2	a_3	a_1
Time period 3	a_2	a_3	a_1	a_3	a_1	a_2

(d) Set of two Latin squares of order 3 × 3

In the above designs, every treatment is followed by the other treatments twice.

The skeleton analysis of variance tables for the above two designs are as displayed below.

(2.12.3) ANOVA for change-over design

Source	d.f.
Blocks	5
Rows	2
Treatments	2
Error	8
Total	17

(2.12.4) ANOVA for two LS – 3

Source	d.f.
Squares	1
Cols. within squares	4
Rows within squares	4
Treatments	2
Treats. × squares	2
Error within squares	4
Total	17

Change-over designs can be extended to 4, 5, 7 treatments and so on, but not for six treatments as the orthogonal Latin squares of order 6 × 6 are not being constructed so far. For vivid discussion of cross-over designs the readers are advised to go through the book by Jones and Kenward (2003).

Merits of cross-over designs

1. Cross-over designs make possible to estimate direct effect as well as indirect effect of the treatments.
2. The comparison of treatments is highly précised.
3. The design is suitable for small number of treatments, generally 2 to 4 treatments.

Demerits of cross-over designs

1. Residual effects are estimated less precisely as compared to direct effects.
2. The analysis of data becomes some what complicated due to estimation of residual effects
3. The design cannot be used if any one of the treatments has a permanent adverse effect.

Analysis of data including residual effects

This has already been discussed that to eliminate carry over effect of one treatment over the next treatment, a rest period is always given between two consecutive treatments applied on the same subject. But to determine the exact length of rest period which fully removes the carry over effect is almost practically impossible due to many compulsions on the part of the experimenter. For instance, an experiment on lactating animals must be completed in the duration of one lactation period, a crop cannot be prolonged beyond its cropping season, a machine in a factory cannot be laid down for long and so on. So the possibility of residual effect of a treatment to be carried over is always there. Hence, a cross-over design is organized in such a manner that the treatments are adjusted for residual effects. The required formulae in the simplest form for the analysis of data of a cross-over design adjusted for residual effects are given in the sequel.

Before providing a general procedure for analysis of variance of a cross-over design estimating the residual effects and eliminating them from treatment effects, certain notations are first decoded. The formulae for estimating direct and residual effects follow. Let,

k – Number of treatments.

s – Number of squares.

T_i – ith treatment total, for $i = 1, 2, ..., k$.

R_i – Total of treatment response occurring in the next period of the treatment i.

S_i – Total of those columns (sequences) in which the treatment i occurs in the last.

P_i – Total of ith period.

G – Grand total.

Further calculate two more quantities.

(*i*) $P_i - k\,G = Q_{1i}$

(*ii*) $k\,P_i - (k + 2)\,G = Q_{2i}$

The formula for estimating direct effect eliminating residual effect is,

$$k \times s(k^2 - k - 2)\hat{t}_i = (k^2 - k - 1)T_i + kR_i + S_i + Q_{1i} \qquad (2.12.1)$$

To obtain the treatment mean adjusted for residual effect, added over all mean to the value of $\hat{t}_i$.

The formula for estimating residual effect r_i of the ith treatment to the next one in the sequence eliminating direct effect is,

$$k \times s\left(k^2 - k - 2\right)\hat{r}_i = kT_i + kR_i + kS_i + Q_{2i} \qquad (2.12.2)$$

Alternative formula: The same value of r_i can also be obtained by the formula,

$$\text{Unadj. mean of } t_i - \frac{k-1}{ks} \text{ (Sum of estimated residual effect other than treatment } i\text{)} \qquad (2.12.2.1)$$

[Check: One should verify that $\sum_i \hat{t}_i = 0$ and $\sum \hat{r}_i = 0$]

It is experienced that direct effects have also some influence over residual effects. As a matter of fact they are interwoven. The treatment sum of square can be splitted up in either of the two parts, i.e. the sum of square due to,

Direct effect (elim. resid. effect) + resid. effects (igno. direct effects)
$$(2.12.3)$$

or Direct effect (igno. resid. effects) + resid. effects(elimin. direct effects)
$$(2.12.3.1)$$

To obtain the sum of squares due to residual effects ignoring direct effects, one has to calculate auxiliary quantities by the formula,

$$\hat{R}_i^* = \hat{R}_i + G - kT_i \qquad (2.12.4)$$

The skeleton analysis of variance table for cross-over design eliminating residual effects will be as delineated below.

Table 2.12.1 Skeleton ANOVA table eliminating residual effect

Source	d.f.
Columns (Sequences)	$ks - 1$
Periods within squares	$s(k-1)$
Direct effects (elim. resid. effect)	$k - 1$
Residual effect (igno. direct effect)	$k - 1$
or	
Direct effect (igno. resid. effect)	$k - 1$
Residual effect (elim. direct effect)	$k - 1$
Error	$(k-1)(ks - s - 2)$
Total	$k^2 s - 1$

The sum of squares for columns, periods within squares and total can be calculated as per common procedure. Of course the sum of squares for other factors given in ANOVA can be worked out by the following formulae.

Direct effects S.S. ignoring residual effects $= \dfrac{1}{ks}\sum_i T_i^2 - \text{C.F.}$ \hfill (2.12.5)

Direct effects S.S. eliminating residual effects $= \dfrac{\sum_i \left\{ ks\left(k^2 - k - 2\right)\hat{t}_i \right\}^2}{ks\left(k^2 - k - 1\right)\left(k^2 - k - 2\right)}$

\hfill (2.12.6)

Residual effects S.S. ignoring direct effects $= \dfrac{\sum_i \hat{R}_i^{*2}}{k^3 s\left(k^2 - k - 1\right)}$ \hfill (2.12.7)

Residual effects S.S. eliminating direct effects $= \dfrac{\sum_i \left\{ ks\left(k^2 - k - 2\right)\hat{r}_i \right\}^2}{k^3 s(k^2 - k - 2)}$

\hfill (2.12.8)

Once the sums of squares for various factors are calculated, it is straight forward to complete the analysis of variance table. If F-value for any factor comes out to be significant, the standard errors of the differences of paired means are required. The same can be worked out by the following formulae.

$$\text{S.E.}\left(\hat{t}_i - \hat{t}_{i'}\right) = \sqrt{\dfrac{2s_e^2\left(k^2 - k - 1\right)}{ks\left(k^2 - k - 2\right)}}, \quad i \neq i' \tag{2.12.9}$$

$$\text{S.E.}\left(\hat{r}_i - \hat{r}_{i'}\right) = \sqrt{\dfrac{2s_e^2 k^2}{ks\left(k^2 - k - 2\right)}}, \quad i \neq i' \tag{2.12.10}$$

where, s_e^2 is the error mean square.

Paired effect means can easily be tested by comparing their differences with lsd or any other critical value of an appropriate test.

The procedure of analysis of data of a cross-over design is further expatiated through the following example.

Example 2.10: An experiment was conducted with twelve crossbred healthy heifers in four Latin squares of order 3×3 to compare three feed mixtures. The periods were taken along rows whereas the columns stood for the heifers. Three treatments were:

T_1 – Wheat straw at lib + conc. mixture.

T_2 – Wheat straw + 15 % water hyacinth + conc. mixture.

T_3 – Wheat straw + 30 % water hyacinth + Conc. mixture.

Adaptation period between treatments was 10 days. The variable measured was digestibility coefficient (Parts/100 parts on dry matter) of organic matter.

Periods	Square I				Square II			
1	T_1 (65.8)	T_2 (65.0)	T_3 (64.4)	Total 195.2	T_1 (60.0)	T_2 (59.2)	T_3 (64.9)	Total 184.1
2	T_2 (59.8)	T_3 (68.7)	T_1 (61.5)	190.0	T_3 (63.2)	T_1 (66.2)	T_2 (61.7)	191.1
3	T_3 (63.0)	T_1 (61.9)	T_2 (58.4)	183.3	T_2 (63.2)	T_3 (62.9)	T_1 (58.1)	184.2
Total	188.6	195.6	184.3	568.5	186.4	188.3	184.7	559.4

Periods	Square III				Square IV			
1	T_2 (63.6)	T_3 (62.4)	T_1 (61.8)	Total 187.8	T_2 (61.5)	T_1 (57.4)	T_3 (64.8)	Total 183.7
2	T_1 (60.1)	T_2 (59.0)	T_3 (68.1)	187.2	T_3 (60.5)	T_2 (56.7)	T_1 (64.1)	181.3
3	T_3 (66.0)	T_1 (65.5)	T_2 (65.1)	196.6	T_1 (59.4)	T_3 (62.4)	T_2 (65.3)	187.1
Total	189.7	186.9	195.0	571.6	181.4	176.5	194.2	552.1

Analysis of data is carried out in the manner this is explicated in theoretical discussion. The method of calculation of T's, R's, S's, Q's and P's is demonstrated step by step in the following part.

In the given example, $k = 3$, $s = 4$.

The procedure of calculation of constants T, R, S and P, as defined on page 86, is further delineated below.

$T_1 = 65.8 + 61.9 + 61.5 + 60.0 + 66.2 + 58.1 + 60.1 + 65.5 + 61.8 +$

$\qquad 59.4 + 57.4 + 64.1$

$\qquad = 741.8$

Similarly T_2 and T_3 are worked out.

$R_1 = 59.8 + 58.4 + 63.2 + 62.9 + 66.0 + 68.1 + 56.7 + 65.3$

$\qquad = 500.4$

Likewise R_2 and R_3 are worked out.

$S_1 = 195.6 + 184.7 + 186.9 + 181.4$

$\qquad = 748.6$

In the same manner, S_2 and S_3 are obtained.

$$P_1 = 65.8 + 65.0 + 64.4 + 60.0 + 59.2 + 64.9 + 63.6 + 62.4$$
$$+ 61.8 + 61.5 + 57.4 + 64.8$$
$$= 750.8$$

In the same way, P_2 and P_3 are computed. Thus,

$T_1 = 741.8,$	$R_1 = 500.4,$	$S_1 = 748.6,$	$P_1 = 750.8$
$T_2 = 738.5,$	$R_2 = 504.5,$	$S_2 = 759.9,$	$P_2 = 749.6$
$T_3 = 771.3,$	$R_3 = 495.9,$	$S_3 = 743.1,$	$P_3 = 751.2$

$$G = 2251.6$$

$Q_{11} = P_1 - 3G = -6004.0,$	$Q_{21} = 3P_1 - 5G = -9005.6$
$Q_{12} = P_2 - 3G = -6005.2,$	$Q_{22} = 3P_2 - 5G = -9009.2$
$Q_{13} = P_3 - 3G = -6003.6,$	$Q_{23} = 3P_3 - 5G = -9004.4$

Estimates of direct effect eliminating residual effect by the formula (2.12.1) are,

$$3 \times 4\left(3^2 - 3 - 2\right)\hat{t}_1 = \left(3^2 - 3 - 1\right) \times 741.8 + 3 \times 500.4 + 748.6 - 6004.0$$

$$48\hat{t}_1 = 3709.0 + 1501.2 + 748.6 - 6004.0 = -45.2$$

$$\hat{t}_1 = -0.942$$

Similarly,

$$48\hat{t}_2 = 5 \times 738.5 + 3 \times 504.5 + 759.9 - 6005.2 = -39.3$$

$$\hat{t}_2 = -0.819$$

Now,

$$48\hat{t}_3 = 5 \times 771.3 + 3 \times 495.9 + 743.1 - 6003.6 = 83.7$$

or $$\hat{t}_3 = 1.744$$

Note: It can be checked that $\sum_i \hat{t}_i = -0.017$, which is almost zero. A slight difference from zero occurs due to rounding of figures.

Estimates of residual effects adjusted for direct effects are worked out with the help of the formula (2.12.2).

$$3 \times 4\left(3^2 - 3 - 2\right)\hat{r}_1 = 3 \times 741.8 + 9 \times 500.4 + 3 \times 748.6 - 9005.6$$

$$\hat{R}_1 = 48\hat{r}_1 = 2225.4 + 4503.6 + 2245.8 - 9005.6$$

$$\hat{R}_1 = 48, \hat{r}_1 = -30.8$$

or $$\hat{r}_1 = -0.6417$$

Similarly,

$$\hat{R}_2 = 48\hat{r}_2 = 3 \times 738.5 + 9 \times 504.5 + 3 \times 759.9 - 9009.2$$

$$\hat{R}_2 = 48\hat{r}_2 = 26.5$$

or $\quad \hat{r}_2 = 0.5520$

$$\hat{R}_3 = 48\hat{r}_3 = 3 \times 771.3 + 9 \times 495.9 + 3 \times 743.1 - 9004.4$$

$$\hat{R}_3 = 48\hat{r}_3 = 1.9$$

$$\hat{r}_3 = 0.0396.$$

Again check that $\sum_i \hat{r}_i = 0.05$ which is quite near to zero.

Calculation of sum of squares,

$$\text{C.F.} = \frac{(2251.6)^2}{36} = 140825.07$$

Total S.S. $= 65.8^2 + 65.0^2 + \ldots\ldots + 62.4^2 + 65.3^2 - \text{C.F.}$

$$= 141136.68 - \text{C.F.} = 311.61$$

Sequence S.S. $= \frac{1}{3}(188.6^2 + 195.6^2 + \ldots\ldots + 176.5^2 + 194.2^2) - \text{C.F.}$

$$= 140942.10 - \text{C.F.} = 117.03$$

S.S. due to periods within squares,

$$= \frac{1}{3}(195.2^2 + 190.0^2 + 183.3^2) - \frac{568.5^2}{9} + \frac{1}{3}(184.1^2 + 191.1^2 + 184.2^2)$$

$$- \frac{559.4^2}{9} + \frac{1}{3}(187.8^2 + 187.2^2 + 196.6^2) - \frac{571.6^2}{9} + \frac{1}{3}(183.7^2 +$$

$$181.3^2 + 187.1^2) - (552.1)^2/9$$

$$= 140909.87 - 140851.29 = 58.58$$

Direct effect S.S. ignoring residual effect,

$$= \frac{1}{12}(741.8^2 + 738.5^2 + 771.3^2) - \text{C.F.}$$

$$= 140879.43 - \text{C.F.} = 54.36$$

Direct effect S.S. eliminating residual effect,

$$= \frac{1}{240}\left\{(-45.2)^2 + (-39.3)^2 + (83.7)^2\right\}$$

$$= 44.14$$

$$\hat{R}_1^* = -30.8 + 2251.6 - 3 \times 741.8 = -4.6$$

$$\hat{R}_2^* = 26.5 + 2251.6 - 3 \times 738.5 = 62.6$$

$$\hat{R}_3^* = 1.9 + 2251.6 - 3 \times 771.3 = -60.4$$

Residual S.S. ignoring direct effects,

$$= \frac{1}{540}\left\{(-4.6)^2 + (62.6)^2 + (-60.4)^2\right\}$$

$$= 14.05$$

Residual S.S. eliminating direct effects,

$$= \frac{1}{432}\left\{(-30.8)^2 + (26.5)^2 + (1.9)^2\right\}$$

$$= 3.83$$

ANOVA table

Source	d.f.	S.S.	M.S.	F-value
Sequences	11	117.03	10.64	1.64
Periods within squares	8	58.58	7.32	1.13
Direct effect (elim.resid.effects)	2 ⎤	44.14 ⎤	22.07 ⎤	3.40 ⎤
Residual effects (igno.direct effects)	2 ⎦	14.05 ⎦	7.02 ⎦	1.08 ⎦
Direct effects (igno.resid.effects)	2 ⎤	54.36 ⎤	27.18 ⎤	4.19 ⎤
Residual effects (elim.direct effects)	2 ⎦	3.83 ⎦	1.92 ⎦	0.30 ⎦
Error	12	77.81	6.48	
Total	35	311.61		

Table values of F-distribution at $\alpha = 0.05$ and respective d.f. required for the above ANOVA from Table B-7 are as given below.

$$F_{.05, 11, 12} = 2.72, \; F_{.05, 8, 12} = 2.85, \; F_{.05, 2, 12} = 3.88$$

Comparing the calculated values of F with the respective tabulated values of F for various effects, it is found that all effects except treatment effects ignoring residual effects are not significant. It reveals that the digestibility of cows is same under three feed mixtures. As a matter of fact, treatment effects adjusted for residual effects and residual effects adjusted for treatment effects are of prime importance as they provide pure effects. So it will be redundant to compare treatment or residual means.

Nota bene: It should always be checked that the aggregate of sum of squares for direct effects eliminating residual effects and residual effects ignoring direct effects is equal to the total sum of squares due to direct effects ignoring residual effects and residual effects eliminating direct effects.

From the above ANOVA, it is easily verifiable that,
$$44.14 + 14.05 = 54.36 + 3.83 = 58.19$$

$$\text{S.E. of treatment mean} = \sqrt{\frac{5 \times 6.48}{3 \times 4 \times 4}} = 0.82$$

$$\text{S.E. of residual effect mean} = \sqrt{\frac{9 \times 6.48}{3 \times 4 \times 4}} = 1.10$$

Standard errors are calculated to show how precise are the estimated means.

Concluding remark

All experiments are conducted with the objective of estimating and testing the treatment effects as precisely as possible. But the interference of nuisance variables cause great problem in meeting our objectives. Hence, various designs are evolved to achieve our goal by controlling or eliminating the influence of nuisance variables for meaningful and reliable conclusions. Soon after collecting and tabulating the data in a proper format, the effects are estimated and analysis of variance is carried out. Subsequently pairwise and non-pairwise contrasts are tested through an appropriate test in case it is required. The matter covered in Chapter 2 embodying the principles given in Chapter 1 will enable the students and researchers to design the experiments and analyze the data without ambiguity. Results should be interpreted within their limits without imposing any preconceived ideas.

The designs discussed in this chapter are basic designs and will continually be used directly or indirectly in one or the other form in the sequel.

Therefore, these designs should thoroughly be studied with all its complexities and complicacies.

QUESTIONS AND EXERCISES

1. Why completely randomized design is called no restrictional design?
2. Skeleton ANOVA of a CRD is given below.

source	d.f.
Treats.	5
Error	25
Total	30

    Can it be concluded that the treatments are equireplicated?
3. Discuss good and bad points of a completely randomized design.
4. Give the expected mean squares for treatments and error in ANOVA in case of Model-I and Model-II of a CRD. To what conclusion do they lead?

5. In what sense randomized block design is better than completely randomized design? Also elucidate its drawbacks.

6. How to carry out the analysis of data of a randomized block design in case of one missing value?

7. Justify that a randomized block design is a balanced and an orthogonal design.

8. Explain iterative method of estimating a number of missing values in a randomized block design.

9. What is the importance of expected mean squares in analysis of variance of data of an experiment?

10. What are the merits and demerits of a randomized block design?

11. When would you prefer to use an RBD, if the block totals are approximately the same or differ substantially?

12. Derive expected mean squares for the components of variation of an RBD assuming random effect model.

13. Suggest what should be the minimum number of replications if a researcher wants to conduct an experiment in RBD with four treatments?

14. Find the number of replication to be taken in an experiment if an investigator desires to detect a difference between treatment means to the extent of 2.3 units. From previous experiment, it is known that error mean square is 4.32 having 14 degrees of freedom for it.

15. What is difference between Subsampling and repeated observations in experimental designs? Also elaborate with the help of examples whether these techniques can be applied in all experimental designs?

16. In what respects analysis of non-orthogonal data of an experiment in randomized block design differs from analysis of orthogonal data? Explain clearly.

17. Discuss a Latin square design giving its important features.

18. Enunciate the following giving their configurations.
    (*i*) Standard Latin square.
    (*ii*) Conjugate Latin square.
    (*iii*) Adjugate Latin Squares.
    (*iv*) Orthogonal Latin squares.

19. In the set of Latin squares of order 3 and 4 how many standard Latin squares in each case can be constructed? Give their configurations as well.

20. Explain the procedure of randomly selecting a Latin square out of a set of Latin squares.

21. Give the statistical model of a Latin square design and underlying assumptions on which the analysis of variance is based?

22. Give the configuration of (*i*) 4 × 4 Graeco-Latin square design (*ii*) set of orthogonal Latin squares of order 4 × 4.

23. Prepare a skeleton analysis of variance table for a 5 × 5 Latin square design and also give expected mean squares assuming fixed effect model.

24. How can the data of a Latin square design having one missing value be analyzed? Explain fully.

25. What are the merits and demerits of a Latin square design?

26. Suggest a design which can isolate three nuisance variables. Give its structure and discuss its advantages and limitations.

27. Give the layout and skeleton ANOVA table of hyper-Graeco-Latin square design.

28. What is a cross-over design and its relevance? Explain the situations in which cross-over designs are most appropriate.

29. Compare the layouts of cross-over designs in RBD and in a set of Latin squares through an illustration.

30. What are other names given to cross-over design?

31. Expatiate the procedure of analysis of data of a cross-over design in which the residual effects are taken into consideration.

32. An experiment was conducted in a 5 × 5 Latin square design to compare the effect of five manures on sugarcane yield. The yield q/ha and layout of the experiment were as given below.

Sugarcane yield (q/ha)
Columns

Rows	1	2	3	4	5
1	A	E	D	C	B
	65.8	40.6	59.4	60.4	51.1
2	D	B	A	E	C
	59.1	50.9	65.4	49.1	59.9
3	B	A	C	D	E
	47.5	61.0	57.6	56.1	49.2
4	C	D	E	B	A
	55.9	53.4	40.6	48.3	60.0
5	E	C	B	A	D
	40.3	56.4	50.9	60.4	50.9

Details of treatments:

A – Inorganic manure (100 kg N/ha)
B – Organic manure (10 tonnes /ha FYM)
C – Organic manure (15 tonnes /ha FYM)
D – Organic manure (20 tonnes /ha FYM)
E – No treatment (Control)

Analyze the experimental data and draw conclusions about the effects of manures. Use proper test for pairwise comparisons and also test the significance of the contrast control versus rest of the treatments.

32. A soil scientist conducted an experiment with twelve treatments in a randomized block design. The yield of wheat grain (q/h) for each plot was recorded as presented below. Anyhow the yield for treatment No. nine (T_9) in second block was missed. Do the complete analysis of data and interpret the results.

Wheat grain yield (q/ha)

Treatments		R_1	R_2	R_3	R_4
100% NPK	T_1	40.50	42.94	35.94	44.88
100% NPK + Zn	T_2	44.69	40.00	43.13	42.25
100% NPK + Zn + S	T_3	41.88	43.13	41.56	46.80
100% NPK + S	T_4	49.06	39.00	40.00	40.94
100% NPK + Azotobactor	T_5	47.50	39.06	44.50	40.63
100% NPK •FYM	T_6	41.56	40.75	47.50	41.13
100% NPK + FYM	T_7	44.38	45.63	50.38	42.50
FYM @ 20 t/ha	T_8	36.25	36.31	32.25	29.06
150% NPK	T_9	45.50	X	46.88	42.31
100% NP	T_{10}	36.25	40.87	41.88	36.31
100% N	T_{11}	36.19	35.25	35.06	35.06
Control	T_{12}	18.44	25.00	20.63	33.44

33. When do you analyze for non-orthogonal data of an experiment? Explain with some practical examples.

34. Give statistical model for an unbalance experiment laid out in randomized block design and normal equations which lead to analysis of non-orthogonal data.

35. The quantity of serum albumin per 100 ml in three groups of lepers of different kinds of leprosy, who were put on four drugs, were as tabulated below. The categories of lepers are considered as blocking factor.

Experimental data

Drugs	Lepromatous leprosy	Tuberculoid leprosy	Intermittent
D_1	3.65	3.20	3.90
	3.60	4.10	3.10
	2.70		

(Contd.)

D_2	3.80 4.00 3.60 2.95	4.20 3.65 4.65	3.15 3.20 4.20
D_3	2.85 3.30 3.80	3.70 3.40 4.80	3.00 3.40
D_4	3.05 3.90 3.75 3.10 2.90	3.20 3.90 3.75	3.45 4.16 3.85

Analyze the experimental non-orthogonal data and interpret the results.

36. In a debate competition, five undergraduate students participated. Five rounds of debate were held to deliver speech on a new but same topic in every round. To provide an equal opportunity to each participant in respect of sequence of topic and order of speech, they were arranged in a Latin square design. Average score of each competitor awarded by a panel of three judges out of 30 are given below along with the layout.

Sequences

Topics	S_1	S_2	S_3	S_4	S_5
T_1	P_1 12.1	P_2 13.7	P_3 12.7	P_4 18.3	P_5 22.6
T_2	P_2 10.6	P_3 14.3	P_4 16.4	P_5 20.3	P_1 14,7
T_3	P_3 15.3	P_4 20.7	P_5 23.7	P_1 16.3	P_2 17.3
T_4	P_4 21.0	P_5 26.3	P_1 15.0	P_2 18.7	P_3 16.7
T_5	P_5 23.0	P_1 19.3	P_2 20.3	P_3 16.0	P_4 21.0

Analyze the data and draw conclusions about the relative performance of the competitors.

37. An experiment was conducted to compare the efficiency of three salesmen S_1, S_2, S_3 on three counters C_1, C_2, C_3 placed in a sequence of a store. Each salesman was allocated a counter for a period of one month randomly and then rotated in order thrice. This experiment was repeated twice. Also it is expected that there remains an impact on sales of salesman even after leaving a particular counter. In this way, the experiment forms a cross-over design with a set of two Latin squares.

(a) The layout of cross-over design with volume of sales in lac rupees in parentheses is given below.

	Square – 1 Counters			Square – 2 Counters		
	C_1	C_2	C_3	C_1	C_2	C_3
Month 1	S_1 (8.3)	S_2 (5.8)	S_3 (6.7)	S_2 (6.8)	S_3 (8.1)	S_1 (7.1)
Month 2	S_2 (7.6)	S_3 (5.0)	S_1 (9.2)	S_3 (5.4)	S_1 (10.6)	S_2 (6.7)
Month 3	S_3 (7.2)	S_1 (9.2)	S_2 (5.6)	S_1 (7.6)	S_2 (5.5)	S_3 (4.8)

Analyze the hypothetical data considering that there is carry over effect of a salesman on a counter even after his leaving.

(b) Consider the layout of the above experiment with volume of sales in a randomized block design with cross-over treatments in six blocks.

	Counters					
	C_1	C_2	C_3	C_4	C_5	C_6
Month 1	S_1 (8.3)	S_2 (5.8)	S_3 (6.7)	S_2 (6.8)	S_3 (8.1)	S_1 (7.1)
Month 2	S_2 (7.6)	S_3 (5.0)	S_1 (9.2)	S_3 (5.4)	S_1 (10.6)	S_2 (6.7)
Month 3	S_3 (7.2)	S_1 (9.2)	S_2 (5.6)	S_1 (7.6)	S_2 (5.5)	S_3 (4.8)

Analyze the data and draw conclusions about the relative performance of the competitors.

38. An experiment was planned to compare the efficacy of five synthetic pesticides against major insect pests of mung bean. Treatment schedules were:

T_1 – Spray of malathion (0.05 %) at 25 DAS and endosulfan (0.07 %) at 55 DAS.

T_2 – Three releases of chrysoperla carnea @ 25000 neonate larvae per hectare at 25, 40, 55 DAS.

T_3 – Three spray of neem oil (0.2%) at 25, 40, 55 DAS.

T_4 – Releases of chrysoperla carnea @ 25000 neonate larvae per hectare at 25 DAS + spray of neem oil (0.2%) at 25 DAS + Spray of malathion (0.05 %) at 55 DAS.

T_5 – Releases of chrysoperla carnea @ 25000 neonate larvae per hectare at 25 DAS + spray of neem oil (0.2%) at 40 DAS + endosulfan (0.07 %) at 55 DAS.

Observations were per cent damaged pods determined after selecting 10 plants randomly and picking 10 pods per plant. The observations were recorded three times a day at regular intervals on days 1, 3, 5, 7 after III application. Percent damaged pods were as tabulated below.

Treatments	Day-1	Day-3	Day-5	Day-7
T_1	12, 10, 12	12, 12, 10	12, 10, 12	14, 13, 14
T_2	8, 9, 8	9, 8, 8	8, 9, 8	10, 11, 11
T_3	10, 11, 10	11, 11, 10	12, 11, 10	13, 14, 14
T_4	6, 5, 6	6, 5, 6	7, 6, 5	9, 9, 8
T_5	20, 18, 20	20, 19, 20	21, 20, 21	21, 20, 21

Since the data are in per cent damaged pods, the values should be transformed using angular transformation. Taking days as block and three observations in percent damaged pods per day as sub-samples, analyze the data and draw inferences.

BIBLIOGRAPHY

1. Agarwal BL. Testing a main effect in a three factor mixed model, *Communication in Statistics*, Vol.19, 1990: 2, 723–38.
3. Klockers AJ, Sax G. Multiple comparisons: Sage university paper series on quantitative applications in the social sciences, *Sage, Newbury Park*, CA, 1986: 07–61.
4. Allan FE, Wishart J. A method of estimating yield of a missing plot in field experimental work. *Journal of Agricultural Sciences*, 1930: 20, 399–406.
5. Atkinson AC, Bailey RA (2001). One hundred years of the design of experiments on and off the pages of Biometrika. *Biometrika*, 88, 53–97.
6. Bartlett MS. Some examples of statistical methods of research in agriculture and applied biology. *Journal of Royal Statistical Society*, 1937: Suppl.4: 137–83.
7. Bose M, Dey A. Some small and efficient cross over designs under an additive model. *Utilities Mathematica*, 2003: 63, 173–82.
8. Bose M, Mukherjee B. Cross over designs in the presence of higher order carry overs. *Australian and New Zealand Journal of Statistics*, 2000: 42, 235–44.
9. Bose SS, Mahalanobis PC. On estimating individual yield in case of mixed up yields of two or more plots in field experiments, *Sankhyā*, 1938: 4, 103–20.
10. Bose RC, Shrikhande SS, Parker ET. Further results on the construction of mutually orthogonal Latin squares and the falsity of Euler's conjecture, *Canadian Journal of mathematics*, 1960: 12, 189–203.
11. Carriere KC. Methods of repeated measures data analysis with missing values, *Journal of Statistical Planning and Inference*, 1999: 77, 221–26.
12. Cochran WG. Testing a linear relation among variances, *Biometrics*, 1951: 7, 17–32.
13. Davenport JM, Webster JT. A comparison of some approximate F–test, *Technometrics*, 1973: Vol. 15, 779–89.
14. Dean A, Voss D. *Design and Analysis of Experiments*, Springer - Verlag, New York, 1999.

15. Delury DB. The analysis of Latin squares when some observations are missing, *Journal of the American Statistical Association*, 1946: 41: 370–89.

16. Duncan, DB. Multiple range and multiple F–test. *Biometrics*, 1955: 11, 1–42.

17. Dunn O.J. Multiple comparisons among means. *Journal of the American Statistical Association*, 1961: 56, 52–64.

18. Dunnet CW. A multiple comparison procedure for comparing several treatments with a control. *Journal of the American Statistical Association*, 1955: 50, 1096–1121.

19. Dunnet CW. New table for multiple comparisons with a control. *Biometrika*, 1964: 31, 249–55.

20. Field AP. *Discovering Statistics using SPSS for Windows: advanced techniques for the beginners*, 2003: Sage, London.

21. Field AP, Hole GJ. *How to design and report experiments*, Sage, London, 2003.

22. Fisher RA, Yates F. The six by six Latin squares. *Proceedings of the Cambridge Philosophical Society*, 1934: 30, 492–507.

23. Fisher RA, Yates F. *Statistical Tables for Biological, Agricultural, and Medical Research*, 6th ed., Edinburgh: Oliver and Boyd, 1963.

24. Friedman M. The use of ranks to avoid the assumptions of normality implicit in the analysis of variance. *Journal of the American Statistical Association*, 1937: 32, 675–701.

25. Giesbrecht FG, Compertz ML. *Planning, Construction, and Statistical Analysis of Comparative Experiments*, John Wiley, Hoboken, NJ, 2004.

26. Grundy, PM. A general technique for the analysis of experiments with incorrectly treated plots. *Journal of Royal Statistical Society*, 1951: B. 13, 272–83.

27. Hartley HO. The maximum F–ratio as short cut for heterogeneity of variance. *Biometrika*, 1950: 37, 308–12.

28. Jackson S, Brashers DE. *Random Factors in ANOVA*. Sage university paper series on quantitative applications in the social sciences, 1994: 07- 098. Sage, Thousand Oaks, CA.

29. Jones B, Michael G. Kenward. *Designs and Analysis of Cross-over Trials*, 2nd ed. Chapman and Hall, New York, 2003.

30. Kempthorne O. *The design and Analysis of Experiments*, John Wiley, New York, 1952.

31. Keselman HJ, Keselman JC. Repeated measures multiple comparison procedures: effects of violating multisample sphericity in unbalance designs. *Journal of Educational Statistics*, 1988: 13(3), 215–26.

32. Kirk RE. *Experimental Design: Procedures for the Behavioral Sciences*, 2nd ed. Pacific Grove, CA: Brooks/ Cole, 1982.

33. Kruskal WH, Wallis WA. Use of ranks in one criterion analysis. *Journal of the American Statistical Association*, 1952: 47, 583–621.

34. Mann HB, Whitney DR. On a test of whether one of two random variables is stochastically larger than the other. *Annals of Mathematical Statistics*, 1947: 18, 50–60.

35. Maxwell SE. Pairwise multiple comparisons in repeated-measures designs. *Journal of Educational Statistics*, 1980: 5(3), 269–287.

36. Maxwell SE, Delaney HD. *Designing experiments and analyzing data*, Wadsworth, Belmont, CA, 1990.

37. Montgomery DC. *Design and analysis of experiments*, 6th ed., John Wiley, Hoboken, New Jersey, 2005.

38. Paul AE. On a preliminary test for pooling mean squares in the analysis of variance. *Annals of Mathematical Statistics*, 1950: 21, 539.

39. Ratkowski DA, Evans MA, Alldredge JR. *Cross Over Experiments: Design, Analysis and Application*, Marcel Dekker, New York, 1993.

40. Sattterthwaite FE. Synthesis of variances. *Psychometrika*, 1941: Vol. 6, 309–16.

41. Sattterthwaite FE. An approximate distribution of estimates of variance components. *Biometrics Bulletin*, 1946: Vol. 2, 110 –14.

42. Scheffe H. A method of judging all contrasts in the analysis of variance. *Biometrika*, 1953: 40, 87–104.

43. Searle SR. *Linear Models for Unbalanced data*, John Wiley, New York, 1987.

44. Siegel S, Castellan NJ. *Nonparametric statistics for behavioral sciences*, 2nd ed. McGraw Hill, New York, 1988.

45. Yates F. The analysis of replicated experiments when field results are incomplete. *Empire Journal of Experimental Agriculture*, 1933: 1, 129–42.

46. Yates F, Hale RW. The analysis of Latin squares with two or more rows, columns or treatments are missing. *Journal of Royal Statistical Society*, 1939: Suppl. 6: 67–79.

Appendix

WINDOWS EXPOSURE FOR ANALYZING DATA THROUGH SPSS 11.2 PACKAGE

Analysis of example 2.1

1. Enter the data as given in example 2.1. While feeding the data, prepare two variables, feeds for treatments and weight gain (wt. gain) for values of dependent variable. Ensure that variable feeds are taken as alphabets value A, B, C, D.

2. Select **Analyze** menu followed by **Compare means** and **One-way ANOVA** as shown in the following window:

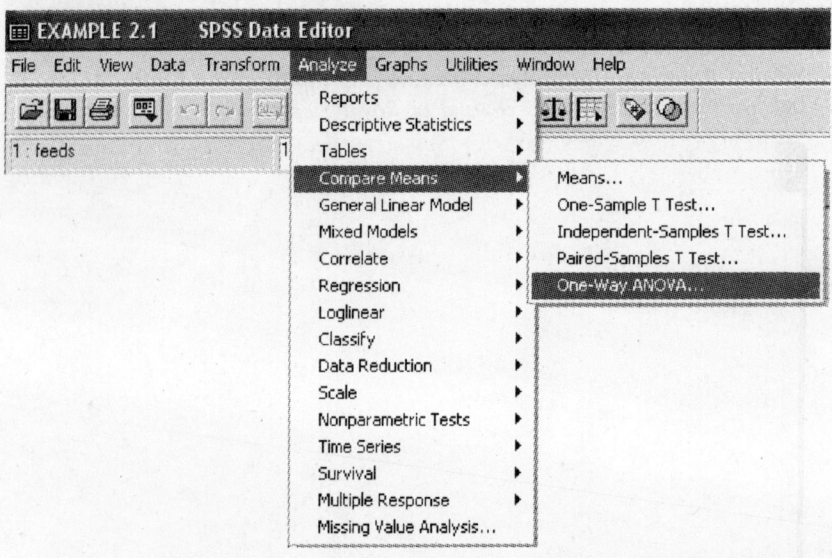

3. When you obtain following window, select each variable and press the corresponding arrow button to shift them to respective panes as shown below.

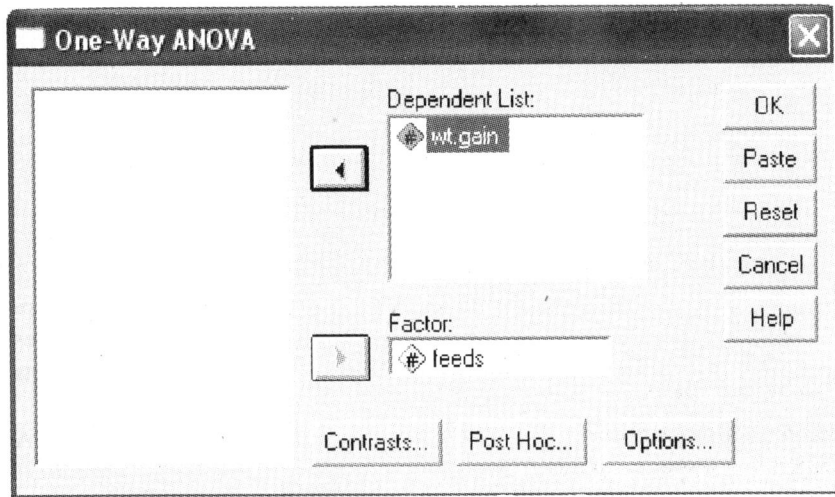

4. Now click on **Post Hoc...** button and in the new window, click on any test you want to apply for comparing the paired treatment means as shown below:

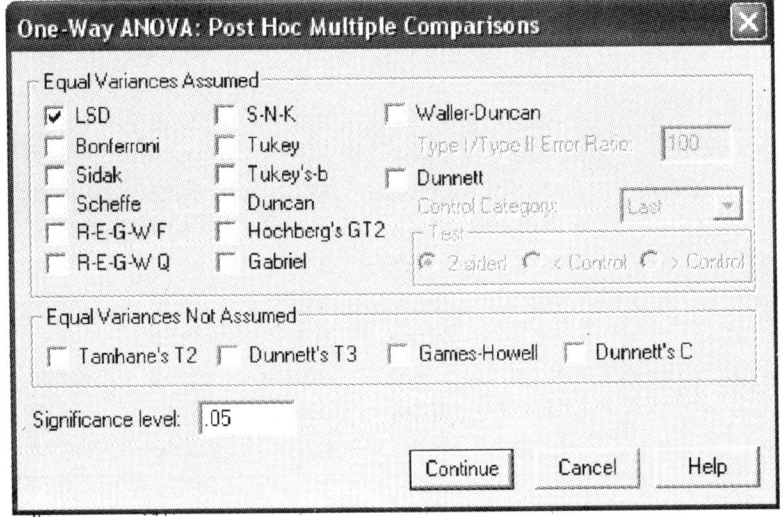

5. Click on **Continue** button to move to the original window. You can click on **Options** button and select the **Descriptive** checkbox to get mean, standard deviations and other relevant statistics.
6. Finally, click on **OK** button to get following output:

One-way

ANOVA

Wt.gain

	Sum of squares	df	Mean square	F	Sig.
Between groups	26320.071	3	8773.357	11.525	.000
Within groups	11418.350	15	761.223		
Total	37738.421	18			

Post-Hoc Tests

Multiple Comparisons
Dependent Variable: Wt.gain
LSD

(I) Feeds	(J) Feeds	Mean Difference (I-J)	Std. error	Sig.	95% confidence interval Lower bound	Upper bound
1	2	−27.20	17.450	.140	−64.39	9.99
	3	−34.95	18.508	.078	−74.40	4.50
	4	−99.00(*)	17.450	.000	−136.19	−61.81
2	1	27.20	17.450	.140	−9.99	64.39
	3	−7.75	18.508	.681	−47.20	31.70
	4	−71.80(*)	17.450	.001	−108.99	-34.61
3	1	34.95	18.508	.078	−4.50	74.40
	2	7.75	18.508	.681	−31.70	47.20
	4	−64.05(*)	18.508	.003	−103.50	−24.60
4	1	99.00(*)	17.450	.000	61.81	136.19
	2	71.80(*)	17.450	.001	34.61	108.99
	3	64.05(*)	18.508	.003	24.60	103.50

* The mean difference is significant at the .05 level.

Analysis of example 2.2

1. Enter the data as given in example 2.2. While feeding the data, prepare two tables, variety and block for independent variable and grain yield (grainyld) for values of dependent variable. Ensure that variable variety is denoted by V's and block by B's.

2. Select **Analyze** menu followed by **General Linear Model** and **Univariate** as shown in the following window:

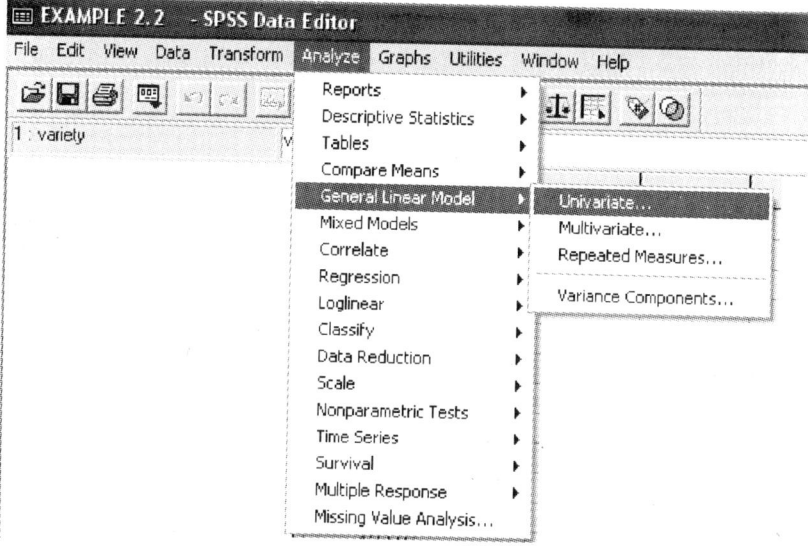

3. When you obtain following window, select each variable and press the corresponding arrow button to shift them to respective panes as shown below.

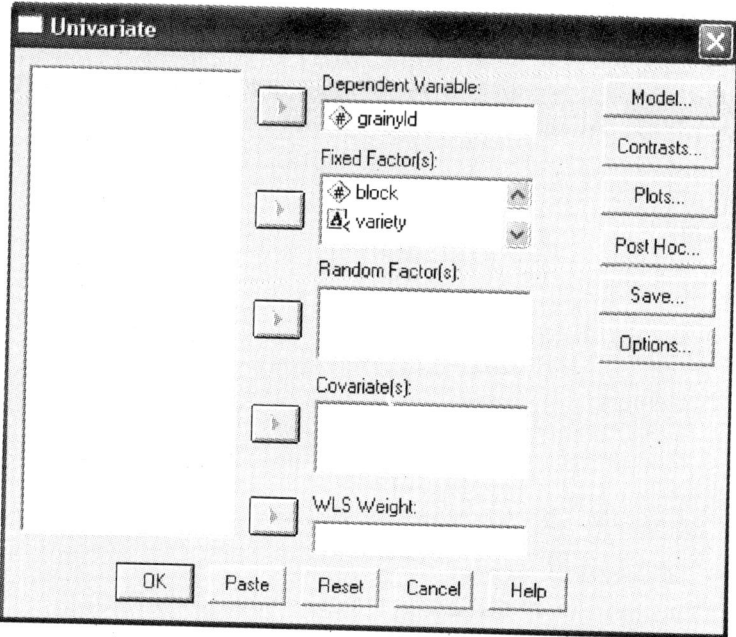

4. Now click on **Model** button and in the new window, specify the model type as **custom** and send **variety and block** variable in model text box as shown below:

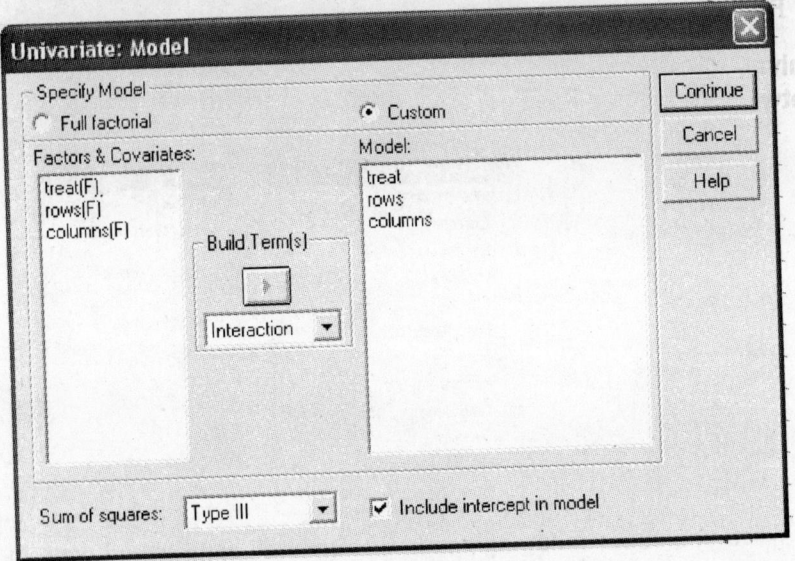

5. Click on Continue button to move to the original window. Now click on **Post Hoc...** button and in the new window, click on Waller-Duncan for comparing the paired treatment means as shown below:

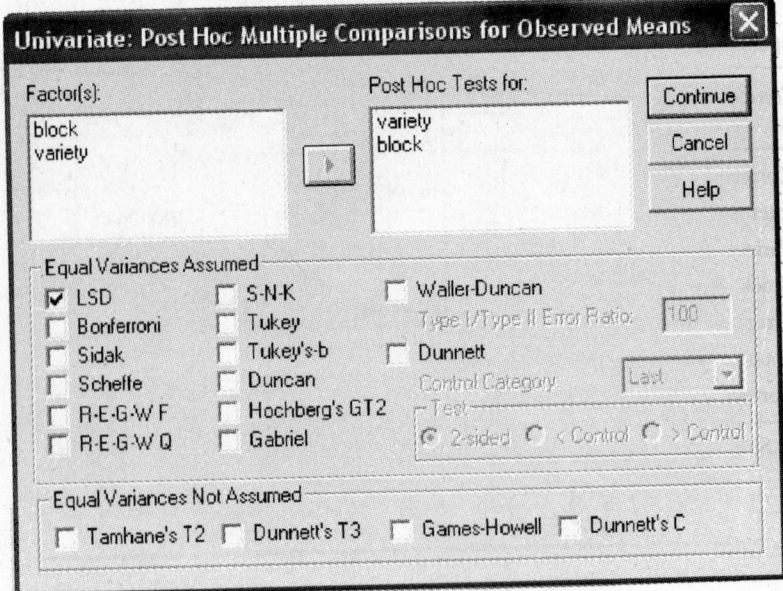

6. Click on Continue button to move to the original window. You can click on **Options** button and select the **Descriptive** check box to get mean, standard deviations and other relevant statistics.
7. Finally, click on **OK** button to get following output:

Univariate Analysis of Variance
Between-Subjects Factors

		N
Variety	*v*1	4
	*v*2	4
	*v*3	4
	*v*4	4
	*v*5	4
	*v*6	4
	*v*7	4
	*v*8	4
	*v*9	4
Block	1	9
	2	9
	3	9
	4	9

Tests of between-Subjects Effects

Dependent Variable: GRAINYLD

Source	*Type III sum of squares*	*df*	*Mean square*	*F*	*Sig.*
Corrected Model	1952.281(*a*)	11	177.480	5.760	.000
Intercept	24289.223	1	24289.223	788.263	.000
Variety	725.060	8	90.633	2.941	.019
Block	1227.221	3	409.074	13.276	.000
Error	739.527	24	30.814		
Total	26981.030	36			
Corrected Total	2691.808	35			

a R Squared = .725 (Adjusted R Squared = .599)

Post-Hoc Tests
VARIETY
Homogeneous Subsets
GRAINYLD

Waller-Duncan

Variety	N	Subset		
		1	2	3
v3	4	16.7500		
v2	4	21.7750	21.7750	
v9	4	24.5750	24.5750	24.5750
v7	4	25.1250	25.1250	25.1250
v6	4	25.3750	25.3750	25.3750
v4	4		28.0750	28.0750
v1	4		29.0500	29.0500
v8	4			31.4500
v5	4			31.6000

Means for groups in homogeneous subsets are displayed. Based on type III sum of squares. Error term is mean square(Error) = 30.814.

a   Uses harmonic mean sample size = 4.000.

b   Type 1/type 2 error seriousness ratio = 100.

Block
Homogeneous Subsets
GRAINYLD

Waller-Duncan

BLOCK	N	Subset		
		1	2	3
3	9	19.5889		
4	9	23.2333	23.2333	
1	9		25.7222	
2	9			35.3556

Means for groups in homogeneous subsets are displayed. Based on type III sum of squares
The error term is mean square(Error) = 30.814.

(a) Uses harmonic mean sample size = 9.000.

(b) Type 1/type 2 error seriousness ratio = 100.

Analysis of Example 2.3

1. Enter the data as given in example 2.3. While feeding the data, prepare two tables, treat and block for independent variable and fresh weight (fresh wt) for values of dependent variable. Ensure that variable variety is denoted by S's and block by roman numbers I, II, III, and IV.

2. Select **Analyze** menu followed by **General Linear Model** and **Univariate** as shown in the following window:

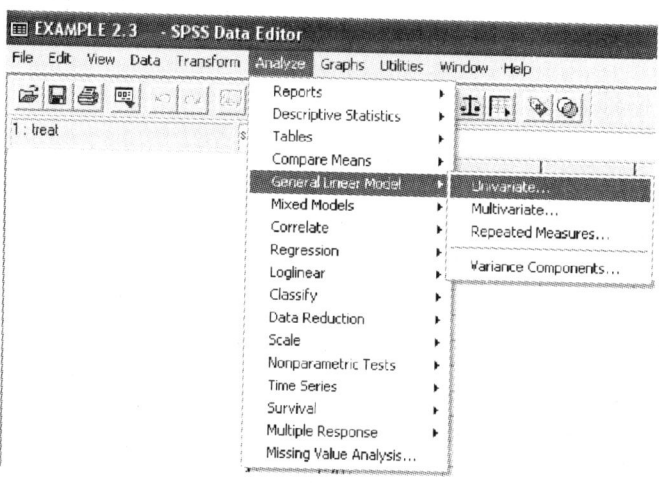

3. On obtaining the following window, select each variable and press the corresponding arrow button to shift them to respective panes as shown below.

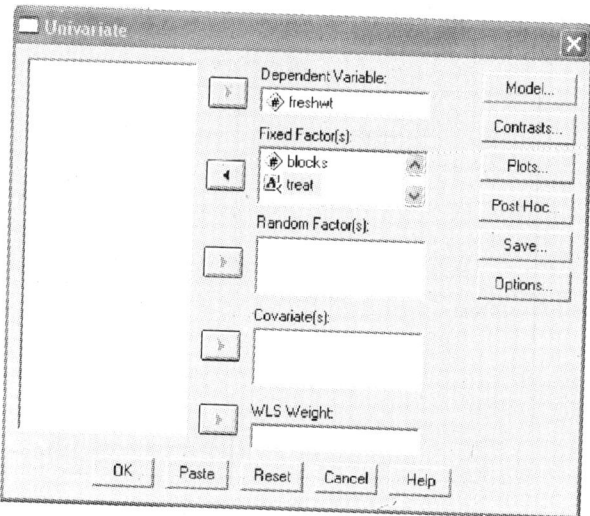

4. Now click on **Post Hoc...** button and in the new window, click on LSD for comparing the paired treatment means as shown below:

Univariate: Post Hoc Multiple Comparisons for Observed Means ☒

Factor(s):
blocks
treat

Post Hoc Tests for:
treat
blocks

[Continue]
[Cancel]
[Help]

Equal Variances Assumed
☑ LSD ☐ S-N-K ☐ Waller-Duncan
☐ Bonferroni ☐ Tukey Type I/Type II Error Ratio: [100]
☐ Sidak ☐ Tukey's-b ☐ Dunnett
☐ Scheffe ☐ Duncan Control Category: [Last ▼]
☐ R-E-G-W F ☐ Hochberg's GT2 ─Test─
☐ R-E-G-W Q ☐ Gabriel ⦿ 2-sided ○ < Control ○ > Control

Equal Variances Not Assumed
☐ Tamhane's T2 ☐ Dunnett's T3 ☐ Games-Howell ☐ Dunnett's C

5. Click on Continue button to move to the original window. You can click on **Options** button and select the **Descriptive** checkbox to get mean, standard deviations and other relevant statistics.

6. Finally, click on **OK** button to get following output:

Univariate Analysis of Variance
Between-Subjects Factors

		N
Blocks	1	5
	2	5
	3	4
	4	5
Treat	s0	4
	s1	3
	s2	4
	s3	4
	s4	4

Tests of Between-Subjects Effects

Dependent Variable: FRESHWT

Source	Type III sum of squares	df	Mean square	F	Sig.
Corrected model	.399(a)	7	.057	18.203	.000
Intercept	11.629	1	11.629	3713.622	.000
BLOCKS	.251	3	.084	26.680	.000
TREAT	.139	4	.035	11.080	.001
Error	.034	11	.003		
Total	12.510	19			
Corrected Total	.433	18			

a R Squared = .921 (Adjusted R Squared = .870)

Post-Hoc Tests
Treat

Multiple Comparisons

Dependent Variable: FRESHWT
LSD

(I) Treat	(J) Treat	Mean Difference (I-J)	Std. error	Sig.	95% Confidence Interval	
					Lower Bound	Upper Bound
s0	s1	−.22058(*)	.042739	.000	−.31465	−.12651
	s2	−.16600(*)	.039569	.001	−.25309	−.07891
	s3	−.25225(*)	.039569	.000	−.33934	−.16516
	s4	−.17400(*)	.039569	.001	−.26109	−.08691
s1	s0	.22058(*)	.042739	.000	.12651	.31465
	s2	.05458	.042739	.228	−.03949	.14865
	s3	−.03167	.042739	.474	−.12574	.06240
	s4	.04658	.042739	.299	−.04749	.14065
s2	s0	.16600(*)	.039569	.001	.07891	.25309
	s1	−.05458	.042739	.228	−.14865	.03949
	s3	−.08625	.039569	.052	−.17334	.00084
	s4	−.00800	.039569	.843	−.09509	.07909
s3	s0	.25225(*)	.039569	.000	.16516	.33934
	s1	.03167	.042739	.474	−.06240	.12574
	s2	.08625	.039569	.052	−.00084	.17334

(Contd.)

	s4	.07825	.039569	.074	−.00884	.16534
s4	s0	.17400(*)	.039569	.001	.08691	.26109
	s1	−.04658	.042739	.299	−.14065	.04749
	s2	.00800	.039569	.843	−.07909	.09509
	s3	−.07825	.039569	.074	−.16534	.00884

Based on observed means.
* The mean difference is significant at the .05 level.

Blocks

Multiple Comparisons

Dependent Variable: FRESHWT

LSD

(I) Treat	(J) Treat	Mean difference (I-J)	Std. error	Sig.	95% confidence interval	
					Lower bound	Upper bound
1	2	-.04340	.035392	.246	−.12130	.03450
	3	.19940(*)	.037539	.000	.11678	.28202
	4	.21880(*)	.035392	.000	.14090	.29670
2	1	.04340	.035392	.246	−.03450	.12130
	3	.24280(*)	.037539	.000	.16018	.32542
	4	.26220(*)	.035392	.000	.18430	.34010
3	1	−.19940(*)	.037539	.000	−.28202	-.11678
	2	−.24280(*)	.037539	.000	−.32542	-.16018
	4	.01940	.037539	.616	−.06322	.10202
4	1	−.21880(*)	.035392	.000	−.29670	−.14090
	2	−.26220(*)	.035392	.000	−.34010	−.18430
	3	−.01940	.037539	.616	−.10202	.06322

Based on observed means.
* The mean difference is significant at the .05 level.

Analysis of Example 2.4

1. Tabulate the data as per following format of the example 2.3. While feeding the data, prepare two tables, variety and block for independent variable and grain yield (grainyld) for values of dependent variable. Ensure that variable variety is denoted by V's and block by numerals 1, 2, 3, and 4.

EXAMPLE 2.4 - SPSS Data Editor

File   Edit   View   Data   Transform   Analyze   Graphs   Ut

	variety	block	grainyld
1	v1	1	20.20
2	v1	2	41.40
3	v1	3	29.00
4	v1	4	25.60
5	v2	1	18.50
6	v2	2	28.80
7	v2	3	15.00
8	v2	4	24.80
9	v3	1	
10	v3	2	25.50
11	v3	3	11.50
12	v3	4	13.50
13	v4	1	27.70
14	v4	2	30.80
15	v4	3	22.50
16	v4	4	31.30
17	v5	1	32.00
18	v5	2	34.70
19	v5	3	29.20
20	v5	4	30.50
21	v6	1	31.30
22	v6	2	35.20
23	v6	3	16.30
24	v6	4	18.70
25	v7	1	25.50
26	v7	2	43.70
27	v7	3	
28	v7	4	18.10
29	v8	1	39.20
30	v8	2	46.30
31	v8	3	20.10
32	v8	4	20.20

2. Select **Analyze** menu followed by **General Linear Model** and **Univariate** as shown in the following window:

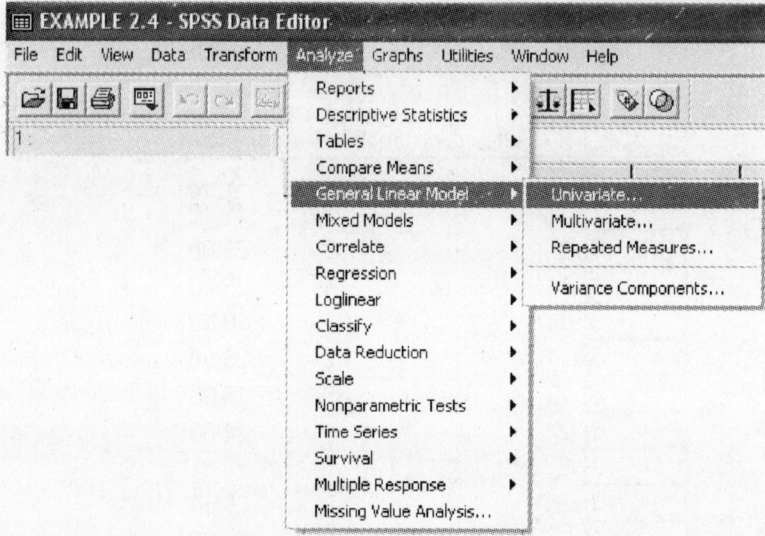

3. On obtaining the following window, select each variable and press the corresponding arrow button to shift them to respective panes as shown below.

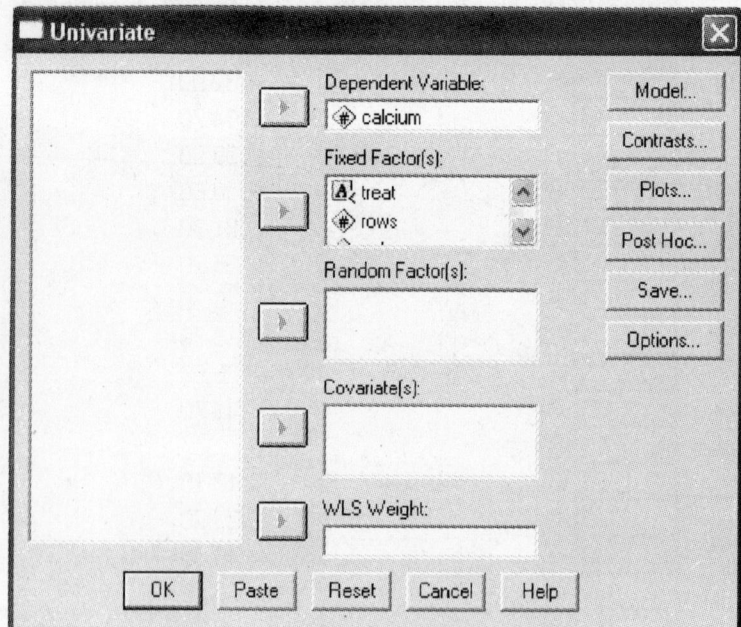

4. Now click on **Model** button and in the new window, specify the model type as **custom** and send **variety and block** variable in model text box as shown below:

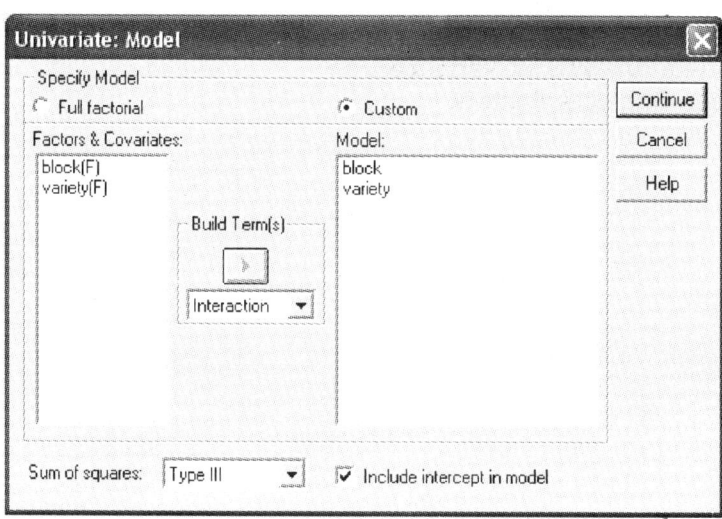

5. Click on Continue button to move to the original window. Now click on **Post Hoc...** button and in the new window, click on LSD for comparing the paired treatment means as shown below:

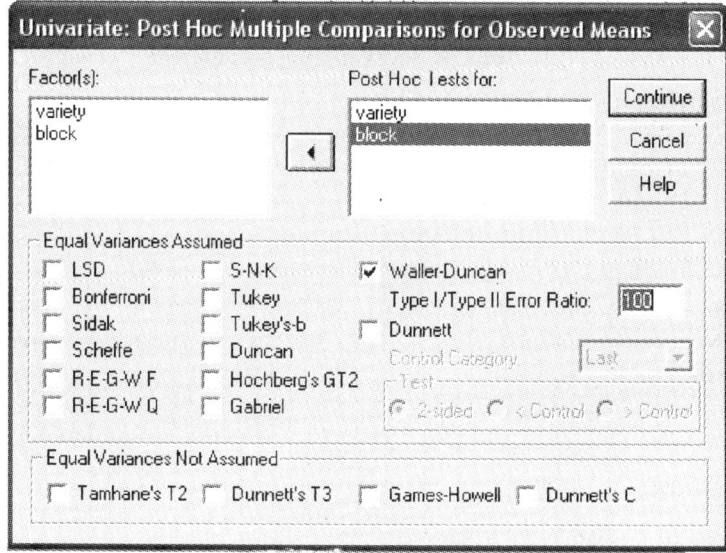

6. Click on Continue button to move to the original window. You can click on **Options** button and select the **Descriptive** checkbox to get mean, standard deviations and other relevant statistics.

7. Finally, click on **OK** button to get following output:

Univariate Analysis of Variance
Between-Subjects Factors

		N
BLOCK	1	8
	2	9
	3	8
	4	9
VARIETY	v_1	4
	v_2	4
	v_3	3
	v_4	4
	v_5	4
	v_6	4
	v_7	3
	v_8	4
	v_9	4

Tests of Between-Subjects Effects

Dependent Variable: GRAINYLD

Source	*Type III sum of squares*	*df*	*Mean square*	*F*	*Sig.*
Corrected model	1730.845(a)	11	157.350	4.992	.001
Intercept	22875.123	1	22875.123	725.749	.000
BLOCK	1083.631	3	361.210	11.460	.000
VARIETY	629.560	8	78.695	2.497	.043
Error	693.426	22	31.519		
Total	26534.540	34			
Corrected total	2424.271	33			

a R Squared = .714 (Adjusted R Squared = .571)

Post-Hoc Tests
Variety

Multiple Comparisons

Dependent Variable: GRAINYLD
LSD

(*I*) Variety	(*J*) Variety	Mean difference (*I-J*)	Std. error	Sig.	95% confidence interval Lower bound	Upper bound
1	2	-.04340	.035392	.246	-.12130	.03450
v1	v2	7.2750	3.96985	.080	-.9580	15.5080
	v3	12.2167(*)	4.28792	.009	3.3241	21.1093
	v4	.9750	3.96985	.808	-7.2580	9.2080
	v5	-2.5500	3.96985	.527	-10.7830	5.6830
	v6	3.6750	3.96985	.365	-4.5580	11.9080
	v7	-.0500	4.28792	.991	-8.9426	8.8426
	v8	-2.4000	3.96985	.552	-10.6330	5.8330
	v9	4.4750	3.96985	.272	-3.7580	12.7080
v2	v1	-7.2750	3.96985	.080	-15.5080	.9580
	v3	4.9417	4.28792	.262	-3.9509	13.8343
	v4	-6.3000	3.96985	.127	-14.5330	1.9330
	v5	-9.8250(*)	3.96985	.022	-18.0580	-1.5920
	v6	-3.6000	3.96985	.374	-11.8330	4.6330
	v7	-7.3250	4.28792	.102	-16.2176	1.5676
	v8	-9.6750(*)	3.96985	.023	-17.9080	-1.4420
	v9	-2.8000	3.96985	.488	-11.0330	5.4330
v3	v1	-12.2167(*)	4.28792	.009	-21.1093	-3.3241
	v2	-4.9417	4.28792	.262	-13.8343	3.9509
	v4	-11.2417(*)	4.28792	.016	-20.1343	-2.3491
	v5	-14.7667(*)	4.28792	.002	-23.6593	-5.8741
	v6	-8.5417	4.28792	.059	-17.4343	.3509
	v7	-12.2667(*)	4.58398	.014	-21.7733	-2.7601
	v8	-14.6167(*)	4.28792	.003	-23.5093	-5.7241
	v9	-7.7417	4.28792	.085	-16.6343	1.1509
v4	v1	-.9750	3.96985	.808	-9.2080	7.2580
	v2	6.3000	3.96985	.127	-1.9330	14.5330

(*Contd.*)

(I) variety	(J) variety	Mean difference (I-J)	Std. error	Sig.	95% confidence interval Lower bound	95% confidence interval Upper bound
v4	v3	11.2417(*)	4.28792	.016	2.3491	20.1343
	v5	−3.5250	3.96985	.384	−11.7580	4.7080
	v6	2.7000	3.96985	.504	−5.5330	10.9330
	v7	−1.0250	4.28792	.813	−9.9176	7.8676
	v8	−3.3750	3.96985	.404	−11.6080	4.8580
	v9	3.5000	3.96985	.387	−4.7330	11.7330
v5	v1	2.5500	3.96985	.527	−5.6830	10.7830
	v2	9.8250(*)	3.96985	.022	1.5920	18.0580
	v3	14.7667(*)	4.28792	.002	5.8741	23.6593
	v4	3.5250	3.96985	.384	−4.7080	11.7580
	v6	6.2250	3.96985	.131	−2.0080	14.4580
	v7	2.5000	4.28792	.566	−6.3926	11.3926
	v8	.1500	3.96985	.970	−8.0830	8.3830
	v9	7.0250	3.96985	.091	−1.2080	15.2580
v6	v1	−3.6750	3.96985	.365	−11.9080	4.5580
	v2	3.6000	3.96985	.374	−4.6330	11.8330
	v3	8.5417	4.28792	.059	−.3509	17.4343
	v4	−2.7000	3.96985	.504	−10.9330	5.5330
	v5	−6.2250	3.96985	.131	−14.4580	2.0080
	v7	−3.7250	4.28792	.394	−12.6176	5.1676
	v8	−6.0750	3.96985	.140	−14.3080	2.1580
	v9	.8000	3.96985	.842	−7.4330	9.0330
v7	v1	.0500	4.28792	.991	−8.8426	8.9426
	v2	7.3250	4.28792	.102	−1.5676	16.2176
	v3	12.2667(*)	4.58398	.014	2.7601	21.7733
	v4	1.0250	4.28792	.813	−7.8676	9.9176
	v5	−2.5000	4.28792	.566	−11.3926	6.3926
	v6	3.7250	4.28792	.394	−5.1676	12.6176
	v8	−2.3500	4.28792	.589	−11.2426	6.5426
	v9	4.5250	4.28792	.303	−4.3676	13.4176
v8	v1	2.4000	3.96985	.552	−5.8330	10.6330
	v2	9.6750(*)	3.96985	.023	1.4420	17.9080
	v3	14.6167(*)	4.28792	.003	5.7241	23.5093

(Contd.)

(I) variety	(J) variety	Mean difference (I-J)	Std. error	Sig.	95% confidence interval	
					Lower bound	Upper bound
v8	v4	3.3750	3.96985	.404	-4.8580	11.6080
	v5	-.1500	3.96985	.970	-8.3830	8.0830
	v6	6.0750	3.96985	.140	-2.1580	14.3080
	v7	2.3500	4.28792	.589	-6.5426	11.2426
	v9	6.8750	3.96985	.097	-1.3580	15.1080
v9	v1	-4.4750	3.96985	.272	-12.7080	3.7580
	v2	2.8000	3.96985	.488	-5.4330	11.0330
	v3	7.7417	4.28792	.085	-1.1509	16.6343
	v4	-3.5000	3.96985	.387	-11.7330	4.7330
	v5	-7.0250	3.96985	.091	-15.2580	1.2080
	v6	-.8000	3.96985	.842	-9.0330	7.4330
	v7	-4.5250	4.28792	.303	-13.4176	4.3676
	v8	-6.8750	3.96985	.097	-15.1080	1.3580

Based on observed means.
* The mean difference is significant at the .05 level.

Block
Multiple Comparisons

Dependent Variable: GRAINYLD

LSD

(I) Block	(J) Block	Mean difference (I-J)	Std. error	Sig.	95% confidence interval	
					Lower bound	Upper bound
1	2	-8.4806(*)	2.72802	.005	-14.1381	-2.8230
	3	6.4875(*)	2.80710	.031	.6659	12.3091
	4	3.6417	2.72802	.196	-2.0159	9.2992
2	1	8.4806(*)	2.72802	.005	2.8230	14.1381
	3	14.9681(*)	2.72802	.000	9.3105	20.6256
	4	12.1222(*)	2.64656	.000	6.6336	17.6109
3	1	-6.4875(*)	2.80710	.031	-12.3091	-.6659
	2	-14.9681(*)	2.72802	.000	-20.6256	-9.3105
	4	-2.8458	2.72802	.308	-8.5034	2.8117

(Contd.)

(I) Block	(J) Block	Mean difference (I-J)	Std. error	Sig.	95% confidence interval Lower bound	95% confidence interval Upper bound
4	1	−3.6417	2.72802	.196	−9.2992	2.0159
	2	−12.1222(*)	2.64656	.000	−17.6109	−6.6336
	3	2.8458	2.72802	.308	-2.8117	8.5034

Based on observed means.
* The mean difference is significant at the .05 level.

Analysis of Example 2.8

Latin square design

1. Arrange the data in the following manner of the example 2.8. While feeding the data, prepare two-way tables taking treatment(treat), rows and columns for independent variable and calcium (mg/gm) for values of dependent variable. Ensure that variable rows and columns are denoted by 1, 2, 3, 4 and 5 and treat is denoted by alphabets a, b, c, d and e.

EXAMPLE 2.8 LSD - SPSS Data Editor

File Edit View Data Transform Analyze Graphs Utilities Window

1 : treat   d

	columns	rows	treat	calcium
1	1	1	d	24.80
2	1	2	b	31.20
3	1	3	c	23.33
4	1	4	e	25.87
5	1	5	a	26.13
6	2	1	b	25.07
7	2	2	a	24.80
8	2	3	d	28.00
9	2	4	c	23.70
10	2	5	e	25.87
11	3	1	a	24.27
12	3	2	c	21.87
13	3	3	e	22.93
14	3	4	d	25.06
15	3	5	b	26.40
16	4	1	e	26.13
17	4	2	d	28.53
18	4	3	a	24.53
19	4	4	b	29.33
20	4	5	c	25.87
21	5	1	c	23.73
22	5	2	e	27.20
23	5	3	b	24.27
24	5	4	a	27.73
25	5	5	d	26.93

2. Select **Analyze** menu followed by **General Linear Model** and **Univariate** as shown in the following window:

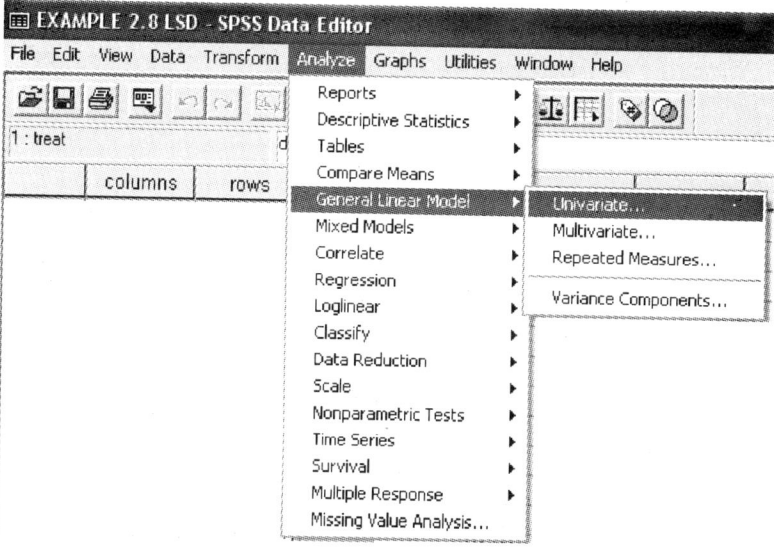

3. When the following window appears, select each variable and press the corresponding arrow button to shift them to respective panes as shown below.

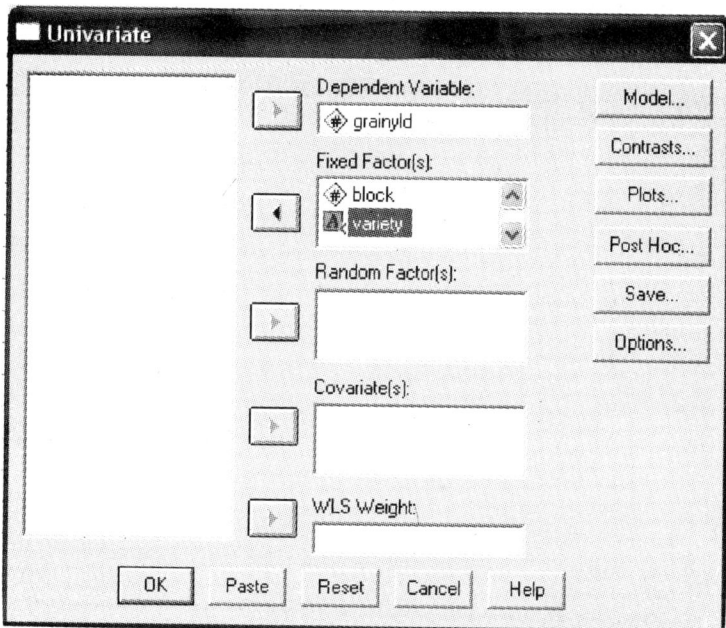

4. Now click on **Model** button and in the new window, specify the model type as **custom** and send **treat, rows and columns** variable in model text box as shown below:

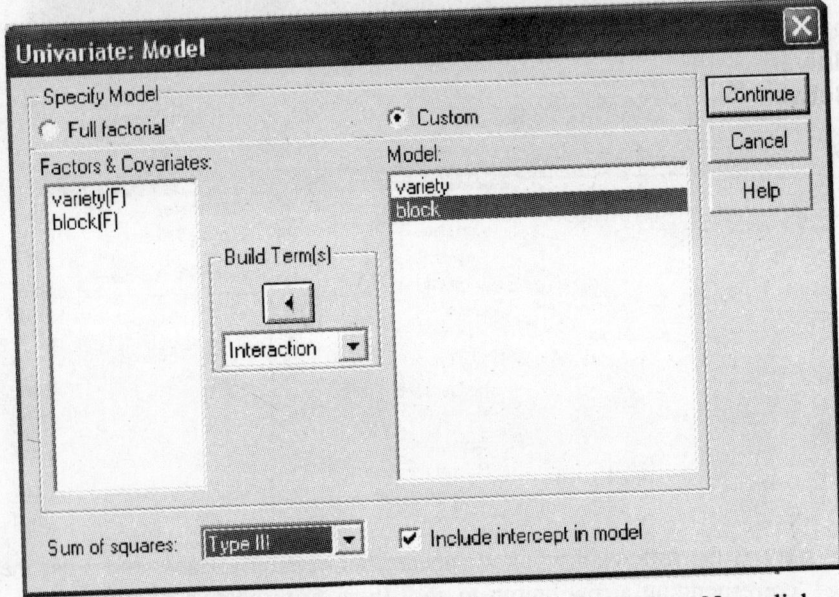

5. Click on Continue button to move to the original window. Now click on **Post Hoc...** button and in the new window, click on S-N-K for comparing the paired treatment means as shown below:

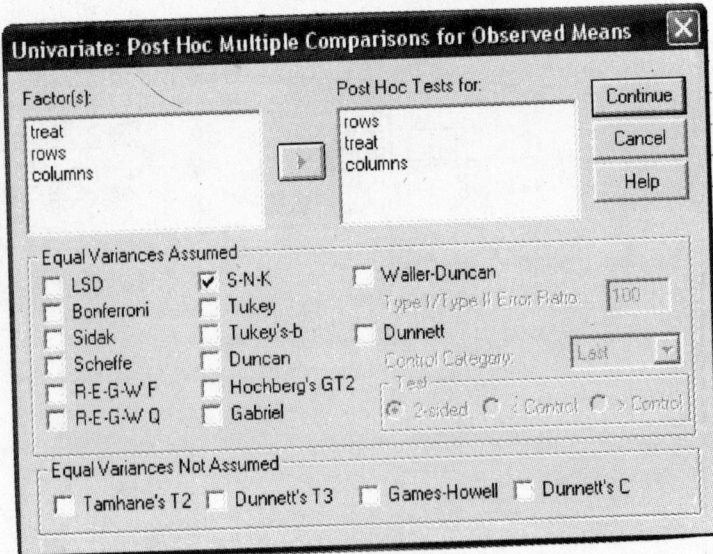

6. Click on Continue button to move to the original window. You can click on **Options** button and select the **Descriptive** checkbox to get mean, standard deviations and other relevant statistics.

7. Finally, click on **OK** button to get following output:

Univariate Analysis of Variance
Between-Subjects Factors

		N
TREAT	a	5
	b	5
	c	5
	d	5
	e	5
ROWS	1	5
	2	5
	3	5
	4	5
	5	5
COLUMNS	1	5
	2	5
	3	5
	4	5
	5	5

Tests of Between-Subjects Effects

Dependent Variable: CALCIUM

Source	Type III sum of squares	df	Mean square	F	Sig.
Corrected Model	77.358(a)	12	6.447	2.427	.069
Intercept	16566.264	1	16566.264	6235.976	.000
TREAT	36.943	4	9.236	3.477	.042
ROWS	18.620	4	4.655	1.752	.203
COLUMNS	21.795	4	5.449	2.051	.151
Error	31.879	12	2.657		
Total	16675.501	25			
Corrected Total	109.237	24			

a R Squared = .708 (Adjusted R Squared = .416)

Post-Hoc Tests
Rows
Homogeneous Subsets

CALCIUM

Student-Newman-Keuls

Rows	N	Subset 1
3	5	24.6120
1	5	24.8000
5	5	26.2400
4	5	26.3380
2	5	26.7200
Sig.		.303

Means for groups in homogeneous subsets are displayed. Based on type III sum of squares
The error term is mean square(Error) = 2.657.

a Uses harmonic mean sample size = 5.000.

b Alpha = .05.

Treat

Homogeneous Subsets

CALCIUM

Student-Newman-Keuls

Treat	N	Subset 1	Subset 2
c	5	23.7000	
a	5	25.4920	25.4920
e	5	25.6000	25.6000
d	5	26.6640	26.6640
b	5		27.2540
Sig.		.059	.361

Means for groups in homogeneous subsets are displayed. Based on type III sum of squares
The error term is mean square (Error) = 2.657.

a Uses harmonic mean sample size = 5.000.

b Alpha = .05.

Columns

Homogeneous Subsets

CALCIUM

Student-Newman-Keuls

Columns	N	Subset
		1
3	5	24.1060
2	5	25.4880
5	5	25.9720
1	5	26.2660
4	5	26.8780
Sig.		.115

Means for groups in homogeneous subsets are displayed. Based on type III sum of squares
The error term is mean square (Error) = 2.657.

a Uses harmonic mean sample size = 5.000.

b Alpha = .05.

3

Non-Parametric Test
Alternatives to ANOVA

Logic: Analysis of variance is based on certain assumptions, particularly the normality of the criterion variable and the random errors. Also the variances are assumed to be homogeneous. At the same time, the tests are sensitive to inaccuracy of measurements as the observed values are used as such. Further in many situations, the observations are on ordinal scale and the data do not follow normal distribution. In such cases, the validity of assumptions is in jeopardy. Often the observations are not as accurate as they ought to be. Under such circumstances it is safe to use non-parametric tests as an alternative to analysis of variance for reliable conclusions. Non-parametric analysis in this chapter is presented for one-way and two-way classifications.

3.1 THE KRUSKAL-WALLIS TEST

The Kruskal-Wallis test (1952) is a non-parametric alternative to one-way ANOVA in the series of parametric tests. This test is often written as K-S test in an abbreviated form and is based on rank data. In this test, the null hypothesis under test is that the factor (treatment) effects are same. If there are k factors or a factor has k levels, then H_0 is that all the factors or factor levels are equally effective.

Assumptions: K-S test is based on the following assumptions.
1. The observations are independent within and between samples.
2. The populations are identical except possibly in respect of median.
3. The data follows continuous distribution.

Test procedure: Independent group wise data (scores) pertaining to a variable are ranked from lowest to highest score ignoring the group identity to which they belong. The lowest value or score is ranked 1, the next higher is given a rank 2 and so on till the highest observation or score receives the rank

N, where N is the total number of values or scores. In this way, the observed values are transmuted to ranks.

Often the situation arises when there is a tie between certain observations (scores). In that case, all tied scores are given the same score which is the average of their ranks treating them in ascending order, no matter which was written first and the other next to it. The average rank is calculated by adding the ranks of tied scores and then dividing the sum of their ranks by the number of tied scores.

Once all data are ranked, separate ranks for each treatment or group and find their totals. Let there be k groups and the sum of ranks for ith group (treatment) be denoted by R_i for $i = 1, 2, \ldots, k$. Also suppose that the number of scores in the ith group is n_i. Thus, $\sum_i n_i = N$. If H_0 is true, the sum of N ranks should be distributed in proportion to the sample size n_i. Thus, for ith sample, the expected sum of ranks should be $n_i (N+1)/2$. The ranked data can be presented in the tabular form as given in Table 3.1.1.

Table 3.1.1 Data table

	Treatments				
	1	*2*	*i*	*k*
	R_{11}	R_{21} .... R_{i1} .... R_{k1}		
	R_{12}	R_{22} .... R_{i2} .... R_{k2}		
	$\vert$			
	$\vert$			
	$\vert$			
	$\vert$			
	$\vert$			
	R_{1n1}	R_{2n2} .... R_{in_i} .... R_{kn_k}		
Sum of ranks	R_1	R_2 .... R_i .... R_k		

Kruskal-Wallis developed a reasonably reliable test statistic H which is based on the function of deviations between observed and expected sum of ranks. The test statistic is,

$$H = \frac{12}{N(N+1)} \sum_{i=1}^{k} \frac{R_i^2}{n_i} - 3(N+1) \tag{3.1.1}$$

Statistic H is distributed as chi-square (χ^2) with $(k-1)$ degrees of freedom.

Correction for ties: If there are substantial number of ties, one must adjust H for tied observation. Suppose there are t_i tied values within the treatment i, the correction factor for H is,

$$C = \sum_i \frac{t_i\left(t_i^2 - 1\right)}{n_i\left(n_i^2 - 1\right)} \tag{3.1.2}$$

Where i varies over all groups having tied observations.
The corrected H statistic is,

$$H_c = H/C \tag{3.1.3}$$

Decision criteria: Let the test be performed at α level of significance. Traditionally α is taken as 5% probability of type I error, i.e. $\alpha = 0.05$. If calculated value of $H \geq \chi^2_{\alpha,(k-1)}$, then reject H_0, otherwise H_0 is accepted.

Post-hoc test

When K-S test results into rejection of null hypothesis H_0, the problem arises of comparing groups in pairs. This is similar to the test of significance of pair wise contrasts. But the methods provided for post-hoc parametric tests cannot be applicable here. So there is a need for searching some suitable non-parametric test. To compare any two groups, Mann-Whitney test (1947) is very suitable to apply following K-S test. To ensure that the level of significance does not build up to more than prefixed level α, one should take help of Bonferroni test, i.e. use α/p as the level of significance instead of α for each individual test where p is the number of pairs of groups which are under test of significance. This book does not cover non-parametric tests. So the readers are referred to the book, *Basic Statistics*, 2009, 5th ed., by B.L.Agarwal, New Age International Publishers, New Delhi.

Example 3.1: A training programme was organized to excel the knowledge of secondary school teachers. Fifteen teachers were randomly selected and divided into three groups consisting of five teachers each. Every group was trained for five weeks under three instructional methods (*a*) traditional method (*b*) traditional method with audio-visual aids (*c*) instructions through computers only. All teachers of the three groups were given a test at an interval of one week. Total scores obtained by each individual teacher were as tabulated below.

Teaching methods	Teacher	Total score
Method (a)	T_1	62
	T_2	60
	T_3	61
	T_4	58
	T_5	48
Method (b)	T_6	66
	T_7	59
	T_8	66
	T_9	70
	T_{10}	65
Method (c)	T_{11}	58
	T_{12}	49
	T_{13}	42
	T_{14}	64
	T_{15}	53

It is assumed that all teachers were equally qualified. To compare the affectivity of three instructional methods, the data can better be analyzed by Kruscal-Wallis test as per the procedure discussed in theoretical description. The hypothesis under test is,

H_0: Instructional methods (a), (b), and (c) are equally effective.

Versus, H_1: All three methods are not equally effective.

First the scores are pooled and arranged in ascending order. Ranks to the ordered scores are given and shown in parentheses below the respective scores

To distinguish scores under three instructional methods, the belonging to method (a) are written as such, under (b) the scores are underlined and under (c) they are superscripted.

$\overline{42}$	48	$\overline{49}$	$\overline{53}$	58	$\overline{58}$	$\underline{59}$	60	61
(1)	(2)	(3)	(4)	(5.5)	(5.5)	(7)	(8)	(9)

62	$\overline{64}$	$\underline{65}$	$\underline{66}$	$\underline{66}$	$\underline{70}$
(10)	(11)	(12)	(13.5)	(13.5)	(15)

Now calculate R_i for $i = 1, 2, 3$.

$$R_1 = 2 + 5.5 + 8 + 9 + 10 = 34.5$$
$$R_2 = 7 + 12 + 13.5 + 13.5 + 15 = 61.0$$
$$R_3 = 1 + 3 + 4 + 5.5 + 11 = 24.5$$

Also, $n_1 = n_2 = n_3 = 5$ and N = 15

By the formula (3.1.1),

$$H = \frac{12}{15 \times 16} \left(\frac{34.5^2}{5} + \frac{61.0^2}{5} + \frac{24.5^2}{5} \right) - 3 \times 16$$

$$= \frac{1}{20} (238.05 + 744.2 + 120.05) - 48$$

$$= 55.115 - 48 = 7.115$$

Tabulated value of chi-square for 2 d.f. and $\alpha = 0.05$ is 5.99. The calculated value of H is 7.115 which is greater than 5.99, i.e. $H > \chi^2_{.05;2}$. Therefore, H_0 is rejected. This leads to the conclusion that all the three instructional methods are not equally effective. Now it becomes imperative to compare three pairs of methods using a post-hoc test. Equality of affectivity of three pairs of methods can be tested utilizing Mann-Whitney test at the level of significance 0.05/3, i.e. 0.0167. For information of the readers, this test can also be performed through SPSS package.

Brief outline of Mann-Whitney test: In this test, one compares the equality of distribution of two series of scores. The procedure is as follows.

Arrange the combined scores of the two methods under test in ascending order.

Denote the scores of one group by X and of other group by Y. Rank the combined scores. Consider the case when Y precedes X. Then, the test statistic U of the Mann-Whiney test can be obtained by the formula,

$$U = n_1 n_2 + \frac{n_2 (n_2 + 1)}{2} - S_2 \qquad (i)$$

where S_2 is the sum of ranks of n_2 Y's in the ordered set.

If the sum of ranks of the variable X in the ordered set are considered, then the formula for Mann-Whitney test statistic U' is,

$$U' = n_1 n_2 + \frac{n_1 (n_1 + 1)}{2} - S_1 \qquad (ii)$$

where S_1 is the sum of ranks of n_1 X's in the ordered set of scores.

Also the relation between U and U' is,

$$U = n_1 n_2 - U' \qquad (iii)$$

Sometimes the probability for n_1, n_2 and U are not available in the tables provided for Mann-Whitney test. In that case, one can consult the table for U'. In that case, formula (*iii*) enables the investigator to obtain the value of U' within no time.

Now making use of the theory of Mann-Whitney U test, the three pairs of instructional methods are compared.

1. Testing the significance of the contrast of methods (a) versus (b). Arrange the combined scores of the two methods in ascending order and rank them as shown below.

Sequence:	X_5	X_4	Y_2	X_2	X_3	X_1	Y_5	Y_3	Y_1	Y_4
Scores :	48	58	59	60	61	62	65	66	66	70
Ranks :	1	2	3	4	5	6	7	8.5	8.5	10

In this case, $n_1 = 5$, $n_2 = 5$, and $S_2 = 3 + 7 + 8.5 + 8.5 + 10 = 37$

Mann-Whitney statistic by the formula (i) is,

$$U = 5 \times 5 + \frac{5 \times 6}{2} - 37$$

$$= 25 + 15 - 37 = 3.0$$

Probability associated with $n_2 = 5$, $U = 3$ and $n_1 = 5$ from the tables meant for Mann-Whitney test is 0.028 which is greater than the level of significance $\alpha/3 = 0.0167$ when $\alpha = 0.05$. This affirms that the instructional methods (a) and (b) are equally effective.

2. Test of significance of methods (b) versus (c).

The value of U is found out in the same manner as in case 1. So details are avoided.

Sequence:	Z_3	Z_2	Z_5	Z_1	Y_2	Z_4	Y_5	Y_3	Y_1	Y_4
Scores :	42	49	53	58	59	64	65	66	66	70
Ranks :	1	2	3	4	5	6	7	8.5	8.5	10

$$S_2 = 1 + 2 + 3 + 4 + 6 = 16$$

$$U = 5 \times 5 + \frac{5 \times 6}{2} - 16$$

$$= 25 + 15 - 16 = 24$$

Tabulated probability associated with $n_2 = 5$, $U = 24$ and $n_1 = 5$ is not available in the tables given for Mann-Whitney test. So it is required to calculate U' by the formula (iii).

$$U' = 5 \times 5 - 24 = 1$$

Theoretical probability from Mann-Whitney table for $U' = 1$, $n_2 = 5$ and $n_1 = 5$, is 0.008. This value of probability is less than 0.0167. Hence, there is a significant difference between the methods (b) and (c) for disseminating matter.

3. Test of significance of the comparison of methods (a) and (c).

Sequence:	Z_3	X_5	Z_2	Z_5	Z_1	X_4	X_2	X_3	X_1	Z_4
Scores:	42	48	49	53	58	58	60	61	62	64
Ranks:	1	2	3	4	5.5	5.5	7	8	9	10

$$S_2 = 1 + 3 + 4 + 5.5 + 10 = 23.5$$

Thus,

$$U = 5 \times 5 + \frac{5 \times 6}{2} - 23.5 = 16.5$$

Probability associated with $n_2 = 5$, $U = 16.5$ and $n_1 = 5$ is not available in the tables given for Mann-Whitney test. Hence, U' is calculated.

$$U' = 5 \times 5 - 16.5 = 8.5$$

From table, the probability associated with $n_2 = 5$, $U' = 8.5$ and $n_1 = 5$ is 0.242 which is greater than 0.0167. Therefore, it leads to the conclusion that the teaching methods (a) and (b) are equally effective.

3.2 FRIEDMAN'S TEST

Friedman's test (1937) is an analogue to two-way analysis of variance. Through this test, the data of a randomized block design can be analyzed. This test becomes more apposite when the data are non-normal or have violated some other assumption(s) of ANOVA. Friedman's test is also based on ranked data and same assumptions as given with Kruskal-Wallis test. In this test there are two independent variables, one the treatments (factors) and the other is conditions (blocking factors) which are likely to affect the treatments. The main interest lies in testing the hypothesis that all treatment effects are equally effective. Also in some trials, the same unit receiving a treatment is measured at regular interval of time for a variable of interest. For instance, the weight of each individual of a group of males put on the same diet is measured initially and after every 30 days interval four times. In this trial, each individual is a block. Then one may be interested whether the mean weight of persons is same. Other problems related to two-way analysis may be seen on the topic randomized block design.

Test procedure: To apply Friedman's test, the data are arranged treatment-wise in as many columns as the number of conditions (number of levels of nuisance variable). Suppose there are v treatments (factor levels) and k columns as per the number of conditions. The scores or observations for each treatment are ranked from 1 to v in each column independently. In this way, the sum of ranks of each block is $v(v + 1)/2$. Also calculate the sum of ranks for every treatment independently. Let the sum of ranks of ith treatment be denoted by R_i for $i = 1, 2, 3, \ldots\ldots, v$. The data and ranks with their totals are as tabulated below.

Table 3.2.1 Data table

Blocks	Treatments					Rank totals
	T_1	T_2	....	T_i	....	T_v	
B_1	$X_{11}(R_{11})$ $X_{21}(R_{21})$		....	$X_{i1}(R_{i1})$	....	$X_{v1}(R_{v1})$	$v(v+1)/2$
B_2	$X_{12}(R_{12})$ $X_{22}(R_{22})$		....	$X_{i2}(R_{i2})$	....	$X_{v2}(R_{v2})$	$v(v+1)/2$
$\vdots$							$\vdots$
$\vdots$							$\vdots$

(Contd.)

B_j	$X_{1j}(R_{1j})$ $X_{2j}(R_{2j})$	....	$X_{ij}(R_{ij})$	....	$X_{vj}(R_{vj})$	$v(v+1)/2$
$\vdots$						$\vdots$
B_k	$X_{1k}(R_{1k})$ $X_{2k}(R_{2k})$	....	$X_{ik}(R_{ik})$	....	$X_{vk}(R_{vk})$	$v(v+1)/2$
Sum of ranks	R_1 R_2	....	R_i	....	R_k	$k\,v(v+1)/2$

The null hypothesis, H_0 that v treatments are equally effective or v factors have identical distributions, can be tested through Friedman's test-statistic F where,

$$F = \frac{12}{kv(v+1)} \sum_{i=1}^{k} R_i^2 - 3k(v+1) \tag{3.2.1}$$

Statistic F is distributed as χ^2 with $(v-1)$ d.f.

Decision criteria: If the calculated value of $F \geq \chi^2_{\alpha:(v-1)}$, where α is the preassigned level of significance of the test, then the null hypothesis H_0 is rejected, otherwise accepted.

Post-hoc tests

When H_0 is rejected, then it becomes evident that all pairwise contrasts of the treatments be tested for their significance. But there is no compulsion that only Mann-Whitney test should only be used as a post-hoc test. Any other suitable test may be applied. The experience tells that Mann-Whitney test is quite laborious particularly when the number of pairs is large. Rather it becomes unpalatable if the number of pairs of treatments is more than 5. So the test described by Siegel and Castellan (1988) is quite convenient to apply. In this test, the difference of the absolute mean ranks of a pair of treatments is compared with a constant critical value based on Z, the standard normal deviate value, at a desired level of significance. This test is parallel to lsd test. In this test also Carlo Bonferroni concept of level of significance is utilized. The table of Z is seen for $\alpha/v(v-1)$ level of significance for one sided right tailed test. The absolute difference between any two treatment rank means is significant if,

$$\left| \bar{R}_i - \bar{R}_{i'} \right| \geq Z_{\alpha/v(v-1)} \times \sqrt{\frac{v(v+1)}{6N}} \tag{3.2.2}$$

For $i \neq i'$ and N is the total number of scores.

The value of Z can be obtained from the table of standard normal distribution given in appendix Table B-5.

Friedman's test is convenient and involves much less calculations as compared to two-way ANOVA and is free from stringent assumptions.

Example 3.2: An experiment was conducted to see the effect of total soluble solids (TSS ^{0} Brix) from 12°B to 20°B at an interval of 2°B on carbon dioxide (CO_2) production in plum juice. The juice samples were fermented and observations were recorded at an interval of six hours regarding production of CO_2 from 0 to 36 hours. The same are tabulated below.

CO_2 (mg/100 ml) produces at various levels of TSS

Time hours	12°B	14°B	16°B	18°B	20°B
0	22	45	89	97	67
6	66	106	106	155	128
12	110	123	133	168	146
18	154	208	226	168	160
24	160	210	234	186	182
30	168	216	230	224	211
36	240	251	399	390	346

The experiment is conducted to compare various levels of TSS°B, so the time periods are considered as blocks. Now rank the observations in each row independently as depicted in the following table. Find the sum of ranks of each TSS°B level.

Table of ranks

Time hours	12°B	14°B	16°B	18°B	20°B
0	1	2	4	5	3
6	1	2.5	2.5	5	4
12	1	2	3	5	4
18	1	4	5	3	2
24	1	4	5	3	2
30	1	3	5	4	2
36	1	2	5	4	3
Total	7	19.5	29.5	29	20
Mean	1.000	2.786	4.214	4.143	2.856

Null hypothesis under test is,

H_0: All levels of TSS°B are equally effective.

Versus H_1: At least two of them differ in their affectivity.
In this trial, $k = 7$, $v = 5$. The value of F by the formula (3.2.1) is,

$$F = \frac{12}{7 \times 5(5+1)} (7^2 + 19.5^2 + 29.5^2 + 29^2 + 20^2) - 3 \times 7(5 + 1)$$

$$= \frac{2}{35} (49 + 385.25 + 870.25 + 841 + 400) - 3 \times 7 \times 6$$

$$= \frac{2 \times 2545.5}{35} - 126$$

$$= 145.46 - 126 = 19.46$$

Critical value of χ^2 for $= 0.05$ and 4 d.f. from the table No. B-4, is 9.49.
Calculated F is greater than the critical value of χ^2. Hence, reject H_0 evincing that the effects of various levels of TSS°B differ significantly.

Now proceed to compare all pairs of treatment means using the inequality (3.2.2). First find the value of the right hand expression of the inequality. For $\alpha = 0.05$, the test is to be performed at the level 0.05/20 = 0.0025.

From the table of the distribution of standard normal deviate given in a statistical book table, read the value of Z by seeing the column of smaller probabilities and searching the value for 0.0025. Z value corresponding to this value is the required value of Z. In this case, $Z_{.0025} = 2.34$.

$$Z_{\alpha/v(v-1)} \sqrt{\frac{v(v+1)}{6N}} = 2.34 \times \sqrt{\frac{5 \times 6}{6 \times 35}}$$

$$= 2.34 \times 0.378 = 0.884$$

Write the mean ranks in order.

12°B	14°B	20°B	18°B	16°B
1.000	2.786	2.856	4.143	4.214

The above diagram reveals that TSS 12°B significantly differs from all other TSS°B. 14°B and 20°B are significantly inferior to 16°B and 18°B, but they are at par with each other. Similarly 16°B and 18°B are equally effective.

QUESTIONS AND EXERCISES

1. When one should use non-parametric tests to analyze experimental data?
2. Kruskal-Wallis test can be applied for how many attributes or measurement variable(s) taken at a time.
3. Whether Kruskal-Wallis test is an alternative to one-way or two-way analysis of variance?
4. Are the non-parametric test based on ranks or measurements as such?

5. Give the full description of post-hoc test after Friedman's test.
6. Friedman's test is applicable to which experimental design?
7. How does Kruskal-Wallis test handle the problem of ties?
8. Whether Friedman's test compares paired or unpaired three or more samples test?
9. What hypothesis is tested by Kruskal-Wallis test?
10. What hypotheses are tested by Friedman's test?
11. How can one perform post-hoc test on Kruskal-Wallis rank sums?
12. Think and write three applied problems in which Friedman's test can be applied.
13. Describe the procedure in which Kruskal-Wallis test can be applied for one-way analysis of data.
14. What statistical distributions are followed by Friedman and Kruskal-Wallis test statistics?
15. Kruskal-Wallis test is used for number of paired samples only. Is this statement correct? Justify your answer by proper reasoning.
16. Following table presents the pulse rates of eight subjects for comparing the effect of strenuous physical exercise on an electrical treadmill as per medical norms. The stages of reading consisted of pretest, study, hyperventila, jogging at varying speeds - 1.7 km/h, 2.5 km/h and post-test stage of rest, i.e. 6 minutes after jogging.

Subjects	Pretest	Study	Hyper-ventila	1.7 km/h	2.5 km/h	After 6 min. rest
S_1	73	79	86	118	105	86
S_2	86	95	114	116	165	106
S_3	88	91	72	127	147	91
S_4	101	93	87	112	123	90
S_5	67	86	88	123	142	83
S_6	109	114	110	120	126	111
S_7	82	86	88	162	135	93
S_8	76	101	102	131	165	90

Test the hypothesis that the distribution of pulse rates under six strenuous stages is same by Friedman's test. In this experiment, each person be considered as a block.

17. A trial was conducted taking twelve entries of groundnut to compare their performance with respect to shelling percentage. The layout of the experiment was randomized block design having four replications. Entry-wise percentage shelling was found as tabulated below:

Entries	Percentage shelling			
	R_1	R_2	R_3	R_4
E_1	62	63	61	62
E_2	61	59	58	59
E_3	63	62	64	62
E_4	64	65	64	62
E_5	62	63	64	62
E_6	63	62	62	61
E_7	62	61	65	64
E_8	65	62	65	64
E_9	67	69	67	65
E_{10}	65	67	68	64
E_{11}	66	62	63	65
E_{12}	60	58	60	63

Using a proper non-parametric test, confirm whether there is a significant difference between entries of groundnut with regard to shelling percentage.

18. Milk yield per lactation in litres of daughters of seven different sire-doe crosses is given below.

Crosses	Milk yield/lactation (litres)
C_1	62, 55, 61, 60, 63
C_2	71, 46, 54, 42, 48
C_3	70, 81, 80, 68
C_4	65, 56, 55, 54
C_5	67, 52, 64, 58
C_6	72, 78, 69, 82, 73
C_7	82, 72, 67

Analyze the data by Kruskal - Wallis test for testing the hypothesis that milk yield of daughters of seven different crosses is same.

BIBLIOGRAPHY

1. Agarwal BL. *Basic Statistics*, 5th ed., New Age International Publishers, New Delhi, 2009.
2. Andy Field. *Discovering Statistics Using SPSS*, 2nd ed. Sage Publication, London, 2005.
3. Andras Vergha, Harold D. Delaney. The Kruskal-Wallis test and stochastic homogeneity. *Journal of Educational and behavioral statistics*, 1998: 23(2), 170–92.

4. Gobo AE. The application of Kruskal-Wallis technique for prediction in Niger Delta, Nigeria. *Management of Environmental Quality: An International Journal,* 2006: 17(3), 275–88.

6. Bernard Ostle and Linda Catron Malone. *Statistics in Research: Basic Concepts and Techniques for Research Workers,* 4th edition, Blackwell Publishing, Oxford, 1988.

7. Carol E. Oshorn. *Applications for Health Information Management.* Jones and Bartlett Publishers, 2005.

8. Calvin Dytham. *Choosing and Using Statistics: A Biologist Guide.* Blackwell Publishing, Oxford, 2003.

9. Conover W. *Practical Nonparametric Statistics.* 2nd ed., John Wiley, New York, 1980.

10. Feir Betty J. The ANOVA F–test versus Kruskal-Wallis test: A robust study. *Paper presented at the American Educational Research Association,* 1974: 59th, Chicago, Illinois.

11. Friedman Milton. The use of ranks to avoid the assumption of normality implicit in the analysis of variance. *Journal of the American Statistical Association,* 1937: 32(200), 675–701.

12. Friedman Milton. A correction: The use of ranks to avoid the assumption of normality implicit in the analysis of variance. *Journal of the American Statistical Association,* 1939: 34(205), 109.

13. Friedman Milton. A comparison of alternative tests of significance for the problem of m rankings. *The Annals of mathematical Statistics,* 1940: 11(1), 86–92.

14. Jean Dickinson Gibbons, Chakraborty S. *Nonparametric Statistical Inference,* 3rd ed., Marcel Dekker, New York, 1992.

15. Leonard J. Kazmier. *Schaum's Outline of Business Statistics,* 4th ed., MacGraw Hill Professional, 2004.

16. Norman Breslow. A generalized Kruskal – Wallis test for comparing k samples subject to unequal paterns of censorship. *Biometrika,* 1970: 57(3), 579–594.

17. O' Gorman TW. A comparison of the F–test, Friedmans test and several rank tests for the analysis of randomized complete blocks. *Journal of Agricultural, Biological and environmental statistics,* 2001: 6(3), 367–378.

18. Siegel S, Castellan NJ. *Nonparametric Statistics for Behavioral Sciences,* 2nd ed. McGraw Hill, New York, 1988.

19. Sprent P. *Applied Nonparametric Statistical Methods.* 2nd edition, Chapman & Hall, London, 1993.

20. Wei LJ. Asymptotic conservasiveness and efficiency of Kruskal-Wallis test for k dependent samples. *Journal of the American Statistical Association,* 1981: 76 (376), 1006–1009.

21. William H. Kruskal. Use of ranks in one-criterion variance analysis. *Journal of the American Statistical Association,* 1952: (Dec. issue), 583–621.

4

Factorial Experiments

4.1 MEANING AND EXPLANATION

A factorial experiment is one in which two or more factors (Treatments) each having two or more levels are considered simultaneously. In such an experiment, each factor level of a factor occurs with all other levels forming all possible treatment combinations. In this way, factors in a factorial experiment are said to be *completely crossed* with each other. In view of this, the term *crossed design* is also used for factorial design. The greatest advantage of factorial experiments is that they not only provide effects of the factors alone but also make it possible to know the influence of the levels of a factor on the performance of the other factor levels. Effect of an individual factor is called *main effect* and the influence of one factor over the other is known as *interaction*.

Definitions and graphical presentation

Main effect: The main effect of a factor can be defined as measure of change in response due to change in the level of that factor averaged over all levels of all other factors involved in the factorial experiment.

Interaction: The interaction between two or more factors may be defined as the failure of a factor to maintain the same response at various levels of the other factor(s). In other words, the effect of a factor depends upon the levels of the other factor.

As a matter of fact, interaction is an additional information which is available due to combined effect of two or more factors.

Two factors interaction is known as *first order interaction*. Three factors interaction is called *second order interaction* and so on.

Mere definitions of main and interaction effects may not give a clear picture to inquisitive minds. Hence, the idea of main and interaction effects has been clarified through graphs for 2 × 2, 2 × 3, 3 × 3 factorials. One will fail

139

to interpret the results of a factorial experiment unless he understands the meaning and practical aspects of main and interaction effects exactly.

To begin with, consider a 2 × 2 factorial experiment with two teachers assigned to two teaching methods. Denote the teachers by T_1 and T_2 and teaching methods by M_1 and M_2. Measurable variable is the average score (O_{ij}) obtained by the group of trainees after an objective test. Let us consider the diagrammatic representation of graphs in different situations as depicted below.

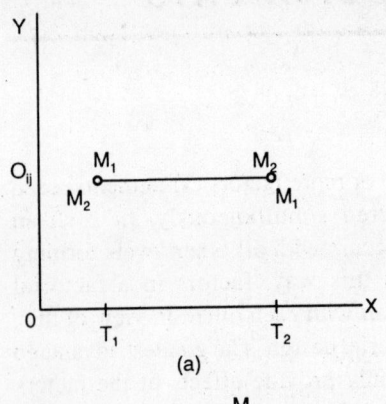

Figure (a) reveals that scores for both the teachers are same. Hence, there is no main effect and no interaction effect.

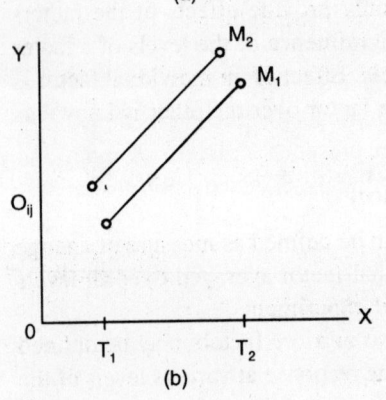

Both teachers have higher scores under method 2 than method 1 and difference is same. Therefore, there is main effect but no interaction.

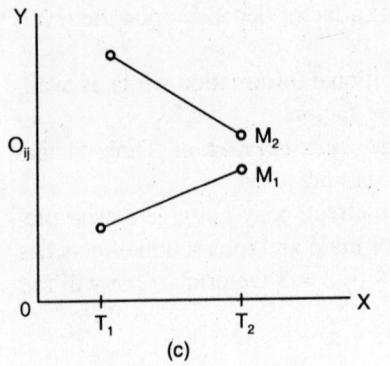

The difference in scores under method T_1 is more than that of T_2. It shows that there exists main effect as well as interaction effect.

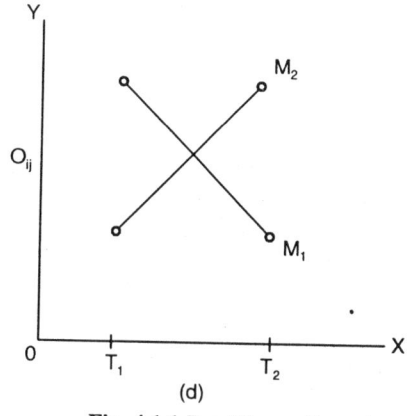

Figure (*d*) depicts a specific situation where students get higher scores under M_1 for T_1 and M_2 for T_2. Also there is a difference in scores under two teachers and two methods as well. This confirms the existence of main effect as well as interaction effect.

Fig. 4.1.1 Possible configurations for main effects and interaction of a 2 × 2 factorial experiment.

Above diagrams lead us to draw following conclusions.

1. The teachers and methods are fixed effects.
2. If the lines for one factor for each level of the other factor are parallel, then this shows the absence of interaction effect.
3. If the lines are coincident, they show the absence of main effects as well.
4. The gap between two parallel lines provides the magnitude of main effect.
5. Any non-parallel lines ensure the presence of interaction effect.
6. The significance or non-significance of the main effects say, A and B has no bearing on the significance or non-significance of the interaction A × B.
7. It strikes to the mind that if a diagram presents a picture of main and interaction effects, where is the need for analysis of data. But this apprehension is not true. The reason being, main and/or interaction effects may be due to chance. Hence, the test of significance ensures with certain risk their presence is real or a matter of chance.

Now consider an experiment in which three genotypes (G) have been grown in two environments (E).

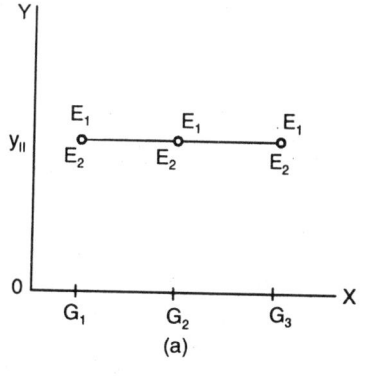

No main effect of G and E. Also no interaction G × E.

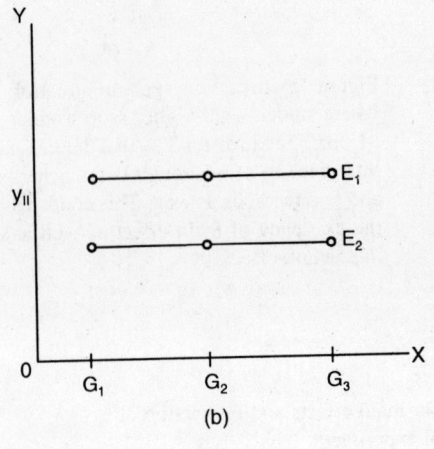

(b)

There is main effect of E but no main effect G. Also there is no interaction G × E.

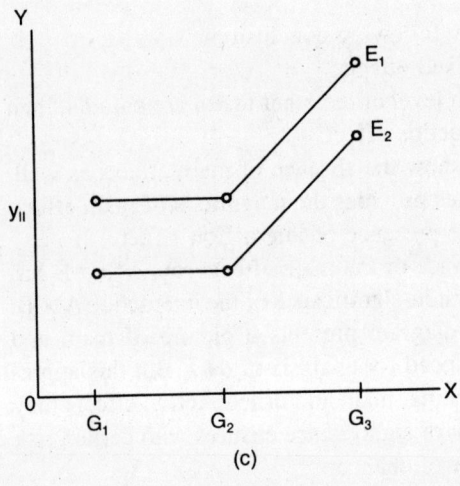

(c)

There is presence of main effects G and E. At the same time, it shows the absence of interaction G × E.

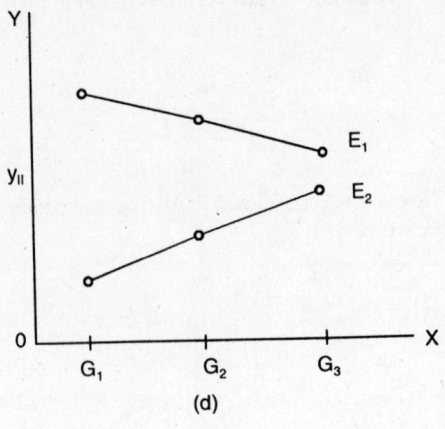

(d)

This diagram depicts the presence of main effects G and E and also the presence of interaction G × E.

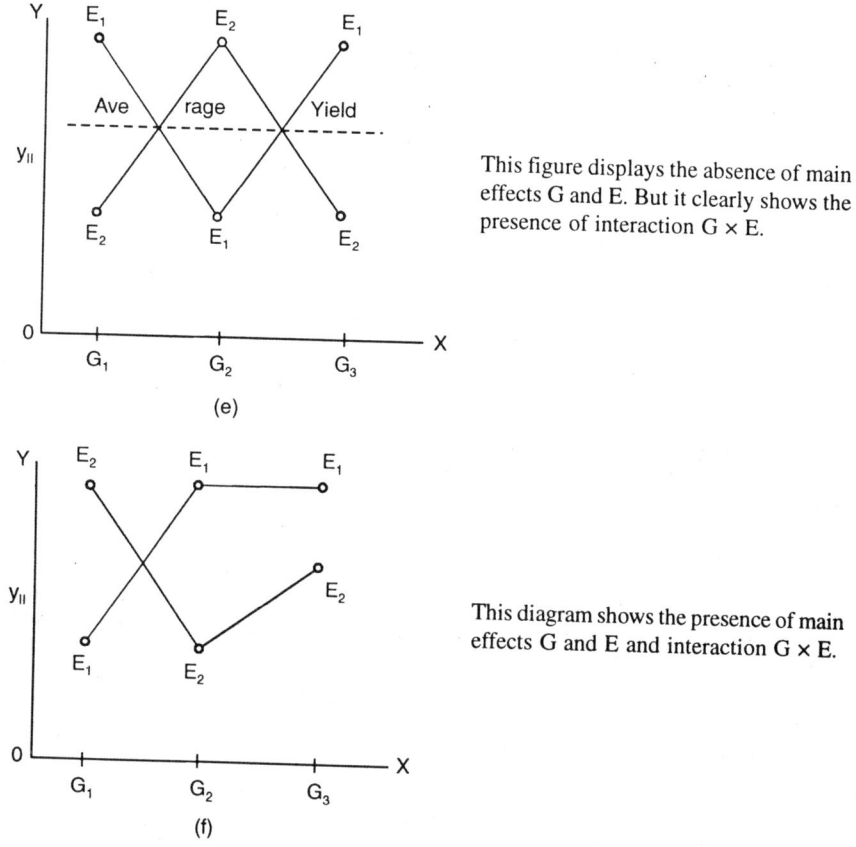

This figure displays the absence of main effects G and E. But it clearly shows the presence of interaction G × E.

This diagram shows the presence of main effects G and E and interaction G × E.

Fig. 4.1.2 Diagrammatic presentation of main and interaction effects of 2 × 3 factorial experiment.

Consider an experiment in which three doses of nitrogen fertilizer (n_0, n_1, n_2) are tried in combination with three doses of phosphorus (p_0, p_1, p_2) to see their main and interaction effects of the yield of a crop.

Possible graphical configurations of 3 × 3 experiment with their average yield are displayed below.

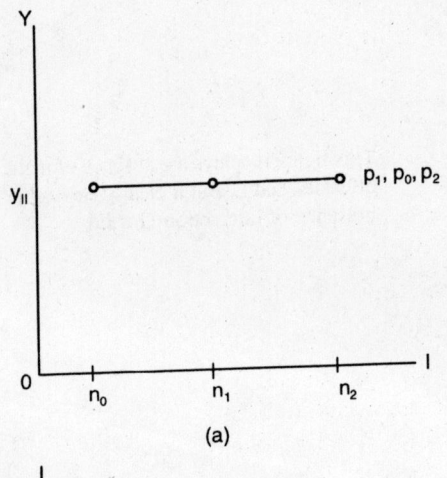

It shows the absence of main effect P and interaction N × P. But the main effect N is present.

(a)

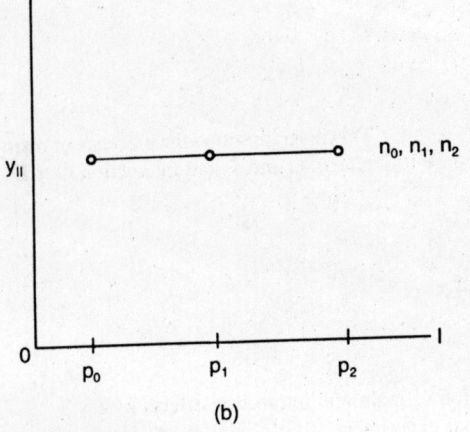

This diagram portray the absence of main effect N and interaction N × P but the presence of main effect of P.

(b)

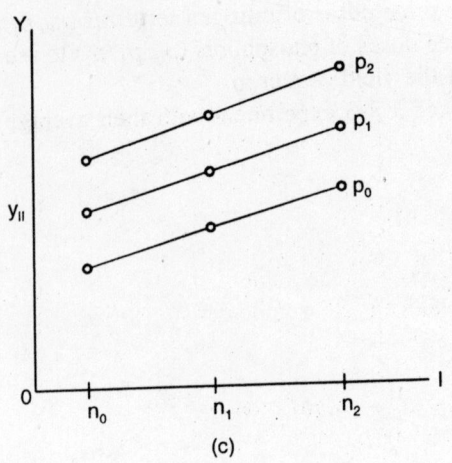

This figure delineate the presence of main effects N and P. At the same time, it shows no interacrtion N × P.

(c)

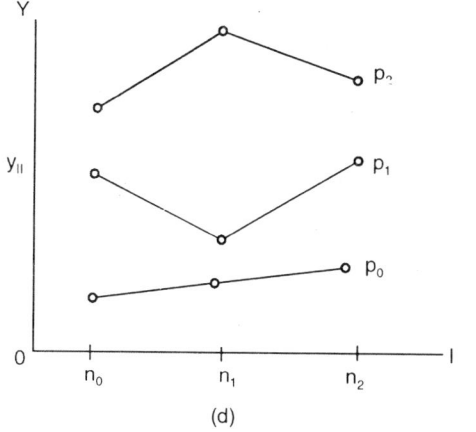

This diagram depicts the presence of main effects N, P and interaction effect N × P.

(d)

Fig. 4.1.3 Graphical presentation of main and interaction effects of a 3 × 3 factorial experiment.

Remark: The significance or non-significance of a second order interaction A × B × C has nothing to do with the significance or non-significance of main effects A, B or C and first order interactions.

A factorial experiment may involve any number of factors having any number of levels. If all factors have same number of levels, then it is said to be a *symmetrical factorial*, otherwise *asymmetrical factorial*. A symmetrical factorial can be denoted by p^n factorial, which means that there are n factors, each at p levels. Whereas a factorial of type $p \times q \times m$ is an asymmetrical factorial in which three factors are at p, q and m levels, respectively. Comparatively it is easy to deal with symmetrical factorials. Factorial 2^n are the simplest one and thus further discussion will be initiated with 2^n factorials. Factorial experiments may be conducted in any of the basic designs or incomplete block designs. But in any experiment there has to be at least two replications to have an estimate of experimental error.

4.2 2^2 AND 2^3-FACTORIAL EXPERIMENTS

This is the simplest form of the factorial experiments. The concept of main effect, interaction and analysis, etc. will first be expatiated for 2^2 factorials which can easily be extended for factorials with larger number of factors.

Consider two factors A and B each at two levels say 0 and 1. The symbol 1 denotes the higher level and 0 the lower level of a factor. Let the factor A with two levels be denoted by a_1 and a_0 and factor B by b_1 and b_0. So there will in all be four treatment combinations a_1b_1, a_1b_0, a_0b_1, a_0b_0. The main effects A and B and the interaction AB can be represented by the **contrasts** as follows.

$$A = a_1(b_1 + b_0) - a_0(b_1 + b_0)$$
$$= a_1 b_1 + a_1 b_0 - a_0 b_1 - a_0 b_0$$

$$= (a_1 - a_0)(b_1 + b_0)$$
$$B = b_1(a_1 + a_0) - b_0(a_1 + a_0)$$
$$= a_1 b_1 + a_0 b_1 - a_1 b_0 - a_0 b_0$$
$$= (a_1 + a_0)(b_1 - b_0)$$
$$AB = a_1(b_1 - b_0) - a_0(b_1 - b_0)$$
$$= a_1 b_1 - a_1 b_0 - a_0 b_1 + a_0 b_0$$
$$= (a_1 - a_0)(b_1 - b_0)$$

Suffixes in the contrasts may be avoided in 2^n factorials by coding a higher level factor by the letter itself and its lower level factor by 1, i.e. a_1 by a, and a_0 by 1, similarly b_1 by b and b_0 by 1. With these notations,

$$A = (a - 1)(b + 1)$$
$$= a b - b + a - (1)$$
$$B = (a + 1)(b - 1)$$
$$= a b - a + b - 1$$
$$A \times B = (a - 1)(b - 1)$$
$$= a b - b - a + 1$$

These contrasts can easily be presented as follows in which only the +ve and −ve signs denote the combinations given in the captions.

Table 4.2.1 Contrasts of a 2^2 factorial

Effects	a b	b	a	(1)
A	+	−	+	−
B	+	+	−	−
A × B	+	−	−	+

One can easily verify that the above three contrasts are mutually orthogonal.

In a 2^3 factorial with three factors A, B, C each at two levels 0 and 1, the contrasts for main and interaction effects can easily be formed and displayed as given below.

Table 4.2.2 Contrasts for a 2^3 factorial

Effects	Contrasts	Expanded contrasts							
		abc	ab	ac	bc	a	b	c	(1)
A	$(a-1)(b+1)(c+1)$	+	+	+	−	+	−	−	−
B	$(a+1)(b-1)(c+1)$	+	+	−	+	−	+	−	−
AB	$(a-1)(b-1)(c+1)$	+	+	−	−	−	−	+	+
C	$(a+1)(b+1)(c-1)$	+	−	+	+	−	−	+	−
AC	$(a-1)(b+1)(c-1)$	+	−	+	−	−	+	−	+
BC	$(a+1)(b-1)(c-1)$	+	−	−	+	+	−	−	+
ABC	$(a-1)(b-1)(c-1)$	+	−	−	−	+	+	+	−

In the like manner, it is trivial to write contrasts for any two 2^n ($n > 3$) factorial.

CALCULATION OF MEAN EFFECTS AND SUM OF SQUARES IN 2^n FACTORIALS

To have the mean effects and sum of squares, there can be three approaches.

I. With the help of two-way and three-way tables.

II. By way of contrasts.

III. By Yates' method.

I. Calculation with the help of two-way table.

Prepare a two-way table putting the total of values aggregated over all replications for various treatment combinations in the respective cells. Consider a 2^2 factorial experiment with factors A and B and r replications. The table will appear as follows:

Factors	B		Total
	b_0	b_1	
A a_0	Y_{00}	Y_{01}	$Y_{0.}$
A a_1	Y_{10}	Y_{11}	$Y_{1.}$
Total	$Y_{.0}$	$Y_{.1}$	$Y_{..}$

$$\text{S.S. due to A} = \frac{Y_{0.}^2}{2r} + \frac{Y_{1.}^2}{2r} - \frac{(Y_{0.} + Y_{1.})^2}{4r}$$

$$= \frac{1}{2r}\left(2Y_{0.}^2 + 2Y_{1.}^2 - Y_{0.}^2 - Y_{1.}^2 - 2Y_{0.}Y_{1.}\right)$$

$$= \frac{1}{2r}\left(Y_{0.} - Y_{1.}\right)^2$$

Similarly,

$$\text{S.S. due to B} = \frac{1}{2r}(Y_{.0} - Y_{.1})^2$$

$$\text{S.S. due to AB} = \frac{1}{r}\left(Y_{00}^2 + Y_{01}^2 + Y_{10}^2 + Y_{11}^2\right) - \frac{Y_{...}^2}{4r} - \text{S.S. due to A}$$

$$- \text{S.S. due to B}$$

$$= \text{Treat. S.S.} - \text{S.S.(A)} - \text{S.S.(B)}$$

II. Sum of squares through contrasts can be obtained by squaring the value of the contrast and dividing it by $2^n \times r$. Thus, the sum of squares are as follows:

$$\text{S.S. due to A} = \frac{(ab + a - b - 1)^2}{4r}$$

$$= \frac{(Y_{11} + Y_{10} - Y_{01} - Y_{00})^2}{4r}$$

$$\text{S.S. due to B} = \frac{(ab + b - a - 1)^2}{4r}$$

$$= \frac{(Y_{11} + Y_{01} - Y_{10} - Y_{00})^2}{4r}$$

$$\text{S.S. due to A} \times \text{B} = \frac{(ab - a - b + 1)^2}{4r}$$

$$= \frac{(Y_{11} - Y_{10} - Y_{01} + Y_{00})^2}{4r}$$

III. Yates' method for calculating sum of squares in a 2^n factorial is a mechanical approach which follows the principle of calculating means and sum of squares through contrasts. In this method, one has not to construct orthogonal contrasts. During the process of Yates' algorithm they are automatically formed and final values of contrasts are obtained after n steps. Stepwise procedure of Yates' method is as follows:

Step 1. Write down the factorial effects in a sequence in column (i).

Step 2. Write the total response value against each treatment combination in column (ii).

Step 3. Add in pairs the values given in column (ii) and place them in column (iii) against first half sequence and then subtract even minus odd serial values in order and place them ahead in column (iii) till all values are exhausted.

Step 4. Repeat addition and subtraction operation on column (iii) and enter these values in column (iv) in the same order.

Step 5. Continue to create as many columns through the operation of addition and subtraction as the number of factors (n). In a 2^2 experiment, there shall be two such columns. In a 2^3 experiment, there shall be three such columns and so on.

The top most value in the last column provides the grand total of responses. The succeeding values turn out to be the factorial effects in the order in which they are written.

To obtain the over all mean value, divide grand total $Y..$ by $2^n r$ and for factorial effect means, divide each effect value of last column by $2^{n-1} r$.

Sum of square due to each factorial effect can be calculated by squaring the factorial effect value and dividing it by $2^n r$.

Note: One may use any of the three approach, the values of sum of squares will always be same.

Total sum of square will in all cases be calculated by taking the sum of each individual squared value and subtracting from it the correction factor.

For illustration Yates' method for 2^2 factorial experiment has been worked out over here.

Treatment combinations *(i)*	*Total response* *(ii)*	*Column* *(iii)*	*Column* *(iv)*
(1)	Y_{00}	$Y_{00} + Y_{10}$	$Y_{00} + Y_{10} + Y_{01} + Y_{11}$
a	Y_{10}	$Y_{01} + Y_{11}$	$Y_{10} - Y_{00} + Y_{11} - Y_{01}$
b	Y_{01}	$Y_{10} - Y_{00}$	$Y_{11} + Y_{01} - Y_{00} - Y_{10}$
ab	Y_{11}	$Y_{11} - Y_{01}$	$Y_{11} - Y_{01} - Y_{10} + Y_{00}$

Comparing entries of column *(iv)* with the contrasts given in approach II, the top entry is equal to grand total and subsequent entries represent the contrasts for the effects A, B and A × B, respectively. Sum of squares will obviously be available by squaring values of column *(iv)* and then dividing by $4r$. This substantiates the statement that Yates' method is a mechanical approach in which contrast method is implied. This method saves lot of time of constructing orthogonal contrasts and reduces the chance of committing error in calculations. Yates' method is applicable to any 2^n factorial experimental data analysis in the manner described above and none else.

Analysis of variance

Factorial experiments can be conducted in any basic design such as CRD, RBD or LSD. But the designs CRD and LSD are seldom used. Therefore, all discussion henceforth shall be given for randomized block design. Whichever may be the design, the analysis of factorial experiment remains same except that in ANOVA for CRD there is no factor as source of variation due to blocks. Also in case of LSD, there is an additional source of variation in ANOVA as rows. Rest of the analysis procedure remains almost same in all designs. The analysis of a factorial experiment in a randomized block design is based on additive linear model,

$$y_{ijk} = \mu + \alpha_i + \beta_j + \rho_k + (\alpha\beta)_{ij} + e_{ijk} \qquad (4.2.1)$$

$$i = 1,2,\ldots\ldots\ldots, p; j = 1,2, \ldots\ldots\ldots, m; k = 1,2,\ldots\ldots\ldots,r.$$

where α_i is the effect of ith level of A, β_j is the effect of jth level of B and $(\alpha\beta)_{ij}$ is the interaction effect of ith level of A on jth level of B. e_{ijk} is the random error independently and identically distributed (i.i.d.) $N(0, \sigma_e^2)$. Under the specifications of the model (4.2.1), the hypotheses that are required to be tested are as follows:

Null hypotheses for main effects are,

H_0: All levels of main effect A are equally effective,

Versus H_1: At least two levels of A differ significantly.

In terms of notation, the above hypotheses can be delineated as follows:

$H_0 : \alpha_1 = \alpha_2 = ... = \alpha_p$ or $\alpha_i = 0$ for all j.

Versus H_1: $\alpha_i \neq \alpha_{i'}$ for some $i \neq i'$ = 1, 2, ..., p.

Similar hypotheses can be established for main effect B.

For interaction AB, the hypotheses are,

H_0: $(\alpha\beta)_{ij} = 0$ for all i and j

Versus H_1: $(\alpha\beta)_{ij} \neq 0$ for some i and j

The analysis of variance table in general will have the following break-up.

Table 4.2.3 Skeleton ANOVA for a p × m factorial

Source	d.f.
Replication	$r - 1$
A	$p - 1$
B	$q - 1$
A × B	$(p - 1)(q - 1)$
Error	$(pm - 1)(r - 1)$
Total	$Pmr - 1$

In ANOVA (4.2.3), all effects are to be tested against experimental error. Hypotheses that can be tested from above ANOVA are as follows:

Analysis of variance for 2^2 experiments with r replications will be of the following form.

Table 4.2.4 ANOVA for a 2^2 factorial

Source	d.f.	S.S.	M.S.	F-value
Replications	$r - 1$	RSS	RMS	RMS/ EMS
A	1	ASS	AMS	AMS/ EMS
B	1	BSS	BMS	BMS/ EMS
A x B	1	ABS	ABMS	ABMS/ EMS
Error	$3(r - 1)$	ESS	EMS	
Total	4r - 1			

All factor effects are to be tested against experimenter error and conclusions can be drawn in the usual manner.

4.3 EXPECTED MEAN SQUARES (EMS)

It has already been emphasized that the choice of an error for testing effect(s) depends on expected mean squares for the effect and other components prevailing in ANOVA. These expected mean squares arise from the model ascribed for the design of the factorial experiment.

General rules for writing expected mean squares

Expected mean squares have been derived in Chapter – 2 for various components of variation appearing in ANOVA table for RBD. It is evident that derivation of expected mean squares is a tedious job. Factorial experiments contain a number of main and interaction effects and this makes the job further exhaustive and time consuming. Rather it is not possible for the scientists from other disciplines to derive expected mean squares. In view of this, general rules are given to write expected mean squares without derivation for any factorial effect. Let any factorial be denoted by U. The factor U may consist of a main effect A, B, C, ..., or two factor interaction AB, AC, BC, ..., or three factor interaction ABC, etc. Suppose that error mean square is denoted by σ_e^2 and due to a factorial effect U by σ_u^2, where U in suffix may represent any main effect or interaction. For example, σ_A^2 is the M.S. due to A, σ_{AB}^2 is the M.S. due to interaction AB and σ_{ABC}^2 is the M.S. for interaction ABC and so on.

It has been enunciated that the statistical model for any factorial design may be fixed effect, random effect or mixed effect model. In each case, expected mean square for an effect will differ in accordance to the specification of the model. In view of models I, II and III, general rules for writing expected mean squares for effects are framed as follows.

Rule 1: Expected value of error mean square is always σ_e^2 irrespective of the type of model and number of factors.

Rule 2: σ_e^2 appears in expected mean square of every factor in addition to other terms.

Rule 3: In the situation of fixed effect model, only two terms appear, i.e. error mean square (σ_e^2) plus [(sum of square due to U)/divided by d.f. for U] × number of replication × levels of all factors except those which appear in U.

Rule 4: In case of random effect model, expected mean square for U consists of σ_e^2 plus all variances involving U with other effects where each of these variance is multiplied by the product of number of replications and the levels of all respective factors which do not appear with that variance.

Table 4.3.1 ANOVA with expected mean squares for 2^2 factorial

Source	d.f.	A and B fixed	A and B random	A fixed, B random	A random, B fixed
Blocks	$r-1$	$\sigma_c^2 + pm \sum\limits_{k=1}^{r} \dfrac{\rho_k^2}{r-1}$			
Treats	$pm-1$				
A	$p-1$	$\sigma_e^2 + rm \sum\limits_{i=1}^{p} \dfrac{\alpha_i^2}{p-1}$	$\sigma_e^2 + r\sigma_{\alpha\beta}^2 + rm\sigma_\alpha^2$	$\sigma_e^2 + rm \sum\limits_{i=1}^{p} \dfrac{\alpha_i^2}{p-1} + r\sigma_{\alpha\beta}^2$	$\sigma_e^2 + rm\sigma_\alpha^2$
B	$m-1$	$\sigma_e^2 + rp \sum\limits_{j=1}^{m} \dfrac{\beta_j^2}{m-1}$	$\sigma_e^2 + r\sigma_{\alpha\beta}^2 + rp\sigma_\beta^2$	$\sigma_e^2 + rp\sigma_\beta^2$	$\sigma_e^2 + rp \sum\limits_{j=1}^{m} \dfrac{\beta_j^2}{m-1} + r\sigma_{\alpha\beta}^2$
A × B	$(p-1)(m-1)$	$\sigma_e^2 + r \sum\limits_{i=1}^{p}\sum\limits_{j=1}^{m} \dfrac{(\alpha\beta)_{ij}^2}{(p-1)(m-1)}$	$\sigma_e^2 + r\sigma_{\alpha\beta}^2$	$\sigma_e^2 + r\sigma_{\alpha\beta}^2$	$\sigma_e^2 + r\sigma_{\alpha\beta}^2$
Error	$(pm-1)(r-1)$	σ_e^2	σ_e^2	σ_e^2	σ_e^2
Total	$pm\,r - 1$				

Rule 5: In case of mixed effect model, expected mean square for any fixed factor U contains σ_e^2 plus the effect U sum of square divided by its d.f. and sum of variances due to all interactions of with random factors each of which is multiplied by the number of replications and levels of factors which do not appear in the interaction.

Following the above rules, one shall be able to write expected mean squares for any factor in case of any type of statistical model. Expected mean squares for factorial experiment in RBD with two treatments A and B having p and m levels respectively and r blocks in view of the model (4.2.1) are given below making use of the above rules.

The above table reveals that the mean square for interaction A × B be tested against error M.S. If it is significant, then the interaction A × B M.S. should be utilized as error for testing the main effects in all situations. If interaction M.S. is non-significant it should be pooled with error M.S. and this pooled M.S. be used as an error to test the main effects A and B.

Expected mean squares for 3-factor factorials

Consider a 3-factor factorial experiment with three factors A, B and C at levels p, m and q, respectively with layout in randomized block design having r blocks. The following table of expected mean squares are displayed taking the factors A, B and C as fixed, random and mixed without explanation. It is left to the reader to examine how well the set rules help to write expected mean squares in any situation. Statistical model for a 3-factor factorial is,

$$y_{ijku} = \mu + \alpha_i + \beta_j + \nu_k + \rho_u + (\alpha\beta)_{ij} + (\alpha\nu)_{ik} + (\beta\nu)_{jk} + (\alpha\beta\nu)_{ijk} + \sigma_{ijku} \qquad (4.3.1)$$

for $i = 1, 2, \ldots\ldots p$; $j = 1, 2, \ldots\ldots, m$; $k = 1, 2, \ldots\ldots, q$; $u = 1, 2, \ldots\ldots, r$

All the notations in the model (4.3.1) can be decoded in the manner similar to model (4.2.1). Expected mean squares in an analysis of variance table for all main and interaction effects in accordance to their nature are produced in table (4.3.2). In the following table, mean square due to a factor U is represented by σ_U^2 for random effects and by κ_U^2 for fixed effects where,

$$\kappa_U^2 = \frac{\text{Adj. S.S. due to effect U}}{\text{d.f. of U}}$$

In analysis of variance, an analyst always prefers an exact F-test. An exact F-test is one in which the expected values of mean squares of F-statistic in the numerator and denominator under null hypothesis (H_0) are same. A critical examination of the three parts of Table 4.3.2 reveals the following facts.

In first column, when all effects are fixed, null hypothesis for any factor is, $H_0: \tau_u = 0$. In this case, all effects can be tested against error mean square providing an exact test.

In second column, when all effects are random, null hypothesis for any effect under test is $\sigma_u^2 = 0$. Expected mean squares for main and interaction

Table 4.3.2 Expected mean squares for 3-factor factorials

Factors	Effects A, B, C fixed	Effects A, B, C random	Estimated M.S.
A	$\sigma_e^2 + rmq\kappa_A^2$	$\sigma_e^2 + r\sigma_{ABC}^2 + rq\sigma_{AB}^2 + rm\sigma_{AC}^2 + rmq\sigma_A^2$	V_A
B	$\sigma_e^2 + rpq\kappa_B^2$	$\sigma_e^2 + r\sigma_{ABC}^2 + rq\sigma_{AB}^2 + rm\sigma_{AC}^2 + rpq\sigma_B^2$	V_B
AB	$\sigma_e^2 + rq\kappa_{AB}^2$	$\sigma_e^2 + r\sigma_{ABC}^2 + rq\sigma_{AB}^2$	V_{AB}
C	$\sigma_e^2 + rpm\kappa_C^2$	$\sigma_e^2 + r\sigma_{ABC}^2 + rm\sigma_{AC}^2 + rp\sigma_{BC}^2 + rpm\sigma_C^2$	V_C
AC	$\sigma_e^2 + rm\kappa_{AC}^2$	$\sigma_e^2 + r\sigma_{ABC}^2 + rm\sigma_{AC}^2$	V_{AC}
BC	$\sigma_e^2 + rp\kappa_{BC}^2$	$\sigma_e^2 + r\sigma_{ABC}^2 + rp\sigma_{BC}^2$	V_{BC}
ABC	$\sigma_e^2 + r\kappa_{ABC}^2$	$\sigma_e^2 + r\sigma_{ABC}^2$	V_{ABC}
Error	σ_e^2	σ_e^2	V_E

Factors	Effect A fixed and effects B, C random
A	$\sigma_e^2 + r\sigma_{ABC}^2 + rq\sigma_{AB}^2 + rm\sigma_{AC}^2 + rmq\kappa_A^2$
B	$\sigma_e^2 + rp\sigma_{BC}^2 + rpq\sigma_B^2$
AB	$\sigma_e^2 + r\sigma_{ABC}^2 + rq\sigma_{AB}^2$
C	$\sigma_e^2 + rp\sigma_{BC}^2 + rpm\sigma_C^2$
AC	$\sigma_e^2 + r\sigma_{ABC}^2 + rm\sigma_{AC}^2$
BC	$\sigma_e^2 + rp\sigma_{BC}^2$
ABC	$\sigma_e^2 + r\sigma_{ABC}^2$
Error	σ_e^2

Where,
$$\kappa_A^2 = \frac{\sum_i \alpha_i^2}{p-1}$$

Note: Expected mean squares for all components in analysis of variance table for any type of factorial effects in an experiment can be written by use of the rules 1 through 5 diligently.

effects divulge that there is no EMS which can be treated as error MS to test any main effect yielding an exact test. Of course, first order interactions can exactly be tested against ABC MS and BC against error MS.

Now the question arises how to test any component factor for which an exact test is not available under various model specifications. To cope with such intrinsic problems, Satterthwaite (1946) put forward an *approximate* F-test. His test is based on the synthesis of expected mean squares. A linear combination of k expected mean squares of the type,

$$L = a_1 \, EMS_1 + a_2 \, EMS_2 + \ldots\ldots\ldots + a_k \, EMS_k \qquad (4.3.2)$$

is used where a's are some constant multiples.

For all practical purposes, EMSs are replaced by their respective estimates values. Thus, linear combination is,

$$L = a_1 \, MS_1 + a_2 \, MS_2 + \text{------} + a_k \, MS_k \qquad (4.3.2.1)$$

where, MS_i is the estimated value of EMS_i for $i = 1, 2, \ldots\ldots, k$.

Degrees of freedom, say, $\hat{\upsilon}$ for L are,

$$\hat{\upsilon} = \cfrac{L^2}{\cfrac{\left(a_1 MS_1\right)^2}{d.f.\,\text{of}\,MS_1} + \cfrac{\left(a_2 MS_2\right)^2}{d.f.\,\text{of}\,MS_2} + \ldots\ldots + \cfrac{\left(a_k MS_k\right)^2}{d.f.\,\text{of}\,MS_k}} \qquad (4.3.3)$$

During the process of analysis, values of EMSs are substituted by their estimates, i.e. by their respective mean square from ANOVA.

Let us denote the expected mean squares for three factor factorial effects by V_U as given in Table 4.3.2, where U = A, B, AB, C, ...

In second column when all effects are random, to test H_0: $\sigma_A^2 = 0$, linear combination of EMS for the denominator is,

$$L = EMS_{AB} + EMS_{AC} - EMS_{ABC} = \sigma_e^2 + r\sigma_{ABC}^2 + rq\sigma_{AB}^2 + rm\sigma_{AC}^2 + rmq\sigma_A^2$$

$$= V_{AB} + V_{AC} - V_{ABC} \qquad (4.3.4)$$

Hence, approximate F-statistic is,

$$F' = \frac{V_A}{V_{AB} + V_{AC} - V_{ABC}} \qquad (4.3.5)$$

F' has υ_1 and υ_2 degrees of freedom.

Note: Similar combination of EMS can be formed for testing the null hypotheses H_0: $\sigma_B^2 = 0$ and $\sigma_C^2 = 0$.

For testing A in the mixed model, F' given (4.3.5) will also work. For all other factors, exact tests are available. In approximate F'-statistic, υ_1 and/or υ_2 are often unknown. So degrees of freedom for any linear combination L can be obtained by the formula given below. Degrees of freedom, say, $\hat{\upsilon}$ for L given by (4.3.4) are,

$$\hat{\upsilon} = \frac{\left(V_{AB} + V_{AC} - V_{ABC}\right)^2}{\dfrac{V_{AB}^2}{d.f.(AB)} + \dfrac{V_{AC}^2}{d.f.(AC)} + \dfrac{V_{ABC}^2}{d.f.(ABC)}} \qquad (4.3.6)$$

$\hat{\upsilon}$ has mostly a fractional value.

Another approximate F-statistic as a ratio of two linear combinations, can be given as,

$$F'' = \frac{V_A + V_{ABC}}{V_{AB} + V_{AC}} \qquad (4.3.7)$$

F'' has υ_3 and υ_4 degrees of freedom. Degrees of freedom υ_3 and υ_4 can be obtained by the formulae,

$$\upsilon_3 = \frac{\left(V_A + V_{ABC}\right)^2}{\dfrac{V_A^2}{d.f.(A)} + \dfrac{V_{ABC}^2}{d.f.(ABC)}} \qquad (4.3.8)$$

$$\upsilon_4 = \frac{\left(V_{AB} + V_{AC}\right)^2}{\dfrac{V_{AB}^2}{d.f.(AB)} + \dfrac{V_{AC}^2}{d.f.(AC)}} \qquad (4.3.9)$$

either of υ_3 and υ_4 or both can be fractional values.

But there is no table which provides critical values of F-distribution for fractional degrees of freedom. Anyhow, Lavbscher (1965) developed *interpolation formulae* to obtain critical values of F-distribution for fractional or untabulated degrees of freedom which are very accurate and easy to apply. These formulae are given below.

In the succeeding interpolation formulae, α is omitted from F as suffix for the sake of convenience and it is assumed to be there.

Case 1: When m appears in the table, but not n. If n' and n'' are the degrees of freedom which immediately precede and follow n in the F-table, the value,

$$F_{m,n} = (1 - A)F_{m,n'} + AF_{m,n''} \qquad (4.3.10)$$

where,

$$A = \frac{n''(n - n')}{n(n'' - n')} \qquad (4.3.10.1)$$

The values of $F_{m,n'}$ and $F_{m,n''}$ for level α are read from F-table and substituted in the above formulae.

Case 2: When n appears in the table, but not m.
The interpolation formula is,

$$F_{m,n} = (1 - B)F_{m',n} + BF_{m'',n} \qquad (4.3.11)$$

In the above formula, all notations can be decoded in the similar manner as in (4.3.10) by replacing n by m and vice-versa. In this case,

$$B = \frac{m''(m - m')}{m(m'' - m')} \qquad (4.3.11.1)$$

Case 3: When neither m nor n appear in F-table. Following the same notations as in cases 1 and 2, the interpolation formula is,

$$F_{m,n} = (1 - A)(1 - B)F_{m',n'} + A(1 - B)F_{m',n''}$$

$$+ (1 - A)BF_{m'',n'} + A \times BF_{m'',n''} \qquad (4.3.12)$$

Constants A and B are as above.
Remark: This topic is pursued in section (7.2) as well. The readers should connect both for better and clear understanding.

Example 4.1: A factorial experiment with 3 factors – two varieties, three nitrogen levels and four spray concentrations was conducted in randomized block design taking three replications to see the effect on nitrogen percentage in wheat grain. Treatments with grain yield are given in the following table:

Table I Grain yield q/ha

Treatments	R_1	R_2	R_3	Total	Grain yield mean
$S_3N_1V_1$	34.6	33.3	35.1	103.0	34.33
$S_3N_2V_1$	41.0	42.6	39.6	123.2	41.07
$S_3N_3V_1$	43.1	52.6	49.6	145.3	48.43
$S_2N_1V_1$	30.7	30.6	36.1	97.4	32.47
$S_2N_2V_1$	40.7	38.0	41.7	120.4	40.13
$S_2N_3V_1$	41.6	43.6	39.7	124.9	41.63
$S_1N_1V_1$	33.0	36.1	34.6	103.7	34.57
$S_1N_2V_1$	43.6	40.2	42.5	126.3	42.10
$S_1N_3V_1$	41.7	47.7	44.3	133.7	44.57
$S_0N_1V_1$	30.0	32.2	36.9	99.1	33.03
$S_0N_2V_1$	42.6	41.2	47.5	131.3	43.77
$S_0N_3V_1$	36.7	45.4	48.7	130.8	43.60

(Contd.)

Treatments	R_1	R_2	R_3	Total	Grain Yield Mean
$S_3N_1V_2$	35.6	38.9	40.2	114.7	38.23
$S_3N_2V_2$	48.6	46.3	45.9	140.8	46.93
$S_3N_3V_2$	48.6	49.7	45.2	143.5	47.83
$S_2N_1V_2$	37.8	47.5	45.2	130.5	43.50
$S_2N_2V_2$	41.7	41.5	46.9	130.1	43.37
$S_2N_3V_2$	43.9	42.0	34.4	120.3	40.10
$S_1N_1V_2$	35.8	45.2	45.0	126.0	42.00
$S_1N_2V_2$	47.6	45.6	41.5	134.7	44.90
$S_1N_3V_2$	56.0	48.9	49.9	154.8	51.60
$S_0N_1V_2$	35.2	32.5	38.7	106.4	35.47
$S_0N_2V_2$	42.0	36.8	43.5	122.3	40.77
$S_0N_3V_2$	42.2	48.9	43.6	134.7	44.90
Total	974.3	1007.3	1016.3	2997.9	999.30

Calculations: $\sum\limits_{i,j,k} y_{ijk} = 2997.90$

$$C.F. = \frac{(2997.90)^2}{72} = 124825.06$$

$$\text{Total S.S.} = (34.6^2 + 33.3^2 + \ldots\ldots + 48.9^2 + 43.6^2) - C.F.$$
$$= 127109.85 - C.F. = 2284.79$$

$$\text{Rep. S.S.} = \frac{1}{24}(974.3^2 + 1007.3^2 + 1016.3^2) - C.F.$$

$$124865.8113 - C.F. = 40.75$$

$$\text{Treat. S.S.} = \frac{1}{3}(103.0^2 + 123.2^2 + \ldots\ldots + 122.3^2 + 134.7^2) - C.F.$$
$$= 126546.4223 - C.F. = 1721.36$$

Table II V × N table

Varieties	N_1	N_2	N_3	Total	Mean
V_1	403.2	501.2	534.7	1439.1	39.98
V_2	477.6	527.9	553.3	1558.8	43.30
Total	880.8	1029.1	1088.0	2997.9	

$$\text{S.S. (V)} = \frac{1}{36}(1439.1^2 + 1558.8^2) - \text{C.F.}$$

$$= 125024.0625 - \text{C.F.} = 199.0012$$

$$\text{S.S. (N)} = \frac{1}{24}(888.8^2 + 1029.1^2 + 1088.0^2) - \text{C.F.}$$

$$= 125774.9771 - \text{C.F.} = 949.9158$$

$$\text{Table-I S.S.} = \frac{1}{12}(403.2^2 + 501.2^2 + \ldots\ldots + 527.9^2 + 553.3^2) - \text{C.F.}$$

$$= 126049.7358 - \text{C.F.} = 1224.6745$$

$$\text{S.S. (V} \times \text{N)} = 1224.6745 - 199.0012 - 949.9158 = 80.7575$$

Table III V × S table

Varieties	S_0	S_1	S_2	S_3	Total
V_1	361.2	363.7	342.7	371.5	**1439.1**
V_2	363.4	415.5	380.9	399.0	**1558.8**
Total	724.6	779.2	723.6	770.5	2997.9
Mean	**40.26**	**43.29**	**40.20**	**42.80**	**41.64**

$$\text{S.S. (S)} = \frac{1}{18}(724.6^2 + 779.2^2 + 723.6^2 + 770.5^2) - \text{C.F.}$$

$$= 124970.2783 - \text{C.F.} = 145.2170$$

$$\text{Table-II S.S.} = \frac{1}{9}(361.2^2 + 363.7^2 + \ldots\ldots + 380.9^2 + 399.0^2) - \text{C.F.}$$

$$= 125242.6989 - \text{C.F.} = 417.6376$$

$$\text{S.S. (V} \times \text{S)} = 417.6376 - 199.0012 - 145.2170 = 73.4194$$

Table IV N × S table

N_2 levels	S_0	S_1	S_2	S_3	Total	Mean
N_1	205.5	229.7	227.9	217.7	880.8	**36.7**
N_2	253.6	261.0	250.5	264.0	1029.1	**42.88**
N_3	265.5	288.5	245.2	288.8	1088.0	**45.33**
Total	724.6	779.2	723.6	770.5	2997.9	

$$\text{Table S.S.} = \frac{1}{6}(205.5^2 + 229.7^2 + \ldots\ldots + 245.2^2 + 288.8^2) - \text{C.F.}$$

$$= 126075.8717 - \text{C.F.} = 1250.8104$$

S.S. (N × S) = 1250.8104 − 949.9158 − 145.2170

= 155.6776

S.S. (V × N × S) = Treat. S.S. − $\sum$(Sum of squares due to main effects)

− $\sum$(Sum of squares due to first order interactions)

= 1721.36 − 199.0012 − 949.9158 − 145.2170

− 80.7575 − 73.4194 − 155.6776

= 117.3715

After calculating the required sum of squares, following analysis of variance table can easily be prepared.

Table V ANOVA table

Source	*d.f.*	*S.S.*	*M.S.*	*F-value*
Replications	2	40.7500	20.3750	1.7932
Treatments	23	1721.36		
V	1	199.0012	199.0012	$F' = 4.39$ NS; $F'' = 3.37$ NS
N	2	949.9158	474.9579	$\dfrac{474.9579}{25.9463} = 18.3054*$
V × N	2	80.7575	40.3788	$\dfrac{40.3788}{19.5619} = 2.0641$
S	3	145.2170	48.4057	$\dfrac{48.4057}{25.4463} = 1.8656$
V × S	3	73.4194	24.4731	$\dfrac{24.4731}{19.5619} = 1.2510$
N × S	6	155.6776	25.9463	$\dfrac{25.9463}{11.3625} = 2.2835$
V × N × S	6	117.3715	19.5619	$\dfrac{19.5619}{11.3625} = 1.7216$
Error	46	522.6787	11.3625	

Note: F-values in the above ANOVA table are calculated by using proper error term as ascertained on the basis of expected mean squares.

* Significant at $\alpha = 0.05$

For $\alpha = 0.05$, critical values of F-distribution are,

$F_{1,46} = 4.056$, $F_{2,46} = 3.206$, $F_{3,46} = 2.816$ and $F_{6,46} = 2.32$.

Now the methods of calculating F′ and F″ and their corresponding degrees of freedom are explicated and used in ANOVA-V.

To apply the statistic F′ from formula (4.3.4), calculate

M.S. (V × N) + M.S. (V × S) − M.S. (V × N × S)

= 40.3788 + 24.4731 − 19.5619

= 45.29

Thus, $$F' = \frac{199.0012}{45.2900} = 4.3939$$

Again for F'', M.S.(V) + M.S.(V × N × S) = 199.0012 + 19.5619

$$= 218.5631$$

M.S (V × N) + M.S. (V × S) = 40.3788 + 24.4731

$$= 64.8519$$

Thus, $$F'' = \frac{218.5631}{64.8519} = 3.37$$

For statistic F', υ_2 by the formula (4.3.5) is,

$$\upsilon_2 = \frac{(45.29)^2}{\dfrac{(40.3788)^2}{2} + \dfrac{(24.4731)^2}{3} + \dfrac{(19.5619)^2}{6}}$$

$$= \frac{2051.1841}{1078.6479} = 1.9016$$

In the like manner, υ_3 and υ_4 are calculated with the help of the formulae (4.3.8) and (4.3.9).

$$\upsilon_3 = \frac{(218.5631)^2}{\dfrac{(199.0012)^2}{1} + \dfrac{(19.5619)^2}{6}}$$

$$= \frac{47769.8287}{39601.4776 + 63.7800} = 1.2043$$

$$\upsilon_4 = \frac{(64.8519)^2}{\dfrac{(40.3788)^2}{2} + \dfrac{(24.4731)^2}{3}}$$

$$= \frac{4205.7689}{815.2237 + 199.6442} = \frac{4205.7689}{1014.8679} = 4.1442$$

Interpolated F-value for $\upsilon_1 = 1$, $\upsilon_2 = 1.9016$ by the formula (4.3.10) can be obtained. In this situation, $n' = 1$ and $n'' = 2$. Let the level of significance of the test be $\alpha = 0.05$. $F_{1,1} = 161.4$, $F_{1,2} = 18.51$.

$$A = \frac{2(1.9016 - 1)}{1.9016(2 - 1)} = \frac{1.8032}{1.9016} = 0.9482$$

$$F_{1,1.9016} = (1 - 0.9482) \times 161.4 + 0.9482 \times 18.51$$
$$= 0.0518 \times 161.4 + 0.9482 \times 18.51 = 25.9117$$

υ_3 and υ_4 both are fractional values. Therefore, formula (4.3.12) will be used to find the critical value of $F_{\upsilon_3, \upsilon_4}$ at $= 0.05$ as follows:

Here, $m' = 1$, $m'' = 2$ and $n' = 4$, $n'' = 5$

For, $F_{1.2043, \, 4.1442}$, $A = \dfrac{2(1.2043 - 1)}{1.2043(2 - 1)} = \dfrac{0.4086}{1.2043} = 0.3393$

$$B = \dfrac{5(4.1442 - 2)}{4.1442(5 - 4)} = \dfrac{0.7210}{4.1442} = 0.1740$$

$F_{1,5} = 6.61$, $F_{2,4} = 6.94$, $F_{1,4} = 7.71$, $F_{2,5} = 5.79$

Thus, from (4.3.11),

$F_{1.2043, \, 4.1442} = (1 - 0.3393)(1 - 0.1740) \times 7.71 + 0.3393 \, (1 - 0.1740) \times 6.61$
$\qquad\qquad + (1 - 0.3393) \times 0.1740 \times 6.94 + 0.3393 \times 0.1740 \times 5.79$

$\qquad\qquad = 0.6607 \times 0.8260 \times 7.71 + 0.3393 \times 0.8260 \times 6.61$
$\qquad\qquad + 0.6607 \times 0.1740 \times 6.94 + 0.3393 \times 0.1740 \times 5.79$

$\qquad\qquad = 7.20$

In ANOVA-V, effects N and S are tested against N × S; V × N and V × S against V × N × S. N × S and V × N × S are tested against experimental error. F-values for all factors are calculated as per given norms and displayed in ANOVA table-V. Critical values of F-distribution are shown below the table and for fractional degrees of freedom, the same have been worked out consecutively. Interpretation of results is left to the reader.

4.4 3^n-FACTORIAL EXPERIMENTS

The number of treatment combinations increases tremendously with the increase in number of levels of factors. Also the main effects and interactions in a 3^n factorial cannot be represented by the contrasts. Hence, a different approach is required. The best approach to deal with the factorials other than 2^n is reduced modulo approach. Though modulo technique is applicable to 2^n also. The sum of squares can also be calculated through 2 or 3 way tables. To begin with, consider 3^2 *factorial* experiment having two factors A and B each with three levels say, 0, 1, 2. There will in all be nine treatment combinations. Proceeding systematically, one can easily write the treatment combinations. Divide 9 by 3, the factor levels. The quotient is 3. Write a_0, a_1 and a_2 in threes. Then divide 3 by 3. The quotient is one. Now write b_0, b_1 and b_2 once against a_0, a_1 and a_2 in order. In this way, all possible treatment combinations are obtained in a sequence. Treatment combinations can be represented through suffixes alone assuming that a is attached to each level in column (i) and b in column (ii).

Treatment combinations	Combinations witwithout letters	Response totals
$a_0\, b_0$	0 0	Y_{00}
$a_0\, b_1$	0 1	Y_{01}
$a_0\, b_2$	0 2	Y_{02}
$a_1\, b_0$	1 0	Y_{10}
$a_1\, b_1$	1 1	Y_{11}
$a_1\, b_2$	1 2	Y_{12}
$a_2\, b_0$	2 0	Y_{20}
$a_2\, b_1$	2 1	Y_{21}
$a_2\, b_2$	2 2	Y_{22}

Modulo technique

Reduced modulo x is defined as $u = x$ mod y, i.e. when u is divided by y, the remainder is x. This approach will now be extensively used to allocate the treatment combinations for main and interaction effects in on going discussion and analysis of variance.

Consider the break-up of degrees of freedom for different factorial effects. This will enable one to know which factor effects and sum of squares have to be worked out. Suppose the responses due to treatments summed over r replications are as displayed above or in a two-way table given below.

Table 4.4.1 Two-way table of responses

Factors	b_0	b_1	b_2	Total	Mean
a_0	Y_{00}	Y_{01}	Y_{02}	$Y_{0.}$	$Y_{0.}/3r$
a_1	Y_{10}	Y_{11}	Y_{12}	$Y_{1.}$	$Y_{1.}/3r$
a_2	Y_{20}	Y_{21}	Y_{22}	$Y_{2.}$	$Y_{2.}/3r$
Total	$Y_{.0}$	$Y_{.1}$	$Y_{.2}$		$Y_{..}/9r$
Mean	$Y_{.0}/3r$	$Y_{.1}/3r$	$Y_{.2}/3r$		$Y_{..}/9r$

Sum of squares due to main effects A and B and interaction A × B can be found with the help of table (4.4.1) in the following manner.

$$\text{S.S. due to A} = \frac{Y_{0.}^2 + Y_{1.}^2 + Y_{2.}^2}{3r} - \frac{Y_{..}^2}{9r} = ASS$$

$$\text{S.S. due to B} = \frac{Y_{.0}^2 + Y_{.1}^2 + Y_{.2}^2}{3r} - \frac{Y_{..}^2}{9r} = BSS$$

S.S. due to A $\times$ B $= \dfrac{1}{r}(Y_{00}^2 + Y_{01}^2 + \ldots\ldots\ldots + Y_{21}^2 + Y_{22}^2) - C.F. - ASS - BSS$

$$= (A \times B)\ SS$$

In this way, one cannot separate the sum of squares due to AB and AB^2. But the sum of squares due to these factors can easily be worked out through modulo approach. Suppose the level of A is denoted by i and that of B by j. Now the treatment combinations associated with main effects and interaction reduced to modulo 3 are listed below giving only the levels in order of a and b.

$$(A)_0 = (A)_{i\,=\,0\,\mathrm{mod}\,3} = 0\ 0 + 0\ 1 + 0\ 2$$
$$(A)_1 = (A)_{i\,=\,1\mathrm{mod}\,3} = 1\ 0 + 1\ 1 + 1\ 2$$
$$(A)_2 = (A)_{i\,=\,2\,\mathrm{mod}\,3} = 2\ 0 + 2\ 1 + 2\ 2$$
$$(B)_0 = (B)_{j\,=\,0\,\mathrm{mod}\,3} = 0\ 0 + 1\ 0 + 2\ 0$$
$$(B)_1 = (B)_{j\,=\,1\,\mathrm{mod}\,3} = 0\ 1 + 1\ 1 + 2\ 1$$
$$(B)_2 = (B)_{j\,=\,2\,\mathrm{mod}\,3} = 0\ 2 + 1\ 2 + 2\ 2$$
$$(AB)_0 = (AB)_{i\,+\,j\,=\,0\,\mathrm{mod}\,3} = 0\ 0 + 1\ 2 + 2\ 1$$
$$(AB)_1 = (AB)_{i\,+\,j\,=\,1\,\mathrm{mod}\,3} = 0\ 1 + 1\ 0 + 2\ 2$$
$$(AB)_2 = (AB)_{i\,+\,j\,=\,2\,\mathrm{mod}\,3} = 0\ 2 + 1\ 1 + 2\ 0$$
$$(AB^2)_0 = (AB^2)_{i\,+2\,j\,=\,0\,\mathrm{mod}\,3} = 0\ 0 + 1\ 1 + 2\ 2$$
$$(AB^2)_1 = (AB^2)_{i\,+2\,j\,=\,1\,\mathrm{mod}\,3} = 0\ 2 + 1\ 0 + 2\ 1$$
$$(AB^2)_2 = (AB^2)_{i\,+2\,j\,=\,2\,\mathrm{mod}\,3} = 0\ 1 + 1\ 2 + 2\ 0$$

Sum of squares for the treatment effects can be obtained by the formulae given below. Let the sum of responses for each effect be denoted by the effect itself.

$$\text{S.S. due to A} = \frac{(A)_0^2 + (A)_1^2 + (A)_2^2}{3r} - \frac{Y..^2}{9r} = ASS$$

$$\text{S.S. due to B} = \frac{(B)_0^2 + (B)_1^2 + (B)_2^2}{3r} - \frac{Y..^2}{9r} = BSS$$

$$\text{S.S. due to AB} = \frac{(AB)_0^2 + (AB)_1^2 + (AB)_2^2}{3r} - \frac{Y..^2}{9r} = ABSS$$

$$\text{S.S. due to AB}^2 = \frac{(AB^2)_0^2 + (AB^2)_1^2 + (AB^2)_2^2}{3r} - \frac{Y..^2}{9r} = AB^2SS$$

Of course, the sum of squares due to replications (blocks), total and error will be obtained in the usual manner.

Nota bene:

1. It can be verified that the sum of square for $A \times B$ with 4 d.f. is equal to the aggregate of sum of squares due to AB and AB^2 each with 2 d.f.
2. One may ask why we have not taken the interaction components $A^2 B$ and $A^2 B^2$. The reason being that $A^2B = A^4 B^2 = A B^2$ reduced mod 3. Similarly $A^2 B^2 = A^4 B^4 = A B$ reduced to modulo 3. Another interesting point to note here is that the treatment combinations belonging to $A^2 B$ are same as those belonging to $A B^2$. To verify,

$$(A^2 B)_0 = (A^2 B)_{2i + j = 0 \, mod \, 3} = 0\,0 + 1\,1 + 2\,2$$
$$(A^2 B)_1 = (A^2 B)_{2i + j = 1 \, mod \, 3} = 0\,1 + 1\,2 + 2\,0$$
$$(A^2 B)_2 = (A^2 B)_{2i + j = 2 \, mod \, 3} = 0\,2 + 1\,0 + 2\,1$$

This confirms that the sum of squares for $A^2 B$ comes out to be the same as for $A B^2$. The same holds true for $A^2 B^2$ and $A B$.

3. Daniel (1976) stressed that the data of a factorial experiment should not be reported in terms of main effects and interaction if an iteraction effect is more than one-third of a main effect.

Analysis of variance

Analysis of variance table with the break up of interaction $A \times B$ can be given as follows.

Table 4.4.2 ANOVA for 3^2 factorials

Source	d.f.	S.S.	M.S.
Replications	$r - 1$	RSS	RMS
A	2	ASS	AMS
B	2	BSS	BMS
A × B	4	(A × B) SS	(A × B)MS
A B $\rceil$	2 $\rceil$	ABSS $\rceil$	ABMS $\rceil$
A B² $\rfloor$	2 $\rfloor$	AB²SS $\rfloor$	AB²MS $\rfloor$
Error	$8(r - 1)$	ESS	EMS
Total	$9r - 1$		

All the effect mean squares are to be tested against error mean square. The significance of any mean square confirms that there is a significant difference among factor level means. If interaction mean square for any component comes out to be significant, it substantiates the assertion that the interaction between two factors A and B meaningfully exists.

The standard error for a mean level of factors A and B in a 3^2 - experiment with r replications $= \sqrt{\dfrac{\text{Error M.S.}}{3r}}$

S.E. for difference between two level means $= \sqrt{\dfrac{2 \, \text{Error M.S.}}{3r}}$

Significance of paired level means can be tested through lsd or some other suitable post-hoc test.

A good analyst always wants to gather as much information from an experimental data as possible. So one has to delve deep by analyzing further and draw conclusions about fortuitous events as well as significant effects.

In discussion of contrasts, it has been revealed that with 3 factor levels there can be two contrasts, one representing the linear effect such as $(a_2 - a_1)$ and other a quadratic effect such as $(a_2 + a_1 - 2 a_0)$ and similar for B. Also it was indicated that interaction of treatment with replication (Treat × Rep.) always provide the error term for that treatment. So an analyst can work out errors for treatment component factors separately. Recall A_l, A_q and B_l, B_q stand for linear and quadratic effects of A and B respectively. Each effect can be test against their genuine error. The skeleton extended analysis of variance table with splitted effects will be as given below. The method of calculating sum of square for every break up component factor has been elucidated through the following example for better understanding. The same procedure can be adapted for higher order factorial experiments.

Table 4.4.3 Skeleton extended ANOVA for 3^2 factorials

Source	d.f.
Replications	$r - 1$
Treatments	8
A	2
$A_l = (a_2 - a_1)$ ⎤	1 ⎤
$A_q = (a_2 + a_1 - 2a_0)$ ⎦	1 ⎦
B	2
$B_l = (b_2 - b_1)$ ⎤	1. ⎤
$B_q = (b_2 + b_1 - 2b_0)$ ⎦	1 ⎦
A × B	4
$A_l B_l$ ⎤	1 ⎤
$A_l B_q$	1
$A_q B_l$	1
$A_q B_q$ ⎦	1 ⎦
Error	$8(r - 1)$
Total	$9r - 1$

4.5 3^3-FACTORIAL EXPERIMENTS

There are many experimental situations in which the researcher wants to investigate 3 treatment factors each having three levels and their interactions. Obviously, 3^3 factorial experiments are the only choice. The experiment can be conducted in randomized block design or any other basic design. Somehow randomized block design is in vogue. So further discussion of 3^3 factorial experiments will confine to randomized block designs. Suppose there are three factors A, B and C, each with three levels 0, 1, and 2. So there will in all be 27 treatment combinations for ABC represented as $a_i b_j c_u$ for $i, j, u = 0$, 1, 2. These treatment combinations can schematically be written as follows:

Schematic display of treatment combinations

0 0 0	1 0 0	2 0 0
0 0 1	1 0 1	2 0 1
0 0 2	1 0 2	2 0 2
0 1 0	1 1 0	2 1 0
0 1 1	1 1 1	2 1 1
0 1 2	1 1 2	2 1 2
0 2 0	1 2 0	2 2 0
0 2 1	1 2 1	2 2 1
0 2 2	1 2 2	2 2 2

All these treatment combinations can be assigned in each of the r blocks randomly and independently.

Table 4.5.1 Skeleton ANOVA for a 3^3 factorials

Source	d.f.
Replications	$r - 1$
Treatments	26
A	2
B	2
A × B	4
AB	2 ⎤
AB^2	2 ⎦
C	2
A × C	4
AC	2 ⎤
AC^2	2 ⎦
B × C	4
BC	2 ⎤
BC^2	2 ⎦

(Contd.)

Source	d.f.
A × B × C	8
ABC $= JJ$	2
$ABC^2 = II$	2
$AB^2C = JI$	2
$AB^2C^2 = IJ$	2
Error	$26\,(r-1)$
Total	$27\,(r-1)$

The notations JJ, II, JI, and IJ were used by Yates (1937)

Sum of squares for main and interaction effects can easily be calculated with the help of modulo technique.

Calculation of sum of squares

$$\text{S.S. due to A} = \frac{(A)_0^2 + (A)_1^2 + (A)_2^2}{9r} - \frac{G^2}{27r}$$

where G = Grand total of all response values.

Similar expressions can be used for main effects B and C.

$$\text{S.S. due to AB} = \frac{(AB)_0^2 + (AB)_1^2 + (AB)_2^2}{9r} - \frac{G^2}{27r}$$

Similar expressions for AC and BC hold good.

$$\text{S.S. due to ABC} = \frac{(ABC)_0^2 + (ABC)_1^2 + (ABC)_2^2}{9r} - \frac{G^2}{27r}$$

Parallel expressions may be used for computing sum of squares due to ABC^2, AB^2C, AB^2C^2.

Denoting the treatment combinations $a_i b_j c_u$ by their suffixes $i\,j\,u$ in order, the treatment combinations comprising the various effects are displayed below. The same may be used to find out the sum of squares due to respective effects.

$(A)_0 = (A)_{i\,=\,0\,mod\,3} = 000 + 001 + 002 + 010 + 011 + 012 + 020 + 021 + 022$

$(A)_1 = (A)_{i\,=\,1\,mod\,3} = 100 + 101 + 102 + 110 + 111 + 112 + 120 + 121 + 122$

$(A)_2 = (A)_{i\,=\,2\,mod\,3} = 200 + 201 + 202 + 210 + 211 + 212 + 220 + 221 + 222$

$(B)_0 = (B)_{j\,=\,0\,mod\,3} = 000 + 001 + 002 + 100 + 101 + 102 + 200 + 201 + 202$

$(B)_1 = (B)_{j\,=\,1\,mod\,3} = 010 + 011 + 012 + 110 + 111 + 112 + 210 + 211 + 212$

$(B)_2 = (B)_{j\,=\,2\,mod\,3} = 020 + 021 + 022 + 120 + 121 + 122 + 220 + 221 + 222$

$(C)_0 = (C)_{u\,=\,0\,mod\,3} = 000 + 010 + 020 + 100 + 110 + 120 + 200 + 210 + 220$

$(C)_1 = (C)_{u\,=\,1\,mod\,3} = 001 + 011 + 021 + 101 + 111 + 121 + 201 + 211 + 221$

$(C)_2 = (C)_{u\,=\,2\,mod\,3} = 002 + 012 + 022 + 102 + 112 + 122 + 202 + 212 + 222$

$(AB)_0 = (AB)_{i + j = 0 \bmod 3} = 000 + 001 + 002 + 120 + 121 + 122 + 210$
$$+ 211 + 212$$

$(AB)_1 = (AB)_{i + j = 1 \bmod 3} = 010 + 011 + 012 + 100 + 101 + 102 + 220$
$$+ 221 + 222$$

$(AB)_2 = (AB)_{i + j = 2 \bmod 3} = 020 + 021 + 022 + 110 + 111 + 112 + 200$
$$+ 201 + 202$$

$(AB^2)_0 = (AB^2)_{i + 2j = 0 \bmod 3} = 000 + 001 + 002 + 110 + 111 + 112 + 220$
$$+ 221 + 222$$

$(AB^2)_1 = (AB^2)_{i + 2j = 1 \bmod 3} = 020 + 021 + 022 + 100 + 101 + 102 + 210$
$$+ 211 + 212$$

$(AB^2)_2 = (AB^2)_{i + 2j = 2 \bmod 3} = 010 + 011 + 012 + 120 + 121 + 122 + 200$
$$+ 201 + 202$$

Similarly one can find the treatment combinations belonging to the effects, $(AC)_0$, $(AC)_1$, $(AC)_2$, $(AC^2)_0$, $(AC^2)_1$, $(AC^2)_2$, and $(BC)_0$, $(BC)_1$, $(BC)_2$, $(BC^2)_0$, $(BC^2)_1$, $(BC^2)_2$.

Second order interaction effects consist of the treatment combinations listed below.

$(ABC)_0 = (ABC)_{i + j + u = 0 \bmod 3} = 000 + 012 + 021 + 102 + 111 + 120 + 201$
$$+ 210 + 222$$

$(ABC)_1 = (ABC)_{i + j + u = 1 \bmod 3} = 001 + 010 + 022 + 100 + 112 + 121 + 202$
$$+ 211 + 220$$

$(ABC)_2 = (ABC)_{i + j + u = 2 \bmod 3} = 002 + 011 + 020 + 101 + 110 + 122 + 200$
$$+ 212 + 221$$

$(ABC^2)_0 = (ABC^2)_{i + j + 2u = 0 \bmod 3} = 000 + 011 + 022 + 101 + 112 + 120 + 202$
$$+ 210 + 221$$

$(ABC^2)_1 = (ABC^2)_{i + j + 2u = 1 \bmod 3} = 002 + 010 + 021 + 100 + 111 + 122 + 201$
$$+ 212 + 220$$

$(ABC^2)_2 = (ABC^2)_{i + j + 2u = 2 \bmod 3} = 001 + 012 + 020 + 102 + 110 + 121 + 200$
$$+ 211 + 222$$

In the like manner, the treatment combinations for the interaction effects (AB^2C) and (AB^2C^2) can be sorted out by taking modulo for $i + 2j + u = x$ mod 3 and $i + 2j + 2u = x$ mod 3 for $x = 0, 1, 2$ in succession. Once the total response using these treatment combinations for each effect is obtained, calculate the sum of squares in the usual manner and prepare the analysis of variance table.

Note: One can break-up the main effects and interactions into linear and quadratic effects. But for second order interactions like $A_lB_lC_l$, $A_lB_qC_l$, etc., it is cumbersome to find their values and sum of squares. Anyhow, even if they are worked out, their practicable interpretation may not be possible.

Example 4.2: The following table presents the data with regard to availability of calcium (mg/g) present in roots of maize from a 3^2 experiment conducted to measure the efficacy of three soil conditioners (Control – C, Phospho – Gypsum – PG and Pyrite – P) under three varying dilutions of effluent (E) discharged from a factory with well water (W) in the ratio of 0 : 1, 1 : 1 and 1 : 0. Let us denote C, PG and P by a_0, a_1 and a_2 and dilution ratios by b_0, b_1 and b_2, respectively. The experimental layout has been randomized block design with four replications.

Table-I Data table

Treat. Comb. (a b)	R_1	R_2	R_3	R_4	Total
0 0	16.0	21.6	18.4	31.2	87.2
0 1	17.6	20.0	23.2	19.2	80.0
0 2	17.6	24.8	28.8	23.2	94.4
1 0	21.5	22.4	16.8	21.6	82.3
1 1	20.8	24.0	16.0	19.2	80.0
1 2	14.4	9.6	12.0	12.8	48.8
2 0	21.6	20.8	21.6	28.8	92.8
2 1	21.6	18.4	17.6	19.2	76.8
2 2	14.4	7.2	13.6	16.0	51.2
Total	165.5	168.8	168.0	191.2	693.5

Detailed analysis of data has been done step by step so as to explicate the application of theory.

Analysis: Sum of squares are calculated through two-way table as well as by modulo technique. It has been shown that same values are obtained under both methods.

Table-II Two-way table

Factors	b_0	b_1	b_2	Total
a_0	87.2	80.0	94.4	261.6
a_1	82.3	80.0	48.8	211.1
a_2	92.8	76.8	51.2	220.8
Total	262.3	236.8	194.4	693.5

$$\text{C.F.} = \frac{(693.5)^2}{36} = 13359.50$$

Treatment S.S. $= \dfrac{1}{4}(87.2^2 + 80.0^2 + \ldots\ldots + 76.8^2 + 51.2^2) - \text{C.F.}$

$\qquad\qquad = 13900.36 - \text{C.F.} = 540.86$

S.S. due to A $= \dfrac{1}{12}(261.6^2 + 211.1^2 + 220.8^2) - \text{C.F.}$

$\qquad\qquad = 13479.20 - \text{C.F.} = 119.70$

S.S. due to B $= \dfrac{1}{12}(262.3^2 + 236.8^2 + 194.4^2) - \text{C.F.}$

$\qquad\qquad = 13555.57 - \text{C.F.} = 196.07$

S.S. due to A $\times$ B $= 540.86 - 119.70 - 196.07$

$\qquad\qquad = 225.09$

Again $\qquad (A)_0 = 00 + 01 + 02 = 87.2 + 80.0 + 94.4 = 261.6$

Similarly, $\qquad (A)_1 = 211.1$ and $(A)_2 = 220.8$

Obviously, the modulo method results into same values and hence sum of square for main effect A is as much through two way table. So is true for other main and interaction effects.

Now the sum of squares for AB and AB^2 are found out by modulo technique. First the component totals are worked out.

$(AB)_0 = 87.2 + 48.8 + 76.8 = 212.8$

$(AB)_1 = 80.0 + 82.3 + 51.2 = 213.5$

$(AB)_2 = 94.4 + 80.0 + 92.8 = 267.2$

$(AB^2)_0 = 87.2 + 80.0 + 51.2 = 218.4$

$(AB^2)_1 = 94.4 + 82.3 + 76.8 = 253.5$

$(AB^2)_2 = 80.0 + 48.8 + 92.8 = 221.6$

S.S. due to (AB) $= \dfrac{1}{12}(212.8^2 + 213.5^2 + 267.2^2) - \text{C.F.}$

$\qquad\qquad = 13521.83 - \text{C.F.} = 162.33$

S.S. due to $(AB^2) = \dfrac{1}{12}(218.4^2 + 253.5^2 + 221.6^2) - \text{C.F.}$

$\qquad\qquad = 13422.28 - \text{C.F.} = 62.78$

It is easy to verify that the aggregate of sum of squares due to AB and AB^2 is 225.11 which is equal to the sum of square due to A $\times$ B.

Rep. S.S. $= \dfrac{1}{9}(165.5^2 + 168.8^2 + 168.0^2 + 191.2^2) - \text{C.F.}$

$\qquad\qquad = 13407.24 - \text{C.F.} = 47.74$

Total S.S. $= 16.0^2 + 21.6^2 + \ldots\ldots + 13.6^2 + 16.0^2 - \text{C.F.}$

$\qquad\qquad = 14276.01 - \text{C.F.} = 916.51$

Error S.S. $= 916.51 - 47.74 - 119.70 - 196.07 - 225.10$

$\qquad\qquad = 327.90$

Table-III ANOVA table

Source	d.f.	S.S.	M.S.	F- value
Rep.	3	47.74	15.78	1.16 NS
A	2	119.70	59.85	4.38*
B	2	196.07	98.04	7.18*
A × B	4	225.11	56.28	4.12*
AB	2 ⎤	162.33	81.16	5.94*
AB^2	2 ⎦	62.78	31.39	2.30 NS
Error	24	327.90	13.66	
Total	35	916.51		

* Significant at 5% level

Table values of F at 5% level of significance: $F_{3,24} = 3.01$, $F_{2,24} = 3.40$, $F_{4,24} = 2.78$.

To compute the effects A_l, B_l, A_q, B_q and their interactions, the following table is to be prepared.

Table-IV Linear and quadratic effects (Ref. Table II)

B	$a_2 - a_1 = A_l$ 1 – 1 – 0	$a_2 + a_1 - 2a_0 = A_q$ 1 + 1 – 2	A	$b_2 - b_1 = B_l$ 1 – 1 – 0	$b_2 + b_1 - 2b_0 = B_q$ 1 + 1 – 2
b_0	92.8 – 82.3 =10.5	92.8 + 82.3 – 2 × 87.2 = 0.7	a_0	94.4 – 80.0 = 14.4	94.4 + 80.0 – 2 × 87.2 = 0
b_1	80.0 – 76.8 = –3.2	76.8 + 80.0 – 2 × 80.0 = –3.2	a_1	48.8 – 80.0 = –31.2	48.0 + 80.0 – 2 × 82.3 = –35.8
b_2	51.2 – 48.8 = 2.4	51.2 + 48.8 – 2 × 94.4 = –88.8	a_2	51.2 – 76.8 = –25.6	51.2 + 76.8 – 2 92.8 = – 57.6
Total	$A_l = 9.7$	$A_q = –91.3$		$B_l = –42.4$	$B_q = –93.4$

Table-V Linear and quadratic effect interactions

		A_l	A_q
B_1	0	2.4 – (–3.2) = 5.6	–88.8 – (–3.2) = –85.6
	–1	A_1B_1	A_qB_1
	1		
	–2	2.4 – 3.2 – 2 × 10.5 = –21.8	–88.8 – 3.2 – 2 × 0.7 = –93.4
B_q	1	A_1B_q	A_qB_a
	1		

The estimates of the effects can be obtained by dividing the effect value by $3 \times r$, *i.e.* $3 \times 4 = 12$.

Sum of squares for the effects are obtained by squaring the effect value and dividing it by proper divisor.

Table-VI Calculation of sum of squares for linear and quadratic effects and interactions

Effects	Divisors	For S.S.	For S.S.
A_l	$3r\{0^2+(-1)^2+1^2\}= 6r$	$A_l^2/6r$	$9.7^2/24 \quad = 3.92$
B_l	$3r\{0^2+(-1)^2+1^2\}= 6r$	$B_l^2/6r$	$(-42.4)^2/24 = 74.91$
A_lB_l	$r\{0^2+(1)^2+1^2\} \times$ $\{0^2+(-1)^2+1^2\} = 4r$	$(A_lB_l)^2/4r$	$(5.6)^2/16 \quad = 1.96$
A_q	$3r\{(-2)^2+1^2+1^2\} =18r$	$A_q^2/18r$	$(-91.3)^2/72 = 115.77$
B_q	$3r\{(-2)^2+1^2+1^2\} = 18r$	$B_q^2/18r$	$(-93.4)^2/72 = 121.16$
A_lB_q	$r\{0^2+(-1)^2+1^2\} \times$ $\{(-2)^2+1^2+1^2\} =12r$	$(A_lB_q)^2/12r$	$(-21.8)^2/48 = 9.90$
A_qB_l	$r\{(-2)^2+1^2+1^2\} \times$ $\{0^2+(-1)^2+1^2\} = 12r$	$(A_qB_l)^2/12r$	$(-85.6)^2/48 \quad = 152.65$
A_qB_q	$r\{(-2)^2+1^2+1^2\} \times$ $\{(-2)^2+1^2+1^2\} = 36r$	$(A_qB_q)^2/36r$	$(-93.4)^2/144 = 60.58$

It is easily verifiable that the aggregate of sum of squares due to A_l and A_q is equal to sum of square due to A and so is true for B. Similarly, the total of sum of squares for A_lB_l, A_lB_q, A_qB_l, A_qB_q is equal to the sum of square due to $A \times B$. Now the error sum of square for each component factor is calculated in the manner given below. For this, two-way tables are being prepared and corresponding error sum of squares are calculated just below each table.

Table-VII $A_l \times$ Replication table

Dilutions ratios	$a_2 - a_1$ Replications				Total
	I	*II*	*III*	*IV*	
b_0	0.1	−1.6	4.8	7.2	10.5
b_1	0.8	−5.6	1.6	0	−3.2
b_2	0	−2.4	1.6	3.2	2.4
Total	0.9	−9.6	8.0	10.4	9.7

$$\text{S.S. due to } (a_2 - a_1) \times \text{Dil.} = \frac{1}{8}\{10.5^2 + (-3.2)^2 + 2.4^2\} - \frac{9.7^2}{24}$$

$$= 15.78 - 3.92 = 11.86$$

Check: The grand total of the above table is 9.7 which is equal to $(a_2 - a_1)$ = 220.8 − 211.1 = 9.7.

This should always be verified as this ensures no calculative error. Also the total represents the contrast value.

$$\text{S.S. due to } (a_2 - a_1) = \frac{9.7^2}{4(1+1) \times 3} = 3.92$$

Table-VIII $A_q \times$ Replication table

Dilutions ratios	$a_2 + a_1 - 2a_0$ Replications				Total
	I	*II*	*III*	*IV*	
b_0	11.1	0	1.6	−12.0	0.7
b_1	7.2	2.4	−12.8	0	−3.2
b_2	−6.4	−32.8	−32.0	−17.6	−88.8
Total	11.9	−30.4	−43.2	−29.6	91.3

$$\text{S.S. due to } (a_2 + a_1 - 2a_0) \times \text{Dil.} = \frac{(-91.3)^2}{4(1+1+4) \times 3} = 115.77$$

Check: Total of sum of squares due to $(a_2 - a_1)$ and $(a_2 + a_1 - 2a_0)$ is 119.69 which is equal to sum of square for A.

The interaction of an effect with replications leads to error term for that effect. In quest of the same, four error terms are calculated below. The terms are Dil. × Rep., Cond. × Rep., A_l × Cond. × Rep., A_q × Cond. × Rep. Each term has 6 d.f. It can be checked that the aggregate of sum of squares of these terms is equal to error sum of square.

Similarly another possible break-up of error into four parts is, Dil. × Rep., Cond. × Rep., B_l × Dil. × Rep., B_q × dil. × Rep., each with six d.f. Again the aggregate of the sum of squares of these four terms equals to error sum of square S.S. due to $(a_2 - a_1) \times$ Dil. × Rep. = Table S.S. − C.F. − S.S. $(a_2 - a_1)$ −S.S. due to b's under A_l for differences.

$$= \frac{1}{2} \left\{ 0.1^2 + (-1.6)^2 + \ldots\ldots\ldots + 1.6^2 + 3.2^2 \right\} - \frac{9.7^2}{24}$$

$$- \frac{1}{6} \left\{ 0.9^2 + (-9.6)^2 + 8.0^2 + 10.4^2 \right\} + \frac{9.7^2}{24}$$

$$- \frac{1}{8} \left\{ 10.5^2 + (-3.2)^2 + 2.4^2 \right\} + \frac{9.7^2}{24}$$

$= 65.28 - 3.92 - 44.19 + 3.92 - 15.78 + 3.92$

$= 9.23$

S.S. due to $(a_2 + a_1 - 2a_0) \times$ Dil. x Rep. = Table S.S. $-$ C.F.

$-$ S.S. $(a_2 + a_1 - 2a_0) -$ S.S. due to b's under A_q

$$= \frac{1}{6}\left\{11.1^2 + 0^2 + - - - - - - + (-32.0)^2 + (-17.6)^2\right\} - \frac{(-91.3)^2}{72}$$

$$- \frac{1}{18}\left\{11.9^2 + (-30.4)^2 + (-43.2)^2 + (-29.6)^2\right\} + \frac{(-91.3)^2}{72}$$

$$- \frac{1}{24}\left\{0.7^2 + (-3.2)^2 + (-88.8)^2\right\} + \frac{(-91.3)^2}{72} -$$

$= 490.30 + 115.77 - 211.56 - 329.01$

$= 65.50$

Table-IX $B_l \times$ Rep. table

Conditioners	$b_2 - b_1$ Replications				Total
	I	*II*	*III*	*IV*	
a_0	0	4.8	5.6	4.0	14.4
a_1	-6.4	-14.4	-4.0	-6.4	-31.2
a_2	-7.2	-11.2	-4.0	-3.2	-25.6
Total	-13.6	-20.8	-2.4	-5.6	-42.4

S.S. due to $(b_2 - b_1) \times$ Cond. $\times$ rep. = Table S.S. $-$ C.F. $-$ S.S. $(b_2 - b_1)$

$-$ S.S. due to a's under b_l for differences.

$= 289.6 - 74.91 - 109.12 + 74.91 - 229.52 + 74.91$

$= 25.87$

Table-X $B_q \times$ Rep. table

Conditioners	$b_2 + b_1 - 2b_0$ Replications				Total
	I	*II*	*III*	*IV*	
a_0	3.2	1.6	15.2	-20.0	0
a_1	-7.8	-11.2	-5.6	-11.2	35.8
a_2	-7.2	-16.0	-12.0	-22.4	-57.6
Total	-11.8	-25.6	-2.4	-53.6	-93.4

S.S. due to $(b_2 + b_1 - 2b_0) \times$ Cond. $\times$ rep.

= Table S.S. $-$ C.F. $-$ S.S. $(b_2 + b_1 - 2b_0)$

$-$ S.S. due to a's under b_q for differences.

= $323.42 - 121.16 - 204.07 + 121.16 - 191.64 + 121.16$

= 48.87

Table-XI Dil. $\times$ rep. table

			Replication			
		I	*II*	*III*	*IV*	*Total*
	b_0	59.1	64.8	56.8	81.6	262.3
Dil.	b_1	60.0	62.4	56.8	57.6	236.8
	b_2	46.4	41.6	54.4	52.0	194.4
Total		165.5	168.8	168.0	191.2	693.5

S.S. due to (Dil. $\times$ Rep.) $= \dfrac{1}{3}$ (Table S.S.) $-$ C.F. $-$ S.S. (b's) $-$ S.S.(Rep.)

$$= \frac{1}{3}(59.1^2 + 64.8^2 + \text{.......} + 54.4^2 + 52.0^2) - \frac{693.5^2}{36} - 196.07 - 47.34$$

$$= 13720.43 - 13359.50 - 196.07 - 47.34 = 117.52$$

Table-XII Conditioners $\times$ Rep. table

			Replication			
		I	*II*	*III*	*IV*	*Total*
	a_0	51.2	66.4	70.4	73.6	261.6
Cond.	a_1	56.7	56.0	44.8	53.6	211.1
	a_2	57.6	46.4	52.8	64.0	220.8
Total		165.5	168.8	168.0	191.2	693.5

S.S. due to (Cond. $\times$ Rep.) $= \dfrac{1}{3}$ (Table S.S.) $-$ C.F. $-$ S.S.(a's) $-$ S.S. Rep

$$= \frac{1}{3}(51.2^2 + 66.4^2 + \text{..........} + 52.8^2 + 64.0^2) - \text{C.F.} - \text{S.S.(a's)} - \text{S.S. Rep.}$$

$$= 13662.99 - 13359.50 - 119.70 - 47.34 = 136.45$$

S.S. due to $(b_2 - b_1) \times$ Cond. $= \dfrac{1}{8}\{14.4^2 + (-31.2)^2 + (-25.6)^2\} - \dfrac{(-42.4)^2}{24}$

$$= 229.52 - 74.91 = 154.61$$

$$\text{S.S. due to } (b_2 + b_1 - 2b_0) = \frac{1}{24}\{0^2 + (-35.8)^2 + (-57.6)^2\} - \frac{(-93.4)^2}{72}$$

$$= 191.64 - 121.16 = 70.48$$

The error sum of square is rived into components to provide proper error for splitted main and interaction effects. After the break-up of experimental error it looks germane to test A_l and A_q against Condition × Replication and B_l and B_q against Dilution × Replication. Also the M.S. $A_l B_l$ be tested against A_l × Dil. × Rep.; $A_q B_l$ against A_q × Dil. × Rep. Similarly for other components proper error M.S. should be used.

Rudimentary approach of analysis in terms of A_l, A_q and B_l, B_q, may be in the following form.

Table-XIV ANOVA with linear and quadratic effects

Source	d.f.	S.S.	M.S.	F-value
Replications	3	47.34	15.78	1.15 NS
Treatments	8	540.87	67.61	4.94*
A_l	1	3.92	3.92	0.28 NS
A_q	1	115.77	115.77	8.46*
B_l	1	74.91	74.91	5.48*
B_q	1	121.16	121.16	8.85*
$A_l B_l$	1	1.96	1.96	0.14 NS
$A_l B_q$	1	9.90	9.90	0.72 NS
$A_q B_l$	1	152.65	152.65	11.16*
$A_q B_q$	1	60.58	60.58	4.43*
Error	24	328.30	13.68	
Total	35	916.51		

* Significant at $\alpha = 0.05$

Table values of F at 5% level of significance from table B-7 are:
$F_{1, 24} = 4.26$, $F_{3, 24} = 3.01$, $F_{8, 24} = 2.36$.

Table-XV Extended ANOVA table

Source	d.f.	S.S.	M.S.	F-value
Blocks	3	47.34	15.78	15.78/13.68 = 1.15 NS
Treatments	8	540.86	67.61	67.61/13.68 = 4.94*
(Cond.) A	2	119.70	59.85	59.85/22.74 = 2.63
$(a_2 - a_1)$	1	3.92	3.92	3.92/22.74 = 0.17 NS

(Contd.)

Source	d.f.	S.S.	M.S.	F-value
$(a_2 + a_1 - 2a)$	1	115.77	115.77	115.77/22.74 = 5.09 NS
(Dil.) B	2	196.07	98.04	98.04/56.28 = 1.74 NS
$(b_2 - b_1)$	1	74.91	74.91	74.91/19.87 = 3.77 NS
$(b_2 - b_1 - 2b_0)$	1	121.16	121.16	121.16/19.87 = 6.10*
A × B	4	225.10	56.28	56.28/13.68 = 4.11*
$(a_2 - a_1)$ × B	2	11.86	5.93	5.93/1.38 = 4.30 NS
$(a_2 + a_1 - 2a_0)$ × B	2	213.23	106.62	106.62/10.92 = 9.76*
or				
$(b_2 - b_1)$ × A	2	154.61	77.30	77.30/4.31 = 17.94*
$(b_2 + b_1 - 2b_0)$ × A	2	70.48	35.24	35.24/8.14 = 4.33 NS
Error	24	328.30	13.68	
A × Rep.	6	136.45	22.74	
B × Rep.	6	117.52	19.87	
$(a_2 - a_1)$ × B × Rep.	6	9.23	1.54	
$(a_2 + a_1 - 2a_0)$ × B × Rep.	6	65.50	10.92	
or				
$(b_2 - b_1)$ × A × Rep.	6	25.87	4.31	
$(b_2 + b_1 - 2b_0)$ A × Rep.	6	48.87	8.14	
Total	35	916.51		

* Significant at $\alpha = 0.05$

Table values of F at 5% level of significance, from table B-7, to test null hypothesis for various effects are:
$$F_{3,24} = 3.01, F_{2,24} = 3.40, F_{4,24} = 2.78, F_{1,6} = 5.99, F_{2,6} = 5.14,$$
$$F_{8,24} = 2.36, F_{4,24} = 2.78$$

Remarks:

1. Sum of squares due to $(a_2 - a_1)$ and $(a_2 + a_1 - 2a_0)$ is equal to the sum of square due to A. So is true for the effect B.

2. The interaction A × B has been splitted up as the interaction of linear effect of A × dilutions and quadratic effect of A × dilutions. It is apparent that the aggregate of sum of squares for these break-up interaction effects is same as that of A × B.

3. The interaction A × B has been splitted up into components, linear effect of B × conditioners and quadratic effect of B × conditioners. Their added sum of squares also constitute the sum of squares of due to A × B.

   The statements 2 and 3 given above substantiate one more concept that the interaction A × B is same as B × A.

4. Error term has been splitted into four components each with 6 d.f. The aggregate of sum of squares of these four components equals to error sum of square. Two sets of four components are constituted. One set contains the terms A × Rep., B × Rep., A_q × B × Rep., and A_l × dil. × Rep. The second set contains A × Rep., B × Rep., B_l × dil. × Rep., B_q × cond. × Rep. Both these sets of error components sum of squares separately add up to error sum of square. Such a break-up facilitates in testing the interaction effects more appropriately.

5. The factor A and its components are tested against A × Rep. term as error while Effect B and its components are tested against B × Rep.

6. Interaction A × B is tested against experimental error. Linear and quadratic components of A × B are tested against $(a_2 - a_1)$ × B × Rep. and $(a_2 + a_1 - 2a_0)$ × B × Rep. respectively. Similarly $(b_2 - b_1)$ × A and $(b_2 + b_1 - 2b_0)$ × A against $(b_2 - b_1)$ × A × Rep. and $(b_2 + b_1 - 2b_0)$ × A × Rep. respectively.

Discussion: For the example (4.2), three analysis of variance tables are prepared to test the significance of main effects and interactions. Table III reveals that the main effects A, B and interaction A × B are significant. Out of three levels of A and B, it can be tested with the help of lsd which of their levels differ significantly and which are at par.

Post-hoc tests

The standard error of A and B is same, i.e.

$$\text{S.E.} = \sqrt{\frac{s_e^2}{r \times 3}} = \sqrt{\frac{13.68}{4 \times 3}} = 1.06$$

And lsd $= \sqrt{2} \times 1.06 \times t_{.05,24} = \sqrt{2} \times 1.06 \times 2.064 = 3.09$

A level means in ascending order are,

a_1	a_2	a_0
17.59	18.40	21.8

The above diagram shows that average yield under control is significantly higher than that of application of PG and P. Hence it is recommended that there is no need of applying Phospho-gypsum and pyrates. It is economical not to apply any soil conditioners.

Similarly for B,

b_2	b_1	b_0
16.20	19.73	21.85

The comparison of E : W shows that yield under irrigation from well water is at par with equal ratio mixture of effluent and well water. Whereas the yield under irrigation from effluent is significantly lower than other two types of irrigations. Interaction A × B is also significant indicating that combination of dilutions with irrigation levels is one that gives higher yield and is most economical. For this calculate the standard error of interaction mean and lsd for comparing interaction means.

$$\text{S.E. of an interaction A} \times \text{B mean} = \sqrt{\frac{13.68}{4}} = 1.85$$

For comparing interaction means, lsd = $\sqrt{2} \times 1.85 \times 2.064 = 5.40$

To select the best combination, start with the highest combination mean and go to the next lower value and see whether the difference between these is more than the critical value 5.40, then there is no need to go for the next lower value. If it is less than the critical value then proceed to the next lower value and compare its difference from first mean value with lsd and decide in the same manner. Continue the process till a non-significant difference occurs or all means are exhausted. Out of all at par values, choose the combination which is most economical. It is worth pointing out that it is not possible to give separate recommendations on the basis of AB and/or AB^2. They are of academic interest and play an important role in many other ways.

Table-XIII presents the analysis of data in a different form. This approach was initiated by F. Yates. Both the factors have three levels and hence they are splitted into linear and quadratic components. These effects and their interactions are tested against experimental error. Significant values are marked with stars. Significance of linear effects shows that the response proportionately increases as the level of a factor increases whereas significance of quadratic effect ensures that there is stagnancy in response after certain increase in levels. Also some idea can be had about the trend from the interaction of linear and quadratic effects but a definite recommendation may not be feasible.

In ANOVA table-XIV, all the effects are tested against experimental error. Its appropriateness is in jeopardy. So an elaborated analysis is carried out and displayed in table-XV in which the error is splitted into components for the linear and quadratic effects and their interactions separately. Also the interaction A × B has been splitted into two sets and each component is tested against its respective error. In calculating the error terms, one basic idea has been utilized, that is, the interaction of a treatment with replication provides the error for that treatment. Conclusions can be drawn in the same way as they have been discussed for AVOVA Table-XIV.

4.6 ASYMMETRICAL FACTORIALS

Explanation: As already defined, if all factors are not having equal number of levels, the experiment is called an *asymmetrical factorial* experiment. A factorial design with differing number of factor levels is also known as *mixed* *factorial.* It needs special attention as one common rule cannot be applied in all cases. The factorials of the type $2 \times 3 \times 4$, $2 \times 2 \times 3$, $2 \times 3 \times 3$ or $2 \times 3 \times 5 \times 4$ are asymmetrical factorials. Generally, asymmetrical factorial experiments contain two or three factors. In such cases, the west approach to work out sum of squares for various factors is through two and three way tables. If there are three factors, second order interactions can be obtained by subtracting the sum of squares for main effects and first order interactions from the treatment sum of square. In case of four factors there is no use of finding the sum of squares for second and third order interactions as it is not possible to draw any conclusions from these interactions of practical relevance.

The skeleton ANOVA table with three factors A, B and C having levels 2, 3 and 4 respectively laid out in randomized block design having r blocks shall be as follows:

Table 4.6.1 Skeleton ANOVA

Source	d.f.
Replications	$r - 1$
Treatments	23
A	1
B	2
A × B	2
C	3
A × C	3
B × C	6
A × B × C	6
Error	$23 (r - 1)$
Total	$24 - 1$

All factors in the above table are tested against error. The analysis procedure is illustrated through examples.

Example 4.3: A field experiment was conducted to study the response of sorghum yield to biofertilizers. Combinations of three nitrogen levels n_0 n_1 and n_2 (0, 30, 60 kg/h) and four biofertilizers b_0, b_1, b_2, and b_3 (Control, Azospirillum 1, Azospirillum 2, Azotobector) were taken and assigned randomly to 12 plots of size 4.5×5.0 Sq.m in a block. In all there were three blocks. The variable measured was 1000 grain weight in grams. The data for the twelve combinations in three replicates are tabulated below along with totals.

Table-I Data table

Treatment combinations	Blocks I	Blocks II	Blocks III	Total
$n_0 b_0$	30.00	29.00	30.70	89.70
$n_0 b_1$	28.50	29.40	30.20	88.10
$n_0 b_2$	28.40	28.18	28.92	85.50
$n_0 b_3$	27.50	27.98	27.02	82.50
$n_1 b_0$	31.74	29.18	31.22	92.14
$n_1 b_1$	30.56	29.92	29.84	90.32
$n_1 b_2$	29.30	30.21	31.94	91.45
$n_1 b_3$	30.42	29.48	29.00	88.90
$n_2 b_0$	30.78	32.64	31.86	95.28
$n_2 b_1$	30.75	30.98	31.35	93.08
$n_2 b_2$	32.42	30.25	29.98	92.65
$n_2 b_3$	31.76	31.03	32.54	95.33
Total	362.13	358.25	364.57	1084.95

The analysis of data has been carried out as asymmetrical factorial in randomized design. A two way table can be prepared as follows.

Table-II Cross-table

N	B b_0	B b_1	B b_2	B b_3	Total
n_0	89.70	88.10	85.50	82.50	345.80
n_1	92.14	90.32	91.45	88.90	362.81
n_2	95.28	93.08	92.65	95.33	376.34
Total	277.12	271.50	269.60	266.73	1084.95

Now sum of squares are worked out using the above table.

$$C.F. = \frac{(1084.95)^2}{36} = 32697.68$$

$$\text{Total S.S.} = 30.00^2 + 28.50^2 + \dots\dots + 29.98^2 + 32.54^2 - C.F.$$
$$= 32769.58 - C.F. = 71.90$$

$$\text{Block S.S.} = \frac{1}{12}\left(362.13^2 + 358.25^2 + 364.57^2\right) - C.F.$$
$$= 32699.37 - C.F. = 1.69$$

$$\text{Treat. S.S.} = \frac{1}{3}\left(89.70^2 + 88.10^2 + \ldots\ldots + 92.65^2 + 95.33^2\right) - \text{C.F.}$$

$$= 32750.66 - \text{C.F.} = 52.98$$

$$\text{Nitrogen (N) S.S.} = \frac{1}{12}\left(345.80^2 + 362.81^2 + 376.34^2\right) - \text{C.F.}$$

$$= 32736.71 - \text{C.F.} = 39.03$$

$$\text{Biofertilizer (B) S.S.} = \frac{1}{9}\left(277.12^2 + 271.50^2 + 269.60^2 + 266.73^2\right) - \text{C.F.}$$

$$= 32704.09 - \text{C.F.} = 6.41$$

$$\text{(N} \times \text{B) S.S.} = \text{Treat.S.S.} - \text{S.S.(N)} - \text{S.S.(B)}$$

$$= 52.98 - 39.03 - 6.41 = 7.54$$

$$\text{Error S.S.} = 71.90 - 1.69 - 52.98 = 17.23$$

Using the above sum of squares, analysis of variance can be carried out as follows.

Table-III ANOVA table

Source	d.f.	S.S.	M.S.	F–value
Blocks	2	1.69	0.845	1.08
Treatments	11	52.98		
Nitrogen (N)	2	39.03	19.52	25.02*
Biofertilizers (B)	3	6.41	2.14	2.74 NS
N × B	6	7.54	1.26	1.62 NS
Error	22	17.23	0.78	
Total	35	71.90		

* Significant at $\alpha = 0.05$

Table values of F at 5% level of significance are:

$$F_{2,22} = 3.44, \quad F_{3,22} = 3.05, \quad F_{6,22} = 2.55$$

On comparison of calculated F-values with respective tabulated values for main effects and interaction, it is found that only the nitrogen levels differ significantly. For comparing the paired nitrogen level means, the standard error of the difference between two nitrogen level means is,

$$\text{S.E.}(d) = \sqrt{\frac{2 \times 0.78}{3 \times 4}} = 0.36$$

and $\qquad$ lsd $= 0.36 \times t_{.05,\,22} = 0.36 \times 2.074 = 0.75$

Nitrogen level means in ascending order are,

n_0	n_1	n_2
28.82	30.23	31.36

All paired differences between mean responses at various levels of nitrogen are greater than lsd value. The figures show that 1000 grain weight of sorghum increases as the application dose of nitrogen fertilizer increases. Biofertilizer show no effect on 1000 grain weight. Also effectiveness of various levels of nitrogen fertilizers is independent of biofertilizers and vice versa.

4.7 FACTORIAL EXPERIMENTS WITH ADDITIONAL TREATMENTS

Orientation: Many times experiments are conducted with treatments which contain a set of factorial combinations and also some non-factorial treatments. Such designs in literature are often called *augmented designs*. Such experiments facilitate to compare individual treatments which sometimes work as a control and also can be compared among themselves. Factorial set provides to estimate and tests main and interaction effects. Further a comparison of non-factorial set versus factorial set is also possible. The analysis procedure remains almost same as for other experiments in general. Anyhow, without going into theoretical aspects of it, the analysis procedure is expatiated through the following solved example.

Example 4.4: A field experiment was laid out in randomized block design with four blocks to investigate the effect of phosphorus on soybean yield by applying two grades of rock phosphates (Low grade and high grade phosphates) incubated for three periods, 20, 30 and 40 days, in combination with two types of solubilising agents (Farm yard manure and phosphobactrine). Four single treatments were taken as control namely, absolute control, SSP (Super phosphate), LGRP (Low grade phosphorus) and HGRP (High grade phosphorus). In this way in all sixteen treatments were taken. This set consists of 12 treatment combinations emerging out of $2 \times 3 \times 2$ factorial treatment combinations and four single treatments. Let the grades of phosphates be denoted by g_1 and g_2, incubation periods by p_1, p_2 and p_3, solubilizing agents by s_1 and s_2.

The yield of soybean (q/h) in four replications was as tabulated below. Treatment and block totals are also given alongside.

Table-I Data table

Treatments	Seed yield				Total
	R_1	R_2	R_3	R_4	
Abs. control	7.44	6.83	7.02	7.01	28.30
SSP unincubated	9.83	10.35	9.05	9.69	38.92
LGRP unincubated	9.15	8.72	8.22	9.38	35.47

(Contd.)

Treatments	Seed yield				Total
	R_1	R_2	R_3	R_4	
HGRP unincubated	9.24	10.58	9.88	10.74	40.44
$g_1 p_1 s_1$	11.38	12.08	10.77	10.44	44.67
$g_1 p_2 s_1$	10.91	10.33	9.58	9.88	40.70
$g_1 p_3 s_1$	11.08	12.41	10.22	10.49	44.20
$g_2 p_1 s_1$	11.24	10.63	11.36	11.44	44.67
$g_2 p_2 s_1$	13.33	13.30	13.88	14.74	55.25
$g_2 p_3 s_1$	9.80	13.30	12.16	12.33	47.59
$g_1 p_1 s_2$	7.55	7.74	6.47	6.88	28.64
$g_1 p_2 s_2$	7.72	8.61	7.91	7.72	31.96
$g_1 p_3 s_2$	11.72	11.94	11.80	11.80	47.26
$g_2 p_1 s_2$	10.52	10.99	9.47	10.66	41.64
$g_2 p_2 s_2$	9.36	11.86	9.33	12.77	43.32
$g_2 p_3 s_2$	11.22	12.16	11.05	11.41	45.84
Total	161.49	171.83	158.17	167.38	658.87

To analyze the experimental data, it seems convenient to give first the analysis of variance table and the components of variation sum of squares are placed after calculating them as per requirement.

Table-II ANOVA table

Source	d.f.	S.S.	M.S.	F-value
Blocks	3	6.94	2.31	4.53*
Treatments	15	195.67	13.04	25.57*
Grades (G)	1	34.82	34.82	69.27*
Solu. bactrine (S)	1	30.75	30.75	60.29*
S × G	1	0.52	0.52	1.02 NS
Periods (P_d)	2	20.00	10.00	19.60*
$P_d \times G$	2	17.94	8.97	17.59*
$P_d \times S$	2	18.77	9.38	18.39*
$P_d \times G \times S$	2	12.12	6.06	11.88*
Nonfact. treat.	3	21.90	7.30	14.31*
Nonfact. Treat. vs. rest	1	38.84	38.84	76.61*
Error	45	22.99	0.51	
Total	63	225.60		

* Significant at p = 0.05

Tabulated F-values at 5 % level of significance, from table B-7 are,
$F_{.05;\ 1,\ 45} = 4.06$, $F_{.05;\ 2,\ 45} = 3.21$, $F_{.05;\ 3,\ 45} = 2.82$, $F_{.05;\ 15,45} = 1.90$
Stepwise calculations of sum of squares are as follows:

$$C.F. = \frac{(658.87)^2}{64} = 6782.96$$

Total S.S. $= 7.44^2 + 9.83^2 + \ldots\ldots\ldots + 12.77^2 + 11.41^2 - C.F.$
$$= 7008.56 - C.F. = 225.60$$

Blocks S.S. $= \frac{1}{16}\left(161.49^2 + 171.83^2 + 156.17^2 + 167.38^2\right) - C.F.$
$$= 6789.90 - C.F. = 6.94$$

To calculate sum of squares due to G, S, P_d, G × S, P_d × G and P_d × S, prepare two way tables. Sum of square due to second order interaction P_d × G × S will be obtained by subtracting the sum of squares due to main effects and first order interactions from treatment sum of square.

Table-III G × S interaction table

Factor	s_1	s_2	Total
g_1	129.57	107.86	237.43
g_2	147.51	130.80	278.31
Total	277.08	238.66	515.74

$$S.S.\ (G) = \frac{1}{24}\left(237.43^2 + 278.31^2\right) - \frac{515.74^2}{48}$$
$$= 5576.23 - 5541.41 = 34.82$$

$$S.S.\ (S) = \frac{1}{24}\left(277.08^2 + 238.66^2\right) - \frac{515.74^2}{48}$$
$$= 5572.16 - 5541.41 = 30.75$$

$$S.S.\ (S \times G) = \frac{1}{12}\left(129.57^2 + 107.86^2 + 147.51^2 + 130.80^2\right) - \frac{515.74^2}{48}$$
$$- 34.82 - 30.75$$
$$= 5607.50 - 5541.41 - 34.82 - 30.75 = 0.52$$

Table-IV $P_d \times G$ interaction table

Factor	g_1	g_2	Totals
p_1	73.31	86.31	159.62
p_2	72.66	98.57	171.23
p_3	91.46	93.43	184.89
Total	237.43	278.31	515.74

$$\text{S.S. } (P_d) = \frac{1}{16}\left(159.62^2 + 171.23^2 + 184.89^2\right) - \frac{515.74^2}{48}$$

$$= 5561.41 - 5541.41 = 20.00$$

$$\text{S.S } (P_d \times G) = \frac{1}{8}(73.31^2 + 86.31^2 + \ldots\ldots + 91.46^2 + 93.43^2) - \frac{515.74^2}{48}$$
$$- 34.82 - 20.00$$
$$= 5614.17 - 5541.41 - 34.82 - 20.00 = 17.94$$

Table-V $P_d \times S$ interaction table

Factor	s_1	s_2	Total
p_1	89.34	70.28	159.62
p_2	95.95	75.28	171.23
p_3	91.79	93.10	184.89
Total	277.08	238.66	515.74

$$\text{S.S. } (P_d \times S) = \frac{1}{8}\left(89.34^2 + 70.28^2 + 91.79^2 + 93.10^2\right) - \frac{515.74^2}{48}$$
$$- 20.00 - 30.75$$
$$= 5610.93 - 5541.41 - 20.00 - 30.75 = 18.77$$

Factorial effects S.S. $= \frac{1}{4}\left(44.67^2 + 40.70^2 + \ldots\ldots + 43.32^2 + 45.84^2\right)$

$$-\frac{515.74^2}{48}$$

$$= 5676.33 - 5541.41 = 134.92$$

S.S. $(P_d \times S \times G) = 134.92 - 34.82 - 30.75 - 20.00 - 0.52 - 17.94 - 18.77$
$$= 12.12$$

$$\text{S.S. (Treatments)} = \frac{1}{4}\left(28.30^2 + 38.92^2 + \ldots\ldots + 43.32^2 + 45.84^2\right) - \frac{658.87^2}{64}$$

$$= 6978.63 - 6782.96 = 195.67$$

S.S. due to non-factorial treatments

$$= \frac{1}{4}\left(28.30^2 + 38.92^2 + 35.47^2 + 40.44^2\right) - \frac{143.13^2}{16}$$

$$= 1302.29 - 1280.39 = 21.90$$

∴ Total of non-factorial effects

$$= 28.30 + 38.92 + 35.47 + 40.44 = 143.13$$

Total of factorial set = 44.67 + 40.70 + ... + 43.32 + 45.84 = 515.74

S.S. due to non-factorial set versus factorial set

$$= \frac{143.13^2}{16} + \frac{515.74^2}{48} - \frac{658.87^2}{64}$$

$$= 1280.39 + 5541.41 - 6782.96 = 38.84$$

Remark: Sum of square due to non-factorial set versus factorial set can also be calculated through contrast in the following manner.

$$\text{S.S.} = \frac{\left[3(28.30 + 38.92 + 35.47 + 40.44) - (44.67 + 40.70 \ldots\ldots\ldots + 45.84)\right]^2}{(3^2 \times 4 + 12) \times 4}$$

$$= \frac{(3 \times 143.12 - 515.74)^2}{192} = 38.84$$

Comparing the calculated F-values for different sources of variation given in the analysis of variance table, it becomes apparent that except the interaction S × G, all factors are significant. Significance of block differences can be construed as a positive contribution towards reducing the experimental error.

Standard errors for different effects are first worked out. Later the same shall be utilized to find out the least significant differences.

$$\text{S.E. of main effects S and } G = \sqrt{\frac{0.51}{4 \times 3 \times 2}} = 0.146$$

$$\text{S.E. of main effect of } P_d = \sqrt{\frac{0.51}{4 \times 2 \times 2}} = 0.178$$

$$\text{S.E. of interactions } P_d \times G \text{ and } P_d \times S = \sqrt{\frac{0.51}{4 \times 2}} = 0.252$$

$$\text{S.E. of interaction } P_d \times S \times G = \sqrt{\frac{0.51}{4}} = 0.357$$

$$\text{lsd for } P_d = \sqrt{2} \times 0.178 \times 2.016 = 0.507$$

Since, $t_{.05,\ 45} = 2.016$

lsd for interactions $P_d \times G$ and $P_d \times S = \sqrt{2} \times 0.252 \times 2.016 = 0.718$

lsd for $P_d \times S \times G$ has not been worked out as it is not practically feasible to recommend a 3-factor combination as best. Significance of second order interaction simply reveals that the presence of one factor does affect the efficacy of other factor when applied together.

lsd for S and G are not required as there have only two levels. Their significance ifso-facto ensures the superiority of higher level of a factor over its lower level.

To compare the performance of three different incubation periods, arrange the mean yields of soybean in descending order and compare pairwise differences of means with lsd 0.507 for periods.

p_3	p_2	p_1
11.56	10.70	9.98

On comparing, it is found that all p_3, p_2, p_1 incubation periods differ from each other significantly. It means that greater incubation period enhances the yield.

For the interaction, $P_d \times G$, start with highest mean value of pairs for

$P_d \times G$, i.e. $\dfrac{1}{8}(98.57 - 93.43) = 0.64$, which is less than lsd value, 0.718.

Now choose the next pairs and find the average mean difference, i.e.

$\dfrac{1}{8}(98.57 - 91.46) = 0.89$ which is greater than 0.718. so rest all differences

will also be significant. It means that one should choose combination either $p_2\, g_2$ or $p_3\, g_2$ whichever is economical. In this case, it should be $p_2\, g_2$.

Similarly proceeding for $P_d \times S$, calculate the mean difference for highest

combination values, i.e. $\dfrac{1}{8}(95.95 - 93.10) = 0.36$, which is less than 0.718.

For the next combination, the difference is, $\dfrac{1}{8}(95.95 - 91.79) = 0.82$ which

is also non-significant. The next difference is, $\dfrac{1}{8}(95.95 - 89.34) = 0.82$,

which is significant. Similarly the difference, $\dfrac{1}{8}(93.10 - 91.79) = 0.16$ is also

non-significant.

The above values of average differences lead to the conclusion that the combinations $p_2\, s_1, p_3\, s_2, p_3\, s_1$ are at par and thus, out of these combinations most suited one be recommended.

$$\boxed{\textbf{QUESTIONS AND EXERCISES}}$$

1. What does factorial experiments mean?
2. Define main effect and interaction. How can they be presented diagrammatically for 2×2 and 2×3 factorial experiments?
3. Differentiate between symmetrical and asymmetrical factorials.
4. How can the main effects and interactions be represented by contrasts.
5. In what manner, the contrasts help in calculating the sum of squares due to main and interaction effects?
6. Describe Yates' method of analysis of 2^n factorials and give the advantages of this method.
7. Write the treatment combinations for factorial experiments of order (a) 2×3 (b) 2^3 (c) $4 \times 3 \times 2$ (d) $2^2 \times 3$.
8. In what situation, it is more appropriate to use pooled error term in testing a hypothesis about treatment means?
9. Write expected mean squares for 3- factor factorial experiment, with factors A, B and C having levels a, b and c respectively, in randomized block design with r blocks. The treatment A is fixed and the treatments B and C are of random nature.
10. What are the advantages of selecting a priori contrasts and testing their significance?
11. An experiment was conducted to compare the effects of three treatments, when applied in combination, on fenugreek yield. There were three row spacing (20, 30 and 40 cm.), denoted by S_1, S_2 and S_3. Also three phosphorus levels were taken, i.e. 20, 40 and 60 kg / ha denoted by P_1, P_1, P_3 and two methods of inoculation were used namely, no inoculation, I_0 and seed inoculation, I_1. In all $3 \times 3 \times 2 = 18$ treatment combinations were applied in a randomized block design taking three replications R_1, R_2 and R_3. Fenugreek yield per hectare was recorded and tabulated as given below.

Treatment combinations	Fenugreek yield (q/ha)		
	R_1	R_2	R_3
$S_1 P_1 I_0$	10.1	10.0	9.9
$S_1 P_1 I_1$	11.5	11.2	10.0
$S_1 P_2 I_0$	10.2	9.8	12.5
$S_1 P_2 I_1$	11.5	12.0	10.5
$S_1 P_3 I_0$	12.0	10.2	11.4
$S_1 P_3 I_1$	10.2	13.0	11.0

(Contd.)

Treatment combinations	Fenugreek yield (q/ha) R_1	R_2	R_3
$S_2 P_1 I_0$	10.2	10.8	11.0
$S_2 P_1 I_1$	11.7	13.0	12.5
$S_2 P_2 I_0$	11.0	11.8	13.4
$S_2 P_2 I_1$	12.7	12.0	13.9
$S_2 P_3 I_0$	13.5	11.8	12.1
$S_2 P_3 I_1$	12.1	14.0	13.0
$S_3 P_1 I_0$	9.8	10.4	11.6
$S_3 P_1 I_1$	12.8	12.1	11.9
$S_3 P_2 I_0$	12.5	11.0	12.0
$S_3 P_2 I_1$	13.4	13.0	12.5
$S_3 P_3 I_0$	13.3	11.9	12.4
$S_3 P_3 I_3$	12.3	13.1	14.0

Analyze the experimental data and comment on the performance of main effects and interactions. Also test the significance of linear and quadratic effects of various factors.

12. A set of 2^3 factorial treatments of three fertilizers N, P and K each at two levels, i.e. nitrogen at 0 and 20 kg/ha, Phosphorus 0 and 40 kg/ha and potash 0 and 20 kg/ha, was applied to experimental plots to test the effects of fertilizers on soybean yield individually as well in combination to others. The lay out of the experiment was randomized block design having four replications. Replicationwise yield (q/ha) of soybean is presented in the following table.

Treatment combinations	Soybean yield q/ha R_1	R_2	R_3	R_4
$N_0 P_0 K_0$	10.10	10.23	10.34	10.25
$N_0 P_0 K_1$	10.45	10.41	10.57	10.68
$N_0 P_1 K_0$	11.25	11.27	11.36	12.45
$N_0 P_1 K_1$	12.31	12.35	12.99	12.76
$N_1 P_0 K_0$	12.27	12.74	12.91	13.02
$N_1 P_0 K_1$	12.94	12.79	13.00	13.41
$N_1 P_1 K_0$	13.43	12.96	14.38	14.90
$N_1 P_1 K_1$	14.96	13.78	14.67	14.62

Analyze the experimental data by Yates' method and interpret the results.

13. Following table presents the data with regard to 1000 – grain weight in grams of sorghum obtained from a 3^2 factorial experiment in which the

combinations of three levels of nitrogen 0, 1, 2 (0, 30,60 kg/ha), three types of biofertilizers 0, 1, 2 (Control, Azotobector, Azospirillum) were tried in a randomized design with three blocks.

Treatment combinations	1000 grain weight (g) R_1	R_2	R_3
0 0	30.00	29.00	30.70
0 1	28.50	29.40	30.20
0 2	28.40	28.18	28.92
1 0	27.50	27.98	27.02
1 1	31.74	29.18	31.22
1 2	30.56	29.92	29.84
2 0	29.30	30.21	31.92
2 1	30.78	32.64	31.86
2 2	30.75	30.90	31.35

Prepare the detailed analysis of variance table for the above experimental data and interpret the results.

14. Under what condition, it would be appropriate to merge 3rd order interaction with experimental error.

15. Following table furnishes the data of 4×5 factorial experiment to assess the effect of four fertilizer levels and five sources of fertilizers on green cob yield of sweet corn. The design of the experiment was randomized block design with three replications. Fertilizer levels: F_1 – 75% of recommended dose, F_2 – 100% of recommended dose, F_3 – 125% recommended dose, F_4 – 150% of recommended dose.

Sources: S_1 – 100% inorganic source, S_2 – 100% organic fertilizer, S_3 – 50% inorganic and 50% organic fertilizer,
S_4 – 40% inorganic + 40% organic fertilizer + Azo + PSB inoculation
S_5 – 25% inorganic + 75% organic fertilizer + Azo + PSB inoculation

Data Table
Green cob yield (q/ha)

Treatments	R_1	R_2	R_3	Treats. contd.	R_1	R_2	R_3
$F_1 S_1$	75.2	70.3	80.4	$F_3 S_1$	76.5	88.3	82.1
$F_1 S_2$	69.9	65.3	60.3	$F_3 S_2$	72.1	65.1	72.6
$F_1 S_3$	64.9	75.4	70.3	$F_3 S_3$	81.9	76.1	70.0
$F_1 S_4$	63.9	74.3	69.5	$F_3 S_4$	80.9	75.1	69.1
$F_1 S_5$	71.3	63.4	77.4	$F_3 S_5$	70.0	84.3	77.3

(Contd.)

Treatments	R_1	R_2	R_3	Treats. contd.	R_1	R_2	R_3
$F_2 S_1$	88.6	80.3	72.5	$F_4 S_1$	83.4	76.5	89.3
$F_2 S_2$	63.9	76.5	70.6	$F_4 S_2$	80.3	78.1	72.2
$F_2 S_3$	75.2	70.3	80.4	$F_4 S_3$	86.1	80.1	70.4
$F_2 S_4$	74.1	69.2	79.4	$F_4 S_4$	85.0	79.1	73.3
$F_2 S_5$	81.9	76.1	70.0	$F_4 S_5$	86.4	81.2	76.3

Analyze the experimental data and draw conclusions about main effects and interactions and relevant contrasts.

16. In an experiment, soil organic carbon was measured at the time of harvesting of wheat crop in all plots of a randomized block design having three replications. The experiment included 24 treatment combinations comprising of four sources of organic nutrients (N), three levels of organic fertilizers (F) and two levels of compost (C).

Details of the treatments

Organic nutrients (N):

N_1 – Azotobacter + Cellulomonous + Trichloderma with high grade rock phosphate.

N_2 – Azotobacter + Cellulomonous + Trichloderma with low grade rock phosphate.

N_3 – Azotobacter + Cellulomonous with high grade rock phosphate.

N_4 – Azotobacter + Cellulomonous with low grade rock phosphate.

Doses of fertilizer (F):

F_1 – No fertilizer NPK

F_2 – 75 % NPK of recommended dose

F_3 – 100 % NPK of recommended dose

Levels of compost (C):

C_1 – 0.5 Ton/ha

C_2 – 1.0 Ton/ha

Tabulated data are as follows:

Data Table

Soil organic percentage

Treatments	R_1	R_2	R_3	Treatments	R_1	R_2	R_3
$N_1 F_1 C_1$	0.70	0.70	0.72	$N_1 F_1 C_2$	0.90	0.80	0.95
$N_1 F_2 C_1$	0.82	0.91	0.83	$N_1 F_2 C_2$	0.90	0.99	1.10
$N_1 F_3 C_1$	0.84	0.97	0.85	$N_1 F_3 C_2$	1.04	1.03	0.98
$N_2 F_1 C_1$	0.65	0.72	0.69	$N_2 F_1 C_2$	0.94	0.86	0.81

(Contd.)

Treatments	R_1	R_2	R_3	Treatments	R_1	R_2	R_3
$N_2F_2C_1$	0.84	0.80	0.91	$N_2F_2C_2$	1.03	0.99	0.91
$N_2F_3C_1$	0.89	0.84	0.88	$N_2F_3C_2$	1.03	1.06	0.95
$N_3F_1C_1$	0.77	0.70	0.70	$N_3F_1C_2$	0.92	0.80	0.90
$N_3F_2C_1$	0.81	0.77	0.84	$N_3F_2C_2$	0.92	0.97	1.05
$N_3F_3C_1$	0.81	0.91	0.80	$N_3F_3C_2$	0.99	1.07	0.95
$N_4F_1C_1$	0.75	0.68	0.75	$N_4F_1C_2$	0.84	0.92	0.86
$N_4F_2C_1$	0.81	0.74	0.87	$N_4F_2C_2$	0.80	0.95	1.08
$N_4F_3C_1$	0.81	0.83	0.95	$N_4F_3C_2$	0.94	0.97	0.90

Analyze the data and write all possible conclusions which can be drawn from the results of analysis of data.

17. Table below provides the data regarding average number of pods of gram per plant (average of five plants per plot chosen randomly) of a 2 × 5 factorial experiment conducted in randomized design with three replications.

Details of treatments:

Treatment A: A_1 – No seed priming, A_2 – Seed priming.

Treatment B: B_1 – No urea spray.

B_2 – Urea spray at vegetative stage.

B_3 – Urea spray at flower initiation stage.

B_4 – Urea spray at pod formation stage.

B_5 – First spray at flower initiation stage and second spray at pod formation stage.

Data Table

Pods per plant (Gram)

Treatment	R_1	R_2	R_3
A_1B_1	58	53	54
A_1B_2	56	60	58
A_1B_3	66	64	68
A_1B_4	59	48	56
A_1B_5	61	52	59
A_2B_1	50	60	41
A_2B_2	55	51	58
A_2B_3	63	61	65
A_2B_4	51	53	51
A_2B_5	60	64	61

Do the detailed analysis of data and interpret the results.

18. What alternative approach can be adapted to deal with analysis of variance when exact F-tests for certain factors are not available?

19. A field trial was carried out to assess the effect of balanced fertilization, varieties and biofertilizers on seed yield of fenugreek. The trial was planned with levels of fertilizers (A_0 – Control, A_1 – N20 + P20, A_2 – N40 + P40); two varieties V_1 and V_2 and four types of biofertilizers (C_0 – Control, C_1 – Rhizobium, C_2 – PSB, C_3 – Rhizobium + PSB) in randomized block design taking three complete blocks. In this trial, it was considered that fertilizer levels are of fixed nature whereas varieties and biofertilizers are of random nature.

Data with regard to seed yield (q/ha) were as tabulated below.

Treatment

Combinations	R_1	R_2	R_3
$A_0V_1C_0$	9.86	9.05	11.50
$A_0V_1C_1$	11.31	12.40	11.13
$A_0V_1C_2$	13.75	9.55	10.86
$A_0V_1C_3$	10.66	11 72	13.72
$A_0V_2C_0$	11.25	15.93	10.50
$A_0V_2C_1$	10.06	13.80	15.97
$A_0V_2C_2$	11.86	14.36	12.80
$A_0V_2C_3$	13.32	14.63	11.66
$A_1V_1C_0$	13.41	13.51	14.00
$A_1V_1C_1$	16.06	14.22	11.38
$A_1V_1C_2$	12.80	14.18	16.11
$A_1V_1C_3$	16.44	13.51	15.05
$A_1V_2C_0$	17.26	14.35	14.18
$A_1V_2C_1$	15.08	14.46	19.13
$A_1V_2C_2$	18.19	15.04	15.77
$A_1V_2C_3$	15.48	16.29	20.43
$A_2V_1C_0$	15.52	12.06	17.13
$A_2V_1C_1$	14.93	18.85	17.43
$A_2V_1C_2$	17.58	15.38	17.82
$A_2V_1C_3$	16.94	20.03	20.36
$A_2V_2C_0$	13.06	18.37	17.67
$A_2V_2C_1$	20.99	18.85	17.63
$A_2V_2C_2$	20.66	19.27	16.97
$A_2V_2C_3$	20.80	19.97	19.47

Do the analysis of data accurately utilizing Satterthwaite approximate F'-statistics. Also discuss the results exhaustively.

20. Can there be fractional degrees of freedom for F-statistics? If so, how to get the critical values of F - distribution?

21. A $2 \times 3 \times 4$ factorial experiment with two varieties: (V_1 - Raj. 1555 and V_2 - HI 8498); three nitrogen levels: (90, 105, 120 kg/ha) and four sprays: (0, 500, 1000, 1500 ppm concentration of Thiourea) was laid out in randomized block design taking three blocks to see the main and interaction effects of the treatments on nitrogen percentage in wheat grain. Replicationwise data for 24 treatment combinations are given below.

Percentage nitrogen in wheat grain

Treatment	R_1	R_2	R_3
$S_3N_1V_1$	1.35	1.50	1.40
$S_3N_2V_1$	1.51	1.68	1.54
$S_3N_3V_1$	1.55	1.72	1.64
$S_2N_1V_1$	1.26	1.42	1.28
$S_2N_2V_1$	1.56	1.55	1.61
$S_2N_3V_1$	1.64	1.55	1.62
$S_1N_1V_1$	1.43	1.39	1.35
$S_1N_2V_1$	1.49	1.42	1.55
$S_1N_3V_1$	1.50	1.63	1.54
$S_0N_1V_1$	1.30	1.32	1.39
$S_0N_2V_1$	1.49	1.45	1.53
$S_0N_3V_1$	1.50	1.54	1.61
$S_3N_1V_2$	1.64	1.55	1.58
$S_3N_2V_2$	1.73	1.76	1.82
$S_3N_3V_2$	1.85	1.84	1.74
$S_2N_1V_2$	1.67	1.52	1.52
$S_2N_2V_2$	1.83	1.76	1.61
$S_2N_3V_2$	1.83	1.84	1.74
$S_1N_1V_2$	1.63	1.50	1.58
$S_1N_2V_2$	1.70	1.72	1.79
$S_1N_3V_2$	1.76	1.85	1.73
$S_0N_1V_2$	1.48	1.62	1.61
$S_0N_2V_2$	1.55	1.69	1.78
$S_0N_3V_2$	1.73	1.79	1.66

Analyze the experimental data with regard to nitrogen percentage and also test some useful contrasts. Interpret the results as vividly as possible.

BIBLIOGRAPHY

1. Agarwal BL Testing a main effect in a three-factor mixed model. *Communication in Statistics—Theory and Methods,* 1990: 19(2), 723–38.
2. Box GEP, Hunter WG, Hunter JS. *Statistics for Experimenters.* John Wiley, New York, 1978.
3. Cobb GW. *Introduction to Design and Analysis of Experiments,* Springer–Verlag, New York, 1998.
4. Daniel C. *Application of Statistics to Industrial Experimentation.* John Wiley, New York, 1976.
5. Hamada M, Balakrishnan N. Analysing unreplicated factorial experiments: A review with some proposals: *Statistica Sinica,* 1998: 8(1), 31–35.
6. Hinkleman K, Kempthorne O. *Design and Analysis of Experiments,* Vol. 2: *Advanced Experimental Design,* John Wiley, Hoboken, New Jersey, 2005.
7. Kao LJ, Notz WI, Dean AM. Efficient block designs for estimating main effects contrasts. *Journal of the Indian Society of agricultural Statistics, Special Golden Jubilee Issue,* 1997: 49, 249–258.
8. Kirk RE. *Experimental Design: Procedures for the behavioral Sciences,* 3rd ed. Belmount, CA: Duxbury Press, 1995.
9. Mead R, Bancroft TA, Han C. Power of analysis variance test procedures for incompletely specified fixed models. *Annals of Statistics,* 1975: 3, 791–808.
10. Montgomery DC. *Design and analysis of experiments,* 6th ed. John wiley, New York, 2004.
11. O' Brien RG. Robust techniques for testing heterogeneity of variance effects in factorial designs. *Psychometrika,* 1978: 43, 327–388.
12. Patterson HD, Bailey RA. Design keys for factorial experiments. *Applied Statistics,* 1978: 27, 335–343.
13. Ryan Thomas P. *Modern Experimental design.* John Wiley, New Jersey, 2007.
14. Stehman SV, Meredith MP. Practical analysis of factorial experiments in forestry. *Canadian Journal of Forest Research,* 1995: 25(3), 446–461.
15. Woodward AJ, Bonett DG. Simple main effects in factorial designs. *Journal of Applied statistics,* 1991: 18, 255–264.
16. Youden WJ. Randomization and experimentation. *Technometrics,* 1972: 14, 13–22.

5

Confounding in Factorial Experiments

Relevance: It is apparent that the number of treatment combinations in factorial experiments increases tremendously as the number of factors or the factor levels increase. In randomized block design, homogeneity of blocks is a prime restriction. Too many plots in a block usually fail to fulfill this condition. Hence, a device known as *confounding* has been advocated to reduce block size. In confounding, a block is splitted into smaller equal size blocks in such a way that the splitted block differences represent a factorial effect, videlicet a factorial effect is inextricably mixed with the block differences. In this way, there is a loss of information about the confounded effect. This leads one to confound an effect with the blocks which is of least interest or no interest. Usually higher order interactions are of least or no interest and thus are confounded with the blocks. Commonly a block of 12 to 15 plots of moderate size is admissible. Therefore, confounding is adapted when the number of treatment combinations is large.

5.1 CONFOUNDING PROCEDURES

There are two approaches of confounding.

1. In case of 2^n factorials, confounding can be carried out through contrasts or modulo technique.
2. For 3^n and asymmetrical factorials, only modulo technique is applicable. A terse description of confounding is given in this chapter.

Definitions

Complete or total confounding: If the same effect is confounded in all replicates in a factorial experiment, it is known as *complete or total confounding*.

Partial confounding: If in a factorial experiment the same effect is not confounded in all replicates, it is called *partial confounding*. It means one effect is confounded in one replicate and other in second replicate and so on.

Partial confounding has an advantage over complete confounding that in the former process complete information is not lost about any effect. The confounded effects can be estimated and tested on the basis of blocks in which they are not confounded. Of course in this case, precision is less for confounded effects as compared to unconfounded effects. But if one is not interested in some higher order interaction or expects some interaction effect to be non-significant then it is preferable to go for complete confounding as an investigator looses information about one factor only and estimates other factors with full precision. In this situation, the confounded effect is merged with experimental error.

In a 2^n experiment, the blocks can be reduced to one-half, one-fourth, one-eighth, or in general to blocks of size 2^{n-k} where k is an integer for $k < n$. If a block is splitted into one-fourth size, then two factorial effects are to be confounded. One factor is confounded to split a block into halves and the second factor to split again each of the half blocks into halves. In this way, a block is divided into four blocks of one-fourth size by confounding two effects. But under this process, a treatment effect is automatically confounded which is the generalized interaction of the two confounded effects.

Generalized interaction: A generalized interaction in a p^n factorial is the product of two effects where ensemble powers are reduced to modulo p. For instance in 2^3 experiment, the generalized interaction of AB and AC is AB $\times$ AC = A^2BC = BC as A^2 reduced to modulo 2 is A^0 which is equal to 1. One should be wary that the selection of two or more effects should not be such that their generalized interaction leads to automatic confounding of a main effect. To illustrate, if we confound the interactions ABC and AC, then their generalized interaction, ABC $\times$ AC = B, which is automatically confounded. It means that the block differences represent the effect B. So two or more effects should always be selected for confounding in such a way that no main effect is automatically confounded. If in a 2^3 factorial, two effects are to be confounded, then preferably select any two first order interactions like, AB and AC. In this situation, the effect that is automatically confounded is AB $\times$ AC = BC which is again a first order interaction. Now the procedure of confounding in specific cases is explained here upon.

5.2 2^4-FACTORIAL EXPERIMENTS IN ONE-HALF REPLICATE

This is an experiment with four factors say, N (nitrogen), P (phosphorus), K (potash) and G (green manure) each at two levels. There will in all be 16 treatment combinations; but only 8 plots in a block are to be accommodated. For this, it is always preferable to confound highest order interaction NPKG.

This interaction effect can be represented by the contrast which shall have eight effects with positive signs and eight effects with negative signs. So in a replicate there will be two blocks of size eight plots. The effects with positive signs will randomly be allocated to the eight plots in one block and the other eight effects with negative signs will independently be assigned to the other block. This process of assigning effects with positive and negative signs in all replications will be carried out independently. The same is elaborated here. The contrast for highest order interaction is,

$$NPKG = (n - 1)(p - 1)(k - 1)(g - 1)$$
$$= npkg - npk - npg - nkg - pkg + np + nk + ng + pk + pg + kg$$
$$- n - p - k - g + (1)$$

In the above contrast, it is noteworthy that all even order combinations are with positive signs and odd order combinations are with negative signs. Assign randomly, (with the help of random number tables) eight combinations with positive signs in half replicate and with negative signs in other half replicate. The entries in two replicates for blocks of size 8-plots are displayed below.

	Replicate – I		Replicate – II	
	Block–1	*Block–2*	*Block–3*	*Block–4*
	n k	n k g	p	n k
	n p	k	k	n p
	n p k g	p	n	p k
	p k	n p k	n p g	p g
	(1)	n p g	n p k	n p k g
	k g	n	p k g	(1)
	p g	p k g	g	k g
	n g	g	n k g	n g

If there are more than two replicates, the respective treatment combinations can be allocated randomly in other replicated too in the same manner.

As regards the analysis of variance of confounded design experimental data, the procedure remains the same as for unconfounded factorial experiment except that the degrees of freedom and sum of square for the confounded effect become the part of error component.

5.2.1. 2^4 factorial in one-fourth replicate

In the present problem, there are 16 treatment combinations of four factors say, N, P, K and G each at two levels. To obtain blocks of size 4 units, firstly the block of size 16 units is decomposed into half-replicate, i.e. into two blocks of size eight units. For this, one effect is to be confounded. Now for further split to reduce 1/2-replicate into 1/4-replicate, another effect is to be confounded. It has already been given that when two effects are confounded in the same replicate, their generalized interaction is automatically confounded. Hence, one should be wary to choose two effects in such a manner that no main effect is confounded in any situation.

In a 2^4 factorial, if a four-factor interaction is confounded in a replication for the first split and any three-factor interaction say, NPK is confounded for the second decomposition, their generalized interaction NPKG × NPK = G will automatically be confounded. That is against the set norms. Therefore, it is easy to contemplate that if two three-factor interactions are selected for confounding whose generalized interaction is not a main effect will be an ideal situation. For NPK and PKG are chosen for confounding, then their generalized interaction, NPK × PKG = NG, is automatically confounded which is not a main effect.

Now the design will be constructed by confounding NPK and PKG in sequence and it will be shown that interaction NG is automatically confounded. Contrast NPK is used for first decomposition.

$$NPK = (n - 1)'(p - 1)(k - 1)(g + 1)$$
$$= npkg + nkp + ng + pg + kg + n + p + k$$
$$- npg - pkg - nkg - np - nk - pk - g - (1)$$

In the above contrast, eight combinations with positive signs will be allocated in one block and the other eight combinations with negative signs in another block of a replicate say blocks I and II as displayed below.

Replication

Block – I

npkg nkp ng pg kg n p k

Block – II

npg pkg nkg np nk pk g (1)

Now blocks I and II will further be splitted into blocks of 4 units by confounding PKG. The contrast for PKG is,

$$PKG = (n + 1)(p - 1)(k - 1)(g - 1)$$
$$= npkg + np + nk + ng + pkg + p + k + g$$
$$-n - pk - kg - pg - npg - nkg - npk - (1)$$

It is interesting to note that always out of eight combinations of blocks I and II, four combinations will have positive signs and four will have negative signs in the contrast of PKG. Again block I will be decomposed into two blocks III and IV and block II into V and VI having four entries with positive and negative signs of blocks I and II, respectively. Complete layout is delineated below.

Replication

Block III	Block IV	Block V	Block VI
Npkg	npk	pkg	npg
p	pg	np	nkg
ng	kg	nk	pk
k	n	g	(1)

Now it is to shown that the interaction NG is automatically confounded.

$$NG = (n - 1)(p + 1)(k + 1)(g - 1)$$
$$= npkg + p + k + ng + pk + nkg + npg + (1)$$
$$- npk - pg - kg - n - pkg - np - nk - g$$

Again interaction NG is represented by the difference between blocks as,

$$III + VI - IV - V$$

Therefore, by definition NG is also confounded.

In case of more replications, the same block entries are allocated randomly and independently within four blocks of the second, third replications, etc.

5.3 CONFOUNDING IN 3^n-FACTORIAL EXPERIMENTS

Confounding in 3^n factorials is carried through modulo technique. In 3^n experiment, one can form blocks of one-third, one-ninth size, i.e. in general the blocks of size 3^{n-k} for $k = 1, 2, 3, \ldots\ldots$ and $k < n$ can be constructed. There is an adage, example is better than percept. So the procedure of confounding for a 3^n factorial is explicated through 3^3 factorial alone. If there are three treatments A, B and C each at three levels, it is mostly preferred to confound highest order interaction viz., A × B × C.

In 3^3 factorial, the blocks of size 27 units are generally not admissible. So three blocks of nine units in a replicate are formed by confounding the interaction A × B × C. The selection of nine treatments combinations in a block will be done on the basis of $(ABC)_0$, $(ABC)_1$ and $(ABC)_2$ where the suffixes 0, 1, 2 are $i + j + k$ reduced to modulo 3. Here i represents the power of A, j the power of B and k the power of C. As a rule, randomization has to be done in all blocks of each replication independently. The layout of a 3^3 factorial in blocks of size nine units on confounding A × B × C for two replications of a randomized block design has been displayed below. Three levels of all the factors A, B, C are taken as 0, 1, 2 and the effects are denoted by factor levels alone in order of A, B, C.

27 treatment combinations of a 3^3 factorial

0 0 0	1 0 0	2 0 0
0 0 1	1 0 1	2 0 1
0 0 2	1 0 2	2 0 2
0 1 0	1 1 0	2 1 0
0 1 1	1 1 1	2 1 1
0 1 2	1 1 2	2 1 2
0 2 0	1 2 0	2 2 0
0 2 1	1 2 1	2 2 1
0 2 2	1 2 2	2 2 2

Using modulo technique, the treatment combinations are sorted out for $(ABC)_0$, $(ABC)_1$ and $(ABC)_2$ by reducing $i + j + k$ to modulo 3. The treatment combinations are shown in random order in the following layout.

	Replication – I			Replication – II	
Block–1	*Block–2*	*Block–3*	*Block–4*	*Block–5*	*Block–6*
1 1 1	1 2 1	1 0 1	1 2 0	0 0 1	2 2 1
0 0 0	0 1 0	2 1 2	0 2 1	2 0 2	0 1 1
2 2 2	2 1 1	0 2 0	0 0 0	1 1 2	1 0 1
0 2 1	1 0 0	2 2 1	0 1 2	2 2 0	2 1 2
2 0 1	2 2 0	1 1 0	1 0 2	0 2 2	0 2 0
0 1 2	0 2 2	2 0 0	2 2 2	1 2 1	0 0 2
2 1 0	0 0 1	0 0 2	1 1 1	2 1 1	1 2 2
1 0 2	2 0 2	1 2 2	2 1 0	0 1 0	2 0 0
1 2 0	1 1 2	0 1 1	2 0 1	1 0 0	1 1 0

A × B × C confounded			A × B × C confounded		
0 mod 3	1 mod 3	2 mod 3	0 mod 3	1 mod 3	2 mod 3

The analysis of data shall be carried in the usual way except that the interaction A × B × C shall be merged with the error. Skeleton ANOVA table for randomized block design having r blocks is displayed below.

Table 5.3.1 Skeleton ANOVA table

Source	d.f.
Replication	$r - 1$
Treatments	26
A	2
B	2
A × B	4
C	2
A × C	4
B × C	4
Error	$26r - 18$
Total	$27r - 1$

Nota bene: Eight degrees of freedom for $A \times B \times C$ are merged with error d.f.

In situations where an investigator does not want to loose complete information about $A \times B \times C$ interaction or any other effect, then partial confounding is more appropriate. Without loosing information about a main effect or an interaction, another suitable approach in 3^n factorials is to confound the component of interaction $A \times B \times C$ like ABC, AB^2 C, ABC^2, AB^2C^2 in different replicates to form blocks of size nine units. In this case, to confound ABC, treatment combinations which satisfy the condition $i + j + k = p$ mod 3 for $i, j, k = 0, 1, 2$ are sorted out and placed in one replicate. Similarly to confound AB^2C, the combination which satisfy the condition $i + 2j + k = p$ mod 3 are sorted out and placed in other replication. In the same manner for confounding ABC^2 and AB^2C^2, the conditions are $i + j + 2k = p$ mod 3 and $i + 2j + 2k = p$ mod 3. In this case, second order interaction can be estimated and tested on the basis of those replicates in which that component is not confounded. The divisors for estimation and sum of squares should be adjusted accordingly.

Example 5.1: An experiment on sorghum was planned to study the response of three fertilizers and a biofertilizer each at two levels, nitrogen (0, 60 kg/ha), phosphorus (0, 40 kg/ha), potash (0, 30 kg/ha) and biofertilizer (No azospirillrum, Azospirillum). Each of the treatment N, P, K and B levels were coded as 0 and 1 in order. It was considered that a block of 16 plots will be too large in view of the condition of the land. Hence, a replicate was divided into two blocks of size of eight plots. Since there were 16 treatment combinations, one treatment contrast NPKB was confounded in each replication. Two replications were taken. Making use of modulo technique, blocks are formed by confounding NPKB. Fodder yield of sorghum plants (kg/plot) for plots of size 4.5×5.0 m^2 was as tabulated below.

| | Replication – I | | | | Replication – II | | |
	Block-1		Block-2		Block-3		Block-4
0000	27.5	0001	29.5	0000	24.5	0001	25.7
0011	23.0	0010	28.0	0011	27.0	0010	27.7
0101	19.0	0100	24.5	0101	21.0	0100	29.0
0110	15.0	0111	27.8	0110	20.5	0111	19.0
1001	31.0	1000	22.2	1001	26.0	1000	26.6
1010	26.0	1011	19.3	1010	29.0	1011	25.0
1100	16.0	1101	16.0	1100	19.5	1101	15.5
1111	19.0	1110	19.0	1111	22.2	1110	19.5

Analysis of data is carried out to assess the effect of main effects and 2-factor and 3-factor interactions.

Calculations:

Block totals, $B_1 = 176.5$, $B_2 = 186.3$, $B_3 = 189.7$, $B_4 = 188.0$

Replication totals, $R_1 = 362.8$, $R_2 = 377.7$

Grand total, $G = 740.5$

   Sum of squares for total, replications, blocks, blocks within replications are calculated first and then the treatment sum of squares are worked out by Yates' method.

$$C.F. = \frac{740.5^2}{32} = 17135.63$$

$$\text{Total S.S.} = 27.5^2 + 23.0^2 + \ldots\ldots + 15.5^2 + 19.5^2 - C.F.$$
$$= 17798.35 - 17135.63 = 662.72$$

$$\text{Rep. S.S.} = \frac{1}{16}(362.8^2 + 377.7^2) - C.F.$$
$$= 17142.57 - 17135.63 = 6.94$$

$$\text{Blocks w. Reps. S.S.} = \frac{1}{8}\left(176.5^2 + 186.3^2\right) - \frac{362.8^2}{16}$$

$$+ \frac{1}{8}\left(189.7^2 + 188.0^2\right) - \frac{377.7^2}{16}$$

$$= 8232.49 - 8226.49 + 8916.26 - 8916.08$$
$$= 6.0 + 0.18 = 6.18$$

Treatments S.S.

Symbols	Treats. combs.	Treat. totals	col. (i)	Col. (ii)	Col. (iii)	Col. (iv)	Treat. S.S.
0000	(1)	52.0	100.8	189.8	374.5	740.5	17135.63
1000	n	48.8	89.0	184.7	366.0	−36.9	42.55
0100	p	53.5	110.7	183.7	−18.9	−95.5	285.00
1100	np	35.5	74.0	182.3	−18.0	−21.3	14.18
0010	k	55.7	112.2	−21.2	−49.5	−6.5	1.32
1010	nk	55.0	71.5	2.3	−47.0	18.9	11.16
0110	pk	35.5	94.3	−6.7	−11.1	9.5	3.44
1110	npk	38.5	88.0	−11.3	−10.2	28.9	26.10
0001	b	55.2	−3.2	−11.8	−5.1	−8.5	2.26
1001	nb	57.0	−18.0	−36.7	−1.4	0.9	0.02
0101	Pb	40.0	−0.7	−40.7	23.5	1.5	0.07
1101	npb	31.5	3.0	−6.3	−4.6	0.9	0.02
0011	kb	50.0	1.8	−14.8	−24.9	3.7	0.43
1011	nkb	44.3	−8.5	3.7	34.4	−28.1	24.68
0111	pkb	46.8	−5.7	−10.3	18.5	59.3	109.89
1111	npkb	41.2	−5.6	0.1	10.4	−8.1	2.05

Sum of square due to treatment combination NPKB is to be merged with experimental error. Thus, the analysis of variance table is as displayed below.

Analysis of variance table of 2^4 confounded design

Source	d.f.	S.S.	M.S.	F-value
Replications	1	6.94	6.94	< 1
Blocks w. reps.	2	6.18	3.09	< 1
Treatments	14	522.55		
N	1	42.55	42.55	4.28
P	1	285.00	285.00	28.70*
N × P	1	14.18	14.18	1.43
K	1	1.32	1.32	< 1
N × K	1	11.16	11.16	1.12
P × K	1	2.82	2.82	< 1
N × P × K	1	26.10	26.10	2.63
B	1	2.26	2.26	< 1
N × B	1	0.02	0.02	< 1
P × B	1	0.07	0.07	< 1
N × P × B	1	0.02	0.02	< 1
K × B	1	0.43	0.43	< 1
N × K × B	1	24.68	24.68	2.48
P × K × B	1	109.89	109.89	11.07*
Error	13	129.10	9.93	
Total	30	662.72		

Critical values of F from Table B-7 at $\alpha = 0.05$ for d.f. (1, 13) and (2, 13) are, $F_{(1, 13)} = 4.67$, $F_{(2, 13)} = 3.81$

F-value with asterisk (*) mark shows significant effect and rest are not significant. There is no need of post-hoc test as each factor has only two levels.

Example 5.2: A 3^3 experiment was conducted with the objective of knowing the effect of three crop geometries (90×22.5 cm, 67.5×30 cm, 45×45 cm), three doses of sulphur (0, 50 kg/ha, 100 kg/ha) and three growth regulators (control, triacontanol @ 2.5 ppm, NAA @ 10 ppm) and their interaction effects on lint yield. Since 27 treatment combinations were too large a number to maintain the homogeneity of blocks of a randomized block design, it was preferred to use a partially confounded design. Three replicates were taken. In first, second and third replicates, the effects CS^2G, CSG^2 and CS^2G^2 were confounded respectively to form blocks of 9 plots in each replicate. The layout and lint yield (q/ha) were as depicted below.

Denote: crop geometry – C, sulphur – S, growth regulators – G.

Replication – I
Confounding $(CS^2G)_{i + 2j + k = x \bmod 3}$

Block-1	Block-2	Block-3
$C_0 S_2 G_2$ - 8.92	$C_0 S_2 G_0$ - 7.63	$C_1 S_2 G_0$ - 6.29
$C_0 S_1 G_1$ - 8.69	$C_2 S_0 G_2$ - 7.00	$C_2 S_0 G_0$ - 5.06
$C_0 S_0 G_0$ - 4.96	$C_1 S_2 G_2$ - 8.15	$C_1 S_1 G_2$ - 8.91
$C_1 S_1 G_0$ - 8.74	$C_0 S_0 G_1$ - 6.43	$C_0 S_1 G_0$ - 6.60
$C_2 S_1 G_2$ - 8.66	$C_0 S_1 G_2$ - 9.08	$C_2 S_1 G_1$ - 7.91
$C_2 S_2 G_0$ - 6.50	$C_1 S_1 G_1$ - 9.59	$C_0 S_0 G_2$ - 7.40
$C_2 S_0 G_1$ - 6.32	$C_1 S_0 G_0$ - 5.14	$C_1 S_0 G_1$ - 7.88
$C_1 S_2 G_1$ - 7.75	$C_2 S_1 G_0$ - 6.40	$C_0 S_2 G_1$ - 10.55
$C_1 S_0 G_2$ - 6.93	$C_2 S_2 G_1$ - 7.57	$C_2 S_2 G_2$ - 9.19
67.47	66.99	69.79
$i + 2j + k = 0 \bmod 3$	$i + 2j + k = 1 \bmod 3$	$i + 2j + k = 2 \bmod 3$

Replication – II
Confounding $(CSG^2)_{i + j + 2k = p \bmod 3}$

Block-1	Block-2	Block-3
$C_1 S_1 G_2$ - 9.79	$C_1 S_1 G_0$ - 5.95	$C_1 S_0 G_0$ - 5.09
$C_0 S_0 G_0$ - 4.18	$C_1 S_2 G_1$ - 9.24	$C_2 S_1 G_2$ - 10.16
$C_2 S_2 G_1$ - 9.05	$C_2 S_0 G_0$ - 5.60	$C_0 S_2 G_1$ - 6.84
$C_0 S_2 G_2$ - 8.47	$C_2 S_1 G_1$ - 9.50	$C_2 S_2 G_0$ - 7.59
$C_2 S_0 G_2$ - 8.04	$C_0 S_1 G_2$ - 8.13	$C_1 S_1 G_1$ - 8.31
$C_0 S_1 G_1$ - 7.43	$C_0 S_2 G_0$ - 5.53	$C_1 S_2 G_2$ - 9.35
$C_2 S_1 G_0$ - 7.37	$C_2 S_2 G_2$ - 10.06	$C_2 S_0 G_1$ - 6.89
$C_1 S_0 G_1$ - 4.97	$C_0 S_0 G_1$ - 6.00	$C_0 S_0 G_2$ - 6.33
$C_1 S_2 G_0$ - 7.60	$C_1 S_0 G_2$ - 7.52	$C_0 S_1 G_0$ - 5.54
66.90	67.53	66.10
$i + j + 2k = 0 \bmod 3$	$i + j + 2k = 2 \bmod 3$	$i + j + 2k = 1 \bmod 3$

Replication – III

Confounding $(CS^2G^2)_{i} + _{2j + 2k = p \bmod 3}$

Block-1	Block-2	Block-3
$C_0 S_0 G_0$ - 3.97	$C_2 S_2 G_2$ - 9.24	$C_1 S_2 G_0$ - 7.22
$C_2 S_0 G_2$ - 8.17	$C_1 S_2 G_1$ - 9.03	$C_0 S_2 G_2$ - 8.19
$C_1 S_1 G_0$ - 6.76	$C_1 S_0 G_0$ - 5.50	$C_2 S_1 G_2$ - 9.69
$C_2 S_2 G_0$ - 7.40	$C_0 S_1 G_1$ - 7.49	$C_2 S_0 G_0$ - 5.75
$C_0 S_1 G_2$ - 8.13	$C_2 S_1 G_0$ - 7.45	$C_2 S_2 G_1$ - 9.46
$C_0 S_2 G_1$ - 7.30	$C_1 S_1 G_2$ - 9.49	$C_1 S_0 G_2$ - 8.34
$C_1 S_2 G_2$ - 10.55	$C_0 S_0 G_2$ - 6.63	$C_0 S_1 G_0$ - 5.74
$C_2 S_1 G_1$ - 9.15	$C_2 S_0 G_1$ - 6.92	$C_0 S_0 G_1$ - 5.81
$C_1 S_0 G_1$ - 6.65	$C_0 S_2 G_0$ - 4.79	$C_1 S_1 G_1$ - 8.26
68.08	66.54	68.46
$i + 2j + 2k = 0 \bmod 3$	$i + 2j + 2k = 1 \bmod 3$	$i + 2j + 2k = 2 \bmod 3$

Now the ANOVA table is prepared so that it is known what has to be worked out for complete analysis and the same values are put in the table. The step wise calculations of sum of squares for all components of analysis of variance table are given just after the table.

Table-I ANOVA

Source	d.f	S.S.	M.S.	F - value
Replications	2	0.27	0.135	< 1
Blocks within rep.	6	0.85	0.142	< 1
Treats. unadjusted for blocks	26	156.87		
C	2	14.15	7.07	< 1
$C_l = C_2 - C_1$	1	0.18	0.18	< 1
$C_q = C_2 + C_1 - 2C_0$	1	13.97	13.97	1.61
S	2	61.02	30.51	3.52*
$S_l = S_2 - S_1$	1	0.005	0.005	< 1
$S_q = S_2 + S_1 - 2S_0$	1	61.014	61.014	7.04
C × S	4	0.23	0. 06	< 1
$C_l S_l$	1	0.0042	0.0042	< 1
$C_l S_q$	1	0.0404	0.0404	< 1

(Contd.)

Source	d.f	S.S.	M.S.	F-value
$C_q S_l$	1	0.1220	0.1220	< 1
$C_q S_q$	1	0.0633	0.0633	< 1
G	2	80.15	40.08	4.62*
$G_l = G_2 - G_1$	1	7.06	7.06	< 1
$G_q = G_2 + G_1 - 2G_0$	1	73.08	73.08	8.43*
$C \times G$	4	0.64	0.16	< 1
$C_l \times G_l$	1	0.0002	0.0002	< 1
$C_l \times G_q$	1	0.0034	0.0034	< 1
$C_q \times G_l$	1	0.2610	0.2610	< 1
$C_q \times G_q$	1	0.3741	0.3741	< 1
$S \times G$	4	0.51	0.13	< 1
$S_l \times G_l$	1	0.0040	0.0040	< 1
$S_l \times G_q$	1	0.0027	0.0027	< 1
$S_q \times G_l$	1	0.3267	0.3267	< 1
$S_q \times G_q$	1	0.1792	0.1792	< 1
$C \times S \times G$	8			
C S G	2	0.03	0.015	< 1
C S^2 G	2	0.11	0.055	< 1
C S G^2	2	0.13	0.065	< 1
C S^2 G^2	2	0.03	0.015	< 1
Error	46	398.96	8.67	
Total	80	557.08		

Calculations:

Replication totals, $R_1 = 67.47 + 66.99 + 69.79 = 204.25$
Similarly, $R_2 = 200.53$, $R_3 = 203.08$
Grand total, $y.. = 204.25 + 200.53 + 203.08 = 607.86$

$$C.F. = \frac{(607.86)^2}{81} = 4561.65$$

Total S.S. $= 8.92^2 + 8.69^2 + \ldots\ldots + 5.81^2 + 8.26^2 - C.F.$
$$= 5118.73 - 4561.65 = 557.08$$

Blocks within replicates S.S. $= \frac{1}{9}\left(67.47^2 + 66.99^2 + 69.79^2\right) - \frac{204.25^2}{27}$

$$+ \frac{1}{9}\left(66.90^2 + 67.53^2 + 66.10^2\right) - \frac{200.53^2}{27}$$

$$+\frac{1}{9}\left(68.08^2+66.54^2+68.46^2\right)-\frac{203.08^2}{27}$$

$$= 1545.61 - 1545.11 + 1489.46 - 1489.34 + 1527.69 - 1527.46$$
$$= 0.50 + 0.12 + 0.23 = 0.85$$

Replications S.S. $=\frac{1}{27}\left(204.25^2+200.53^2+203.08^2\right)-\frac{607.86^2}{81}$

$$= 4561.92 - 4561.65 = 0.27$$

To obtain the sum of squares for main effects and first order interactions, three two-way tables are prepared. The required sum of squares are calculated with the help of these tables.

Table-II C × S table

Factors	C_0	C_1	C_2	Total
S_0	51.71	58.02	59.75	169.48
S_1	66.83	75.80	76.29	218.92
S_2	68.22	75.18	76.06	219.46
Total	186.76	209.00	212.10	607.86

Table-III S × G table

Factors	S_0	S_1	S_2	Total
G_0	45.25	60.55	60.55	166.35
G_1	57.87	76.33	76.79	210.99
G_2	66.36	82.04	82.12	230.52
Total	169.48	218.92	219.46	607.86

Table-IV C × G table

Factors	C_0	C_1	C_2	Total
G_0	48.94	58.29	59.12	166.35
G_1	66.54	71.68	72.77	210.99
G_2	71.28	79.03	80.21	230.52
Total	186.76	209.00	212.10	607.86

$$\text{S.S.(C)} = \frac{1}{27}\left(186.76^2+209.00^2+212.10^2\right)-\frac{607.86^2}{81}$$

$$= 4575.80 - 4561.65 = 14.15$$

$$S.S.(S) = \frac{1}{27}\left(169.48^2 + 218.92^2 + 219.46^2\right) - C.F.$$

$$= 4622.67 - 4561.65 = 61.02$$

$$S.S.(C \times S) = \frac{1}{9}\left(51.71^2 + 58.02^2 + \ldots\ldots + 75.18^2 + 76.06^2\right)$$

$$-C.F. - S.S.(C) - S.S.(S)$$

$$= 4637.05 - 4561.65 - 14.15 - 61.02 = 0.23$$

$$S.S.(G) = \frac{1}{27}\left(166.35^2 + 210.99^2 + 230.52^2\right) - C.F.$$

$$= 4641.80 - 4561.65 = 80.15$$

$$S.S(S \times G) = \frac{1}{9}\left(45.25^2 + 60.55^2 + \ldots\ldots\ldots + 82.04^2 + 82.12^2\right)$$

$$- C.F. - S.S.(S) - S.S.(G)$$

$$= 4703.33 - 4561.65 - 61.02 - 80.15 = 0.51$$

$$S.S(C \times G) = \frac{1}{9}\left(48.94^2 + 58.29^2 + ---- + 79.03^2 + 80.21^2\right)$$

$$- C.F. - S.S.(C) - S.S.(G)$$

$$= 4656.59 - 4561.65 - 14.15 - 80.15 = 0.64$$

Unadjusted treatment totals

000 - 13.11	100 - 15.73	200 - 16.41
001 - 18.24	101 - 19.50	201 - 20.13
002 - 20.36	102 - 22.79	202 - 23.21
010 - 17.88	110 - 21.45	210 - 21.22
011 - 23.61	111 - 26.16	211 - 26.56
012 - 25.34	112 - 28.19	212 - 28.51
020 - 17.95	120 - 21.11	220 - 21.49
021 - 24.69	121 - 26.02	221 - 26.08
022 - 25.58	122 - 28.05	222 - 28.49

$$\text{Treat.S.S.} = \frac{1}{3}(13.11^2 + 18.24^2 + \ldots\ldots + 28.08^2 + 28.49^2) - C.F.$$

$$= 4718.52 - 4561.65 = 156.87$$

Using the calculations so far, linear and quadratic components of main effect are estimated and their sum of squares are also worked out.

Lin. effect, $C_2 - C_1 = 212.10 - 209.00 = 3.10$

$$\text{S.S. due to linear effect (C)} = \frac{(3.10)^2}{(1+1) \times 3 \times 9} = 0.18$$

Quad. effect, $C_2 + C_1 - 2C_0 = 212.10 + 209.00 - 2 \times 186.76 = 47.58$

S.S. due to quad. effect (C) $= \dfrac{(47.58)^2}{(1+1+4) \times 3 \times 9} = 13.97$

Lin. effect, $S_2 - S_1 = 219.46 - 218.92 = 0.54$

S.S. due to lin. effect (S) $= \dfrac{(0.54)^2}{(1+1) \times 3 \times 9} = 0.0054$

Quad. effect, $S_2 + S_1 - 2S_0 = 219.46 + 218.92 - 2 \times 169.48 = 99.42$

S.S due to quad. effect (S) $= \dfrac{(99.42)^2}{(1+1+4) \times 3 \times 9} = 61.014$

Lin. effect, $G_2 - G_1 = 230.52 - 210.99 = 19.53$

S.S due to lin. effect (C) $= \dfrac{(19.53)^2}{(1+1) \times 3 \times 9} = 7.06$

Quad. effect, $G_2 + G_1 - 2G_0 = 230.52 + 210.99 - 2 \times 166.35 = 108.81$

S.S. due to quad. effect (G) $= \dfrac{(108.81)^2}{(1+1+4) \times 3 \times 9} = 73.08$

For estimating the effects of the components; $C_l S_l, C_l S_q, C_q S_l, C_q S_q$ of the first order interaction C × S, following table is prepared.

Table-V C × S components table

	$C_l = C_2 - C_1$ 0 −1 1	$C_l = C_2 + C_1 - 2C_0$ −2 1 1		$S_l = S_2 - S_1$ 0 −1 1	$S_q = S_2 + S_1 - 2S_0$ −2 1 1
S_0	1.73	14.35	C_0	1.39	31.63
S_1	0.49	18.43	C C_1	− 0.62	34.94
S_2	0.88	14.80	C_2	− 0.23	32.86
Total	$C_l = 3.10$	47.58	Total	$S_l = 0.54$	$S_q = 99.42$
0 S_l −1 1	0.88 − 0.49 = 0.39 $C_l S_l$	14.80 − 18.43 = − 3.63 $C_q S_l$	0 C_l −1 1	− 0.23 + 0.62 = 0.39 $S_l C_l$	32.86 − 34.94 = − 2.08 $C_l S_q$
− 2 S_q 1 1	−2 × 1.73 + 0.49 + 0.88 = −2.09 $C_l S_q$	−2 × 14.35 + 18.43 + 14.80 = 4.53 $C_q S_q$	−2 C_q 1 1	−2 × 1.39 − 0.62 − 0.23 = −3.63 $S_l C_q$	−2 × 31.63 + 34.94 + 32.86 = 4.54 $C_q S_q$

Sum of squares due to four components of the interaction C × S is obtained by using the values of the components given in the table-V.

$$\text{S.S. due to } C_l S_l = \frac{(0.39)^2}{(1+1)(1+1)\times 9} = 0.0042$$

$$\text{S.S. due to } C_l S_q = \frac{(-2.09)^2}{(1+1)(1+1+4)\times 9} = 0.0404$$

$$\text{S.S. due to } C_q S_l = \frac{(-3.63)^2}{(1+1+4)(1+1)\times 9} = 0.1220$$

$$\text{S.S. due to } C_q S_q = \frac{(4.53)^2}{(1+1+4)(1+1+4)\times 9} = 0.0633$$

Note: The total of these components S.S. = 0.2299, which is almost same as found earlier from two-way table. A slight difference occurs due to rounding of figures.

Similar to C × S, linear and quadratic interaction components of S × G are estimated and their sum of squares are calculated in the same manner.

Table-VI Linear and quadratic components of S × G interaction

	$G_l = G_2 - G_1$	$G_q = G_2 + G_1 - 2G_0$		$S_l = S_2 - S_1$	$S_q = S_2 + S_1 - 2S_0$
	0 −1 1	−2 1 1		0 −1 1	−2 1 1
S_0	8.49	33.73	G_0	0	30.60
S_1	5.71	37.27	G_1	0.46	37.38
S_2	5.33	37.81	G_2	0.08	31.44
	$G_l = 19.53$	108.81		$S_l = 0.54$	$S_q = 99.42$
0	(5.33 − 5.71)	(37.81 − 37.27)	0	(0.08 − 0.46)	(31.44 − 37.38)
S_l−1	= −0.38	= 0.54	G_l −1	= −0.38	= −5.94
1	$S_l G_l$	$S_l G_q$	1	$S_l G_l$	$S_q G_l$
−2	(−2 × 8.49 + 5.71 + 5.33)	(−2 × 33.73 + 37.27 + 37.81)	−2	(2 × 0 + 0.46 + 0.08)	(−2 × 30.60 + 37.38 + 31.44)
S_q 1	= −5.94	= 7.62	G_q 1	= 0.54	= 7.62
1	$S_q G_l$	$S_q G_q$	1	$S_l G_q$	$S_q G_q$

$$\text{S.S. due to } S_l G_l = \frac{(-0.38)^2}{(1+1)(1+1)\times 9} = 0.0040$$

$$\text{S.S. due to } S_q G_l = \frac{(-5.94)^2}{(1+1+4)(1+1)\times 9} = 0.3267$$

$$\text{S.S. due to } S_l G_q = \frac{(0.54)^2}{(1+1)(1+1+4)\times 9} = 0.0027$$

$$\text{S.S. due to } S_q G_q = \frac{(7.62)^2}{(1+1+4)(1+1+4)\times 9} = 0.1792$$

Total of components of $S \times G = 0.5126$ which is almost the same as was calculated for $S \times G$.

In the like manner, linear and quadratic component sums of squares for the interaction $C \times G$ are obtained by creating the following table.

Table-VII Linear and quadratic components of $C \times G$ interaction

	$G_l = G_2 - G_1$ 0 -1 1	$G_q = G_2 + G_1 - 2G_0$ -2 1 1		$C_l = C_2 - C_1$ 0 -1 1	$C_q = C_2 + C_1 - 2C_0$ -2 1 1
C_0	4.74	39.94	G_0	0.83	19.53
C_1	7.35	34.13	G_1	1.09	11.37
C_2	7.44	34.74	G_2	1.18	16.68
Total	$G_l = 19.53$	108.81	Total	$C_l = 3.10$	$C_q = 47.58$
0 $C_l -1$ 1	7.44 − 7.35 = 0.09 $C_l G_l$	34.74 − 34.13 = 0.61 $C_l G_q$	0 $G_l -1$ 1	1.18−1.09 = 0.09 $C_l G_l$	16.68−11.37 = 5.31 $C_q G_l$
−2 C_q 1 1	−2 × 4.74 + 7.35 + 7.44 = 5.31 $C_q G_l$	2 × 39.94 + 34.13 + 34.74 = −11.01 $C_q G_q$	−2 G_q 1 1	−2 × 0.83 + 1.09 + 1.18 = 0.61 $C_l G_q$	−2 × 19.53 + 11.37 + 16.68 = −11.01 $C_q G_q$

$$\text{S.S. due to } C_l G_l = \frac{(0.09)^2}{(1+1)(1+1)\times 9} = 0.0002$$

$$\text{S.S. due to } C_q G_l = \frac{(5.31)^2}{(1+1+4)(1+1)\times 9} = 0.2610$$

$$\text{S.S. due to } C_l G_q = \frac{(0.61)^2}{(1+1)(1+1+4)\times 9} = 0.0034$$

$$\text{S.S. due to } C_q G_q = \frac{(-11.01)^2}{(1+1+4)(1+1+4)\times 9} = 0.3741$$

Total of sum of squares due components of C × G = 0.6353, which is almost the same as it was found earlier. This works as a check of calculations. The components of second order interactions will be calculated by modulo technique. The effects which are not confounded with the blocks shall be worked out from all replications whereas second order interaction component which is confounded in a replicate shall be obtained from the replicates in which this is not confounded. Some portion of sum of squares due to confounded effects will automatically contribute towards experimental error.

Sum of square due to unconfounded component CSG is estimated from all replicates. Their treatment combinations and total responses are worked out as given below.

$(CSG)_0$ = 000, 012, 021, 102, 111, 120, 201, 210, 222
   = 13.11 + 25.34 + 24.69 + 22.79 + 26.16 + 21.11 + 21.13 + 21.22
   + 28.49 = 203.04

$(CSG)_1$ = 001, 010, 022, 100, 112, 121, 202, 211, 220
   = 18.24 + 17.88 + 25.58 + 15.73 + 28.19 + 26.02 + 23.21 + 26.56
   + 21.49 = 202.90

$(CSG)_2$ = 002, 011, 020, 101, 110, 122, 200, 212, 221
   = 20.36 + 23.61 + 17.95 + 19.50 + 21.45 + 28.05 + 16.41 + 28.51
   + 26.08 = 201.92

$$\text{S.S. due to CSG} = \frac{1}{27}(203.04^2 + 202.90^2 + 201.92^2) - \frac{607.86^2}{81}$$

$$= 4561.68 - 4561.65 = 0.03$$

In the like manner, sum of square for the interaction CS^2G has been obtained from the replications II and III.

$(CS^2G)_0$ = 022, 011, 000, 110, 212, 220, 201, 121, 102
   = 16.66 + 14.92 + 8.15 + 12.71 + 19.85 + 14.99 + 13.81 + 18.27
   + 15.86 = 135.22

$(CS^2G)_1$ = 020, 202, 122, 001, 012, 111, 100, 210, 221
   = 10.32 + 16.21 + 19.90 + 11.81 + 16.26 + 16.57 + 10.59 + 14.82
   + 18.51 = 134.99

$(CS^2G)_2$ = 120, 200, 112, 010, 211, 002, 101, 021, 222
   = 14.82 + 11.35 + 19.28 + 11.28 + 18.65 + 12.96 + 11.62 + 14.14
   + 19.30 = 133.40

$$\text{S.S. due to } CS^2G = \frac{1}{18}(135.22^2 + 134.99^2 + 133.40^2) - \frac{403.61^2}{54}$$

$$= 3016.79 - 3016.68 = 0.11$$

The interaction CSG^2 sum of square has been worked out from I and III replications in the same way as for CS^2G.

$(CSG^2)_0$ = 112, 000, 221, 022, 202, 011, 210, 101, 120
$\qquad$ = 18.40 + 8.93 + 17.03 + 17.11 + 15.17 + 16.18 + 13.85 + 14.53
$\qquad$ + 13.51 = 134.71

$(CSG^2)_1$ = 100, 212, 021, 220, 111, 122, 201, 002, 010
$\qquad$ = 10.64 + 18.35 + 17.85 + 13.90 + 17.85 + 18.70 + 13.24
$\qquad$ + 14.03 + 12.34 = 136.90

$(CSG^2)_2$ = 110, 121, 200, 211, 012, 020, 222, 001, 102
$\qquad$ = 15.50 + 16.78 + 10.81 + 17.06 + 17.21 + 12.42 + 18.43
$\qquad$ + 12.24 + 15.27 = 135.72

$$\text{S.S due to } CSG^2 = \frac{1}{18}(134.71^2 + 136.90^2 + 135.72^2) - \frac{407.33^2}{54}$$

$$= 3072.68 - 3072.55 = 0.13$$

Interaction component CS^2G^2 sum of square shall be obtained from replications I and II in the manner similar to the components CS^2G and CSG^2.

$(CS^2G^2)_0$ = 000, 202, 110, 220, 012, 021, 122, 211, 101
$\qquad$ = 9.14 + 15.04 + 14.69 + 14.09 + 17.21 + 17.39 + 17.50 + 17.41
$\qquad$ + 12.85 = 135.32

$(CS^2G^2)_1$ = 222, 121, 100, 011, 210, 112, 002, 201, 020
$\qquad$ = 19.25 + 16.99 + 10.23 + 16.12 + 13.77 + 18.70 + 13.73
$\qquad$ + 13.21 + 13.16 = 135.16

$(CS^2G^2)_2$ = 120, 022, 212, 200, 221, 102, 010, 001, 111
$\qquad$ = 13.89 + 17.39 + 18.82 + 10.66 + 16.62 + 14.45 + 12.14
$\qquad$ + 12.43 + 17.90 = 134.30

$$\text{S.S. due to } CS^2G^2 = \frac{1}{18}(135.32^2 + 135.16^2 + 134.30^2) - \frac{404.78^2}{54}$$

$$= 3034.23 - 3034.20 = 0.03$$

An examination of the analysis of variance Table-I reveals that there is a significant role of sulphur and growth regulators in increasing the cotton lint yield. Further it is interesting to note that the linear effect of fertilizers and growth regulators are non-significant whereas their quadratic effects are significant. This is an interesting result which signifies that the effect of sulphur fertilizer and growth regulators increase lint yield to certain extent with increasing dose but it becomes stagnant there after.

The standard error of main effect $= \sqrt{\dfrac{9.07}{3 \times 3 \times 3}} = 0.58$

Least significant difference $= \sqrt{2} \times 0.58 \times t_{.05:46}$

$$= \sqrt{2} \times 0.58 \times 2.015 = 1.65$$

Mean yields of S in descending order are:

S_2	S_1	S_0
8.14	8.11	6.28

Mean yields of G in descending order are:

G_2	G_1	G_0
8.54	7.81	6.16

On comparing the mean lint yields of sulphur levels and levels of growth regulators, it is evident that 50 kg/ha and 100 kg/ha of sulphur fertilization result into same yield. Also the effect of Triacontanol @ 2.5 ppm contribute equally towards enhancing lint yield. But both the levels of sulphur and growth regulators significantly increase lint yield as compared to control.

5.4 CONFOUNDING IN ASYMMETRICAL FACTORIALS

Concept: In the preceding section, confounding in symmetrical factorials is explicated adequately. The methods applicable to symmetrical factorials are not suitable for confounding in asymmetrical factorials. They need special adroitness. Various methods for confounding have to be adapted for different asymmetrical factorials. But certain principle ideas hold well in all cases. Firstly, it is pertinent to understand the meaning of confounding in asymmetrical factorials. Method of contrasts or modulo technique cannot be applied as such. One has to stick to an approach which protects the main effects to have maximum precision, i.e. they should not be confounded in any case. A main effect is not confounded means that every level of the factor occurs in a block equal number of times. If this condition does not hold good, the effect is confounded with the blocks. In view of this, the methods of confounding are so evolved that main effects remain devoid of block differences and two or more factor interactions are confounded with blocks. This imposes a restriction on block size. With this restriction in case of $2 \times 2 \times 3$ or $2 \times 3 \times 3$ factorials, the block size has to be of six units. Otherwise it will not be possible to keep main effects uncoufounded. In brief, block size should be the least common factor of the factor levels or its multiple of two or more factors. So fix the size of block, say it is M. Let the asymmetrical factorial in general be $p_1 \times p_2 \times \ldots\ldots\ldots \times p_t = P$. Then divide P by M, i.e. P/M = N (say). Here N is either a prime number or a prime power, i.e. $N = s^k$, where s is a prime number and k is an integer. So out of t factors, those factors whose number of levels is s are called *real factors* and rest of the factors are known as *factors of asymmetry*.

Confounding in asymmetric factorial is usually carried out with the help of *pseudo factors*. Corresponding to factors of asymmetry, pseudo factors are those which are used as dummy factors to convert the asymmetric factorials into symmetric factorials by the combination of requisite number of factors each at s levels. Suppose that a factor X has four levels, 0, 1, 2, 3. These levels of X can also be represented by two factors X_1 and X_2 each at two levels 0, 1 by establishing a correspondence between them as given below:

X	X_1	X_2
0	0	0
1	0	1
2	1	0
3	1	1

Thus, the equivalence of levels is, $00 \equiv 0$, $01 \equiv 1$, $10 \equiv 2$, $11 \equiv 3$.

The factors X_1 and X_2 are called pseudo factors in this example. In this way, any experiment of the type 4^t in blocks of size 4^{t-m} can be represented by a 2^{2t} experiment in 2^{2m} blocks of size $2^{(t-m)}$. Often one comes across the factorial experiments of the type $2^u \times 4^v$ or $2^4 \times 3^v$ or $2^4 \times w \times 3^v$ where w is a prime number and u and v are integers. Confounding for some particular factorials has been discussed hereafter.

5.5 CONFOUNDING IN A $2^3 \times 4$ FACTORIAL IN BLOCKS OF SIZE EIGHT UNITS

The given factorial can be converted into a symmetrical factorial as $2^3 \times 2^2$, i.e. 2^5 factorial. The new factorial has three real factors say, A, B, C each with levels 0, 1 and two pseudo factors X_1 and X_2 of X each at two levels 0, 1. In this way, there are 32 treatment combinations and hence to obtain blocks of size 8 units, two factorial effects will be confounded and one factor as their generalized interaction will automatically be confounded. On confounding interactions say, ACX_1, BCX_2, the interaction ABX_1X_2 will automatically be confounded. Choice of second order interactions saves all main effects and two factor interactions.

Making use of modulo technique, confound first ACX_1 to obtain blocks of size 16 units. Then again confound BCX_2 to get blocks of 8 units. In this way, there are five factor combinations in the order $ABCX_1X_2$. X_1 and X_2 are pseudo factors. So putting $00 \equiv 1$, $01 \equiv 1$, $10 \equiv 2$, $11 \equiv 3$, the layout of confounded design in blocks of size eight is obtained.

Confounding ACX_1, following two blocks are obtained.

$(ACX_1)_{0 \bmod 2}$ = 00000, 00001, 00110, 00111, 01000, 01001, 01110, 01111,
10010, 10011, 10100, 10101, 11010, 11011, 11100, 11101

$(ACX_1)_{1 \bmod 2}$ = 00010, 00011, 00100, 00101, 01010, 01011, 01100, 01101,
10000, 10001, 10110, 10111, 11000, 11001, 11110, 11111

Confounding BCX_2 in the above two blocks of size 16 units in succession, four blocks of required size of 8 units are obtained by modulo technique as follows:

Block-1

$(BCX_2)_{0 \bmod 2}$ = 00000, 00111, 01001, 01110, 10010, 10101, 11011, 11100

Block-2

$(BCX_2)_{1 \bmod 2}$ = 00001, 00110, 01000, 01111, 10011, 10100, 11010, 11101

Block-3

$(BCX_2)_{0 \bmod 2}$ = 00010, 00101, 01011, 01100, 10000, 10111, 11001, 11110

Block-4

$(BCX_2)_{1 \bmod 2}$ = 00011, 00100, 01010, 01101, 10001, 10110, 11000, 11111

Putting for the last two digits which stand for $X_1 X_2$ into their equivalent levels of the factor X, confounded factorial design for asymmetrical factorial in blocks of size eight is obtained as displayed below.

Table 5.5.1 Layout of $2^3 \times 4$ factorial in blocks of 8 units

Block–1	*Block–2*	*Block–3* .	*Block–4*
0000	0001	0002	0003
0013	0012	0011	0010
0101	0100	0103	0102
0112	0113	0110	0111
1002	1003	1000	1001
1011	1010	1013	1012
1103	1102	1101	1100
1110	1111	1112	1113

Nota bene:

1. The above approach is applicable only if the asymmetrical factorial can be converted into a symmetrical factorial as power of the same prime number.
2. It is trivial to verify that each factor level occurs in each column of any block equal number of times.
3. Further, all two factor combinations occur twice in each block which ensures that first order interactions also remain unconfounded.

General rule: If b blocks are constructed in a confounded design, then always $(b - 1)$ degrees of freedom for factorial effects will be confounded. In the above example one d.f. is lost due to confounding of ACX_1 and 1 d.f. for BCX_2 and 1 d.f. for ABX_1X_2 which is equivalent to ABX.

5.6 CONFOUNDING IN A $2^U \times 3^V$ FACTORIAL EXPERIMENT

By collapsing of blocks : Consider a practical situation in which there is a factorial experiment with two factors A and B each at two levels 0, 1 and two factors C and D each at three levels 0, 1, 2. In this experiment, only a block of size six units is admissible. So this leads to an experiment of the type $2^2 \times 3^2$. In all there are 36 treatment combinations of ABCD. A block can contain only six units such that the main effects remain unconfounded. This can be achieved by converting the asymmetrical factorial into a symmetrical factorial by a simple approach. In this endeavor, confound AB in 2^2 factorial to obtain two blocks of size 2 units and also obtain three blocks of size 3 units for 3^2 factorial by confounding $C^2 D$ or $C D^2$ by modulo technique. The confounded designs I and II after confounding are as follows:

Design-I	*Design-II*
$(AB)_{0 \bmod 2}$ 00 11	$(C^2 D)_{0 \bmod 3}$ 00 11 22
$(AB)_{1 \bmod 2}$ 01 10	$(C^2 D)_{1 \bmod 3}$ 01 12 20
	$(C^2 D)_{2 \bmod 3}$ 02 10 21

Now collapse the blocks of design-I with the blocks of design-II systematically. This provides six blocks of size six units. Thus the combined treatment combinations of designs I and II after collapsing result into six blocks of size 6 units providing the required design.

Blocks	*Treatment combinations*					
1.	0000	0011	0022	1100	1111	1122
2.	0001	0012	0020	1101	1112	1120
3.	0002	0010	0021	1102	1110	1121
4.	0100	0111	0122	1000	1011	1022
5.	0101	0112	0120	1001	1012	1020
6.	0102	0110	0121	1002	1010	1021

In the above layout the confounded effects including their generalized interactions with factors A and B reduced to modulo 2 and factors C and D reduced to modulo 3 are: AB, $C^2 D$, $C D^2$, ABC^2D, $ABCD^2$.

5.7 BALANCED CONFOUNDING IN A 3 × 3 × 2 FACTORIAL IN BLOCKS OF SIZE SIX UNITS

Let there be two factors A and B each having 3 levels 0, 1, 2 and a factor C with two levels 0 and 1. The experiment has 18 treatment combinations. Blocks of size 6 units have to formed such that main effects and first order interactions which are not confounded be saved. This can easily be achieved by confounding AB separately by modulo approach to yield design-I. Then combining three blocks of design-I with two levels of factor C, the final layout of confounded design is obtained.

Blocks	Design-I	Design-II	Final confounded design					
1	(AB)$_0$: 00 12 21	0 1	000	001	120	121	210	211
2	(AB)$_1$: 01 10 22		010	011	100	101	220	221
3	(AB)$_2$: 02 11 20		020	021	110	111	200	201

In the above layout, the confounded effects are: AB and ABC.

In the above design, it is easy to verify that the main effects A, B, C and first order interactions AC and BC are not confounded.

In this manner, confounding in any asymmetrical factorial experiments can be carried out to obtain blocks of desired size within its feasibility. Required number of replications can be had in an experiment by doing independent randomization of entries within each block.

5.8 CONSTRUCTION OF A CONFOUNDED DESIGN IN BLOCKS OF SIZE 12 IN ASYMMETRICAL FACTORIAL 4 × 6 × 3

The design is constructed by the method of collapsing the columns of symmetrical factorial designs. Let the three factors be A, B and C at levels 0, 1, 2, 3 ; 0, 1, 2, 3, 4, 5 and 0, 1, 2, respectively. Convert four levels of factor A as 2^2 experiment with pseudo factors X_1 and X_2 each at two levels 0, 1 and six levels of B as 2 × 3 experiment with pseudo factors Y_1 and Y_2 at levels 0, 1 and 0, 1, 2, respectively. Combining two levels each of X_1 and X_2 with levels of Y_1 results into a 2^3 experiment whereas combining of 3 levels of Y_2 with levels of C forms 3^2 factorial experiment. Confound the interaction $X_1X_2Y_1$ in 2^3 experiment to form a confounded design-I consisting of blocks of size four units. Also construct a design-II having blocks of size three units by confounding the effect Y_2C in 3^2 experiment.

Following mapping will be used in levels of factors and pseudo factors.

A	X_1	X_2	B	Y_1	Y_2	C
0	0	0	0	0	0	0
1	0	1	1	0	1	1
2	1	0	2	0	2	2
3	1	1	3	1	0	
			4	1	1	
			5	1	2	

The confounded designs I and II are as given below.

Blocks	Design-I				Blocks	Design-II		
1.	000	011	101	110	1.	00	12	21
2.	001	010	100	111	2.	01	10	22
					3.	02	11	20
	Effect confounded $X_1X_2Y_1$					Effect confounded Y_2C		

Now collapsing designs I and II, the design having blocks of size 12 units with five digit treatment combinations in the order $X_1X_2Y_1Y_2 C$ will be constructed. In this design, replace the two digit levels of X_1X_2 by 0, 1, 2, 3 and of Y_1Y_2 by 0, 1, 2, 3, 4, 5 as per above mapping to obtain the required design.

The effects which are confounded in the construction of this design are $X_1X_2Y_1$, Y_2C and $X_1X_2Y_1Y_2C$. In this way, the degrees of freedom for confounded effects are five, i.e. 1 d.f. for $X_1X_2Y_1$, 2 d.f. for Y_2C and 2 d.f. for $X_1X_2Y_1Y_2C$.

The design with blocks of size 12 units for the treatment combinations $X_1X_2Y_1Y_2C$ by collapsing designs I and II is as displayed below.

Blocks

1	2	3	4	5	6
00000	00001	00002	00100	00101	00102
00012	00010	00011	00112	00110	00111
00021	00022	00020	00121	00122	00120
01100	01101	01102	01000	01001	01002
01112	01110	01111	01012	01010	01011
01121	01122	01120	01021	01022	01020
10100	10101	10102	10000	10001	10002
10112	10110	10111	10012	10010	10011
10121	10122	10120	10021	10022	10020
11000	11001	11002	11100	11101	11102
11012	11010	11011	11112	11110	11111
11021	11022	11020	11121	11122	11120

Through mapping, replace the levels of X_1X_2 and Y_1Y_2 by the original levels. On doing so, required confounded design is obtained as displayed below.

Layout of final confounded design

Blocks

1	2	3	4	5	6
000	001	002	030	031	032
012	010	011	042	040	041
021	022	020	051	052	050
130	131	132	100	101	102
142	140	141	112	110	111
151	152	150	121	122	120
230	231	232	200	201	202
242	240	241	212	210	211
251	252	250	221	222	220
300	301	302	330	331	332
312	310	311	342	340	341
321	322	320	351	352	350

The entries of each block should be allocated randomly within the blocks and independently in any number of replications.

5.9 BALANCED CONFOUNDING IN A 3 × 2 × 2 FACTORIAL IN BLOCKS OF SIZE SIX UNITS

The factorial under consideration has 12 treatment combinations of the effects say, A, B and C in order. A has three levels 0, 1, 2 and each of B and C has two levels 0, 1. If one confounds either of the main effects B or C, or the interaction B × C, blocks of size six units will be obtained. The problem is quite simple. But such a design is categorized as split plot design which the readers will study in the next chapter. All the more, main effects are usually protected from confounding. So confounding in a 3 × 2 × 2 factorial has been carried out by converting it to a 2^4 symmetrical factorial by introducing two pseudo factors X_1 and X_2 for A each at two levels as discussed in beginning of this topic. This gives rise to four factors each at two levels 0 and 1. Now creating equivalence among the levels of A and X_1X_2 combination, it is evident that the level combination 11 does not exist as shown below.

A	X_1X_2
0	00
1	01
2	10
	11

In this way, out of sixteen treatment combinations of X_1X_2 BC, four combinations containing levels 11 will be rejected leaving only twelve combinations as depicted below.

Level combinations of X_1X_2 BC: 0000, 0001, 0010, 0011, 0100, 0101, 0110, 0111, 1000, 1001, 1010, 1011, <u>1100, 1101, 1110, 1111</u>

To be rejected

Unrejected twelve treatment combinations will be utilized in the construction of the design. To obtain a balanced confounded design with blocks of size six units, partial confounding will be carried out by confounding X_1BC, X_2BC and X_1X_2 BC in first, second and third replications, respectively. The layout of the required experiment is as delineated below.

Replication - I		Replication - II		Replication - II.	
Bl.-1	Bl.-2	Bl.-3	Bl.-4	Bl.-5	Bl.-6
X_1BC confounded		X_2BC confounded		X_1X_2BC confounded	
$(X_1BC)_0$	$(X_1BC)_1$	$(X_2BC)_0$	$(X_2BC)_1$	$(X_1X_2BC)_0$	$(X_1X_2BC)_1$
0000	0001	0000	0001	0000	0001
0011	0010	0011	0010	0011	0010
0100	0101	0101	0100	0101	0100
0111	0110	0110	0111	0110	0111
1001	1000	1000	1001	1001	1000
1010	1011	1011	1010	1010	1011

Reverting the level of X_1X_2 to the original levels of A by equivalence, the entries of the above design result into the final layout as follows:

Replication-I		Replication-II		Replication-III	
Bl.-1	Bl.-2	Bl.-3	Bl.-4	Bl.-5	Bl.-6
X_1BC confounded		X_2BC confounded		X_1X_2BC confounded	
000	001	000	001	000	001
011	010	011	010	011	010
100	101	101	100	101	100
111	110	110	111	110	111
201	200	200	201	201	200
210	211	211	210	210	211

Blocks within replications and entries within blocks should be assigned randomly and independently.

5.10 ANALYSIS OF BALANCED CONFOUNDED DESIGNS IN CASE OF ASYMMETRICAL FACTORIALS

The analysis discussed in this section is for randomized block design. The analysis of factorial experiments has already been discussed in adequate details. In balanced confounded asymmetrical factorial block designs also, the sum of squares for all effects except confounded effects are calculated in the usual way. Only problem lies with the effects confounded with the blocks. Due to confounding, the data are non-orthogonal. The analysis of data can be carried out by two approaches. One simple approach is to adjust each observation for the block effects by subtracting the average of the block entries to

which they belong. That is $y_{iju} - \dfrac{\beta_j}{k} = x_{iju}$, where k is the block size and x_{iju}

is the adjusted observation. In this situation, the analysis of data is performed in the usual manner using x_{iju} values. The analysis provides exactly the same sum of squares of all main effects and interactions which are not confounded as would be obtained with unadjusted data. But sum of square for confounded interaction(s) has to be divided by a fraction of information recovered for it.

Definition: The fraction of information is the ratio of the variance of the contrast for the effect without confounding to the variance with confounding.

For certain standard designs, the value of fraction of information is known. That value can be used as such. This makes the analysis very handy. In case, fraction of information is not known, one would have to obtain it by utilizing the method of analysis of non-orthogonal data given in section (2.7).

Analysis by second approach: Balanced confounded designs which are under consideration have blocks of equal size, say k. Also each treatment combination is replicated equal number of times, say r. Based on these ideas, method of finding the fraction of information for the confounded effects in case of $3 \times 2 \times 2$ confounded design in which the interactions ABC and BC are confounded is delineated here.

For each confounded interaction, a linear function of treatment combinations using the coefficients 1, −1 and 0 has been formed in such a way that they represent the contrasts for the confounded interactions. Such linear functions are called reparametrized treatments or in short R-treatments. For the interaction ABC, consider the R-treatments t_0, t_1 and t_2 as a function of treatment combinations as follows:

$$t_0 = a_0 (b_1c_1 + b_0c_0 - b_1c_0 - b_0c_1)$$
$$t_1 = a_1 (b_1c_1 + b_0c_0 - b_1c_0 - b_0c_1)$$
$$t_2 = a_2 (b_1c_1 + b_0c_0 - b_1c_0 - b_0c_1)$$

Above R-treatments are formed by multiplying the contrast for BC by a_0, a_1 and a_2, respectively.

R-treatments, say v_0 and v_1 for BC as linear functions can be defined as,

$(BC)_{0 \bmod 2} = v_0 = a_0 b_1 c_1 + a_0 b_0 c_0 + a_1 b_1 c_1 + a_1 b_0 c_0 + a_2 b_1 c_1 + a_2 b_0 c_0$

$(BC)_{1 \bmod 2} = v_1 = a_0 b_1 c_0 + a_0 b_0 c_1 + a_1 b_1 c_0 + a_1 b_0 c_1 + a_2 b_1 c_0 + a_2 b_0 c_1$

The above functions for v_0 and v_1 are developed in contrast BC having positive and negative signs with three levels of A, respectively.

For convenience, the layout of the design is reproduced.

Replication-I		Replication-II		Replication-III	
Block-1	*Block-2*	*Block-3*	*Block-4*	*Block-5*	*Block-6*
000	001	000	001	000	001
011	010	011	010	011	010
100	101	101	100	101	100
111	110	110	111	110	111
201	200	200	201	201	200
210	211	211	210	210	211

Reduced normal equations (2.7.6) and (2.7.7) are repeated below to have them at a glance.

$$\left(n_{i.} - \sum_j \frac{n_{ij}^2}{n_{.j}} \right) \hat{\tau}_i - \sum_{i \neq l} \left(\sum_j \frac{n_{ij} n_{lj}}{n_{.j}} \right) \hat{\tau}_l = Q_i \qquad (5.10.1)$$

Where, $$Q_i = y_{i..} - \sum_j \frac{n_{ij} y_{.j.}}{n_{.j}} \qquad (5.10.2)$$

Here it is worth emphasizing that our treatments are t_0, t_1 and t_2 for ABC and v_0, v_1 for BC. To write the reduced normal equations, the values of frequencies for the given design are:

	Bl.-1	*Bl.-2*	*Bl.-3*	*Bl.-4*	*Bl.-5*	*Bl.-6*
t_0	$n_{01} = 2$	$n_{02} = -2$	$n_{03} = 2$	$n_{04} = -2$	$n_{05} = 2$	$n_{06} = -2$
t_1	$n_{11} = 2$	$n_{12} = -2$	$n_{13} = -2$	$n_{14} = 2$	$n_{15} = -2$	$n_{16} = 2$
t_2	$n_{21} = -2$	$n_{22} = 2$	$n_{23} = 2$	$n_{24} = -2$	$n_{25} = -2$	$n_{26} = 2$

Also $n_{0.} = n_{1.} = n_{2.} = 12$

$\sum_j n_{0j}^2 = \sum_j n_{1j}^2 = \sum_j n_{2j}^2 = 24$

and $n_{.j} = 6$ for all blocks. Similarly, when $\hat{\tau}_i = t_0$ and $\hat{\tau}_l = t_1$

	Bl.-1	*Bl.-2*	*Bl.-3*	*Bl.-4*	*Bl.-5*	*Bl.-6*
t_0	$n_{01} = 2$	$n_{02} = -2$	$n_{03} = 2$	$n_{04} = -2$	$n_{05} = 2$	$n_{06} = -2$
t_1	$n_{11} = 2$	$n_{12} = -2$	$n_{13} = -2$	$n_{14} = 2$	$n_{15} = -2$	$n_{16} = 2$

$$\sum_j n_{ij} n_{lj} = 2 \times 2 + (-2) \times (-2) + 2 \times (-2) + (-2) \times 2 + 2 \times (-2) + (-2) \times 2$$

$$= 4 + 4 - 4 - 4 - 4 - 4 = -8$$

Similarly, when $\hat{\tau}_i = t_0$ and $\hat{\tau}_l = t_2$

$$\sum_j n_{ij} n_{lj} = 2 \times (-2) + (-2) \times 2 + 2 \times 2 + (-2) \times (-2) + 2 \times (-2) + (-2) \times 2$$

$$= -4 - 4 + 4 + 4 - 4 - 4 = -8$$

The reduced normal equation for $i = 0$ and $l = 1, 2$ is,

$$\left(12 - \frac{24}{6}\right) t_0 - \left(-\frac{8}{6} t_1 - \frac{8}{6} t_2\right) = Q_{t_0} \tag{5.10.3}$$

or $\quad 8t_0 + \dfrac{4}{3}(t_1 + t_2) = Q_{t_0}$ (5.10.3.1)

where $\quad Q_{t_1} = T_{t_0} - \dfrac{1}{6}(2\,B_1 - 2\,B_2 + 2\,B_3 - 2\,B_4 + 2\,B_5 - 2\,B_6)$ (5.10.4)

where, B_j represents the jth block total and T_{t_0}, the total for treatment combinations belonging to t_0.

Similarly, the reduced normal equations when $i = 1$ and l takes the values 0, 2 and when $i = 2$ and l takes on the valued 0, 1 are,

$$8t_1 + \frac{4}{3}(t_0 + t_2) = Q_{t_1} \tag{5.10.5}$$

Where $\quad Q_{t_1} = T_{t_1} - \dfrac{1}{6}(2B_1 - 2B_2 - 2B_3 + 2B_4 - 2B_5 + 2B_6)$ (5.10.6)

Again, $\quad 8t_2 + \dfrac{4}{3}(t_0 + t_1) = Q_{t_2}$ (5.10.7)

where, $\quad Q_{t_2} = T_{t_2} - \dfrac{1}{6}(-2B_1 + 2B_2 + 2B_3 - 2B_4 - 2B_5 + 2B_6)$ (5.10.8)

In the similar manner, the reduced normal equations for the interaction BC can be obtained as follows:

	Bl.-1	Bl.-2	Bl.-3	Bl.-4	Bl.-5	Bl.-6
v_0	$n_{01} = 4$	$n_{02} = 2$	$n_{03} = 4$	$n_{04} = 2$	$n_{05} = 2$	$n_{06} = 4$
v_1	$n_{11} = 2$	$n_{12} = 4$	$n_{13} = 2$	$n_{14} = 4$	$n_{15} = 4$	$n_{16} = 2$

$n_{.j} = 6$, $n_{0.} = n_{1.} = 18$, and $\displaystyle\sum_j n_{0j}^2 = \sum_j n_{1j}^2 = 60$

$$\sum_j n_{0j} n_{1j} = 4 \times 2 + 2 \times 4 + 4 \times 2 + 2 \times 4 + 2 \times 4 + 4 \times 2 = 48$$

when $i = 0$ and $1 = 1$, the reduced normal equation is,

$$\left(18 - \frac{60}{6}\right) v_0 - \frac{48}{6} v_1 = Q_{v0} \qquad (5.10.9)$$

Where, $Q_{v0} = T_{v0} - \frac{1}{6}(4 B_1 + 2 B_2 + 4 B_3 + 2 B_4 + 2 B_5 + 4 B_6)$ (5.10.10)

Similarly, when $i = 1$ and $1 = 0$, the reduced normal equation is,

$$\left(18 - \frac{60}{6}\right) v_1 - \frac{48}{6} v_0 = Q_{v1} \qquad (5.10.11)$$

where, $Q_{v1} = T_{v1} - \frac{1}{6}(2 B_1 + 4 B_2 + 2 B_3 + 4 B_4 + 4 B_5 + 2 B_6)$ (5.10.12)

Reduced normal equations in t's and v's are again written with slight modifications.

$$\frac{20}{3} t_0 + \frac{4}{3} \sum_{i=1}^{3} t_i = Q_{t0} \qquad (5.10.13)$$

$$\frac{20}{3} t_1 + \frac{4}{3} \sum_{i=1}^{3} t_i = Q_{t1} \qquad (5.10.14)$$

$$\frac{20}{3} t_2 + \frac{4}{3} \sum_{i=1}^{3} t_i = Q_{t2} \qquad (5.10.15)$$

And

$$8v_0 - 8v_1 = Q_{v0} \qquad (5.10.16)$$

$$- 8v_0 + 8v_1 = Q_{v1} \qquad (5.10.17)$$

Under the assumption $\sum_{i=1}^{3} t_i = 0$, equations (5.9.13) through (5.9.15) come

out to be,

$$t_0 = \frac{3}{20} Q_{t0} \qquad (5.10.18)$$

$$t_1 = \frac{3}{20} Q_{t1} \qquad (5.10.18)$$

$$t_2 = \frac{3}{20} Q_{t2} \qquad (5.10.19)$$

Now two rules are enunciated which shall be useful in further analysis of data.

Rule 1: The variance of a linear function of the true treatment effec, say,
$Z = \alpha_1 \tau_1 + \alpha_2 \tau_2 + \ldots + \alpha_v \tau_v$ which is estimated by $\lambda_1 Q_{t1} + \lambda_2 Q_{t2} + \ldots$
$+ \lambda_v Q_{tv}$, where Q_{ti} is the expression for treatment i, ($i = 1, 2, \ldots, v$), Z is
given by $(\alpha_1 \lambda_1 + \alpha_2 \lambda_2 + \ldots + \alpha_v \lambda_v) \sigma^2$.

Rule 2: If there is no confounding, the relation between t's and Q_i's from

(5.10.1) shall be $n_i t_i = Q_{ti}$ for $i = 0, 1, 2$ or $t_i = \dfrac{1}{n_{i.}} Q_{ti}$. The variance of any

contrast among t_i's is easily obtainable by the correspondence among Q_{ti}'s by
rule-1.

Making use of the above rules, it is trivial to obtain the variances of the
contrast $(t_0 - t_1)$ with confounding and without confounding. In the same
manner, the variances of the contrast $(v_0 - v_1)$ can be obtained. The ratio of the
variance with no confounding to the variance with confounding will provide
the fraction of information which is used to divide the sum of square of the
confounded effect.

The contrast $(t_0 - t_1)$ is estimated by the contrast $\dfrac{3}{20}(Q_{t0} - Q_{t1})$ (5.10.20)

Thus,

$$\text{Var}(t_0 - t_1) = \left\{ 1 \times \frac{3}{20} + (-1) \times \left(-\frac{3}{20} \right) \right\} \sigma^2 = \frac{6}{20} \sigma^2 \quad (5.10.21)$$

If there were no confounding, following relations hold.

$$12\, t_0 = Q_{t0} \text{ and } 12\, t_1 = Q_{t1}$$

or
$$(t_0 - t_1) = \frac{1}{12}(Q_{t0} - Q_{t1}) \quad (5.10.22)$$

Thence,

$$\text{Var}(t_0 - t_1) = \left\{ 1 \times \frac{1}{12} + (-1)\left(-\frac{1}{12} \right) \right\} \sigma^2 = \frac{2}{12} \sigma^2 \quad (5.10.23)$$

The fraction of information for the interaction ABC as per definition is,

$$\frac{\left(\dfrac{2}{12} \sigma^2 \right)}{\left(\dfrac{6}{20} \sigma^2 \right)} = \frac{5}{9} \quad (5.10.24)$$

Similarly, the variance of the contrast $(v_0 - v_1)$ can be estimated by the
contrast in Q_{tv}'s with the help of (5.10.16) and (5.10.17) as,

$$v_0 - v_1 = \frac{1}{16}(Q_{v0} - Q_{v1}) \qquad (5.10.25)$$

The variance of the contrast, when BC is confounded, by the rule-1 is,

$$\text{Var}\,(v_0 - v_1) = \left\{ 1 \times \frac{1}{16} + (-1)\left(-\frac{1}{16}\right) \right\} \sigma^2 = \frac{2}{16}\sigma^2 \qquad (5.10.26)$$

When BC is not confounded,
$$18\,v_0 = Q_{v0} \text{ and } 18\,v_1 = Q_{v1}$$

Or
$$(v_0 - v_1) = \frac{1}{18}(Q_{v0} - Q_{v1}) \qquad (5.10.27)$$

Thus,
$$\text{Var}\,(v_0 - v_1) = \left\{ 1 \times \frac{1}{18} + (-1)\left(-\frac{1}{18}\right) \right\} \sigma^2 = \frac{2}{18}\sigma^2 \qquad (5.10.28)$$

The fraction of information for the interaction BC is,

$$\frac{\dfrac{2}{18}\sigma^2}{\dfrac{2}{16}\sigma^2} = \frac{8}{9} \qquad (5.10.29)$$

As regards the analysis of variance, all the sum of squares will be calculated in the usual way except the sum of squares for the confounded effects. The break up of degrees of freedom is also simple. For $3 \times 2 \times 2$ example discussed herewith, sum of square for the interaction ABC is,

$$= \frac{3}{20}\left\{ \sum_{i=1}^{3} Q_{ti}^2 - \frac{\left(\sum_i Q_{ti}\right)^2}{3} \right\} \qquad (5.10.30)$$

And sum of square for BC is,

$$= \frac{1}{16}\sum_{i=1}^{2} Q_{vi}^2 \qquad (5.9.31)$$

In the calculation of sum of square due to BC, correction factor $\dfrac{\left(\sum\limits_{i=1}^{2} Q_{vi}\right)^2}{2}$ has not been subtracted as $\sum\limits_{i} Q_{vi} = 0$. Analysis of variance table can be prepared in the usual manner and conclusions should be drawn by following the same procedure as for any factorial experiments. Adjusted sum of squares for ABC and BC will be obtained by dividing their sum of squares by 5/9 and 8/9, respectively.

The method of analysis of variance of $3 \times 2 \times 2$ mixed factorial would be utilized in the following example as such.

Example 5.2: An experiment was conducted to study the response of fertilizers and biofertilizers on sorghum crop. The treatments consisted of three levels of nitrogen (N), (0, 30, 60 kg/ha) coded as 0, 1, 2; two levels of biofertilizers, i.e. no azospirillium, azospirillium (F) coded as 0, 1; two doses of phosphorus, (P), (0 and 40 kg/ha) coded as 0, 1. Due to field conditions, it was decided to take blocks of six plots of size 4.50×5.00 m^2. The design was a $3 \times 2 \times 2$ asymmetrical confounded randomized block design in which the effects X_1FP, X_2FP and X_1X_2FP were confounded. [X_1 and X_2 are pseudo factors for N). The layout with grain yield (kg/plot) is as tabulated below. Block totals and block means also displayed for the sake of brevity.

Table-I

Replication - I		*Replication - II*		*Replication – III*	
Block-1	*Block-2*	*Block-3*	*Block-4*	*Block-5*	*Block-6*
000-4.60	001-4.00	000-4.10	001-6.10	000-4.20	001-5.20
011-5.60	010-6.50	011-5.30	010-6.00	011-6.00	010-5.60
100-5.70	101-7.90	101-7.00	100-4.80	101-6.80	100-5.90
111-8.40	110-8.20	110-7.60	111-10.20	110-6.80	111-8.90
201-7.60	200-5.90	200-4.80	201-6.90	201-7.00	200-5.40
210-8.00	211-10.00	211-9.60	210-7.60	210-7.50	211-8.80
X_1FP confounded		X_2 FP confounded		X_1X_2 FP confounded	

Total	39.90	42.50	38.40	41.60	38.30	39.80
Mean	6.65	7.08	6.40	6.93	6.38	6.63

Also $Y_{.1.} = 82.4$, $Y_{.2.} = 80.0$, $Y_{.3} = 78.1$, $Y_{...} = 240.50$

The analysis of data will be carried out by the procedure described in the preceding section without any explanation.

Table-II contains adjusted observation, x_{iju}, for NFP combinations given in the preceding Table-I.

Table-II

Replication – I		Replication –II		Replication – III	
Bl.-1 x_{iju}	Bl.-2 x_{iju}	Bl.-3 x_{iju}	Bl.- 4 x_{iju}	Bl.-5 x_{iju}	Bl.-6 x_{iju}
000 - -2.05	001 - -3.08	000 - -2.30	001 - -0.83	000 - -2.18	001 - -1.43
011 - -1.05	010 - -0.58	011 - -1.10	010 - -0.93	011 - -0.38	010 - -1.03
100 - -0.95	101 - 0.82	101 - 0.60	100 - -2.13	101 - 0.42	100 - -0.73
111 - 1.75	110 - 1.12	110 - 1.20	111 - 3.27	110 - 0.42	111 - 2.27
201 - 0.95	200 - -1.18	200 - -1.60	201 - -0.03	201 - 0.62	200 - -1.23
210 - 1.35	211 - 2.92	211 - 3.20	210 - 0.67	210 - 1.12	211 - 2.17

Firstly calculate sum of squares due to blocks, replications and blocks within replications and total from unadjusted observations.

$$\text{Total S.S.} = 4.60^2 + 5.60^2 + \ldots\ldots + 5.40^2 + 8.80^2 - \frac{Y^2_{\ldots}}{36}$$

$$= 1702.95 - \frac{(240.5)^2}{36}$$

$$= 1702.95 - 1606.67 = 96.28$$

$$\text{Blocks S.S.} = \frac{1}{6}(39.90^2 + 42.50^2 + \ldots\ldots + 38.30^2 + 39.80^2) - \text{C.F.}$$

$$= 1609.05 - 1606.67 = 2.38$$

$$\text{Rep. S.S.} = \frac{1}{12}\left(82.4^2 + 80.0 + 78.1^2\right) - \text{C.F.}$$

$$= 1607.75 - \text{C.F.}$$

$$= 0.78$$

Blocks within replication S.S.

$$= \frac{1}{6}(39.90^2 + 42.50^2) - \frac{82.40^2}{12} + \frac{1}{6}(38.40^2 + 41.60^2) - \frac{80.00^2}{12}$$

$$+ \frac{1}{6}(38.30^2 + 39.80^2) - \frac{78.10^2}{12}$$

$$= 566.3766 - 565.8133 + 534.1866 - 533.3333 + 508.4883 - 508.3008$$

$$= 0.5633 + 0.8533 + 0.1875 = 1.6041 \approx 1.60$$

For calculating the sum of squares for main effects and interactions three two-way tables and one three-way tables have been prepared as displayed ahead.

Table-III N × F table

Levels N / F	0	1	2	Total
0	− 11.87	−1.97	−2.47	−16.31
1	− 5.07	10.03	11.43	16.39
Total	− 16.94	8.06	8.96	0.08

Table-IV F × P table

Levels F / P	0	1	Total
0	− 14.35	3.34	− 11.01
1	− 1.96	13.05	11.09
Total	− 16.31	16.39	0.08

Table-V N × P table

Levels N / P	0	1	2	Total
0	− 9.07	− 1.07	− 0.87	− 11.01
1	−7.87	9.13	9.83	11.09
Total	16.94	8.06	8.96	0.08

Table-VI N × F × P table

Levels N	0		1		2	
Levels F	0	1	0	1	0	1
Levels P						
0	− 6.53	− 2.54	− 3.81	2.74	− 4.01	3.14
1	− 5.34	− 2.53	1.84	7.29	1.54	8.29

Note: The sum of all adjusted observations comes out to be 0.08 due to rounding of blocks average values. Actually, it should have been zero. Hence, it will be treated as zero and no correction factor will appear in the calculations of sum of squares.

$$\text{S.S. (F)} = \frac{1}{18}\{(-16.31)^2 + (16.39)^2\} = 29.70$$

S.S. (N) = = 36.02

$$S.S.(N \times F) = \frac{1}{6}\{(-11.87)^2 + (-1.97)^2 + \ldots\ldots + (10.03)^2 + (11.43)^2\}$$

$$- 29.70 - 36.02$$
$$= 67.97 - 29.70 - 36.02 = 2.25$$

$$S.S.(P) = \frac{1}{18}\{(-11.01)^2 + (11.09)^2\} = 13.57$$

$$S.S.(N \times P) = \frac{1}{6}\{(-9.07)^2 + (-1.07)^2 + \ldots\ldots + (9.13)^2 + (9.83)^2\}$$

$$- 36.02 - 13.57$$
$$= 54.35 - 36.02 - 13.57 = 4.76$$

$$S.S.(F \times P) = \frac{1}{9}\{(-14.35)^2 + (3.34)^2 + (-1.96)^2 + (13.05)^2\} - 29.70 - 13.57$$

$$= 43.47 - 29.70 - 13.57 = 0.20$$

$$S.S.(N \times F \times P) = \frac{1}{3}\{(-6.35)^2 + (-2.54)^2 + - - - - - + (1.54)^2 + (8.29)^2\}$$

$$- 29.70 - 36.02 - 2.25 - 13.57 - 4.76 - 0.20$$
$$= 86.53 - 86.50 = 0.03$$

The fractions of information for F × P and N × F × P have already been derived in the preceding section are 8/9 and 5/9, respectively. Hence, the adjusted sum of squares due to F × P and N × F × P are,

$$\text{Adj. S.S. } (F \times P) = \frac{0.20}{8/9} = 0.225$$

$$\text{Adj. S.S. } (N \times F \times P) = \frac{0.03}{5/9} = 0.054$$

Adjusted sum of squares will be used in the following ANOVA table.

Now it seems interesting to show empirically that sum of squares for F × P and N × F × P worked out by above method are same as would be obtained through R-treatments.

Calculate first the R-treatments for the interaction N × F × P and F × P.

$$T_{r0} = 011 + 000 - 010 - 001$$
$$= 16.9 + 12.9 - 18.1 - 15.3 = -3.60$$
$$T_{r1} = 111 + 100 - 110 - 101$$
$$= 27.5 + 16.4 - 22.6 - 21.7 = -0.04$$
$$T_{r2} = 211 + 200 - 210 - 201$$
$$= 28.4 + 16.1 - 23.1 - 21.5 = -0.10$$

$$v_0 = 011 + 000 + 111 + 100 + 211 + 200$$
$$= 16.9 + 12.9 + 27.5 + 16.4 + 28.4 + 16.1 = 118.2$$
$$v_1 = 010 + 001 + 110 + 101 + 210 + 201$$
$$= 18.1 + 15.3 + 22.6 + 21.7 + 23.1 + 21.5 = 122.3$$

From the formulae (5.10.4) through (5.10.12), the values of Q_{pi} ($i = 1, 2, 3$), Q_{v0} and Q_{v1} are,

$$Q_{t0} = -3.60 - \frac{2}{6}(39.90 - 42.5 + 38.40 - 41.60 + 38.30 - 39.80)$$

$$= -3.60 + \frac{2}{6} \times 7.30 = -3.60 + 2.43 = -1.17$$

$$Q_{t1} = -0.40 - \frac{2}{6}(39.90 - 42.50 - 38.40 + 41.60 - 38.30 + 39.80)$$

$$= -0.40 - \frac{2}{6} \times 2.10 = -0.40 - 0.70 = -1.10$$

$$Q_{t2} = -0.10 - \frac{2}{6}(-39.90 + 42.50 + 38.40 - 41.60 - 38.30 + 39.80)$$

$$= -0.10 - \frac{2}{6} \times 0.9 = -0.10 - 0.30 = -0.40$$

$$Q_{v0} = 118.2 - \frac{1}{6}(4 \times 39.9 + 2 \times 42.5 + 4 \times 38.4$$
$$+ 2 \times 41.6 + 2 \times 38.30 + 4 \times 39.8)$$

$$= 118.2 - \frac{1}{6} \times 717.2 = 118.2 - 119.53 = -1.33$$

$$Q_{v1} = 122.3 - \frac{1}{6}(2 \times 39.9 + 4 \times 42.5 + 2 \times 38.4$$
$$+ 4 \times 41.6 + 4 \times 38.3 + 2 \times 39.8)$$

$$= 122.3 - \frac{1}{6} \times 725.82 = 122.3 - 120.97 = 1.33$$

$$\text{S.S. } (N \times F \times P) = \frac{3}{20}\left\{(-1.17)^2 + (-1.10)^2 + (0.40)^2 - \frac{(-2.67)^2}{3}\right\}$$

$$= \frac{3}{20}(2.7389 - 2.3763) = 0.054$$

$$\text{S.S. } (F \times P) = \frac{1}{16}\left\{(-1.33)^2 + (1.33)^2\right\} = 0.221$$

A slight difference in third decimal place of S.S. (F × P) under two approach may be attached to rounding of figures. So this purports that one may find the sum of squares for confounded effects from either of the two approaches.

ANOVA table

Source	d.f.	S.S.	M.S.	F-value
Between blocks	5	2.38	0.476	1.236 NS
Replications	2 ⎤	0.78	0.390	1.028 NS
Blocks within replications	3 ⎦	1.60	0.533	1.384 NS
Treatments	11			
N	2	36.02	18.01	46.779*
F	1	29.70	29.70	77.143*
N × F	2	2.25	1.125	2.922 NS
P	1	13.57	13.57	35.247*
N × P	2	4.76	2.38	6.182*
F × P	1'	0.225	0.225	0.584 NS
N × F × P	2'	0.054	0.027	0.070 NS
Error	19	7.321	0.385	
Total	35	96.28		

Table values of F: $F_{0.05; 1,19} = 4.38$, $F_{0.05; 2, 19} = 3.52$, $F_{0.05; 3,19} = 3.13$, $F_{0.05; 5,19} = 2.74$.

Significant at 5% level of significance, NS – Non-significant

Comparing calculated F-values for main effects and interactions with corresponding tabulated F-values, it is inferred that all main effects are significant. Also interaction F × P is significant. Individual level comparisons for N and N × P can be done by using fisher's lsd. This needs no elaboration.

Concluding remarks

Confounding is often used in factorial experiments to reduce block size so as to maintain homogeneity of blocks in respect of various nuisance variables. Confounding in experiments is mostly applied in the areas of field experiments, decontamination experiments, and industrial, medical science researches. Those who are involved in research work will find that confounded designs are very helpful when they have factorial treatments good in number and each having many levels. The matter covered in this chapter provides enough knowledge about confounded designs so as to understand and utilize the concepts and analysis procedures in the right manner.

QUESTIONS AND EXERCISES

1. Write an annotation on the importance of confounding in factorial experimental designs.
2. Define the following terms:
   (a) Complete confounding.
   (b) Partial confounding.
   (c) Generalized interaction.
   (d) Fraction of information.
   (e) M = x Modulo p.
   (f) Balanced confounding
3. Differentiate between the following:
   (a) Symmetrical and asymmetrical factorials.
   (b) Real and pseudo factors.
   (c) Contrast method and modulo technique of confounding.
4. For confounding in asymmetrical factorials in what way one has to decide about the size of blocks.
5. In what situations contrasts can be used for confounding various effects in different blocks?
6. Give the layout of a partial confounded design for a 2^5 factorial experiment in which the blocks of size eight are admissible. The layout is in a randomized block design with two replications. In one block effect ABC and in other block effect ACD are confounded. Show that the effect BD is automatically confounded.
7. Give the layout of a 2^5 factorial experiment of totally confounded design in RBD dividing the blocks into 1/2-replicate.
8. In what manner, confounding can be done by converting asymmetrical factorials into symmetrical factorials?
9. What is the basis of analysis of data of asymmetrical factorial confounded designs? Give the outline of the same.
10. How the confounded designs in asymmetrical factorials be constructed by collapsing the blocks? Explain by choosing a suitable example.
11. Construct a partially confounded design for a 2^4 factorial design in which the effects ABC and BDE are confounded in two replications, respectively. Also give the skeleton analysis of variance table for this design.
12. What are the merits and demerits of partial and total confounding?
13. Give the layout of 2^5 factorial design having blocks of size 8 units. Explain main features of the design.
14. Determine the treatment combinations of a 3^3 factorial design to obtain blocks of one-third replicate.

15. Display the layout of an asymmetrical confounded design for a $2^2 \times 3^2$ factorial in blocks of nine units. Also give the skeleton ANOVA table for the said design.
16. What is the basis of analysis of data of confounded designs?
17. Can Yates' method of analysis of data be used in confounded designs for 2^n factorials? Justify your statement by taking a suitable example.
18. An experiment on soybean was conducted to see the effect of three fertilizers in combination with sulphur each at two levels coded as 0, 1. The levels of nitrogen (N) were 0 and 20 kg/ha, phosphorus (P) 0 and 40 kg/ha, potash (K) 0 and 20 kg/ha, sulphur (S) 0 and 20 kg/ha. It was considered that a plot of 16 plots will be too large to maintain homogeneity. Therefore, partially confounded design was used. In one replication, the interaction NPK and in the second replication PKS was confounded. The layout and yield q/ha were as displayed below.

| Replication – I | | | | Replication – II | | | |
Block-1		Block-2		Block-3		Block-4	
p	11.27	n p	13.40	p k	12.34	n p k s	15.38
n s	15.38	n k s	13.43	n p k	13.78	n p	14.34
k	10.70	n p s	14.38	k s	13.00	n s	14.60
n p k	13.41	p k s	12.93	(1)	10.23	p	11.36
k s	12.94	p k	12.76	n	11.91	s	10.01
n p k s	15.38	s	10.23	n p s	14.67	n k	12.94
n	12.74	n k	12.66	n k s	14.90	k	10.57
p s	14.90	(1)	9.65	p s	14.34	p k s	13.66
Confounded NPK				Confounded PKS			

Analyze the experimental data and interpret the results as vividly as possible.

BIBLIOGRAPHY

1. Bailey RA. Patterns of confounding in factorial designs. *Biometrika*, 1977: 64, 579–603.
2. Kempthorne O. A simple approach to confounding and fractional replication in factorial experiments. *Biometrika*, 1947: 34, 255–72.
3. Kirk RE. *Experimental Designs: Procedures for Behavioral Science*, 2nd ed. Pacific Grove, CA: Brooks/Cole, 1982.
4. Kirk RE. *Confounded Factorial Designs. In: Edwards LK. (Ed.), Applied Analysis of Variance in Behavioral Science.* Marcel Dekker, New York, 1993.

6

Fractional Factorial Designs

6.1 SIGNIFICANCE

Confounding has been a good device to reduce the block size in factorial experiments. But confounded designs make use of all treatment combinations and require more than one replication in every experiment. But as number of factors and their levels increase, the set of treatment combinations becomes gigantic and it is not possible to conduct an experiment even with a single replication due to paucity of material and resources. For example, in an experiment with eight factors each at two levels, 256 treatment combinations are to be accommodated in one replicate which is quite cumbersome and unsuitable for a number of reasons. Similarly taking seven factors each at 3 levels leads to 2187 treatment combinations which are too large to handle in any trial. Such situation often arises in experiments concerning chemical industry, breeding of cattles, effect of nutrients on crops, formation of alloys, etc. To tackle such experiments, Finney (1945, 1946) developed the concept of *fractional factorial experiments* or *fractional replications*. These experiments include a subset of the set of treatment combinations chosen skillfully. Later it was developed by Plackett and Burman (1946), Kempthorne (1947). The name fractional replication is used in the sense that instead of using a full replicate only a fraction of it is used, e.g. in 2^5 experiment, one uses 1/2 or 1/4 replicate. In general, in 2^n experiment, one uses $\dfrac{1}{2^s}$ replicate where $s < n$.

In 3^5 experiment, one uses 1/9 or 1/27 replicate. In general, in 3^n experiment, an investigator utilizes $\dfrac{1}{3^s}$ replicate for $s < n$. The author has kept out of scope of this book the case of asymmetrical factorial experiments. Before describing the particular situations, it is relevant to define two terms namely, defining contrasts and aliases.

Defining contrast: Any treatment effect used to split the complete factorial into divisible fractional replicate is called a *defining contrast*. In some situations, there can be more than one defining contrasts. Mostly the interaction(s) of third or higher order is/are chosen as defining contrast(s). Such interaction is said to form the identity.

To obtain a one-half fractional factorial, only one effect is to be confounded. So there is only one defining contrast. For example, in a 2^3 factorial usually ABC is confounded to obtain blocks of four units, so called a 2^{3-1} fractional factorial. Let us write,

$$I = ABC$$

This relation is called *defining relation*. Here I is the identity. For one-fourth fraction of 2^n design, one has to confound two effects and so on.

General rule: In general, $\dfrac{1}{2^i}$ -fractional factorial of a 2^n factorial denoted as FF-2^{n-i} where, i is the key to identify the design fraction, i.e. i defining contrasts are said to used to obtain the required fractional factorial FF-2^{n-i}, then their generalized interactions also form the part of defining contrasts.

Aliases: Any two or more treatment effects which are represented by the same contrast are called *aliases*. The name aliases was given by Finney (1945). It is easy to identify an alias of any main effect or interaction. The generalized interaction of any effect with the defining contrast is its alias.

6.2 ONE-HALF REPLICATE OF 2^4-FACTORIAL IN FRACTIONAL REPLICATION

Though it is not an appropriate example for a fractional factorial, still it has been chosen for better understanding. Let there be four factors A, B, C, and D each at two levels say, 0 and 1. Evidently higher order interactions are either negligible or irrelevant. So, highest order interaction ABCD is chosen as defining contrast. Eight treatment combinations of the contrast of ABCD with positive signs are,

(1), *ab, ac, ad, bc, bd, cd, abcd*

and those with negative signs are,

a, b, c, d, abc, abd, acd, bcd

Now one can form the half replicate for all the main effects and interactions out of eight treatment combinations given above either with positive (+) signs or with negative (−) signs. Here treatment combinations having positive signs have been chosen. Thus, half-replicate with main effects and interactions will be estimated as follows through the following contrasts. The total responses (y) for all the eight treatment combinations are also given alongside in Table 6.2.1.

Table 6.2.1 Table of contrasts

Treat. combn.	A	B	C	D	AB	AC	AD	BC	BD	CD	ABC	ABD	ACD	BCD	ABCD
(1) y_1	–	–	–	–	+	+	+	+	+	+	–	–	–	–	+
(ab) y_2	+	+	–	–	+	–	–	–	–	+	+	–	–	+	+	+
(ac) y_3	+	–	+	–	–	+	–	–	+	–	–	+	–	+	+
(ad) y_4	+	–	–	+	–	–	+	+	–	–	+	–	–	+	+
(bc) y_5	–	+	+	–	–	–	+	+	–	–	–	+	+	–	+
(bd) y_6	–	+	–	+	–	+	–	–	+	–	+	–	+	–	+
(cd) y_7	–	–	+	+	+	–	–	–	–	+	+	+	–	–	+
(abcd) y_8	+	+	+	+	+	+	+	+	+	+	+	+	+	+	+

Note: Here it should be noted that all contrasts in the above scheme are not orthogonal.

It is trivial to identify the effects and their aliases in the above scheme. Aliases can also be obtained by taking the generalized interaction of a factor effect with the defining contrast. Both the ways one gets the same set of aliases. The list of effects and aliases is known as *aliasing scheme*. The same is presented below.

Table 6.2.2 Aliasing scheme

Main effects	Aliases	First order interactions	Aliases
A	A × ABCD = BCD	AB	AB × ABCD = CD
B	B × ABCD = ACD	AC	AC × ABCD = BD
C	C × ABCD = ABD	AD	AD × ABCD = BC
D	D × ABCD = ABC		

In the above example, ABCD is used to obtain $\frac{1}{2}$-fractional factorial. Also,

$$D = ABC.$$

This relation is known as the *generating relation*. The effects D and ABC are same. On multiplying the respective columns of D and ABC, one gets a column of positive signs only identified as I. In nomenclature,

$$D = ABC$$
$$D \times D = ABC \times D$$
$$D^2 = ABCD$$
$$I = ABCD$$

Since, D^2 reduced to modulo 2 is I. Similarly, $A^2 = B^2 = C^2 = I$. The relation I = ABCD is called the *defining relation*.

It can be noticed that the contrast of A is same as for BCD, contrast for B is same as for ACD and so on. Similarly the contrasts for AB and CD, AC and BD, AD and BC are same. So any main effect or interaction cannot be estimated individually. In fact the effects which can be estimated or tested are the combined effects of (A + BCD), (B + ACD), (C + ABD), (D + ABC), (AB + CD), (AC + BD) and (AD + BC). Considering the second order interactions as negligible, the combined effect (A + BCD) may be attributed to A alone. Similarly this hold true for other main effects too. But for first order interaction, one is unable to estimate the individual effect and test them. The only way to interpret whether the interaction AB is effective or CD is to see the significance of the main effects. If A and B are significant and C and D are not, then one may infer that AB is likely to be significant and CD is not. But such a conclusion is not based on any strong grounds. In many cases, main effects may be nonsignificant but their interaction may be significant. The same discussion holds for other first order interactions.

The estimates of combined effects will be as follows:

$(A + BCD)$ is estimated by $\dfrac{1}{4}(- y_1 + y_2 + y_3 + y_4 - y_5 - y_6 - y_7 + y_8)$

$(B + ACD)$ is estimated by $\dfrac{1}{4}(- y_1 + y_2 - y_3 - y_4 + y_5 + y_6 - y_7 + y_8)$

$(C + ABD)$ is estimated by $\dfrac{1}{4}(- y_1 - y_2 + y_3 - y_4 + y_5 - y_6 + y_7 + y_8)$

$(D + ABC)$ is estimated by $\dfrac{1}{4}(- y_1 - y_2 - y_3 + y_4 - y_5 + y_6 + y_7 + y_8)$

Similarly the first order interaction effects are estimated as,

$(AB + CD)$ is estimated by $\dfrac{1}{4}(y_1 + y_2 - y_3 - y_4 - y_5 - y_6 + y_7 + y_8)$

$(AC + BD)$ is estimated by $\dfrac{1}{4}(y_1 - y_2 + y_3 - y_4 - y_5 + y_6 - y_7 + y_8)$

$(AD + BC)$ is estimated by $\dfrac{1}{4}(y_1 - y_2 - y_3 + y_4 + y_5 - y_6 - y_7 + y_8)$

The idea of fractional replication of 2^4-factorial into one-half-replicate can be extended to any higher order 2^n factorial ($n > 4$) splitted into $\dfrac{1}{2}$-replicate.

6.3 ONE-FOURTH REPLICATE OF A 2^5-FACTORIAL IN FRACTIONAL REPLICATION

In an experiment with five factors A, B, C, D and E each at two levels 0 and 1, the situation calls for a block of eight units only. To save main effects and first order interactions, it is preferable to choose a second order interaction to decompose the 32 combinations block into halves containing treatment effects with positive and negative signs. Let us consider the block of 16 treatment combinations having positive signs. Again select another second order interaction to split the selected block of 16 units into half replicate in such a way that the generalized interaction of these two defining treatment effects is a third order interaction. To be more realistic, let us take two interactions ABC and CDE as defining interaction effects. It is evident that their generalized interaction ABDE will automatically be confounded. Set of 16 treatment combinations in terms of a, b, c, d, and e with positive signs and denoted by their levels will be as follows. These treatment combinations will easily be available by the modulo technique as $i + j + k = 1$ mod 2 for the interaction ABC, where i, j, k, l, u denote the levels 0 and 1 of A, B, C, D and E, respectively.

$$(\text{ABC})_{1 \; mod \; 2} = 00100 \quad 00101 \quad 00110 \quad 00111 \quad 01000 \quad 01001 \quad 01010 \quad 01011$$
$$c \qquad ce \qquad cd \qquad cde \qquad b \qquad be \qquad bd \qquad bde$$
$$10000 \quad 10001 \quad 10010 \quad 10011 \quad 11100 \quad 11101 \quad 11110 \quad 11111$$
$$a \qquad ae \qquad ad \qquad ade \qquad abc \qquad abce \qquad abcd \qquad abcde$$

Now the eight treatment combinations with positive signs out of these sixteen on confounding CDE will be obtained by modulo technique $k + l + u = 1$ mod 2. Thus, the group of interactions ABC, CDE, and ABDE forms the identity group of interactions. The estimating contrasts for the main effects and interactions will be as follows:

Table 6.3.1 Chart of contrasts for one-fourth fractional replication

Treatment combinations		A	B	C	D	E	AB	AC	AD	AE	BC	BD	BE
00100	c	−	−	+	−	−	+	−	+	+	−	+	+
00111	cde	−	−	+	+	+	+	−	−	−	−	−	−
01001	be	−	+	−	−	+	−	+	+	−	−	−	+
01010	bd	−	+	−	+	−	−	+	−	+	−	+	−
10001	ae	+	−	−	−	+	−	−	−	+	+	+	−
10010	ad	+	−	−	+	−	−	−	+	−	+	−	+
11100	abc	+	+	+	−	−	+	+	−	−	+	−	−
11111	abcde	+	+	+	+	+	+	+	+	+	+	+	+

Columns continued,

CD	CE	DE	ABC	ABD	ABE	ACD	ACE	ADE	BCD	BCE	BDE	CDE
−	−	+	+	−	−	+	+	−	+	+	−	+
+	+	+	+	+	+	−	−	−	−	−	−	+
+	−	−	+	+	−	−	+	+	+	−	−	+
−	+	−	+	−	+	+	−	+	−	+	−	+
+	−	−	+	+	−	+	−	−	−	+	+	+
−	+	−	+	−	+	−	+	−	+	−	+	+
−	−	+	+	−	−	−	−	+	−	−	+	+
+	+	+	+	+	+	+	+	+	+	+	+	+

Columns continued−

ABCD	ABCE	ABDE	ACDE	BCDE	ABCDE
−	−	+	−	−	+
+	+	+	−	−	+
−	+	+	−	+	−
+	−	+	−	+	−
−	+	+	−	−	−
+	−	+	+	−	−
−	−	+	+	+	+
+	+	+	+	+	+

With the help of the above chart of contrasts for all factorial effects, the effects and aliases are listed below.

Table 6.3.2 Aliasing scheme

Defining contrasts: $I = ABC = CDE = ABDE$

Main effects	Aliases
A	BC, ACDE, BDE
B	AC, BCDE.ADE
C	AB, DE, ABCDE
D	ABCD, CE, ABE
E	ABCE, CD, ABD
First order interactions	*Aliases*
AB	C, ABCDE, DE
AC	B, ADE, BCDE

(Contd.)

First order interactions	Aliases
AD	BCD, ACE, BE
AE	BCE, ACD, BD
BC	A, BDE, ACDE
BD	ACD, BCE, AE
BE	ACE, BCD, AD
CD	ABD, E, ABCE
CE	ABE, D, ABCD
DE	ABCDE, C, AB

Second order interactions	Aliases
ABD	CD, ABCE, E
ABE	CE, ABCD, D
ACD	BD, AE, BCE
ACE	BE, AD, BCD
ADE	BCDE, AC, B
BCD	AD, BE, ACE
BCE	AE, BD, ACD
BDE	ACDE, BC, A

Third order interactions	Aliases
ABCD	D, ABE, CE
ABCE	E, ABD, CD
ACDE	BDE, A, BC
BCDE	ADE, B, AC

Note: Factorial effects and their aliases could be obtained by finding the generalized interaction of effects with the identity group of interactions. Still the contrasts are necessitated to estimate main and interaction effects. Sum of squares of various treatment effects can also be calculated through them.

After going through the aliasing schemes of two fractional factorials, it looks germane to discuss the salient features of the aliasing scheme.

1. The column(s) corresponding to any interaction included in any defining contrast(s) is/are not a contrast. They all have their either positive or negative signs. As a matter of fact that is a multiple of mean and is sum of observations.

2. The coefficients of contrasts of more than one factorial are identical. Such effects are called aliases and none of them can be estimated or tested individually.

3. Least square estimate of an effect for an additive model with single observation per cell (cell mean model) will have to be obtained jointly for four effects in case of one-fourth fractional replication. This is due to the fact that each factorial effect has three aliases. To estimate these effects, each contrast value should be divided by $\dfrac{\upsilon}{s}$, υ where is the number of treatment combinations in the fractional replication and s is the number of aliased effects. For example, the estimate of four aliased effects (A + BC + BDE + ACDE) will be estimated by

$$\frac{1}{2} (- y_1 - y_2 - y_3 - y_4 + y_5 + y_6 + y_7 + y_8).$$ Where, $y_1, y_2, y_3, y_4, y_5, y_6,$ y_7, y_8 are responses due to treatment combinations 00100, 00111, 01001, 01010, 10001, 10010, 11100, 11111, respectively.

4. As regards the interpretation of results, it is not possible to draw a clear conclusion. It is based on unqualified assumptions. If the effect of factor A on testing of hypothesis comes out to be non-significant, still one can not conclude that A is not significant. One possibility is that A, BC, BDE and ACDE are zero. Another eventuality is that some of them may have opposite affects. Because of this lacuna, fractional replications are run as screening experiments for some rough idea rather than solid results.

6.4 FRACTIONAL REPLICATION IN 3^n FACTORIALS

The fraction of replication will be of the type 3^{n-s} experiment where s is an integer $(s < n)$. 3^{n-s} are quite unpopular because of complications in analysis of data and more so difficulty of interpretation. Consider a 3^3 experiment with 1/3-replication. There shall be nine treatment combinations in a replicate which are easily obtained through modulo technique. Let there be three factors A, B, and C each with levels 0, 1and 2. One would always prefer to use ABC, the highest order interaction in this case, as defining contrast. So the entries in a replication be either a set of treatment combinations obtained from $i + j + k$ = 0 or 1 or 2 mod 3. It hardly matters which set out these three is chosen. When ABC is confounded, $A^2B^2C^2$ is automatically confounded as the treatment combinations which satisfy the condition $i + j + k = x$ mod 3 will also satisfy the condition $2i + 2j + 2k = x$ mod 3 where x is either of 0, 1 or 2.

Randomization: Once the entries of a fractional replication are finalized, they should be randomly assigned within the replication.

The aliasing scheme will be as follows:

Table 6.4.1 Aliasing scheme

Defining contrasts, I = ABC, $A^2B^2C^2$

Main effect	Aliases to ABC	Aliases to $A^2B^2C^2$
A	A^2BC	$B^2C^2 = BC$
B	$A\,B^2C$	$A^2C^2 = AC$
C	ABC^2	$A^2B^2 = AB$

It can easily be verified that AB = C, AC = B, BC = A.

Further on considering analysis of variance, total degrees of freedom will be 8. Out of 8 degrees of freedom, 6 d.f. shall be consumed by main effects. Any first order interaction has four d.f. which is not available. This ensures that main effects can be estimated and tested in this type of fractional replication only if the two factor interactions are likely to be negligible or assumed to be negligible. Because of the meager utility and seldom use of such experiments, further discussion is omitted in this volume.

6.5 RESOLUTION OF FRACTIONAL FACTORIAL DESIGNS

From the preceding discussion, it is evident that identity groups of interactions play an important role in classification of fractional factorial designs. Box and Hunter (1961 a, b) coined the term *resolution* by classifying fractional factorials. If the smallest order interaction in the identity group consists of R-factors, then the fractional design is said to be of resolution-R. As a result of this definition in a resolution-R design, no p-factor interaction ($p <$ R) is aliased with any other effect containing less than (R – p) factor interactions. Whereas all interactions involving (R + 1)-factors or more are assumed to be negligible.

On the basis of the above definitions fractional factorials are classified into three types of resolutions namely, III, IV, V. The factorial effects in the identity group of interactions are called *words*. A simple approach to identification of resolution of fractional factorial designs may be given as, 'the number of letters in the shortest *word* in the identity group of a 2^4-factorial, only one word ABCD is used as defining contrast which consists of four letters. Hence, such a design is of resolution-IV.

In the example of one-fourth replicate of 2^5-factorial in fractional replication, there are three interactions in the identity group of interactions namely, ABC, CDE, ABDE. The shortest words in the group consist of three letters. Hence the given design is of resolution-III.

6.6 CONFOUNDING IN FRACTIONAL FACTORIAL REPLICATION

When experiments with large number of factors having 2, 3 or more levels are conducted into 1/2 or 1/4-fractional replicate in completely randomized block design, then often homogeneity condition appears to be violated. In such a situation, reduction of a fractional replicate into smaller blocks through confounding seems much better *alternative* as it provides more precise results. In this method, besides identity group used in a 2^n factorial to decompose into 1/2 or 1/4-replicate, one has to select diligently one interaction to break into two blocks or two interactions to break it into four blocks and so on. Actually care has to be taken in selection of effects to be confounded that no main effects or first order interactions are confounded with blocks including generalized interaction(s) which are of interest of the investigator. For instance, in a 2^6 factorial experiment into half replicate, one chooses a 6-factor interaction ABCDEF to divide the experiment into half-replicate to contain the following 32 treatment combinations with negative signs. Thus, 32 treatment combinations of *a, b, c, d, e, f* with odd number of letters are as follows.

Half-replicate of a 2^6 factorial with I = ABCDEF

a	*b*	*c*	*d*	*e*	*f*	*abc*	*abd*
abe	abf	acd	ace	acf	ade	adf	aef
bcd	bce	bcf	bde	bdf	bef	cde	cdf
cef	def	abcde	abcdf	bcdef	acdef	abdef	abcef

Note: It is ones own choice whether he selects the above 32 treatment combinations or the remaining 32 treatment combinations having positive signs, i.e. with even number of letters.

Now to divide it into two blocks of 16 units, it is advisable to select a 3-factor interaction say, ABC or DEF. This will save all main effects and 2-factor interactions. If one uses ABC as defining contrast, DEF will automatically be confounded and vice-versa.

The disposition of 1/2-fractional replicate into two blocks with 16 entries per block on confounding ABC in the above block of negative sign treatment combinations is as given below.

Table 6.6.1 1/2-fractional replication in two blocks taking
I = ABCDEF and confounding the effect, ABC = DEF

Block-1	*Block-2*
a	d
b	e

(Contd.)

Block-1	*Block-2*
c	f
abc	abd
ade	abe
adf	abf
aef	acd
bde	ace
bdf	acf
bef	bcd
cde	bce
cdf	bcf
cef	def
abcde	bcdef
abcdf	acdef
abcef	abdef

Before proceeding to analysis of data, it seems necessary to prepare and understand aliasing scheme.

Table 6.6.2 Aliasing scheme for 1/2-fractional replication of a 2^{6-1} fractional factorial in two blocks with I = ABCDEF = ABC = DEF

ABCDE × ABCDEF = F
ABCDF × ABCDEF = E
ABCEF × ABCDEF = D
BCDEF × ABCDEF = A
ACDEF × ABCDEF = B
ABDEF × ABCDEF = C
A × ABC = BC
B × ABC = AC
C × ABC = AB
D × DEF = EF
E × DEF = DF
F × DEF = DE

ABC × DEF = ABCDEF = I	DEF × ABC = ABCDEF = I
ADE × DEF = AF	ABD × ABC = CD
ADF × DEF = AE	ABE × ABC = CE
AEF × DEF = AD	ABF × ABC = CF
BDE × DEF = BF	ACD × ABC = BD
BDF × DEF = BE	ACE × ABC = BE
BEF × DEF = BD	ACF × ABC = BF
CDE × DEF = CF	BCD × ABC = AD
CDF × DEF = CE	BCE × ABC = AE
CEF × DEF = CD	BCF × ABC = AF

The aliasing scheme given in Table 6.6.2 clearly reveals that 32 entries of Table 6.6.1 in two blocks are equivalent to 6 main effects and 15 two-factor interactions. This aliasing scheme leads to proper analysis of data.

Skeleton ANOVA for the confounded fractional replication in two blocks shall be as displayed below.

Table 6.6.3 Skeleton ANOVA

Source	d.f.
Blocks	1
Main effects	6
Two-factor interactions	15
Error (residual)	9
Total	31

In case one has to resort to the blocks of size 8 units only, he should choose another interaction, preferably 3-factor interaction, and form four blocks of 8 units out of the above two blocks. Suppose the interaction ABD is selected for confounding. Then, the effect CEF, CD and ABDE are automatically confounded as generalized interactions with the above defining contrasts. The entries of the four blocks are as displayed below.

Table 6.6.4 Layout plan

Block-1	Block-2	Block-3	Block-4
a	c	d	e
b	abc	abd	f
aef	ade	ace	abe
bef	adf	acf	abf
cde	bde	bce	acd
cdf	bdf	bcf	bcd
abcde	cef	def	bcdef
abcdf	abcef	abdef	acdef

Skeleton analysis of variance table for 1/2-fractional replication having four blocks is given in Table 6.6.5.

Table 6.6.5 Skeleton ANOVA

Source	d.f.
Blocks	3
Main effects	6
Two-factor interactions	14
Error (residual)	9
Total	31

Note: 1. In the above ANOVA, main effect has 6 d.f. and two-factor interactions have 14 degrees of freedom as it has been proved in Table 6.6.2 that 32 entries of the design tantamount to 6 main effects and 15 two-factor interactions.

2. In the layout given in Table 6.6.4, two-factor interaction CD has been confounded in spite of best efforts to avoid this situation. Hence, the degrees of freedom for two-factor interactions are 14 instead of 15.

As regards computation of sum of squares for main effects and interactions, this can be done by any of the usual methods. Namely, (*i*) by preparing two way tables, (*ii*) by forming orthogonal contrasts, (*iii*) by Yates' method.

The author prefers the contrasts method as it is totally mechanical and fool proof. All the more, error sum of square can easily be obtained by subtraction in the usual way.

Example 6.1: An experiment was planned on soybean for knowing the affect of 6 nutrients each having two levels namely, nitrogen (N) - 0 and 20 kg/ha; phosphorus (P) - 0 and 40 kg/ha; potash (K) - 0 and 20 kg/ha; zinc (Z) - 0 and 5 kg/ha; sulphur (S) - 0 and 20 kg/ha; iron (F) - 0 and 2.5 kg/ha. Grain yield (q/ha) of soybean was recorded. The data have been recasted into $\frac{1}{2}$-fractional replication with four blocks for the purpose of showing the method of analysis of data.

Layout and grain yield (q/ha) of $\frac{1}{2}$-fractional replication in 4 blocks with effects possessing negative signs has been given below.

The given design is constructed taking, I = NPKZSF

Defining contrasts, NPK = ZSF, NPZ = KSF, KZ = NPSF.

Block-1		Block-2		Block-3		Block-4	
n	12.70	k	10.41	z	10.22	s	10.34
p	11.25	npk	14.96	npz	13.46	f	10.25

(Contd.)

Block-1		Block-2		Block-3		Block-4	
nsf	13.02	nzs	13.48	nks	12.85	nps	14.38
psf	12.45	nzf	12.04	nkf	12.70	npf	12.96
kzs	10.80	pzs	12.88	pks	12.99	nkz	12.70
kzf	10.74	pzf	12.49	pkf	12.35	pkz	12.34
npkzs	15.38	ksf	10.68	zsf	10.39	pkzsf	13.65
npkzf	15.00	npksf	14.62	npzsf	14.95	nkzsf	13.22
Blocks Total	101.34		101.56		99.91		99.84

Precise skeleton ANOVA

Source	d.f.
Blocks	3
Main effects	6
Two-factor interactions	14
Error (Residual)	9
Total	31

Next table provides the chart of contrasts for main effects and first order interactions to calculate the sum of squares due to them.

Nota bene:

1. Writing of contrasts listed in table I is very simple. For the contrast of an effect (main effect or interaction), a treatment combination possesses a positive sign if it contains all the letter(s) of the effect on no letter. Otherwise the sign of the treatment combination will be negative. This thumb rule will enable the investigator to write all contrasts without stress and strain.
2. Second approach may be to write contrast for all main effects. The contrasts for interactions can then be obtained by multiplying the signs of the main effect factors involved in the interaction.
3. The interaction KZ has not been included for testing its effect as it is confounded in second split with the blocks. So it is merged with the error. Calculation of sum of squares:

$$C.F. = \frac{(402.65)^2}{32} = 5066.47$$

Total S.S. $= 12.70^2 + 11.25^2 + \ldots\ldots\ldots + 13.65^2 + 13.22^2 - C.F.$

$$= 5140.58 - 5066.47 = 74.11$$

Table-I

Combi-nations	Yield (q/ha)	N	P	K	Z	S	F	NP	NK	NZ	NS	NF	PK	PZ	PS	PF	KS	KF	ZS	ZF	SF
n	12.70	+	−	−	−	−	−	−	−	−	−	−	+	+	+	+	+	+	+	+	+
p	11.25	−	+	−	−	−	−	−	+	+	+	+	−	−	−	−	+	+	+	+	+
k	10.41	−	−	+	−	−	−	+	−	+	+	+	−	+	+	+	−	−	+	+	+
npk	14.96	+	+	+	−	−	−	+	+	−	−	−	+	−	−	−	−	−	+	+	+
z	10.22	−	−	−	+	−	−	+	+	−	+	+	+	−	+	+	+	+	−	−	+
npz	13.46	+	+	−	+	−	−	+	−	+	−	−	−	+	−	−	+	+	−	−	+
nkz	12.70	+	−	+	+	−	−	−	+	+	−	−	−	−	+	+	−	−	−	−	+
pkz	12.34	−	+	+	+	−	−	−	−	−	+	+	+	+	−	−	−	−	−	−	+
s	10.34	−	−	−	−	+	−	+	+	+	−	+	+	+	−	+	−	+	−	+	−
nps	14.38	+	+	−	−	+	−	+	−	−	+	−	−	−	+	−	−	+	−	+	−
nks	12.85	+	−	+	−	+	−	−	+	−	+	−	−	+	−	+	+	−	−	+	−
nzs	13.48	+	−	−	+	+	−	−	−	+	+	−	+	−	−	−	−	+	+	−	−
pks	12.99	−	+	+	−	+	−	−	−	+	−	+	+	−	+	−	+	−	−	+	−
pzs	12.88	−	+	−	+	+	−	−	+	−	−	+	−	+	+	−	−	+	+	−	−
kzs	10.80	−	−	+	+	+	−	+	−	−	−	+	−	−	−	+	+	−	+	−	−
f	10.25	−	−	−	−	−	+	+	+	+	+	−	+	+	+	−	+	−	+	−	−
npf	12.96	+	+	−	−	−	+	+	−	−	−	+	−	−	−	+	+	−	+	−	−

(Contd.)

Combinations (q/ha)	Yield	N	P	K	Z	S	F	NP	NK	NZ	NS	NF	PK	PZ	PS	PF	KS	KF	ZS	ZF	SF
nkf	12.70	+	−	+	−	−	+	−	+	−	−	+	−	+	+	−	−	+	+	−	−
pkf	12.35	−	+	+	−	−	+	−	−	+	+	−	+	−	−	+	−	+	+	−	−
pzf	12.49	−	+	−	+	−	+	−	+	−	+	−	−	+	−	+	+	−	+	+	−
nsf	13.02	+	−	−	−	+	+	−	−	−	+	+	+	+	−	−	−	−	−	−	+
psf	12.45	−	+	−	−	+	+	−	+	+	−	−	−	−	+	+	−	−	−	−	+
nzf	12.04	+	−	−	+	−	+	−	−	+	−	+	+	−	+	−	+	−	−	+	−
ksf	10.68	−	−	+	−	+	+	+	−	+	−	−	−	+	−	−	−	+	−	−	+
kzf	10.74	−	−	+	+	−	+	+	−	−	+	−	−	−	+	−	−	+	−	+	−
zsf	10.39	−	−	−	+	+	+	+	+	−	−	−	+	−	−	−	+	−	+	+	+
npkzs	15.38	+	+	+	+	+	−	+	+	+	+	+	+	+	+	−	+	+	+	−	−
npkzf	15.00	+	+	+	+	−	+	+	+	+	−	+	+	+	−	+	−	+	−	+	−
npksf	14.62	+	+	+	−	+	+	+	+	−	+	+	+	−	+	+	+	−	−	−	+
npzsf	14.95	+	+	−	+	+	+	+	−	+	+	+	−	+	+	+	−	+	+	+	+
nkzsf	13.22	+	−	+	+	+	+	−	+	+	+	+	−	−	−	−	+	+	+	+	+
pkzsf	13.65	−	+	+	+	+	+	−	−	−	−	−	+	+	+	+	+	+	+	+	+
Effect	**402.65**	34	29	8	4	9	0	−3	0	1	1	−3	4	3	3	1	−4	0	5	2	−0
Totals		.19	.57	.13	.83	.51	.37	.57	.75	.25	.25	.17	.81	.55	.47	.29	.21	.69	.39	.07	.61

Blocks S.S. $= \dfrac{1}{8}(101.34^2 + 101.56^2 + 99.91^2 + 99.84^2) - \text{C.F.}$

$= 5066.78 - 5066.47 = 0.31$

Main effects S.S. = Aggregate of sum of squares of each individual Main effect contrast value divided by 32.

$$= \frac{1}{32}\sum_{i=1}^{6}\phi_i^2$$

$$= \frac{1}{32}\left(34.19^2 + 29.57^2 + 8.13^2 + 4.83^2 + 9.51^2 + 0.37^2\right)$$

$$= 69.48$$

Similarly 2-factor interaction S.S.

$$= \frac{1}{32}\{(-3.57)^2 + 0.75^2 + \ldots\ldots + 2.07^2 + (-0.61)^2\}$$

$$= 3.99$$

Each individual main effect and interaction sum of squares have been obtained by squaring their respective contrast value and dividing each of it by 32 and displayed in ANOVA table. The null hypotheses that the effect is zero are tested against experimental error and a decision about their significance is taken by comparing the calculated F-value with the tabulated critical value of F corresponding to their degrees of freedom and predecided level of significance. For example,

$$\text{S.S.}(\text{N}) = \frac{(34.19)^2}{32} = 36.5298$$

$$\text{S.S.}(\text{P}) = \frac{(29.57)^2}{32} = 27.3245$$

$$\text{S.S.}(\text{NP}) = \frac{(-3.57)^2}{32} = 0.3983$$

$$\text{S.S.}(\text{NK}) = \frac{(0.75)^2}{32} = 0.1758$$

Similarly all other sum of squares for main effects and first order interactions are calculated.

Table-II ANOVA table

Source	d.f	S.S.	M.S.	F-value
Blocks	3	0.31	0.1033	< 1
Main effects	6	69.48	11.5800	281.06*
N	1	36.5298	36.5298	886.64*
P	1	27.3245	27.3245	663.22*
K	1	2.0655	2.0655	50.13*
Z	1	0.7290	0.7290	17.69*
S	1	2.8262	2.8262	68.60*
F	1	0.0043	0.0043	0.10 NS
Two factor interactions	14	3.99	0.2850	6.92*
N × P	1	0.3983	0.3983	9.67*
N × K	1	0.0175	0.0175	0.43 NS
N × Z	1	0.0488	0.0488	1.18 NS
N × S	1	0.0488	0.0488	1.18 NS
N × F	1	0.3140	0.3140	7.62*
P × K	1	0.7230	0.7230	17.55*
P × Z	1	0.3938	0.3938	9.56*
P × S	1	0.3763	0.3763	9.13*
P × F	1	0.0520	0.0520	1.26 NS
K × S	1	0.5539	0.5539	13.44*
K × F	1	0.0149	0.0149	0.36 NS
Z × S	1	0.9079	0.9079	22.04*
Z × F	1	0.1339	0.1339	3.25 NS
S × F	1	0.0116	0.0116	0.28 NS
Error	8	0.33	0.0412	
Total	31	74.11		

Tabulated F-values from table B-7 at P = 0.05 and (υ_1, υ_2) d.f. are,
$$F_{1,8} = 5.32, F_{3,8} = 4.07, F_{6,8} = 3.58, F_{14,8} = 3.24.$$

On comparing calculated F-values with the respective tabulated F-values, it is concluded that all main effects except iron show a significant effect on the yield of soybean. Also the interactions N × P, N × F, P × K, P × Z, K × S and Z × S in ANOVA table are significant. Hence, their joint effects further boost the grain yield of soybean.

Merits and demerits of fractional factorial designs

Merits:

1. Fractional factorial designs enable the investigator to evaluate a large number of factorial treatments in spite of curtailing the experiment size appreciably.
2. Fractional factorial designs technique can be utilized in any of the completely randomized, randomized bocks and Latin square designs. But it is most frequently used in completely randomized design.
3. Usually fractional replication experiments are performed in 2^n factorials and sometimes in 3^n factorials reduced to any fraction of replication 2^{n-i} or 3^{n-i} where n and i are integers such that $i < n$.
4. For a large number of factorial treatments to be evaluated simultaneously, fractional factorial designs are quite satisfactory alternative.
5. Analysis of data is simple particularly in case of 2^{n-i} fractional factorial replications.

Demerits:

1. Main drawback of fractional factorial design lies with the interpretation of results. A large number of main effects and interactions are aliased with each other. Hence, it is not possible to evaluate the influence of an individual factor. The only option is to assume that the interaction effects are zero in such cases. But this has no sound logic or basis.
2. Except for 2^n factorials, the layout and analysis of 3^n and other factorials in fractional replication becomes very complicated. More over interpretation becomes problematic.
3. Fractional factorial designs are not applicable for asymmetrical designs. Hence, their utility is limited.
4. In general, layout and analysis of fractional factorial designs are more complex than any other parallel conventional designs.
5. Concrete conclusions can never be derived from fractional factorial experiments. They only provide some vague idea about the influence of various factors on response variable.

Concluding remarks

The experience tells that factorial fractional experiments are sometimes conducted in agricultural research. There are also examples of their application in engineering and industrial experiments. Mostly the experiments are confined to the fractions of 2^n factorials. In theory, a large number of fractional factorial designs are discussed in advance books on experimental designs. The construction of factorial fractional designs through orthogonal arrays is also available in literature. Resolution of FF-designs is also discussed at large. But the matter in this chapter is confined to the extent subject to their applications in general.

QUESTIONS AND EXERCISES

1. Under what situations fractional factorial design are preferred over confounded designs?
2. Explain clearly the terms, (*i*) defining contrast, (*ii*) Aliases.
3. In a 2^4 factorial, construct a $\frac{1}{2}$-fractional factorial designs taking the defining contrast ABCD and choosing the treatment combinations possessing negative signs. Also list the effect and aliases by preparing a list of contrasts.
4. What problem is caused in interpretation of results due to aliases?
5. Construct the fractional factorial design of size $\frac{1}{4}$-replicate in case of 2^5 factorial experiment. Also display the aliasing scheme for this design.
6. How can you estimate various effects from fractional factorial design given in question-5?
7. How can you construct a $\frac{1}{3}$-fractional replication design in case of 3^3 factorial experiments? What problems a researcher faces in construction and analysis of 3^n factorials in fractional replications?
8. What do you understand by the resolution of fractional factorial designs? Give definitions as well.
9. Clarify the concept of resolutions III, IV and V.
10. Give the layout of a 2^6 factorial into $\frac{1}{4}$-fractional replication. Also display the aliasing scheme.
11. Give the method of analysis of data for a $\frac{1}{2}$-fractional factorial design of a 2^5 factorial taking the highest order interaction as defining contrast.
12. Give the break-up of degrees of freedom for various factors in analysis of variance table for a $\frac{1}{4}$-fractional factorial design in case of 2^5 factorial experiments.

13. In what manner the sum of squares for various effects and interactions can be calculated for fractional factorial designs of 2^n factorial? Which approach would you prefer and why?

14. How can you detect the effects and aliases when a complete list of contrasts is given for a fractional factorial designs?

15. How can you interpret the results of a fractional factorial design? What are the assumptions underlying these interpretations?

16. Comment on the complexities of inferences drawn from a fractional factorial designs.

17. What points are in favor and against fractional factorial designs?

18. Are fractional factorial designs applicable to all types of factorial experiments? Justify your answer through practical illustrations.

19. What are the scientific areas in which fractional factorial designs are in use?

20. Give the layout and indicate the aliases of treatments for the following fractional factorial designs.

   (*i*) CRFF-2^{5-2} with AB and CDE as defining contrasts.

   (*ii*) CRFF-3^{4-2} with ABC and BCD as defining contrasts.

BIBLIOGRAPHY

1. Bisgaard S. A method for identifying defining contrasts for 2^{k-p} experiments. *Journal of Quality Technology*, 1993: 26(4), 288–96.

2. Block R, Mee RW. Resolution IV designs with 128 runs. *Journal of Quality Technology*, 2005: 37(4), 282–293.

3. Bonett DG, Woodward AJ. Analysis of simple main effects in fractional factorial experimental designs of Resolution V. *Communication in Statistics*, 1993: 22, 1585–1593.

4. Box GEP, Hunter JS. The 2^{k-p} fractional factorial designs. *Technometrics*, 1961a: 3, 311–352.

5. Box GEP, Hunter JS. The 2^{k-p} fractional factorial designs, II. *Technometrics*, 1961b: 3, 449–458.

6. Box GEP, Meyer RD. An analysis of unreplicated fractional factorials. *Technometrics*, 1986: 28(1), 11–18.

7. Connor WS, Zelen M. *Fractional factorial experimental designs for experiments with factors at two and three levels*. National Bureau of Standards, Applied Mathematics Series, 1959: 54.

8. Dean A, Voss D. *Design and analysis of experiments*. Springer-Verlag, New York, 1999.

9. Dey A. *Orthogonal fractional factorial designs*. Wiley Eastern Ltd., New Delhi, 1985.

10. Finney DJ. The fractional replication of factorial arrangements. *Ann. Eugen.*, 1945: 12, 291–301.

11. Finney DJ. Recent developments in the design of field experiments, III. Factorial replications. *Journal of Agricultural Sciences*, 1946: 36, 184–191.

12. Kempthorne O. A simple approach to confounding and fractional replication in factorial experiments. *Biometrika*, 1947: 34, 255–272.

13. Lewis SM, Dean AM. Fractional experiments in resolvable generalized cyclic designs. *Bulletin in Applied Statistics*, 1980: 7, 159–167.

14. Plackett RL, Burman JP. The designs of multifactorial experiments. *Biometrika*, 1946: 33, 305–325.

15. Webb SR. Non-orthogonal designs of even resolution. *Technometrics*, 1968a: 10, 291–300.

7

Split Plot Designs

7.1 CONCEPTUAL IDEAS AND LAYOUT

Split plot designs were originally evolved for agricultural research by necessity. In many experiments, a factor A with a number of levels was such that it could not be applied in small size plots whereas another factor B with various levels could be managed conveniently in small plots. For instance, in experiments with factors irrigation and fertilizers having two or more levels of both the factors, irrigation levels cannot be managed in small plots. On the other hand, fertilizer levels can efficiently be applied in small size plots. For such experiments, split plot design is ideal one. Later these designs have been immensely used in green house studies, industrial research, dairy science investigations, storage management studies, baking experiments, education and behavioral research, etc.

In short, a factorial experiment requiring large and small size plots was laid out in split plot design. The levels of a factor may be allocated in randomized blocks design, Latin square design or any other basic design. Split plot design is an incomplete block design. The larger plots are known as *main plots (MP)* or *whole plots* and the treatments assigned to the main plots are called *main plot or whole plot treatments*. Whereas the smaller plots within the whole plot are known as *sub-plots (SP)* and the treatment levels allotted to these plots are called *sub-plot treatments*. As a matter of fact, split plot design is a factorial confounded design in which the variation between the levels of the main treatment is inextricably mixed with the whole plot differences. Another interesting point with regard to split plot design is that this design greatly resembles to partially hierarchical design. This point will further be discussed in the sequel.

Consider a split plot design in r randomized blocks, an SP-p, m design, i.e. the layout has r blocks each with p main plots and in every main plot there are m sub-plots. p levels of the factor A are randomly allocated to main plots in each block and m factor levels of B are assigned to sub-plots in each main plot randomly and independently. Given below is the layout for an SP-3, 4 design with $r = 3$.

Layout   7.1.1

Table 7.1.1 SP-design with main plots and split plots in randomized blocks

Blocks-I			Block-II			Block-III		
MP-1	MP-2	MP-3	MP-4	MP-5	MP-6	MP-7	MP-8	MP-9
a_2	a_1	a_0	a_2	a_1	a_0	a_0	a_1	a_2
b_3	b_0	b_1	b_2	b_2	b_0	b_0	b_3	b_0
b_1	b_2	b_3	b_1	b_1	b_3	b_1	b_1	b_3
b_2	b_3	b_2	b_3	b_3	b_2	b_2	b_2	b_2
b_0	b_1	b_0	b_0	b_0	b_1	b_3	b_0	b_1

Nota bene:

1. Plot effects are nested within the levels of factor A whereas the factors A and B are crossed. Hence, split plot design is a partially hierarchical design.
2. Differences between the levels of factor A are inseparably mixed with main plot differences.
3. Comparison of levels of factor B is confined to within plot variation as they are free from the variation due to main plots.
4. Interaction effect A × B is also measurable as factors A and B are crossed.
5. An experimenter can take repeated measurements on each subject if need arises.
6. Main plot and sub-plot treatments are not necessarily the levels of single factor. But they can be the set of factorial combinations of different factors if required.
7. Sub-plot treatments may be arranged in various ways to control natural variation between sub-plots within a whole plot.
8. Randomization procedure for allotting factor levels of treatments depends on the type of design chosen for a particular trial.
9. Split plot design is the only design which accommodates different size of experimental units in the same trial as per the requirement of treatments.

7.2 ANALYSIS OF VARIANCE OF SPLIT PLOT DESIGN

Spit plot design is a special design because it involves different size experimental units for various factors. The analysis of spit plot design consists of two parts. One part is the analysis of main plots receiving p levels of A independently in each replication which involves the comparison of levels of factor A based on the total of responses from sub-plots within main plot. Second part consists of the analysis for comparison of m levels of factor B and interaction A × B which solely depend on the comparisons of responses from split plots within the whole plots. So variation due to levels of factor A is confined to the differences between whole plots whereas variation due to levels of factor B and interaction A × B are restricted to the sub-plot variations within whole plots. These estimates are free from the effect of whole plot differences. Hence, the factors belonging to main plots and sub-plots will have their own errors.

Statistical model for the p, m split plot in r replicates is,

$$y_{iju} = \mu + \alpha_i + \beta_j + (\alpha\beta)_{ij} + \Pi_{u\,(i)} + \Pi'_{j\,(ui)} + \varepsilon_{iju} \qquad (7.2.1)$$

where,

$$i = 1, 2, \ldots\ldots, p \; ; j = 1, 2, \ldots\ldots, m \; ; u = 1, 2, \ldots\ldots, r$$

and

y_{iju} = the response in block u for the treatment combination $a_i\,b_j$.

μ = over all mean effect.

α_i = the effect of ith level of factor A.

β_j = the effect of jth level of factor B.

$(\alpha\beta)_{ij}$ = the interaction effect of ith level of factor A and jth level of factor B.

$\Pi_{u\,(i)}$ = the effect of the uth block for the ith level of A and is NID $(0, \sigma_\Pi^2)$.

$\Pi'_{j\,(ui)}$ = joint effect of jth level of B and block u receiving the ith level of A independent of $\Pi_{u\,(i)}$ and is NID $(0, \sigma_{\Pi'}^2)$.

ε_{iju} = a random component of error associated with ijuth sub-plot and each ε_{iju} is NID $(0, \sigma_\varepsilon^2)$.

Prior to proceeding for the analysis of data, it is pertinent to tabulate the experimental data in the following format.

Table 7.2.1 Data table

Factors	Rep.-1 $a_0\,a_1 \quad \ldots\ldots \quad a_{p-1}$	Rep.-2 $a_0\,a_1 \quad \ldots\ldots \quad a_{p-1}$	.....	Rep.-r $a_0\,a_1 \quad \ldots\ldots \quad a_{p-1}$
b_0	$y_{001}\,y_{101} \ldots\ldots y_{(p-1)01}$	$y_{002}\,y_{102} \ldots\ldots y_{(p-1)02}$	.....	$y_{00r}\,y_{10r} \ldots\ldots y_{(p-1)0r}$
b_1	$y_{011}\,y_{111} \ldots\ldots y_{(p-1)11}$	$y_{012}\,y_{112} \ldots\ldots y_{(p-1)12}$	.....	$y_{01r}\,y_{11r} \ldots\ldots y_{(p-1)1r}$
I	⋮	⋮	⋮	⋮
b_m	$y_{0m1}\,y_{1m1} \ldots\ldots y_{(p1)m1}$	$y_{0m2}\,y_{1m2} \ldots\ldots y_{(p-1)m2}$		$y_{0mr}\,y_{1mr} \ldots\ldots y_{(p-1)mr}$
Total	$y_{0.1}\,y_{1.1} \ldots\ldots y_{(p-1).1}$	$y_{0.2}\,y_{1.2} \ldots\ldots y_{(p-1).2}$		$y_{0.r}\,y_{1.r} \ldots\ldots y_{(p-1).r}$

Totals of levels of b,
$$b_0 = y.0., \ b_1 = y.1., \ ...., \ b_m = y.m.$$
Replication totals,
$$R_1 = y..1, \ R_2 = y..2, \ ...., \ R_r = y..r \ ; \ \text{Grand total} = y...$$
Two way table for calculating the main effect and interaction sum of squares is to be prepared as follows:

Table 7.2.2 A × B interaction table

Factors	a_0	a_1	a_2 .... a_i .... a_{p-1}	Total
b_0	$y_{00}.$	$y_{10}.$	$y_{20}.$ .... $y_{i0}.$ .... $y_{(p-1)0}.$	$y._0.$
b_1	$y_{01}.$	$y_{11}.$	$y_{21}.$ .... $y_{i1}.$ .... $y_{(p-1)1}.$	$y._1.$
b_j	$y_{0j}.$	$y_{1j}.$	$y_{2j}.$ .... $y_{ij}.$ .... $y_{(p-1)j}.$	$y._j.$
b_m	$y_{0m}.$	$y_{1m}.$	$y_{2m}.$ .... $y_{im}.$ .... $y_{(p-1)m}.$	$y._m.$
Total	$y_0..$	$y_1..$	$y_2..$ .... $y_i..$ .... $y_{(p-1)}..$	$y...$

Sum of squares for all the factors with the help of the above tables can be calculated by the following formulae.

$$\text{C.F.} = \frac{y^2...}{pmr}$$

$$\text{S.S. (Rep.)} = \frac{1}{pm}(y^2.._1 + y^2.._2 + ... + y^2.._i + y^2.._m) - \text{C.F.}$$

$$\text{S.S. (A)} = \frac{1}{mr}(y^2_0.. + y^2_1.. + ... + y^2_i.. + y^2_{(p-1)}..) - \text{C.F.}$$

$$\text{S.S. (MP)} = \frac{1}{m}\{(y^2_0._1 + y^2_1._1 + ... + y^2_i._1 + y^2_{(p-1)._1})$$

$$+ (y^2_{0.2} + y^2_{1.2} + ... + y^2_{i.2} + y^2_{(p-1).r})$$

$$+ ...... + (y^2_{0.r} + y^2_{1.r} + ... + y^2_{i.r} + y^2_{(p-1).r})\} - \text{C.F.}$$

Main plot error S.S. = S.S.(MP) – S.S.(A) – S.S.(Rep.) = Error (MP)

$$\text{S.S.(B)} = \frac{1}{pr}(y^2._0. + y^2._1. + ... + y^2._j. + y^2._m.) - \text{C.F.}$$

$$\text{S.S.(AB)} = \frac{1}{r}(y^2_{00}. + y^2_{10}. + ... + y^2_{im}. + y^2_{(p-1)m}.) - \text{C.F.} - \text{S.S.(A)} - \text{S.S.(B)}$$

Table 7.2.3 ANOVA table for split plot design

Source	d.f.	S.S.	M.S.	Expected mean squares Model-II
Main plot				
Replication	υ_1	SS (Rep.)	$\dfrac{SS(\text{Rep.})}{\upsilon_1} = MS(\text{Rep.})$	
A	υ_2	SS (A)	$\dfrac{SS(A)}{\upsilon_2} = MS(A) = V_1$	$\sigma_\varepsilon^2 + \left(1-\dfrac{r}{R}\right)\left(1-\dfrac{m}{M}\right)\sigma_{\pi'}^2 + r\left(1-\dfrac{m}{M}\right)\sigma_{\alpha\beta}^2 + m\left(1-\dfrac{r}{R}\right)\sigma_\pi^2 + rm\sigma_\alpha^2$
Error (MP)	$\upsilon_1 \times \upsilon_2$	Error (MP)	$\dfrac{SSEMP}{\upsilon_1 \times \upsilon_2} = MSEMP$ $= V_2$	$\sigma_\varepsilon^2 + \left(1-\dfrac{m}{M}\right)\sigma_{\pi'}^2 + m\sigma_\pi^2$
Sub-plot				
B	υ_3	SS (B)	$\dfrac{SS(B)}{\upsilon_3} = MS(B) = V_3$	$\sigma_\varepsilon^2 + \left(1-\dfrac{r}{R}\right)\sigma_{\pi'}^2 + r\left(1-\dfrac{p}{P}\right)\sigma_{\alpha\beta}^2 + rp\sigma_\beta^2$
AB	$\upsilon_2\upsilon_3$	SS (AB)	$\dfrac{SS(AB)}{\upsilon_2 \times \upsilon_3}$ $= V_4$	$\sigma_\varepsilon^2 + \left(1-\dfrac{r}{R}\right)\sigma_{\pi'}^2 + r\sigma_{\alpha\beta}^2$
Error (SP)	$p\upsilon_2\upsilon_3$	SS(ESP)	$\dfrac{SS(ESP)}{p\upsilon_1\upsilon_3} = MS(ESP)$ $= V_5$	$\sigma_\varepsilon^2 + \sigma_{\pi'}^2$
Total	prm−1			

Total S.S. $= (y_{001}^2 + y_{101}^2 + \ldots\ldots + y_{(p-1)m1}^2 + y_{002}^2 + y_{102}^2 + \ldots\ldots +$

$$y_{(p-1)m2}^2 + y_{00r}^2 + y_{10r}^2 + \ldots\ldots + y_{(p-1)mr}^2) - \text{C.F.}$$

Sub-plot error S.S. = Total S.S. $-$ S.S.(MP) $-$ S.S.(B) $-$ S.S.(AB) = SS(ESP)

Situation (i): Analysis of variance table along with the expected mean squares considering that A, B and block effects are random in nature.

where, $r - 1 = \upsilon_1$, $p - 1 = \upsilon_2$, $m - 1 = \upsilon_3$,

In the table (7.2.3), expected mean squares are considered to be randomly selected from a population of R blocks. In this situation, $r/R = 0$ and thus, $(1 - r/R) = 1$. Also the fraction p/P is equal to 1 if the effect A is fixed and $m/M = 1$, if the effect B is fixed, whereas for effects A and B to be random effects they are zero.

If the effect A is fixed, σ_α^2 in the expression of expected mean square be replaced by $\sum_i \alpha_i^2 / (p-1)$ and in case B is a fixed effect, σ_β^2 be replaced by $\sum_j \beta_j^2 / (m-1)$. Again if A and B both are fixed effects, then $\sigma_{\alpha\beta}^2$ be replaced by $\sum_i \sum_j (\alpha\beta)_{ij}^2 / (p-1)(m-1)$. If either of the effects A and B is fixed and other is random, then AB is considered to be a random factor and $\sigma_{\alpha\beta}^2$ is retained as such. With the above rules, it is simple to give the expected mean squares for model III as usually one can not assume block effects as fixed.

Situation (i): In view of the above norms, when A, B and blocks are all random effects, then $p/P = m/M = r/R = 0$. Substituting these values in table (7.2.3), the expected mean squares in simplified form will be as follows:

Table 7.2.4 Abridged table

Source	Expected mean squares
Main plot *Replication*	
A	$\sigma_\varepsilon^2 + \sigma_{\pi'}^2 + r\sigma_{\alpha\beta}^2 + m\sigma_\pi^2 + rm\sigma_\alpha^2 = V_1$
Error (MP)	$\sigma_\varepsilon^2 + \sigma_{\pi'}^2 + m\sigma_\pi^2 = V_2$
Sub-plot	
B	$\sigma_\varepsilon^2 + \sigma_{\pi'}^2 + r\sigma_{\alpha\beta}^2 + rp\sigma_\beta^2 = V_3$
AB	$\sigma_\varepsilon^2 + \sigma_{\pi'}^2 + r\sigma_{\alpha\beta}^2 = V_4$
Error (SP)	$\sigma_\varepsilon^2 + \sigma_{\pi'}^2 = V_5$

From the Tables 7.2.3 and 7.2.4, another important point to note is that no proper error mean square is available to test the significance of effect A in case of model II. As a remedy, an analyst should make use of Satterthwait's (1946) approximate F'-statistics. For the convenience of the readers, a little portion of the matter given in section (4.3) is repeated so as to maintain the continuity in reading. He proposed to synthesize the mean squares in the form of linear combination such that the numerator and denominator under H_0 have the same expected values. The ratio of such synthesized variances does not follow exact F-test but an approximate F which is denoted as F'. The degrees of freedom of linear combination of mean squares say,

$$L = a_1 \, MS_1 + a_2 \, MS_2 + \dots\dots\dots + a_k \, MS_k \qquad (7.2.2)$$

where, $a_1, a_2, \dots\dots\dots, a_k$ are properly chosen coefficients and MS_1, MS_2, $\dots\dots\dots, MS_k$ are the estimated mean squares for error and other components involved in the linear combination which are obtained from ANOVA table.

The degrees of freedom for L are obtained by the formula,

$$\hat{\upsilon} = \frac{L^2}{\dfrac{\left(a_1 MS_1\right)^2}{\upsilon_1} + \dfrac{\left(a_2 MS_2\right)^2}{\upsilon_2} + \dots\dots\dots + \dfrac{\left(a_k MS_k\right)^2}{\upsilon_k}} \qquad (7.2.3)$$

where, $\upsilon_1, \upsilon_2, \dots\dots\dots, \upsilon_k$ are the degrees of freedom associated with MS_1, $Ms_2, \dots\dots\dots, MS_k$ respectively. The value of L is easily available by (7.2.2). It is well known that the ratio of two mean squares follows F-distribution. Let it be denoted by $F'_{m, n; \alpha}$ where α is the level of significance of the test. The values of m and n are often a fraction as obtained by the formula (7.2.3) or the values which are not tabulated. To solve this problem Lavscher (1965) provided explicit formulae to facilitate interpolation of F-values. In the succeeding interpolation formulae α is omitted from F for the sake of convenience and it is assumed to be there.

Case 1: When m appears in the table, but not n. If n' and n'' are the degrees of freedom which immediately precede and follow n in the F-table, the value,

$$F_{m, n} = (1 - A) \, F_{m, n'} + A \, F_{m, n''} \qquad (7.2.4)$$

where,
$$A = \frac{n''(n - n')}{n(n'' - n')} \qquad (7.2.4.1)$$

The values of $F_{m, n'}$ and $F_{m, n''}$ for α level are read from F-table and substituted in the above formulae.

Case 2: When n appears in the table, but not m.
The interpolation formula is,

$$F_{m, n} = (1 - B) \, F_{m', n} + B \, F_{m'', n} \qquad (7.2.5)$$

In the above formula, all notations can be decoded in the similar manner as in (7.2.4) by replacing n by m and vice-versa. In this case,

$$B = \frac{m''(m - m')}{m(m'' - m')} \qquad (7.2.5.1)$$

Case 3: When neither m nor n appear in F-table. Following the same notations as in cases 1 and 2, the interpolation formula is,

$$F_{m, n} = (1 - A)(1 - B) F_{m', n'} + A(1 - B) F_{m', n''}$$
$$+ (1 - A) B F_{m'', n'} + A \times B F_{m'', n''} \qquad (7.2.6)$$

In situation (i), all factors were considered to be random. Now other situations are considered and only the expected mean squares for various factors are given. Because the contents of all other columns of the ANOVA Table 7.2.3 remain same.

Resuming back to table (7.2.4) of expected mean squares (EMS), it is evident that no EMS is there which can be used to exactly test the effect A. For the sake of brevity, EMS are denoted by V_1, V_2, V_3, V_4, V_5. To test V_1, synthesize other EMS which results into the same terms as in V_1 except $r\, m\, \sigma_\alpha^2$. Under H_0, $\sigma_\alpha^2 = 0$. Thus, the linear combination $V_2 + V_4 - V_5$ provides the required error mean square. Degrees of freedom for $V_2 + V_4 - V_5$ can be obtained by the formula (7.2.3). Further, table value of quasi-F can be interpolated by a suitable Lavbcher's formula.

To test the effect B, it is proposed that one should use preliminary test of significance PTS). To apply PTS, first test the significance of V_4 against V_5. In this test, one should choose a lower level of significance than the traditional $\alpha = 0.05$. May it be $\alpha = 0.20$. If AB is significant, then test V_3 against V_4. Again if AB is non-significant, then pool V_4 and V_5 to get pooled error V_{45} where,

$$V_{45} = \frac{\upsilon_2 \upsilon_3 V_4 + p \upsilon_1 \upsilon_3 V_5}{\upsilon_2 \upsilon_3 + p \upsilon_1 \upsilon_3} \qquad (7.2.7)$$

For testing the significance of AB, test V_3 against V_{45} using traditional 5 % or 1 % level of significance. For further understanding of PTS, one should read the research papers by B. L. Agarwal (1990), Agarwal and Gupta (1993). **Situation (ii):** When the effects A and B are fixed and block effects are random. In this situation $r/R = 0$ and $m/M = 1$, $p/P = 1$. Expected mean squares shall be as displayed in the following table.

Table 7.2.5

Source	Expected mean squares (EMS)
Main plot Replication	
A	$\sigma_\varepsilon^2 + m\sigma_\pi^2 + rm\sum_i \alpha_i^2 \Big/ (p-1) = V_1$
Error (MP)	$\sigma_\varepsilon^2 + m\sigma_\pi^2 = V_2$
Sub-plot	
B	$\sigma_\varepsilon^2 + \sigma_{\pi'}^2 + rp\sum_j \beta_j^2 \Big/ (m-1) = V_3$
AB	$\sigma_\varepsilon^2 + \sigma_{\pi'}^2 + r\sum_i\sum_j (\alpha\beta)_{ij}^2 \Big/ (p-1)(m-1) = V_4$
Error (SP)	$\sigma_\varepsilon^2 + \sigma_{\pi'}^2 = V_5$

Nota bene:

From Table 7.2.4, it is apparent that error mean square of main plot treatments is always greater than the error mean square of sub-plot treatments. That is why main plot treatments are tested less precisely than sub-plot treatments. In case for some experimental anomaly, SP-error mean square is greater than MP-error mean square, the two errors MS be pooled and all effects be tested against pooled error.

Expected mean squares in the Table 7.2.5 clearly reveal that V_2 provides an exact F-statistic for testing the effect A and for testing the effects B and AB, V_5 is an exact error.

Situation (iii): When A is fixed, B and block effect are random in nature. In

Table 7.2.6 Abridged table

Source	Expected mean squares (EMS)
Main plot Replication	
A	$\sigma_\varepsilon^2 + \sigma_{\pi'}^2 + r\sigma_{\alpha\beta}^2 + m\sigma_\pi^2 + rm\sum_i \alpha_i^2 \Big/ p-1 = V_1$
Error (MP)	$\sigma_\varepsilon^2 + \sigma_{\pi'}^2 + m\sigma_\pi^2 = V_2$
Sub-plot	
B	$\sigma_\varepsilon^2 + \sigma_{\pi'}^2 + rp\sigma_\beta^2 = V_3$
AB	$\sigma_\varepsilon^2 + \sigma_{\pi'}^2 + r\sigma_{\alpha\beta}^2 = V_4$
Error(SP)	$\sigma_\varepsilon^2 + \sigma_{\pi'}^2 = V_5$

this situation $p/P = 1$, $m/M = r/R = 0$. Expected mean squares for different factors are as displayed in Table 7.2.6.

In case A is fixed and B is random, there is no exact EMS which can be taken as error to provide an exact F-test for testing the significance of the effect A. So it is preferable to make use of a linear combination of mean squares to obtain a proper error MS. The linear combination $(V_2 + V_4 - V_5)$ serves as an error for testing the effect A. This is same as in situation (i). Whereas V_3 and V_4 are to be tested against V_5 resulting into an exact F-test.

Situation (iv): When B is fixed, effect A and block effects are taken to be random. This condition leads to $m/M = 1$, $p/P = 0$ and $r/R = 0$. Expected mean squares for all the factors of ANOVA (7.2.3) are depicted in the following table.

Table 7.2.7 Abridged ANOVA table

Source	*Expected mean squares (EMS)*
Main plot	
Replication	
A	$\sigma_\varepsilon^2 + m\sigma_\pi^2 + rm\sigma_\alpha^2 = V_1$
Error (MP)	$\sigma_\varepsilon^2 + m\sigma_\pi^2 = V_2$
Sub-plot	
B	$\sigma_\varepsilon^2 + \sigma_{\pi'}^2 + r\sigma_{\alpha\beta}^2 + rp\sum_j \beta_j^2 \Big/ m-1 = V_3$
AB	$\sigma_\varepsilon^2 + \sigma_{\pi'}^2 + r\sigma_{\alpha\beta}^2 = V_4$
Error (SP)	$\sigma_\varepsilon^2 + \sigma_{\pi'}^2 = V_5$

From the Table 7.2.7, it is evident that V_2 is an exact error for testing A. V_5 is an appropriate error for testing AB. But for testing the effect B, use PTS as discussed in situation (i).

Standard errors for the differences of paired means

Let us call the error mean square for main plot treatment as E_a, and for testing the sub-plot treatments as E_b. The standard errors for the differences of paired means can be obtained by the following formulae.

(i) Standard error (S.E.) of the difference between two main plot treatment means, i.e. difference between two A level means is,

$$= \sqrt{\frac{2E_a}{rm}} \qquad (7.2.8)$$

(*ii*) S.E. of a difference between two sub-plot treatment plot treatment means, i.e. difference between two A level means is,

$$= \sqrt{\frac{2E_b}{rp}}$$ (7.2.9)

(*iii*) S.E. of difference between two B level means at the same level of A is,

$$= \sqrt{\frac{2E_b}{r}}$$ (7.2.10)

(*iv*) S.E. of difference between two A level means at the same level of B or at different levels of sub-plot treatments, e.g. S.E. of $(a_1b_1 - a_0b_1)$ or $(a_1b_1 - a_0b_0)$ is,

$$= \sqrt{\frac{2(m-1)E_b + E_a}{rm}}$$ (7.2.11)

Formula (7.2.11) does not provide an exact *t*-test. Some approximations are used which are omitted here.

Uppers and downers of split plot design

Every design has some qualities which induce for its usage and possesses some drawbacks which restrict its use. The same have been delineated here.

Uppers

1. This is the only design which accommodates treatments requiring large and small size experimental units.
2. Sub-plot treatment and interaction effects have greater precision as compared to randomized block design (RBD).
3. Main plot or sub-plot treatments can be arranged in Latin square design which increases the precision over RBD.
4. Main plot and sub-plot treatments need not necessarily be the single treatment levels, but they can be a set of factorial combinations.
5. Many variations in split plot design are possible to fulfill the requirement of an experiment under variety of situations. The same have been given subsequently.
6. Analysis of data is usually simple except for some situations.
7. Interpretation of results can be given without ambiguity.

Downers

1. Main plot treatments have less precision as compared to a factorial experiment with same number of treatment combinations in RBD.
2. A missing value in split plot design causes much more complication in analysis of data as compared to RBD.

3. Comparison of paired treatment means requires various standard errors which need greater dexterity.

Example 7.1: A researcher in agronomy conducted an experiment in split plot design to assess the effect of sulphur on green pod of peas. He included nine treatments in main plots as elemental sulphur (E) at four levels, 100, 150, 200 and 250 kg/ha and four levels of gypsum 1, 2, 3, 4 and no treatment as a control (C_0). Three doses of foliar spray (F) in sub-plots in each main plot namely, plain water as control (F_0), 0.1% $FeSO_4$ (F_1) and 0.2% of $FeSO_4$ (F_2) were applied.

The layout and data pertaining to green pod yield (q/ha) along with main plot totals for four replications were as given below. In the following layout, elemental sulphur levels are denoted by E_1, E_2, E_3, E_4 and gypsum levels by G_5, G_6, G_7, G_8 and control C_0.

Rep.-1

E_3	C_0	E_1	G_5	E_2	G_8	G_7	G_6	E_4
F_1	F_0	F_0	F_2	F_0	F_0	F_1	F_0	F_1
59.6	54.2	58.2	56.8	50.9	69.6	59.6	55.8	75.3
F_2	F_2	F_1	F_0	F_1	F_1	F_0	F_1	F_0
68.2	55.4	58.4	54.3	55.6	74.2	59.2	59.2	76.2
F_0	F_1	F_2	F_1	F_2	F_2	F_2	F_2	F_2
57.8	56.2	57.2	57.5	60.7	76.6	74.2	60.0	76.2
185.6	165.8	173.8	168.6	167.2	220.4	193.0	175.0	227.7

Rep.-2

G_5	G_6	E_1	C_0	E_4	G_8	E_3	G_7	E_2
F_2	F_1	F_2	F_2	F_1	F_0	F_2	F_2	F_2
57.8	59.9	57.8	51.2	74.6	69.2	61.4	68.4	65.1
F_0	F_0	F_1	F_1	F_0	F_1	F_1	F_1	F_1
57.2	54.1	61.2	55.4	72.3	74.9	55.4	58.4	60.2
F_1	F_2	F_0	F_0	F_2	F_2	F_0	F_0	F_0
60.4	61.3	60.9	53.1	69.8	73.4	60.1	64.5	51.7
175.4	175.3	179.9	159.7	216.7	217.5	176.9	191.3	177.0

Rep.-3

E_2	E_3	G_7	G_6	E_1	C_0	G_5	E_4	G_8
F_2 63.2	F_2 60.9	F_0 68.3	F_0 59.6	F_1 60.4	F_2 57.4	F_1 58.9	F_0 65.3	F_2 69.9
F_0 53.3	F_0 71.4	F_2 66.6	F_2 57.2	F_2 64.3	F_1 56.2	F_0 59.1	F_1 68.6	F_0 72.4
F_1 66.1	F_1 63.9	F_1 61.8	F_1 56.8	F_0 54.3	F_0 59.6	F_2 60.1	F_2 74.2	F_2 70.9
182.6	196.2	196.7	173.6	179.0	173.2	178.1	208.1	213.2

Rep.-4

C_0	E_1	E_3	E_4	G_7	E_2	G_8	G_5	G_6
F_2 56.3	F_2 61.6	F_2 70.2	F_2 72.6	F_2 70.3	F_0 52.4	F_2 73.4	F_1 60.0	F_0 64.3
F_1 58.5	F_1 62.5	F_0 66.2	F_0 69.2	F_0 67.2	F_2 64.4	F_0 69.2	F_2 62.5	F_1 61.2
F_0 63.2	F_0 56.4	F_1 74.4	F_1 70.8	F_1 70.3	F_1 67.3	F_1 74.9	F_0 56.4	F_2 58.5
178.0	180.5	210.8	212.6	207.8	184.1	217.5	178.9	184.0

Step by step procedure for analysis of experimental data has been presented here. For the sake of convenience, first prepare the skeleton ANOVA so as to know what factors' sum of squares are to be worked out. Then put in the calculated values in this analysis of variance table for final results. In this way, calculations follow analysis of variance table.

Table-I ANOVA

Source	d.f.	S.S.	M.S.	F-value
Main plot				
Replications	3	162.22	54.07	3.58*
Between M.P. treats	8	3443.33	430.54	28.51*
Bet. levels within E	3	1304.16	434.72	28.79*
Bet. levels within G	3	1553.82	517.94	34.30*

(Contd.)

Source	d.f.	S.S.	M.S.	F-value
Ele. sulphur vs. gypsum	1	0.60	0.60	0.04 NS
Control vs. sulphur source	1	585.75	585.75	38.79*
Error (a)	24	362.53	15.10	
Sub – plot				
Between sprays F	2	168.27	84.14	8.10*
Linear effect $(F_1 - F_2)$	1	17.60	17.60	1.70 NS
Quadratic effect $F_1 + F_2 - 2F_0$	1	150.67	150.67	14.52*
Main plot treats. × F	16	361.11	22.57	2.17*
Ele. sulphur × F	6	175.07	29.18	2.81*
Gypsum × F	6	94.50	15.75	1.52 NS
Two sulphur sources × F	2	24.53	12.26	1.18 NS
Sulphur and no sulphur × F	2	67.00	33.50	3.23*
Error (b)	54	560.80	10.38	
Total	107	5059.26		

Critical values of F-distribution at = 0.05 for required d.f. are:

$F_{1, 24} = 4.26$; $F_{3,24} = 3.01$; $F_{8,24} = 2.36$; $F_{1, 54} = 4.02$; $F_{2, 54} = 3.18$;

$F_{6, 54} = 2.27$; $F_{16, 54} = 1.77$

* Significant at 5 % level of significance.

Calculations

Replication totals,

$R_1 = 1677.1$, $R_2 = 1669.7$, $R_3 = 1700.7$, $R_4 = 1754.2$.

Grand total, $G = 6801.7$ and C.F. $= \dfrac{(6801.7)^2}{108} = 428362.25$

Total S.S. $= (59.6^2 + 54.2^2 + \ldots\ldots + 56.4^2 + 58.5^2) -$ C.F.

$= 433421.51 -$ C.F. $= 5059.26$

Rep. S.S. $= \dfrac{1}{27}(1677.1^2 + 1669.7^2 + 1700.7^2 + 1754.2^2) -$ C.F.

$= 428524.47 -$ C.F. $= 162.22$

Main plot S.S. $= \dfrac{1}{3}(185.6^2 + 165.8^2 + \ldots\ldots\ldots + 178.9^2 + 184.0^2) -$ C.F.

$= 432331.33 -$ C.F. $= 3969.08$

Main plot treatment totals,

$C_0 = 676.7$, $E_1 = 713.2$, $E_2 = 710.9$, $E_3 = 769.5$, $E_4 = 865.1$, $\displaystyle\sum_{i=1}^{4} E_i = 3058.7$

$$G_5 = 701.0, \ G_6 = 707.9, \ G_7 = 788.8, \ G_8 = 868.6, \ \sum_{j=5}^{8} G_j = 3066.3$$

Main plot treat. S.S. $= \dfrac{1}{12} (676.7^2 + 713.2^2 + \ldots + 788.8^2 + 868.6^2) - \text{C.F.}$

$$= 431806.58 - \text{C. F.} = 3444.33$$

$$\text{Error } (a) = 3969.08 - 162.22 - 3444.33 = 362.53$$

Ele. sulphur S.S. $= \dfrac{1}{12} (713.2^2 + 710.9^2 + 769.5^2 + 865.1^2) - \dfrac{(3058.7)^2}{48}$

$$= 196213.44 - 194909.28 = 1304.16$$

Gypsum S.S. $= \dfrac{1}{12} (701.0^2 + 707.9^2 + 788.8^2 + 868.6^2) - \dfrac{(3066.3)^2}{48}$

$$= 197432.90 - 195879.08 = 1553.82$$

Ele. sulphur vs. gypsum S.S. $= \dfrac{(\sum E - \sum G)^2}{2 \times 48} = \dfrac{(3058.7 - 3066.3)^2}{2 \times 48} = 0.6017$

Control vs. source S.S.

$$= \dfrac{\{8C_0 - (E_1 + E_2 + E_3 + E_4 + G_5 + G_6 + G_7 + G_8)\}^2}{12 \times (8^2 + 8)}$$

$$= \dfrac{(8 \times 676.7 - 6125.0)^2}{864} = \dfrac{(5413.6 - 6125.0)^2}{864}$$

$$= \dfrac{506089.96}{864} = 585.75$$

Check: Aggregate of sum of squares due to main plot components,

$$= 1304.16 + 1553.82 + 0.6017 + 585.75$$

$$= 3444.3317 = \text{Main plot S.S.}$$

Sub-plot analysis:

Totals of foliar spray levels,

$$F_0 = 2207.1, \ F_1 = 2279.5, \ F_2 = 2315.1$$

Foliar spray S.S. $= \dfrac{1}{36} (2207.1^2 + 2279.5^2 + 2315.1^2) - \text{C.F.}$

$$= 428530.52 - 428362.25 = 168.27$$

$(F_1 \text{ vs. } F_2) \text{ S.S.} = \dfrac{(2279.5 - 2315.1)^2}{2 \times 36} = 17.60$

$$(F_1 + F_2 - 2\ F_0)\ \text{S.S.} = \frac{(2279.5 + 2315.1 - 2x2207.1)^2}{6x36} = 150.67$$

Prepare the following two-way table to compute the interaction, main plot treats. × foliar spray.

Table-II Sulphur × Foliar spray table

Foliar sprays	Main plot treatments									Total
	C_0	E_1	E_2	E_3	E_4	G_5	G_6	G_7	G_8	
F_0	230.1	229.8	208.3	255.5	283.0	227.0	233.8	259.2	280.4	2207.1
F_1	226.3	242.5	249.2	253.3	289.3	236.8	237.1	250.1	294.9	2279.5
F_2	220.3	240.9	253.4	260.7	292.8	237.2	237.0	279.5	293.3	2315.1
Total	676.7	713.2	710.9	769.5	865.1	701.0	707.9	788.8	868.6	6801.7

$(S \times F)$ S.S. = Main plot × spray = Table S.S. – C.F. – S.S (F) – S.S. (S)

$$= \frac{1}{4}(230.1^2 + 229.8^2 + \ldots\ldots + 279.5^2 + 293.3^2) - \text{C.F.} - 168.7 - 3444.3$$

$$= 432335.96 - 428362.25 - 168.27 - 3444.33$$

$$= 361.11$$

Prepare the following tables to obtain the components sum of squares of the interaction S × F.

Table-III Ele. sulphur × sprays

Foliar sprays	Ele. Sulphur levels				Total
	E_1	E_2	E_3	E_4	
F_0	229.8	208.3	255.5	283.0	976.6
F_1	242.5	249.2	253.3	289.3	1034.3
F_2	240.9	253.4	260.7	292.8	1047.8
Total	713.2	710.9	769.5	865.1	3058.7

$(\text{Ele. sulphur} \times F)$ S.S. $= \frac{1}{4}(229.8^2 + 208.3^2 + \ldots\ldots + 260.7^2 + 292.8^2)$

$$+ \frac{(3058.7)^2}{48} - \frac{1}{16}(976.6^2 + 1034.3^2 + 1047.8)^2$$

$$- \frac{1}{12}(713.2^2 + 710.9^2 + 769.5^2 + 865.1^2)$$

$$= 196567.29 + 194909.28 - 195088.06 - 196213.44$$

$$= 175.04$$

Table-IV Gypsum levels × sprays

Foliar sprays	Gypsum levels				Totals
	G_5	G_6	G_7	G_8	
F_0	227.0	233.8	259.2	280.4	1000.4
F_1	236.8	237.1	250.1	294.9	1018.9
F_2	237.2	237.0	279.5	293.3	1047.0
Total	701.0	707.9	788.8	868.6	3066.3

$$(\text{Gypsum} \times F) \text{ S.S.} = \frac{1}{4}(227.0^2 + 233.8^2 + \ldots\ldots + 279.5^2 + 293.3^2)$$

$$+\frac{(3066.3)^2}{48} - \frac{1}{16}(1000.4^2 + 1018.9^2 + 1047.0^2)$$

$$-\frac{1}{12}(701.0^2 + 707.9^2 + 788.8^2 + 868.6^2)$$

$$= 197596.22 + 195879.08 - 195947.90 - 197432.90$$
$$= 94.50$$

To calculate source × F sum of square, prepare the following table.

Table-V Source × Foliar spray table

Foliar spray	Sources of sulphur	Ele. sulphur gypsum	Total
F_0	976.6	1000.4	1977.0
F_1	1034.3	1018.9	2053.2
F_2	1047.8	1047.0	2094.8
Total	3058.7	3066.3	6125.0

$$(\text{Source} \times F) \text{ S.S.} = \frac{1}{16}(976.6^2 + 1000.4^2 + \ldots\ldots + 1047.8^2 + 1047.0^2)$$

$$+\frac{(6125.0)^2}{96} - \frac{1}{32}(1977.0^2 + 2053.2^2 + 2094.8^2)$$

$$-\frac{1}{48}(3058.7^2 + 3066.3^2)$$

$$= 391035.95 + 390787.76 - 391010.82 - 390788.36$$
$$= 24.53$$

Interaction (S and no S) × F sum of square can be calculated with the help of the following table.

Table-VI (Sulphur and no sulphur) x Foliar spray

Foliar sprays	Control and sulphur application C_0	S-application	Total
F_0	230.1	1977.0	2207.1
F_1	226.3	2053.2	2279.5
F_2	220.3	2094.8	2315.1
Total	676.7	6125.0	6801.7

$$(S \text{ and no } S) \times F \text{ S.S.} = \frac{1}{4}(230.1^2 + 226.3^2 + 220.3^2)$$

$$+ \frac{1}{32}(1977.02 + 2053.2^2 + 2094.8^2) - C.F.$$

$$- S.S.(F) - \frac{1}{12}(676.7^2) - \frac{1}{96}(6125.0^2) + C.F.$$

$$= 38172.45 + 391010.82 - 168.27 - 38160.24 - 390787.76$$
$$= 67.00$$

Standard errors of differences between two means

S.E. of the difference between two main plot means by (7.2.8),

$$= \sqrt{\frac{2 \times 15.10}{4 \times 3}} = 1.59$$

Least significant difference for comparing paired levels of elementary sulphur and gypsum.

$$\text{lsd} = 1.59 \times t_{24, \, 0.05}$$
$$= 1.59 \times 1.711 = 2.72$$

Ele. sulphur means in ascending order are,

E_2	E_1	E_3	E_4
59.24	59.43	64.12	72.09

This shows that 250 kg/ha dose of elemental sulphur is best as it gives significantly higher yield than all other lower levels of application. E_3 is meaningfully better than E_1 and E_2 in respect of yield whereas E_1 and E_2 are at par.

Gypsum means in ascending order,

G_5	G_6	G_7	G_8
58.42	58.99	65.73	72.38

Gypsum follows the same pattern as elemental sulphur.

Elemental sulphur versus gypsum is non-significant. This leads to conclude that it makes no difference whether elemental sulphur is applied or gypsum. Anyhow application of sulphur definitely contributed to increase green pod yield of peas as the yield after application of sulphur is significantly higher as compared to control, i.e. no application of sulphur.

There is only quadratic effect of foliar spray since F_1 versus F_2 is not significant. Further there is a significant interaction effect of elemental sulphur with foliar spray but not of gypsum. So foliar spray should only be applied in case of elemental sulphur.

S.E. of the difference between two sub-plot treatment means by the formula (7.2.9) is,

$$= \sqrt{\frac{2 \times 10.38}{4 \times 9}} = 0.76$$

lsd to compare any two levels of foliar spray is,

$$= 0.76 \times t_{54, \, 0.05}$$

$$= 0.76 \times 2.00 = 1.52$$

Significant difference between foliar spray level means is depicted below.

F_0	F_1	F_2
61.31	63.32	64.31

It is evident from the above diagram that F_1 and F_2 are significantly superior than water spray whereas the spray of H_2SO_4 and $FeSO_4$ are at par. Therefore, it is recommended that a scientist should spray either of these which is less costly.

S.E. of the difference between two main plot treatment means at a level of spray by the formula (7.2.11) is,

$$= \sqrt{\frac{2 \times 2 \times 10.38 + 15.10}{4 \times 3}} = 2.17$$

S.E. of the difference between two spray levels at the same level of main plot treatment is,

$$= \sqrt{\frac{2 \times 10.38}{4}} = 2.28$$

Variations in spit plot design

Several variations in split plot design are diligently exploited by the researchers. They are succinctly discussed here one by one.

7.3 SUB-PLOT TREATMENTS IN LATIN SQUARE ARRANGEMENT

To increase the precision of the sub-plot treatments, one may prefer to assign sub-plot treatments in Latin square arrangement provided the row effects are suspected. This is possible only when the number of replications is equal to the number of sub-plot treatments (treatment levels). In this arrangement, sub-plot treatments under each level of main plot treatment form a Latin square. Given below is the layout of an experiment for an SP-3,4 design having sub-plot treatment levels in Latin square arrangement.

Layout 7.3.1

Rep.-1			Rep.-2			Rep.-3			Rep.-4		
a_2	a_1	a_0	a_2	a_1	a_0	a_0	a_1	a_2	a_1	a_2	a_0
b_0	b_1	b_0	b_1	b_0	b_3	b_1	b_2	b_2	b_3	b_3	b_2
b_1	b_2	b_3	b_3	b_1	b_2	b_0	b_3	b_0	b_0	b_2	b_1
b_2	b_3	b_2	b_0	b_2	b_1	b_3	b_0	b_3	b_1	b_1	b_0
b_3	b_0	b_1	b_2	b_3	b_0	b_2	b_1	b_1	b_2	b_0	b_3

This type of layout further controls one more nuisance parameter, i.e. row to row variation.

Skeleton analysis of variance table for SP-p, m design with m replications is produced below. It is worth noting that in this situation $r = m$.

Expressions for sum of squares have been avoided as the calculations can be done in the wonted manner.

Table 7.3.2 Skeleton ANOVA

Source	d.f	SP-3, 4 and r = 4 d.f.
Main plot		
$\mathcal{K}$ Replications	$m - 1$	3
Treatment A	$p - 1$	2
Error MP	$(p - 1)(m - 1)$	6
Sub-plots		
Sub-plot		
Treat. B	$m - 1$	3
A × B	$(p - 1)(m - 1)$	6
Rows within MP	$p(m - 1)$	9
Error SP	$p(m - 1)(m - 2)$	18
Total	$p\,m^2 - 1$	47

The sum of squares for different factors can be calculated in the usual manner as are being worked out in example 7.1. F-tests in ANOVA and post-hoc test can also be applied as per requirement on the basis of F-tests. Therefore, they have not been repeated here.

7.4 USE OF FACTORIAL TREATMENTS IN SPLIT PLOT DESIGN

This is not necessary that an investigator should use single treatments with a number of levels in main plots and sub-plots. One can use combinations of more than one factor having various levels. For instance, an experimenter may apply the combinations of levels of irrigation and row to row distances in main plots and combination of two or more fertilizers levels in sub-plots. But in general practice, factorial treatments are mostly taken as sub-plots treatments whereas a single treatment at different levels is preferred in main plots. Anyhow this should not be considered as a rule. In many situations, it may be otherwise also. The layout of the experiment will be same as with single treatments except that one uses the treatment combinations in place of single factors.

Consider an experiment with three levels of irrigation (I) which are randomly signed to main plots and the combinations of 3-levels of nitrogen (N), 2-levels of phosphorus (P), 2-levels of potash (K) are randomly allotted to sub-plots. The layout of the experiment in split plot design will of the kind given below.

Layout 7.4.1

Rep.-I			Rep.-II			Rep.-III		
I_2	I_1	I_0	I_2	I_1	I_0	I_1	I_0	I_2
010	000	000	101	011	111	101	011	201
111	100	001	000	200	200	110	111	100
101	001	211	110	000	101	200	010	010
200	200	110	210	201	000	011	201	011
201	110	111	011	001	211	201	000	200
001	201	210	010	100	100	210	200	210
210	111	010	001	111	210	000	100	000
000	010	100	100	010	010	100	101	211
011	210	201	200	211	110	211	110	111
110	101	011	201	110	201	001	211	001
100	211	200	211	210	001	111	210	110
211	011	101	111	101	011	010	001	101

Skeleton analysis of variance table is displayed below for the above experimental design.

Table 7.4.1 ANOVA Table

Source	d.f.
Main plot	
Replications	2
Irrigation (I)	2
Error MP (a)	4
Sub-plot	
Treat. combinations (B)	11
N	2
P	1
N × P	2
K	1
N × K	2
P × K	1
N × P × K	2
I × B	22
I × N	4
I × P	2
I × N × P	4
I × K	2
I × N × K	4
I × P × K	2
I × N × P × K	4
Error SP (b)	66
Total	107

The method of calculating sum of squares is not given here as these values can be obtained in same way as in factorial experiments. All the more, the same will further be explicated through the solved example given immediately following the ANOVA.

Example 7.2: An experiment on fenugreek was conducted to assess the effect of balanced fertilization, varieties and biofertilization on seed yield of fenugreek. The layout of the experiment was in split plot design taking four biofertilizers in main plots and six combinations of two varieties and three fertilizers in sub-plots.

Treatments were as follows:
Biofertilizers (B): B_1 - Control, B_2 - Rhizobium inoculation, B_3 - PSB inoculation, B_4 - Rhizobium + PSB.

Varieties (V): V_1 - RMT 1, V_2 - RMT 301.
Fertilizers (F): F_1 - Control, F_2 - N 20 kg/ha + P_2O_5 20 kg/ha,
$\qquad$ F_3 - N 40 kg/ha + 40 kg/ha.
Layout of the trial and plotwise data regarding 1000 seed weight (Test weight) were as follows:

Test weight in grams

Rep.-1

B_4	B_1	B_3	B_2
V_2F_2 12.0	V_2F_3 13.3	V_1F_3 12.8	V_2F_1 11.8
V_1F_1 11.8	V_2F_2 11.9	V_2F_3 13.4	V_1F_3 13.6
V_1F_3 13.8	V_1F_2 11.4	V_1F_1 11.5	V_1F_1 11.6
V_1F_2 11.4	V_2F_1 11.2	V_2F_1 9.9	V_1F_2 12.9
V_2F_1 10.7	V_1F_1 10.4	V_1F_2 12.1	V_2F_3 15.3
V_2F_3 15.7	V_1F_3 13.9	V_2F_2 12.1	V_2F_2 11.9
75.4	72.1	71.8	77.1

Rep.-2

B_1	B_4	B_2	B_3
V_2F_2 11.9	V_2F_3 13.9	V_2F_2 12.5	V_2F_3 13.1
V_1F_3 11.5	V_2F_1 11.6	V_2F_1 9.8	V_1F_3 13.0
V_1F_1 10.6	V_1F_2 13.1	V_1F_1 10.6	V_1F_2 12.1
V_2F_1 12.2	V_1F_1 11.3	V_1F_3 13.0	V_2F_2 12.9
V_1F_2 12.9	V_1F_3 14.5	V_1F_2 12.5	V_1F_1 11.0
V_2F_3 13.6	V_2F_2 13.0	V_2F_3 13.3	V_2F_1 10.9
72.7	77.4	71.7	73.0

Rep.-3

B₂	B₃	B₁	B₄
V_2F_1 11.2	V_2F_2 12.6	V_2F_2 12.0	V_2F_3 12.2
V_2F_2 12.2	V_1F_3 13.3	V_1F_1 11.1	V_2F_2 13.4
V_1F_3 12.6	V_1F_2 11.5	V_1F_2 11.9	V_2F_1 11.6
V_1F_2 12.4	V_2F_1 11.1	V_2F_1 11.2	V_1F_1 9.5
V_1F_1 11.1	V_2F_3 13.3	V_1F_3 12.3	V_1F_2 12.2
V_2F_3 13.1	V_1F_1 11.7	V_2F_3 13.5	V_1F_3 12.6
72.6	73.5	72.0	71.5

Note: Main plot totals are worked out and given in the last rows of the replicationwise data table.

Analysis of experimental data is carried out as per procedure delineated in section (7.4).

Replication totals:

R_1 = 75.4 + 72.1 + 71.8 + 77.1 = 296.4, similarly, R_2 = 294.8, R_3 = 289.6.

Grand total: $\qquad$ G = 296.4 + 294.8 + 289.6 = 880.8

Correction factor: C.F. $= \dfrac{(880.8)^2}{72} = 10775.120$

$$\text{Rep. S.S.} = \frac{1}{24}(296.4^2 + 294.8^2 + 289.6^2) - \text{C.F.}$$

$$= \frac{1}{24} \times 258628.16 - \text{C.F.}$$

$$= 10776.173 - 10775.120 = 1.053$$

Main plot treatment totals: B_1 = 72.1 + 72.7 + 72.0 = 216.8,

Similarly, $\qquad B_2$ = 221.4, B_3 = 218.3, B_4 = 224.3

$$\text{Biofertilizers S.S.} = \frac{1}{18}(216.8^2 + 221.4^2 + 218.3^2 + 224.3^2) - \text{C.F.}$$

$$= \frac{1}{18}(193985.58) - C.F.$$

$$= 10776.98 - C.F. = 1.857$$

Main plot S.S. $= \frac{1}{6}(75.4^2 + 72.1^2 + \ldots\ldots\ldots + 72.0^2 + 71.5^2) - C.F.$

$$= \frac{1}{6}(64698.42) - C.F.$$

$$= 10783.07 - C.F. = 7.950$$

Error (a) S.S. $= 7.950 - 1.053 - 1.857 = 5.040$

For sub-plot analysis, a 2×3 table for varieties and fertilizers is prepared to calculate the sum of squares for main effects and interaction.

Table-I

	F_1	F_2	F_3	Total
V_1	132.2	146.4	156.9	435.5
V_2	133.2	148.4	163.7	445.3
Total	265.4	294.8	320.6	880.8

(Table I) S.S. $= \frac{1}{12}(132.2^2 + 146.4^2 + \ldots\ldots + 148.4^2 + 163.7^2) - C.F.$

$$= \frac{1}{12}(130089.9) - C.F.$$

$$= 10840.82 - C.F. = 65.700$$

Varieties (V) S.S. $= \frac{1}{36}(435.5^2 + 445.3^2) - C.F.$

$$= \frac{1}{36}(387952.34) - C.F.$$

$$= 10776.45 - C.F. = 1.334$$

Fertilizers (F) S.S. $= \frac{1}{24}(265.4^2 + 294.8^2 + 320.6^2) - C.F.$

$$= \frac{1}{24}(260128.56) - C.F.$$

$$= 10838.69 - C.F. = 63.570$$

Interaction (V $\times$ F) S.S. = (Table-I) S.S. $-$ S.S.(V) $-$ S.S.(F)

$$= 65.700 - 1.334 - 63.570 = 0.796$$

To calculate interaction B × F, following 3 × 4 table is prepared.

Table-II

	B_1	B_2	B_3	B_4	Total
F_1	66.7	66.1	66.1	66.5	265.4
F_2	72.0	74.4	73.3	75.1	294.8
F_3	78.1	80.9	78.9	82.7	320.6
Total	216.8	221.4	218.3	224.3	880.8

In the above table, $B_1F_1 = 11.2 + 10.4 + 10.6 + 12.2 + 11.1 + 11.2 = 66.7$
Similarly, other B_pF_q values are calculated.

$$\text{(Table-II) S.S.} = \frac{1}{6}(66.7^2 + 66.1^2 + \dots + 78.9^2 + 82.7^2)$$

$$= \frac{1}{6}(65050.74) = 10841.790$$

Interaction (B × F) S.S. = (Table II) S.S. – C.F. – S.S.(B) – S.S. (F)
$$= 10841.790 - 10775.120 - 1.857 - 63.570$$
$$= 1.243$$

To calculate interaction B × V, following 2 × 4 table is prepared.

Table-III

	B_1	B_2	B_3	B_4	Total
V_1	106.0	110.3	109.0	110.2	435.5
V_2	110.8	111.1	109.3	114.1	445.3
Total	216.8	221.4	218.3	224.3	880.8

In the above table,
$B_1V_1 = 11.4 + 10.4 + 13.9 + 11.5 + 10.6 + 12.9 + 11.1 + 11.9 + 12.3 = 106.0$
Similarly, other entries of table-III are worked out.
Interaction (B × V) S.S. = (Table-III) S.S. – C.F. – S.S.(V) – S.S. (B)

$$\text{(Table-III) S.S.} = \frac{1}{9}(106.0^2 + 110.3^2 + \dots + 109.3^2 + 114.1^2)$$

$$= \frac{1}{9}(97012.28) = 10779.142$$

(B × V) S.S. = $10779.142 - 10775.120 - 1.334 - 1.857 = 0.831$

To calculate sum of square due to B × V × F interaction, a three-way table is prepared as given below.

Table-IV

	V₁				V₂				Total
	B_1	B_2	B_3	B_4	B_1	B_2	B_3	B_4	
F_1	32.1	33.3	34.2	32.6	34.6	32.8	31.9	33.9	265.4
F_2	36.2	37.8	35.7	36.7	35.8	36.6	37.6	38.4	294.8
F_3	37.7	39.2	39.1	40.9	40.4	41.7	39.8	41.8	320.6
Total	106.0	110.3	109.0	110.2	110.8	111.1	109.3	114.1	880.8

Note: In Table IV, marginal totals have been worked out to verify that all entries are correct.

Interaction $(B \times V \times F)$ S.S. = (Table-IV) S.S. − C.F. − S.S.(B) − S.S.(V)
$$- \text{S.S.(F)} - \text{S.S. } (B \times V) - \text{S.S. } (B \times F) - \text{S.S. } (V \times F)$$

$$\text{(Table-IV) S.S.} = \frac{1}{3} (32.1^2 + 33.3^2 + \ldots\ldots\ldots + 39.8^2 + 41.8^2)$$

$$= \frac{1}{3} (32543.58) = 10847.86$$

$(B \times V \times F)$ S.S. = 10847.86 − 10775.120 − 1.857 − 1.334
$$- 63.570 - 0.831 - 1.243 - 0.796$$
$$= 3.109$$

Total S.S. = Sum of squares of each individual 72 values − C.F.
$$= 12.0^2 + 13.3^2 + \ldots\ldots\ldots + 13.5^2 + 12.6^2 - \text{C.F.}$$
$$= 10875.14 - 10775.12 = 100.02$$

Once sum of squares for all factors are calculated, following analysis of variance table can easily be prepared.

ANOVA

Source	d.f.	S.S.	M.S.	F-value
Main plot				
Replications	2	1.053	0.5265	0.627
Biofertilizers (B)	3	1.857	0.6177	0.735
Error (*a*)	6	5.040	0.8400	
Split plot				
Sub-plot treats.	5			
Varieties (V)	1	1.334	1.334	2.517
Fertilizers (F)	2	63.570	31.785	59.972*

(Contd.)

Source	d.f.	S.S.	M.S.	F-value
V × F	2	0.796	0.398	0.751
B × V	3	0.831	0.277	0.523
B × F	6	1.243	0.207	0.390
B × V × F	6	3.109	0.518	0.977
Error (b)	40	21.187	0.530	
Total	71	100.02		

Tabulated values of F at 5% level of significance for various degrees of freedom required in the above ANOVA are:

$F_{1,40} = 4.08$, $F_{2,40} = 3.23$, $F_{3,40} = 2.84$, $F_{6,40} = 2.34$

*Significant at 5% level of significance. All other effects are non-significant.

Results: On comparing the calculated F-values with the corresponding tabulated F-values, it is inferred that only fertilizer levels differ significantly and all other main and interaction effects are non-significant. Therefore, post-hoc test will be applied for comparing the paired fertilizer levels only. Since there are three levels of fertilizer, lsd test will be most appropriate.

Standard error of the difference between two level means of fertilizer by the formula (7.2.9) is,

$$S.\ E. = \sqrt{\frac{2E_b}{rp}}$$

From ANOVA table, $E_b = 0.530$, $r = 3$ and $p = 4$.

$$\therefore \quad S.E. = \sqrt{\frac{2 \times 0.530}{3 \times 4}} = 0.297$$

$$lsd = S.E. \times t_{.05;\ 40}$$
$$= 0.297 \times 2.021 = 0.600$$

Fertilizer level means are,

$$\overline{F}_1 = 11.058,\ \overline{F}_2 = 12.283,\ \overline{F}_3 = 13.358$$

Fertilizer level means in ascending order are,

$\overline{F}_1$	$\overline{F}_2$	$\overline{F}_3$
11.058	12.283	13.358

The differences between all fertilizer level means are more than least significant difference. Hence, it is recommended that one should use F_3 fertilizer dose as it gives highest test weight.

7.5 SPLIT-SPLIT PLOT DESIGN

A split design contains two factors, one associated with main plots and the other with sub-plots. Their interaction is assessed from sub-plots. But split plot design can be extended to more than two factors, may be three or more. In a split plot design with three factors at various levels, one factor is allotted to main plots as per requirement and second factor to sub-plots. Again each split plot is further decomposed into as many plots as the number of levels of third factor. These further splitted plots are known as *sub-sub plots*. Theoretically there is no restriction on the number of factors to be included in split plot design. Still more than three or four factors are practically unmanageable. Hence, such designs are avoided.

Given below is the layout of a split plot design with 3 levels of factor A, 4 levels of factor B and 2 levels of factor C having three replications.

Layout 7.5.1

Rep. I

a_1		a_0		a_2	
c_0	c_1	c_1	c_0	c_1	c_0
	b_0		b_2		b_1
c_1	c_0	c_0	c_1	c_0	c_1
	b_2		b_1		b_0
c_0	c_1	c_0	c_1	c_1	c_0
	b_1		b_3		b_2
c_0	c_1	c_1	c_0	c_0	c_1
	b_3		b_0		b_3

Rep.2

a_2		a_1		a_0	
c_0	c_1	c_1	c_0	c_0	c_1
	b_1		b_2		b_0
c_1	c_0	c_0	c_1	c_0	c_1
	b_0		b_0		b_2
c_0	c_1	c_1	c_0	c_0	c_1
	b_3		b_1		b_1
c_1	c_0	c_1	c_0	c_1	c_0
	b_2		b_3		b_3

Rep. 3

a_1		a_2		a_0	
c_1	c_0	c_0	c_1	c_0	c_1
	b_3		b_0		b_3
c_0	c_1	c_1	c_0	c_0	c_1
	b_1		b_2		b_1
c_0	c_1	c_0	c_1	c_0	c_1
	b_0		b_1		b_2
c_1	c_0	c_1	c_0	c_1	c_0
	b_2		b_3		b_0

The design needs no additional clarification as it possesses all properties of split plot design. Analysis of variance for the split-split plot design with p levels of factor A, m levels of factor B and q levels of factor C can be carried out according to the following break-up. The design has r replications. Side by side the break-up of degrees of freedom for the above design has also been displayed.

Table 7.5.1 Skeleton ANOVA

Source	d.f.	d.f.
Main plots		
Replications	$r - 1$	2
A	$p - 1$	2
$\text{Error}_{MP}(a)$	$(p - 1)(r - 1)$	4
Sub-plot		
B	$m - 1$	3
A × B	$(p - 1)(m - 1)$	6
$\text{Error}_{SP}(b)$	$p(r - 1)(m - 1)$	18
Sub-sub-plot		
C	$q - 1$	1
A × C	$(p - 1)(q - 1)$	2
B × C	$(q - 1)(m - 1)$	3
A × B × C	$(p - 1)(m - 1)(q - 1)$	6
$\text{Error}_{SSP}(c)$	$pm(q - 1)(r - 1)$	24
Total	$rpmq - 1$	

Sum of squares for all the sources of variations given in ANOVA Table 7.5.1 can be calculated in the same manner as for any other factorial experiment. Analysis of variance table can be completed as usual and the significance of all main effects and interactions can be tested by F-test using appropriate errors. Of course standard errors of the difference between two levels of treatment means of a split-split plot design are not so simple and direct. Therefore, the formulae for these standard errors are given below on ultimate unit basis.

Let the error mean squares for error (a), error (b) and error (c) are denoted by E_a, E_b and E_c, respectively.

S.E. for the comparison of a pair of levels of factor A,

$$= \sqrt{\frac{2E_a}{rmq}} \tag{7.5.1}$$

S.E. for the difference of a pair of levels of A at the same level of B,

$$= \sqrt{\frac{2[(m-1)E_b + E_a]}{rmq}} \tag{7.5.2}$$

and at the same level of C,

$$= \sqrt{\frac{2[(q-1)E_c + E_a]}{rmq}} \tag{7.5.3}$$

S.E. for the comparison of a pair of levels of factor B,

$$= \sqrt{\frac{2E_b}{rpq}} \tag{7.5.4}$$

S.E. for the difference of a pair of levels of B at the same level of A,

$$= \sqrt{\frac{2E_b}{rq}} \tag{7.5.5}$$

and the same level of factor C,

$$= \sqrt{\frac{2[(q-1)E_c + E_b]}{rpq}} \tag{7.5.6}$$

S.E. for comparison of a pair of levels of treatment C,

$$= \sqrt{\frac{2E_c}{rpm}} \tag{7.5.7}$$

S.E. for the difference of a pair of levels of C at the same level of A,

$$= \sqrt{\frac{2E_c}{rm}} \tag{7.5.8}$$

and at the same level of factor B,

$$= \sqrt{\frac{2E_c}{rp}} \qquad (7.5.9)$$

These formulae for standard errors are of prime importance for comparing any two levels of a factor at various levels of the other factor(s). Hence, they should be understood carefully.

Remark: It is further added that expected mean squares, not given here, indicate that the respective errors for main plot, sub-plot and sub-sub plot treatments are appropriate to test the effects providing exact F-tests.

In split-split plot design, treatments assigned to sub-sub plot are tested more precisely than main plot and sub-plot effects. Obviously sub-plot treatments are tested more precisely than main plot treatment. In the analysis, one will always find that $E_c < E_b < E_a$. If this inequality does not hold, then it may be possible that either the experiment has not been properly planned or some other gross mistake has occurred. In this situation, the errors be pooled as per need and effects be tested against pooled error.

7.6 SPLIT BLOCK DESIGN

Sometimes in split plot design, both the factors are such they require large size plots. For instance, one treatment is depth of ploughing and other is irrigation regime. In such a case, a split block design is better than a simple split plot design. Split block design is also named as *strip plot design*. In this design, both the factors are allocated to main plots. One factor is taken in horizontal direction and the other in vertical direction. Their interaction is assessed from sub-plots. The layout of a split block design with 4 levels of a factor A and 3 levels of a factor B having 3 replicates is as portrayed below.

Layout 7.6.1

Rep.-1

	b_0	b_2	b_1
a_1			
a_0			
a_3			
a_2			

Rep.-II

	b_2	b_1	b_0
a_0			
a_1			
a_3			
a_2			

Rep.-III

	b_2	b_1	b_0
a_1			
a_3			
a_0			
a_2			

Skeleton analysis of variance table for a split plot design in which there are p levels of factor A and m levels of factor B and r replications is given below. Degrees of freedom for the above design are also displayed for quick understanding.

Table 7.6.1 Skeleton ANOVA table

Source	d.f	d.f.
Replications	$r - 1$	2
A	$p - 1$	3
Error (a)	$(p - 1)(r - 1)$	6
B	$m - 1$	2
Error (b)	$(r - 1)(m - 1)$	4
A × B	$(p - 1)(m - 1)$	6
Error (c)	$(r - 1)(p - 1)(m - 1)$	12
Total	$rpm - 1$	35

In this design factors A and B are on equal footing from the angle of precision whereas interaction AB is tested more precisely than the effects A and B.

Let the error mean squares for error (a), error (b) and error (c) be denoted by E_a, E_b and E_c, respectively.

Standard errors of the difference of pairs of various factor levels in split block design for p levels of A, m levels of B and r replications are as given below.

S.E. of the difference of pairs of means of levels of A on sub-unit basis is,

$$= \sqrt{\frac{2E_a}{rm}} \qquad (7.6.1)$$

S.E. of the difference of pairs of means of levels of B on sub-unit basis is,

$$= \sqrt{\frac{2E_b}{rp}} \qquad (7.6.2)$$

Standard error of the difference of treatment combinations of the type $(a_1b_1 - a_0b_1)$ is,

$$= \sqrt{\frac{2[(m-1)E_c + E_a]}{rm}} \qquad (7.6.3)$$

Standard error of the difference of treatment combinations of the type $(a_1b_1 - a_1b_0)$ is,

$$= \sqrt{\frac{2[(p-1)E_c + E_b]}{rp}} \qquad (7.6.4)$$

By using appropriate standard error, a suitable post-hoc test can be applied to know about the significance of differences between paired factor level means or the significance of a contrast.

Example 7.3: An experiment was conducted to know the impact of three population densities on five varieties of maize with regard to grain yield (q/ha). The experiment was laid out in strip plot design with three replicates due to operational convenience. The yield data were recorded as per plan given below. Block totals are also displayed alongside for the sake of brevity.

Details of the treatments:

Varieties (V): V_1 - Ganga-2, V_2 - Ganga-11, V_3 - VL-54, V_4 - PHEM-1, V_5 - PHEM-2.

Spacing (crop geometry): S_0 - 60 × 25 cm, S_1 - 80 × 25 cm, S_2 - 50 × 20 cm.

It should be kept in mind that different spacing lead to change in population density per hectare.

Layout and data table

Rep. I

	S_0	S_1	S_2	Total
V_3	27.39	29.30	30.97	87.66
V_2	26.41	28.25	29.31	83.97
V_1	26.25	27.30	28.10	81.65
V_5	34.62	36.29	34.80	105.71
V_4	28.40	30.29	32.68	91.37
Total	143.07	151.43	155.86	450.36

Rep. II

	S_2	S_0	S_1	Total
V_2	30.25	27.55	26.98	84.78
V_4	33.85	26.24	31.65	91.74
V_3	31.67	28.38	31.28	91.33
V_5	33.25	33.11	29.75	96.11
V_1	29.31	27.01	29.01	85.33
Total	158.33	142.29	148.67	449.29

Rep. III

	S_0	S_2	S_1	Total
V_2	28.48	28.65	28.64	85.77
V_4	27.00	31.44	32.75	91.19
V_3	26.75	32.85	32.65	92.25
V_1	28.40	31.30	27.80	87.50
V_5	32.88	34.68	33.40	100.96
Total	143.51	158.92	155.24	457.67

Analysis of experimental data can be carried out in the following manner and inferences can be drawn on the basis of the results.

Prior to calculating the main plot treatments and interaction sum of squares, it seems germane to calculate the following sum of squares which shall be required for ANOVA table and are directly calculable from the given data.

Grand total: $G = 450.36 + 449.29 + 457.67 = 1357.32$

Correction factor: C.F. $= \dfrac{(1357.32)^2}{45} = 40940.39$

Total S.S. $= (27.39^2 + 29.30^2 + \ldots\ldots\ldots + 34.68^2 + 33.40^2) - $ C.F.
$= 41268.39 - 40940.39 = 328.00$

Rep. S.S. $= \dfrac{1}{15}(450.36^2 + 449.29^2 + 457.67^2) - $ C.F.

$= 40943.16 - $ C.F. $= 2.77$

Main plot S.S. for varieties $= \dfrac{1}{3}(87.66^2 + 83.97^2 + \ldots\ldots$

$+ 87.50^2 + 100.96^2) - $ C.F.
$= 41139.92 - $ C.F. $= 199.33$

Main plot S.S. for spacing $= \dfrac{1}{5}(143.07^2 + 151.43^2 + \ldots\ldots$

$+ 158.92^2 + 155.24^2) - $ C.F.
$= 41012.03 - $ C.F. $= 71.64$

To work out the sum of squares for varieties, spacing and their interaction, it is pertinent to prepare a two-way table as follows:

Table-I V × S table

Spacing(S) variety(V)	S_0	S_1	S_2	Total	Varity means
V_1	81.66	84.11	88.71	254.48	28.276
V_2	82.44	83.87	88.21	254.52	28.280
V_3	82.52	93.23	95.49	271.24	30.138
V_4	81.64	94.69	97.97	274.30	30.478
V_5	100.61	99.44	102.73	302.78	33.642
Total	428.87	455.34	473.11	1357.32	
Spacing mean	28.591	30.356	31.540		

Variety S.S. $= \dfrac{1}{9}(254.48^2 + 254.52^2 + 271.24^2 + 274.30^2 + 302.78^2) - $ C.F.

$= 41114.21 - 40940.39 = 173.82$

Spacing S.S. $= \dfrac{1}{15}(428.87^2 + 455.34^2 + 473.11^2) - $ C.F.

$= 41006.47 - $ C.F. $= 66.08$

$(V \times S)$ S.S.= Table S.S. $-$ C.F. $-$ S.S.(V) $-$ S.S.(S)

$$= \frac{1}{3}(81.66^2 + 84.11^2 + \ldots\ldots + 99.44^2 + 102.73^2) - 40940.39$$

$$- 173.82 - 66.08$$

$$= 41212.37 - 40940.39 - 173.82 - 66.08$$

$$= 32.08$$

Error S.S. (V) = $199.33 - 2.77 - 173.82 = 22.74$

Error S.S. (S) = $71.64 - 2.77 - 66.08 = 2.79$

Error S.S. (V $\times$ S) = $328.00 - 199.33 - 66.08 - 2.79 - 32.08 = 27.72$

Finally, the analysis of variance table can be prepared without difficulty as given below.

Table-II ANOVA table

Source	d.f.	S.S.	M.S.	F-value
Replications	2	2.77	1.385	0.487 NS
Varieties (V)	4	173.82	43.455	15.290*
Error$_1$ (V)	8	22.74	2.842 (E$_1$)	
Spacing (S)	2	66.08	33.04	47.335*
Error$_2$ (S)	4	2.79	0.698 (E$_2$)	
V $\times$ S	8	32.08	4.01	2.315 NS
Error$_3$ (V $\times$ S)	16	27.72	1.732 (E$_3$)	
Total	44	328.00		

Tabulated F-values at 5% level of significance:

$$F_{2,8} = 4.46, F_{4,8} = 3.84, F_{2,4} = 6.94, F_{8,16} = 2.59$$

On comparing the calculated F-values with the respective tabulated F-values as given below the table, the effects which are marked with a star show the significant differences and those which are non-significant are marked NS.

Calculation of standard errors of differences:

S.E. of the difference of any two varietal means on sub-unit basis by the formula (7.6.1) is,

$$S.E._{\cdot 1} = \sqrt{\frac{2 \times 2.842}{3 \times 3}} = 0.795$$

S.E. of the difference of any two spacing means on sub-unit basis by the formula (7.6.2) is,

$$S.E._{\cdot 2} = \sqrt{\frac{2 \times 0.698}{3 \times 5}} = 0.305$$

S.E. of a mean difference between varieties at the same level of spacing by the formula (7.6.3) is,

$$S.E._3 = \sqrt{\frac{2 \times 2 \times 1.732 + 2.842}{3 \times 3}} = 1.042$$

S.E. of a mean difference among spacings at the same level of varieties by the formula (7.6.4) is,

$$S.E._4 = \sqrt{\frac{2 \times 4 \times 1.732 + 0.698}{3 \times 5}} = 0.985$$

Comparison of calculated F-values with the respective tabulated F-values for various factors reveals that there is a significant difference among varieties as well as spacings whereas the interaction V × S is non-significant. Therefore, critical differences (CD) for varieties and spacings will be worked out at 5% level of significance for comparing all pairs of variety means and spacing means.

$$CD_1 = S.E._1 \times t_{error1 \, d.f.; \, .05} = 0.795 \times t_{8; \, .05}$$
$$= 0.795 \times 2.306 = 1.833$$
$$CD_2 = S.E._2 \times t_{error \, 2 \, d.f.; \, .05} = 0.305 \times t_{4;.05}$$
$$= 0.305 \times 2.776 = 0.847$$
$$CD_3 = SE_3 \times t_1^*$$

where,

$$t_1^* = \frac{(p-1)E_3 \times t_{error3d.f.;\alpha} + E_1 \times t_{error1d.f.;\alpha}}{(p-1)E_3 + E_1}$$

$$= \frac{4 \times 1.732 \times t_{16;.05} + 2.842 \times t_{8;.05}}{4 \times 1.732 + 2.842}$$

$$= \frac{4 \times 1.732 \times 2.120 + 2.842 \times 2.306}{4 \times 1.732 + 2.842} = 2.174$$

Thus, $\quad CD_3 = 1.042 \times 2.174 = 2.265$

Similarly, $\quad CD_4 = SE_4 \times t_2^*$

where, $\quad t_2^* = \dfrac{(q-1)E_3 \times t_{error3d.f.;\alpha} + E_2 \times t_{error2d.f.;\alpha}}{(q-1)E_3 + E_2}$

$$= \frac{2 \times 1.732 \times t_{16;.05} + 0.698 \times t_{4;.05}}{2 \times 1.732 + 0.698}$$

$$= \frac{2 \times 1.732 \times 2.120 + 0.698 \times 2.776}{2 \times 1.732 + 0.698} = 2.230$$

$$CD_4 = 0.985 \times 2.230 = 2.196$$

To compare pairs of variety means, write them in ascending order and compare their paired differences with $CD_1 = 1.833$.

V_1	V_2	V_3	V_4	V_5
28.276	28.280	30.138	30.487	33.642

In the above diagram, non-significant differences are underlined. The comparison of paired means reveals that variety V_5 has a significant higher yield than all other varieties. V_3 and V_4 are better than V_1 and V_2. On the whole V_1 is at par with V_2 and V_3 is at par with V_4. Similarly for comparing spacing means, write them in ascending order and compare the differences with $CD_2 = 0.847$.

S_0	S_1	S_2
28.591	30.356	31.540

Above diagram manifests that S_1 and S_2 are significantly better than S_0 and S_2 is superior to S_1 in respect of their yields.

To compare variety means at the same level of spacing say S_0, write variety means at S_0 level only as follows and compare their paired differences with $CD_3 = 2.265$.

Means at S_0

V_1	27.22	a
V_2	27.48	a c
V_3	27.50	a c e
V_4	27.21	a c e g
V_5	33.54	b d f h

This shows that at spacing S_0, V_5 is significantly better than all other varieties which are at par among themselves. Similarly comparison of varieties at the spacing levels S_1 and S_2 can be done.

To compare spacings at the same level of a variety, write the spacing means say at level V_1 and compare the paired mean differences with $CD_4 = 2.196$.

Spacing means at V_1

S_0	27.22	a
S_1	28.04	a c
S_2	29.57	b c

Above diagram depicts that S_2 is better than S_0 whereas S_0 and S_1 are at par and so are S_1 and S_2.

7.7 REPEATED EXPERIMENTS

Whatever research is being conducted, it has to be applied in real life. Large numbers of designs of experiments are extensively used in agricultural research and its findings are supposed to be widely adapted at farmer's fields in many regions and for long periods. This is possible only if the experiments are conducted at a number of distant locations and for a number of years. Such experiments testify whether the performance of treatments is uniform over the number of locations and years. This is indicated by the interaction of treatments with locations or years or both. Repeated experiment implies that the treatments and the design of experiment remains exactly the same at all locations and for all years. This depends on the purpose of study whether the experiment is spread over different locations or is conducted for a number of years at the same location or one includes a number of locations and years simultaneously.

7.8 POOLED ANALYSIS

To have an overall picture of the repeated experimental results, it becomes necessary to do combined analysis of data. For this purpose, firstly the data are analyzed for each location or year independently. For such an analysis, there may emerge two situations, (*i*) the treatments have same performance on all the locations or years, (*ii*) the responses differ at different locations or in various years. In situation (*i*), one may feel that when the performance of treatments is same at all locations or over all the years, then there is no need of doing pooled analysis. But in such a situation too, pooled analysis helps to know treatment and location (year) interaction persists. In situation (*ii*), an investigator is in dilemma to decide what conclusion can be drawn from varying results. Thus, to make a candid recommendation, pooled analysis is the only procedure at the disposal of the experimenter. The method of *pooled* or *combined analysis* is expatiated herein.

Consider an experiment which is conducted in randomized block design with t treatments, r replications and at u locations. Let an observation per unit for i-th treatment, jth block and uth location be represented by y_{iju}, where $i = 1, 2, \ldots\ldots, v$; $j = 1, 2, \ldots\ldots, r$ and $u = 1, 2, \ldots\ldots, k$. Procedure for pooled analysis is explicated in a realistic manner.

Firstly, analysis of variance for every location is to be carried out separately in the usual manner. ANOVA table for any uth location will be of the type given below.

Table 7.8.1 ANOVA

Source	d.f.	S.S.	M.S.	F-value
Blocks	$r-1$	B_{yy}	$\dfrac{B_{yy}}{r-1} = B_y$	$\dfrac{B_y}{s_u^2} = F_{r-1,\,\eta e}$
Treatments	$v-1$	T_{yy}	$\dfrac{T_{yy}}{v-1} = T_y$	$\dfrac{T_y}{s_u^2} = F_{v-1,\,\eta e}$
Error	$(r-1)(v-1) = n_{eu}$	$E_{yy} = s_u^2$		
Total	$rv-1$			

In this way for u locations, there shall be u ANOVA tables resulting into u error mean squares, say, $s_1^2, s_2^2, \ldots\ldots, s_u^2$. Prior to pooled analysis, it becomes necessary to test the homogeneity of these error mean squares as the further procedure of pooled analysis will be subjected to the result of the test. Bartlett's (1937) test is one of the popular tests commonly used to test the equality of three or more error mean squares (variances). Bartlett's test statistic to test H_0: $\sigma_1^2 = \sigma_2^2 \ldots\ldots = \sigma_u^2$ versus, H_1: At least two of them differ from one another significantly, is

$$\chi^2 = \cfrac{1}{1 + \cfrac{1}{3(k-1)}\left(\sum_u \dfrac{1}{n_{eu}} - \dfrac{1}{\sum_u n_{eu}}\right)}\left[\log_e \overline{s}_u^2 \times \sum_u n_{eu} - \sum_u n_{eu}\log_e s_u^2\right] \qquad (7.8.1)$$

where $\overline{s}_u^2 = \dfrac{\displaystyle\sum_{u=1}^{k} n_{eu}s_u^2}{k \times n_{eu}}$ $\qquad\qquad (7.8.2)$

Statistic χ^2 has $(k-1)$ d.f. But soon after Hartley (1940, 1950) developed a much simpler test for testing the homogeneity of several variances simultaneously, i.e. to test, H_0: $\sigma_1^2 = \sigma_2^2 \ldots\ldots = \sigma_u^2$ versus, H_1: At least two of them differ from one another significantly. His test is based on the theme that it is enough to test the hypothesis between the largest and the smallest of the u variances, i.e. test H_0: $\sigma_{max}^2 = \sigma_{min}^2$ versus, H_1: $\sigma_{max}^2 \neq \sigma_{min}^2$. Equality of two variances can be tested by F-test. The test statistic is,

$$F = \frac{s_{max}^2}{s_{min}^2} \qquad (7.8.3)$$

F has υ_1 and υ_2 degrees of freedom, where υ_1 are d.f. of s_{max}^2 and υ_2 are the d.f. of s_{min}^2, respectively. The decision about H_0 can be taken in the usual manner. If H_0 is rejected, it means u error mean squares are heterogeneous and in case of acceptance of H_0 it is affirmed that they are homogeneous.

In a situation when H_0 is accepted, skeleton analysis of variance table for pooled analysis shall be as displayed below.

Table 7.8.2 Pooled ANOVA

Source	d.f.
Locations	$k - 1$
Treatments	$v - 1$
Treat. × locations	$(v - 1)(k - 1)$
Polled error	$k (v - 1)(k - 1)$

Sum of squares and mean squares are calculated in the wonted manner. Treatment × location interaction is tested first against pooled error. If it comes to be significant, then treatment mean square is tested against interaction mean square.

Pooled error is obtained by the formula,

$$s_p^2 = \frac{\sum_u n_{eu} s_u^2}{\sum_u n_{eu}} \tag{7.8.4}$$

Pooled error is obtained from (7.8.4) is a biased estimate of error variance. Thus, an improved estimate can be obtained by the formula,

$$s_p^2 = \frac{\sum_u r_u s_u^2}{\sum_u r_u} \tag{7.8.5}$$

But if all r_u are same or not, in both the situations it is simply the average of locationwise error variances.

Again when it is confirmed through Bartlett's χ^2-test or Hartley's F-test that error mean squares for various locations are heterogeneous, analysis of variance cannot be carried out as such since homogeneity of error variances is a prime assumption. Hence, pooled analysis becomes somewhat complicated. The procedure for testing the significance of treatment effect differences is based on the significance / non-significance of treatment × location interaction.

Case-1: When the error variances are homogeneous. In this situation, set up a two-way table of treatment means and carry out the usual analysis of variance and test the interaction mean square against error mean square. This procedure is called *unweighted analysis of variance*.

Case-2: When the error variances are heterogeneous, then *weighted analysis* *of variance* is to be carried out. Stepwise procedure for weighted analysis is as follows:

(*i*) Calculate the weight for each experiment, where $w_u = \dfrac{r}{s_u^2}$ (7.8.6)

(*ii*) Work out the treatment mean $\bar{x}_{iu}$ for each location separately.

(*iii*) Find $P_u = \sum\limits_{i=1}^{v} \bar{x}_{iu}$ (7.8.7)

(*iv*) Compute the quantities, $\mu_u = \sum\limits_{i} \bar{x}_{iu}^2$ (7.8.8)

Sum of squares for ANOVA can be obtained through the following expressions.

(*v*) Grand total $= \sum\limits_{i}\sum\limits_{u} w_u \bar{x}_{iu}$ (7.8.9)

(*vi*) Correction factor $= \dfrac{\left(\sum\limits_{i}\sum\limits_{u} w_u \bar{x}_{iu}\right)^2}{v\sum\limits_{u} w_u}$ (7.8.10)

(*vii*) Location S.S. $= \dfrac{1}{v}\sum\limits_{u} w_u P_u^2 - \text{C.F.}$ (7.8.11)

(*viii*) Treatment S.S. $= \dfrac{\sum\limits_{i}\left(\sum\limits_{u} w_u \bar{x}_{iu}\right)^2}{\sum\limits_{u} w_u} - \text{C.F.}$ (7.8.12)

(*xi*) Total S.S. $= \sum w_u \mu_u - \text{C.F.}$ (7.8.13)

(*xii*) Treat. × location S.S. = Total S.S. – Treat. S.S.   (7.8.14)
                                      – Location S.S.

(*xiii*) Significance of interaction S.S. can be tested by χ^2 test by the formula,

$$\chi^2 = \frac{(n_e - 4)(n_e - 2)}{n_e(n_e + v - 3)} \times \text{Interaction S.S.} \qquad (7.8.15)$$

where, $n_e = (k - 1)(v - 1)$
χ^2 has v d.f.

where $v = \dfrac{(k-1)(v-1)(n_e - 4)}{(n_e + v - 3)}$ (7.8.16)

If the critical value of $\chi^2_{\upsilon,0.05} \leq \chi^2_{cal}$, then interaction is present, otherwise interaction is absent.

Degrees of freedom υ for χ^2 may be a fraction or an untabulated value. In this situation, critical value of χ^2 can be interpolated. If interaction is present, carry out unweighted analysis of mean data.

When it is affirmed that the interaction is non-significant, then it is not feasible to perform an overall test for treatment mean differences. For detailed study, readers are advised to go through the papers of Cochran (1937) and Yates and Cochran (1938).

Example 7.4: An experiment was conducted to test the performance of seven genotypes of maize at four locations in respect of the days to 50% silking. The experiment was in randomized block design with three replications at all locations. Locationwise data were as presented below. Totals have also been shown. ANOVA table for each location is prepared and displayed alongside of the data table.

Location-1   Days to 50% silking

Genotypes	Rep.-1	Rep.-2	Rep.-3	Total	ANOVA			
					Source	d.f.	S.S.	M.S.	F-value
G-1	56	55	53	164	Source	d.f.	S.S.	M.S.	F-value
G-2	57	56	54	167					
G-3	50	54	52	156	Reps.	2	9.24	4.62	2.18
G-4	53	54	55	162					
G-5	52	54	53	159	Geno.	6	28.00	4.67	2.20
G-6	51	55	52	158					
G-7	52	54	55	161	Error	12	25.43	2.12	
Total	371	382	374	1127	Total	20	62.67		

Location-2   Days to 50% silking

Genotypes	Rep.-1	Rep.-2	Rep.-3	Total	ANOVA			
					Source	d.f.	S.S.	M.S.	F-value
G-1	62	57	60	179	Source	d.f.	S.S.	M.S.	F-value
G-2	58	64	60	182					
G-3	60	57	58	175	Reps.	2	16.29	8.14	1.40
G-4	65	56	60	181					
G-5	62	59	61	182	Geno.	6	18.26	3.04	0.52
G-6	62	60	62	184					
G-7	60	61	62	183	Error	12	69.74	5.81	
Total	429	414	423	1266	Total	20	104.29		

Location-3   Days to 50% silking

Genotypes	Rep.-1	Rep.-2	Rep.-3	Total	ANOVA			
G-1	62	58	61	181	Source	d.f.	S.S.	M.S.	F-value
G-2	62	60	61	183					
G-3	59	56	60	175	Reps.	2	23.52	11.76	11.95
G-4	61	58	60	179					
G-5	61	57	58	176	Geno.	6	25.62	4.27	4.33*
G-6	61	60	62	183					
G-7	61	61	62	184	Error	12	11.81	0.984	
Total	427	410	424	1261	Total	20	60.95		

Location-4   Days to 50% silking

Genotypes	Rep.-1	Rep.-2	Rep.-3	Total	ANOVA			
G-1	60	61	63	184	Source	d.f.	S.S.	M.S.	F-value
G-2	60	57	58	175					
G-3	62	58	60	180	Reps.	2	4.10	2.05	< 1
G-4	57	61	59	177					
G-5	58	54	60	172	Geno.	6	56.57	9.43	2.43
G-6	57	58	56	171					
G-7	63	61	60	184	Error	12	46.57	3.88	
Total	417	410	416	1243	Total	20	107.24		

Critical value of F-distribution at = 0.05 and (6, 12) d.f. is,

$$F_{.05;\ (6,\ 12)} = 3.00$$

Locationwise analysis reveals that at location 3, genotypes are significantly different whereas at locations 1,2 and 4, they are not. So the scientist cannot draw a definite conclusion. To draw a firm and reliable result, the only way is to do pooled analysis.

As per procedure described in theory, first test the homogeneity of error variances by Bartlett's test.

Test the hypothesis H_0: $\sigma_{1e}^2 = \sigma_{2e}^2 = \sigma_{3e}^2 = \sigma_{4e}^2$ versus, H_1: At least any two of them differ significantly.

From ANOVA tables, $s_{1e}^2 = 2.12$, $s_{2e}^2 = 5.81$, $s_{3e}^2 = 0.984$, $s_{4e}^2 = 3.88$.

$\bar{s}_e^2 = 3.198$, $\log_{10} \bar{s}_e^2 = 0.5045$, $\log_e \bar{s}_e^2 = 2.3026 \times 0.5045$

$$= 1.1617$$

$$\sum_{u=1}^{4} n_{eu} \log_e s_{eu}^2$$

$$= 12 \times (0.3263 + 0.7642 - 0.0070 + 0.5888) \times 2.3026$$
$$= 12 \times 1.6725 \times 2.3026 = 46.2132$$

By the formula (7.8.1),

$$\chi^2 = \frac{1}{1 + \frac{1}{3 \times 3}\left(\frac{3}{12} - \frac{1}{48}\right)}(48 \times 1.1617 - 46.2132)$$

$$= \frac{1}{1.0254}(55.7616 - 46.2132) = 9.3119$$

From Table B-6, critical value of $\chi^2_{.05,3} = 7.815$. Since the calculated value of χ^2 is greater than the critical value, it affirms that the error variances are heterogeneous. Thus, this leads to weighted pooled analysis.

Prepare the following table of locationwise genotypes means and other factors.

Genotypes	Locationwise treatment means				Weighted total$_i$
	1	*2*	*3*	*4*	$\sum \omega_u \bar{x}_{iu} = T_i$
G_1	54.67	59.67	60.33	61.33	339.50
G_2	55.67	60.67	61.00	58.33	341.16
G_3	52.00	58.33	58.33	60.00	327.91
G_4	54.00	60.33	59.67	59.00	335.08
G_5	53.00	60.67	58.67	57.33	329.50
G_6	52.67	61.33	61.00	57.00	336.22
G_7	53.67	61.00	61.33	61.33	341.82
w_u	1.415	0.516	3.049	0.773	$5.753 = \sum w_u$
P_u	375.68	422.00	420.33	414.32	1632.33
$w_u P_u$	531.59	217.75	1281.58	320.27	$2351.19 = G$

$w_1 P_1^2 = 199706.68$, $w_2 P_2^2 = 91891.34$, $w_3 P_3^2 = 538689.11$,
$w_4 P_4^2 = 132694.00$, $\sum w_u P_u^2 = 962981.14$

$\mu_1 = 20171.56$, $\mu_2 = 25446.67$, $\mu_3 = 25248.14$, $\mu_4 = 24541.86$

$$\text{C.F.} = \frac{(2351.19)^2}{7 \times 5.753} = 137272.34$$

Geno. S.S. $= \dfrac{1}{5.753}(339.50^2 + 341.16^2 + \ldots\ldots + 336.22^2 + 341.82^2) - \text{C.F.}$

$$= 137303.84 - \text{C.F.} = 31.50$$

Total S.S. $= \sum_u w_u \mu_u$ – C.F.

$= (1.415)(20171.56) + (0.516)(25446.67) + (3.049)(25248.14)$
$+ (0.773)(24541.86)$

$= 137625.68 - \text{C.F.} = 353.34$

Location S.S. $= \frac{1}{7}(1.415 \times 375.68^2 + 0.516 \times 422.00^2$
$+ 3.049 \times 420.33^2 + 0.773 \times 414.32^2) - \text{C.F.}$

$= 137568.73 - \text{C.F.} = 296.39$

Genotypes × locations = 353.34 – 31.50 – 296.39 = 25.45

Error d.f., $n_e = 12$.

Statistic χ^2 for testing the significance of genotype × locations interaction by the formula (7.8.15) is,

$$\chi^2 = \frac{(12-4)(12-2)}{12 \times (12+7-3)} \times 25.45$$
$$= 10.61$$

Degrees of freedom for χ^2,

$$\upsilon = \frac{(4-1)(7-1)(12-4)}{(12+7-3)} = 9$$

Tabulated value of χ^2 at = 0.05 and 9 d.f. = 16.92. Hence, the interaction is not significant. This leads to the conclusion that performance of genotypes does not vary from location to location. In this situation, an analyst should study the papers referred in theory and then proceed for further analysis.

Note: If per chance the value of comes out to be a fraction, then the value of χ^2 should be interpolated by the formula (A.) given in Appendix-A.

Undoutedly an inquisitive mind would be eager to know, if the interaction is present, how to proceed for further analysis. In this case, carry out unweighted analysis. In example (7.4), interaction is absent. But for the time being it assumed to be present. Procedure for unweighted pooled analysis is as given below.

Two-way table of means with additional available information.

Genotypes	Locationwise treatment means				Total
	1	*2*	*3*	*4*	
G_1	54.67	59.67	60.33	61.33	236.00
G_2	55.67	60.67	61.00	58.33	235.67
G_3	52.00	58.33	58.33	60.00	228.66

(Contd.)

Contd. table

Genotypes	Locationwise treatment means				Total
	1	2	3	4	
G_4	54.00	60.33	59.67	59.00	233.00
G_5	53.00	60.67	58.67	57.33	229.67
G_6	52.67	61.33	61.00	57.00	232.00
G_7	53.67	61.00	61.33	61.33	237.33
Total	375.68	422.00	420.33	414.32	1632.33
Error S.S	25.43	69.74	11.81	46.57	153.55
Error d.f.	12	12	12	12	

$$C.F. = \frac{(1632.33)^2}{28} = 95160.76$$

Genotype S.S. $= \frac{1}{4}(236.00^2 + 235.67^2 + \dots\dots + 232.00^2 + 237.33^2) - C.F.$

$$= 16.38$$

Locations S.S. $= \frac{1}{7}(375.68^2 + 422.00^2 + 420.33^2 + 414.32^2) - C.F.$

$$= 204.64$$

Table S.S. $= 54.67^2 + 55.67^2 + \dots\dots + 57.00^2 + 61.33^2) - C.F.$

$$= 247.47$$

Genotype × location S.S. = Table S.S. − Genotype S.S. − Location S.S.

$$= 247.47 - 16.38 - 204.64 = 26.45$$

Pooled error S.S. $= 25.43 + 69.74 + 11.81 + 46.57 = 153.55$

ANOVA

Source	d.f.	S.S.	M.S.	F-value
Genotypes	6	16.38	2.73	0.85 NS
Locations	3	204.64	68.21	21.31*
Geno. × Loc.	18	26.45	1.47	0.46 NS
Pooled error	48	153.55	3.20	

- * - Significant at 5% level of significance.
- NS - Not significant

Critical value of $F_{.05, (3, 48)}$ can be obtained by the interpolation formula. From F-table, at $= 0.05$, $F_{3, 40} = 2.84$, $F_{3, 60} = 2.76$.

In this case, $m = 3$, $n = 48$, $n' = 40$ and $n'' = 60$.

$$A = \frac{60(48 - 40)}{48(60 - 40)} = 0.5$$

$$F_{3,48} = (1 - 0.5) \times 2.84 + 0.5 \times 2.76 = 2.80$$

The above analysis reveals that there is a significant difference between locations and locations × genotypes interaction is absent.

Epitome

In experimental research, split-plot designs are very popular and frequently used designs in various fields. They are used because of their peculiar merits which no other design possesses. They accommodate any number of treatments which require large as well as small size experimental units. Their layout and analysis of data is not tedious. There are no complexities in interpretation of results. Also a large number of variations in their layout can be incorporated. They contain all characters of factorial experiments and are efficient as well. Treatments of greater interest can be estimated and tested more precisely than others. Theory of split-plot designs covered in this chapter will benefit the students, teachers and researchers to a large extent.

QUESTIONS AND EXERCISES

1. In what respects, split plot designs differ from factorial experiments?
2. What are the criteria choosing main-plot and sub-plot treatments in split-plot designs?
3. What manipulations are possible in split-plot design to increase precision?
4. Give the statistical model of a split plot design with two factors at various levels. Also give the expressions for expected mean squares taking the factors as fixed, random and mixed.
5. When do you require the use of Satterthwaite's approximate F-statistics?
6. In split-plot design whether main-plot treatment is nested in split-plots or vice-versa. Explain clearly.
7. Justify that main plot-error mean square is greater than split-plot error mean square. Can it happen otherwise also? If so, suggest the proper solution.
8. Give the formulae for standard error of differences between different paired means of treatments in a $p \times m$ split-plot design having r replications.
9. Give the layout of a split-split-plot design and its merits and demerits.
10. Explain the method of analysis of data of a split-split-plot design.
11. When do you need a split-block design? Give the skeleton analysis of variance table for this design.

12. What do you understand by pooled analysis of a group of experiments?
13. What are the advantages and problems of pooled analysis of group of experiments?
14. When should one use weighted analysis of variance while analyzing data of a location trial?
15. In which areas of research split-plot designs are more commonly used?
16. How can one make use of Latin square design pattern in split-plot design? Explain by taking a suitable example. Also mention its advantages.
17. Explain the differences between split-plot and strip-plot design. Also discuss the differences occurring in analysis of variance tables.
18. Consider a split-plot design with single main-plot (MP) treatment A with four levels and another sub-plot (SP) factor B with three levels. There are two replications. How many degrees of freedom shall be there for the SP error?
19. An experiment was conducted in split-plot design taking four replications for testing the differences between five levels of sulphur (0, 20, 40, 60, and 80 kg/ha) in main-plots and five levels of zinc (0, 2.5, 5, 7.5, and 10 kg/ha) in sub-plots. The crop was mustard and the character under study was siliqua/plant. Tabulated data are presented below.

Siliqua/plant

TREAT.	R_1	R_2	R_3	R_4
S_1Zn_1	272	217	230	252
S_1Zn_2	307	245	209	308
S_1Zn_3	287	275	267	308
S_1Zn_4	270	300	302	295
S_1Zn_5	317	266	287	267
S_2Zn_1	275	260	307	265
S_2Zn_2	296	315	353	252
S_2Zn_3	282	354	313	344
S_2Zn_4	314	376	318	319
S_2Zn_5	325	366	328	279
S_3Zn_1	300	318	296	302
S_3Zn_2	345	353	295	340
S_3Zn_3	388	321	376	328
S_3Zn_4	320	375	364	384
S_3Zn_5	352	332	397	337
S_4Zn_1	307	334	329	328
S_4Zn_2	349	374	300	400
S_4Zn_3	413	312	425	357

S₄Zn₄	383	419	399	349
S₄Zn₅	391	383	417	329
S₅Zn₁	298	300	332	320
S₅Zn₂	399	304	306	361
S₅Zn₃	391	302	371	392
S₅Zn₄	403	335	404	357
S₅Zn₅	344	369	400	364

Analyze the experimental data and give your recommendations.

20. A trial was planned to assess the impact of three irrigation levels (1, 2, 3 irrigations) and six weed control methods on sorghum. Both of these factors are such that they require large size plots from operational point of view. Hence, the layout of the experiment was split-block design. The experiment was carried taking three replications.

The layout along with measurements pertaining to 1000 grain weight (gm) is displayed below.

1000 grain weight (gm)

Rep.-1

	I_1	I_2	I_3
W_1	30.40	31.60	32.30
W_2	29.50	32.45	32.60
W_3	30.15	32.28	31.78
W_4	30.79	32.45	34.06
W_5	31.65	33.05	32.70
W_6	30.80	31.95	31.78

Rep.-2

	I_1	I_2	I_3
W_1	30.75	31.95	33.00
W_2	31.12	31.20	32.10
W_3	29.60	30.80	32.10
W_4	31.30	33.70	32.70
W_5	30.70	32.35	33.72
W_6	31.10	32.85	32.18

Rep.-3

	I_1	I_2	I_3
W_1	30.95	32.90	33.40
W_2	29.92	32.05	32.50
W_3	29.50	31.18	32.66
W_4	32.20	31.95	33.08
W_5	30.95	31.80	32.16
W_6	30.35	31.65	33.24

Analyze the data of the trial and interpret the results.

21. Following the layout plan given in section (7.5), a trial was planned in split-split-plot design. Treatments in this trial were 3 dates of sowing at an interval of 15 days (D_1, D_2, D_3) allocated to main-plots; 4 levels of nitrogen – 50, 100, 150, 200 kg/ha (N_1, N_2, N_3, N_4) assigned to sub-plots; 3 varieties of paddy namely, Sughand-1, Sughand-2 and Mahi Sughand (V_1, V_2, V_3). There were four replications.

Replicationwise tabulated data are presented below.

Grain yield (q/ha)

	R_1	R_2	R_3	R_4
$D_1N_1V_1$	51.24	42.14	41.92	44.36
$D_1N_1V_2$	40.15	48.13	43.36	40.48
$D_1N_1V_3$	39.59	33.60	31.72	36.60
$D_1N_2V_1$	58.22	65.87	65.43	55.78
$D_1N_2V_2$	62.55	53.23	53.01	51.68
$D_1N_2V_3$	44.36	49.91	51.57	36.04
$D_1N_3V_1$	74.86	66.98	73.19	63.99
$D_1N_3V_2$	62.10	70.20	65.43	62.21
$D_1N_3V_3$	55.56	50.90	46.58	55.45
$D_1N_4V_1$	75.86	77.63	77.85	68.09
$D_1N_4V_2$	72.09	69.65	68.20	65.99
$D_1N_4V_3$	52.23	56.78	61.00	43.81
$D_2N_1V_1$	51.46	40.15	45.03	51.01
$D_2N_1V_2$	44.36	41.48	48.02	42.70
$D_2N_1V_3$	39.92	28.06	36.38	41.59
$D_2N_2V_1$	55.01	71.64	68.76	63.21
$D_2N_2V_2$	59.89	57.89	56.00	61.77

(Contd.)

	R_1	R_2	R_3	R_4
$D_2N_2V_3$	39.37	54.12	54.67	47.02
$D_2N_3V_1$	80.62	68.76	70.98	82.18
$D_2N_3V_2$	69.98	66.54	74.30	64.21
$D_2N_3V_3$	61.22	50.13	57.89	61.00
$D_2N_4V_1$	73.42	83.18	87.83	77.63
$D_2N_4V_2$	70.98	73.42	71.64	77.63
$D_2N_4V_3$	51.35	63.66	63.88	54.90
$D_3N_1V_1$	47.24	39.70	35.82	44.91
$D_3N_1V_2$	36.49	44.80	38.48	39.92
$D_3N_1V_3$	35.93	31.05	26.62	35.93
$D_3N_2V_1$	53.01	60.66	59.55	49.91
$D_3N_2V_2$	53.79	49.46	49.24	51.12
$D_3N_2V_3$	42.14	47.02	48.35	37.71
$D_3N_3V_1$	70.64	63.32	57.11	68.87
$D_3N_3V_2$	55.34	64.32	57.67	60.44
$D_3N_3V_3$	55.67	46.36	39.92	55.45
$D_3N_4V_1$	66.76	70.98	74.30	58.11
$D_3N_4V_2$	67.76	62.10	61.77	64.32
$D_3N_4V_3$	49.13	54.01	55.89	43.25

Analyze the experimental data and draw conclusions.

BIBLIOGRAPHY

1. Agarwal BL. Testing a main effect in a three factor mixed model. *Communication in Statistics – Theory and Methods*, 1990: 19(2), 723–38.
2. Agarwal BL, Gupta VP. Approximate power of test procedures based on two preliminary tests in the mixed model. *Journal of the Indian Society of Agricultural Statistics*, 1983: 2, 56–68.
3. Bartlett MS. Some examples of statistical methods of research in agriculture and applied biology. *Journal of the Royal Statistical Society* (Suppl.), 1937: 4, 137.
4. Box GEP, Hunter JS, Hunter WG. *Statistics for Experimenters: Design, Innovation and Discovery*. Hoboken, John Wiley, New Jersey, 2005.
5. Cochran WG. Problems arising in the analysis of a series of similar experiments. *Journal of the Royal Statistical Society*, 1937: 4, 102–118.
6. Guseo R. Split and strip plot configurations of two-level fractional factorials: A review. *Journal of the Italian Statistical Society*, 2000: 9, 85–86.
7. Hartley HO. Testing of homogeneity of a set of variances. *Biometrika*, 1940: 31, 249–255.
8. Hartley HO. The maximum F-ratio as a short cut test for heterogeneity of variances. *Biomerika*, 1950: 37, 308–312.

9. Lavbscher NF. Interpolation in F-tables. *The American Statistician*, 1965.
10. Miller A. Strip-plot configuration of fractional factorials. *Technometrics*, 1997: 39(2), 153–160.
11. Satterthwaite FE. An approximate distribution of estimates of variance components. *Biometrics*, 1946: 2, 110.
12. Taguchi G. *System of experimental design*. Kraus International Publications, White Plain, New York, 1987.
13. Vivacqua CA, Bisgaard S. Strip-block experiments for process improvement and robustness. *Quality Engineering*, 2004: 16(3), 495–500.
14. Wooding WN. The split-plot design. *Journal of Quality Technology*, 1973: 5, 16–33.
15. Yates F, Cochran WG. The analysis of group of experiments. *Journal of the Agricultural Sciences*, 1938: 28, 556–580.

8

Nested or Hierarchical Classification

8.1 STRUCTURE OF THE NESTED CLASSIFICATION

Factor B with a number of levels is said to be nested within a second factor A if each level of B is woven with just one level of the second factor A and it is denoted as B(A), i.e. the factor B is nested in A. The process of nesting can be extended to a third factor C such that each level of C is woven with just a single level of factor B. Obviously, C is also nested in A and it is denoted by C(BA). Thus, a process of repeated sampling and sub-sampling from each sampling unit is known as nested or hierarchical classification. Such type of classification is often used in a variety of research areas. To name a few are: agricultural sciences, behavioral sciences, education, industries, animal science, medical science, etc. Experiments following *nested classification* can be cited as follows:

(*i*) An experiment on cultivators' field was planned to know the effect of certain treatments on a crop. For this, p villages and in each village q farms were randomly selected. On each farm, v treatments were applied and m samples for each treatment were randomly selected and observations taken in all villages. Such an experiment paves the way to a nested design.

(*ii*) Suppose the affectivity of two methods of instructions namely, traditional method a_1 and audio-visual aid a_2 are to be evaluated on students of ninth standard in two schools, each having three sections. Also it has been assumed that number of students in all sections is same. Here three sections b_1, b_2 and b_3 of one school are allotted to a_1 and sections b_4, b_5 and b_6 of the other school are assigned to a_2. Diagramatically, this is

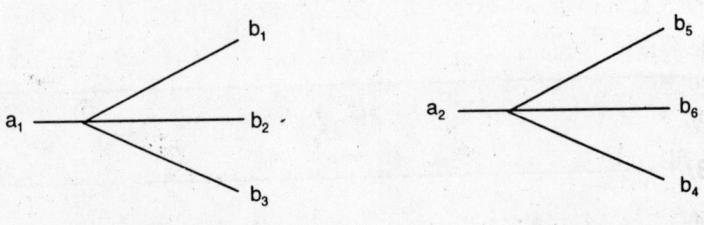

Fig. 8.1   **Fig. 8.2**

a problem which leads to hierarchical classification in educational research.

(*iii*) In a spare-part producing factory, three types of machines which produce a particular type of spare-part are to be tested. There are six mechanics that can operate these machines. Two mechanics are allocated to each machine and their six hourly outputs are being recorded to compare the efficiency of machines as well as operators in conjunction to machines. Such an experimental plan is nothing but is a nested design.

(*iv*) In animal breeding each of s_i sires is mated to d_j dams and the milk yield of their r_k daughters is recorded. Such a mating plan leads to hierarchical classification.

(*v*) A new drug is introduced to replace the current drug for treating the same disease. To compare their efficacy, four hospitals each having two wards were selected for administering the drug. It is supposed that each ward has the same number of patients. To compare the effect of drugs, two hospitals were allotted to each drug and two wards in each hospital. The nested classification can diagrammatically be shown as,

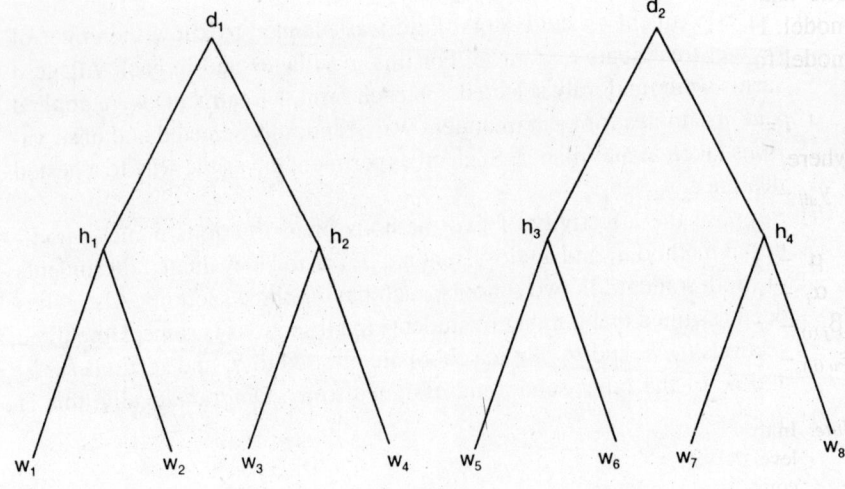

Fig. 8.3

8.2 PARTIAL HIERARCHICAL CLASSIFICATION

In a design, if a treatment B is nested in A, i.e. B (A), but a third factor C crossed with A and B(A), then such a design is called a partial hierarchical design.

8.3 BALANCED HIERARCHICAL DESIGN

A hierarchical design is said to be *balanced* if it contains equal number of experimental units for each treatment combination and an equal number of levels of the nested treatment B in each level of treatment A and so on. Otherwise the design is considered to be *unbalanced*.

8.4 LAYOUT OF HIERARCHICAL DESIGNS

Generally, hierarchical designs are conducted in completely randomized design or randomized complete block design and they are abbreviated as CRH-*pm* (A) and RBH-*pm* (A) respectively, where treatment B with *m* levels is nested in each of the *p* levels of A.

In case there are three factors A, B and C, factor B is nested in A and C is nested in A and C is nested in A(BA) then, in similar manner as above, they are denoted as CRH-*pmq* (AB) and RBH-*pmq* (AB), where *q* is the number of levels of C nested in BA.

Note: A hierarchical design may be conducted in any other design but they are not considered in this book.

8.5 STATISTICAL MODELS AND ANALYSIS OF HIERARCHICAL DESIGNS

The analysis of data of a design of experiment is based on its statistical model. Hence, firstly the model of the design is given prior to ANOVA. The model for CRH-*pm* (A) in general is.

$$Y_{uij} = \mu + \alpha_i + \beta_{j(i)} + \varepsilon_{u(ij)} \qquad (8.5.1)$$

$$i = 1, 2, \ldots\ldots\ldots, p; j = 1, 2, \ldots\ldots\ldots, m; u = 1, 2, \ldots\ldots\ldots, r.$$

where,

y_{uij} – the response for the *u*th experimental unit receiving *ij*th treatment combination.

μ – grand mean.

α_i – effect of the *i*th level of treatment A.

$\beta_{j(i)}$ – effect of the *j*th level of treatment B nested in the *i*th level of A.

$\varepsilon_{u(ij)}$ – error due to *u*th unit receiving the treatment combination *ij* and is always distributed as N $(0, \sigma_\varepsilon^2)$ and is independent of α_i and $\beta_{j(i)}$.

Note: In the above model, it is just possible that B has different number of levels for each level of A and also there may be a varied number of determinations for the treatment combinations *ji* say, r_{ij} instead of constant *r*.

(*i*) Assuming the model (8.5.1) to be a fixed effect model, the restrictions on α_i and $\beta_{j(i)}$ are $\sum_{i=1}^{p} \alpha_i = 0$ and $\sum_{j=1}^{m} \beta_j = 0$ for all *i*.

(*ii*) If the treatments A and B(A) are random effects, then model (8.5.1) is a random effect model and it follows the conditions; $\alpha_i \sim NID(0, \sigma_\alpha^2)$ and $\beta_{j(i)} \sim NID(0, \sigma_\beta^2)$.

(*iii*) In case the treatment A is random and B(A) is of fixed nature, then model (8.5.1) represents a mixed model and the assumptions under this model are: $\alpha_i \sim NID(0, \sigma_\alpha^2)$ and $\beta_{j(i)}$ is subject to the restriction $\sum_{j=1}^{m} \beta_{j(i)} = 0$ for all *i*.

The calculations of mean sum of squares for CRH-*pm* (A) with $u = 1, 2, \ldots, r$ remains the same for the three types of models considered above. The difference for the three types of models lies in their expected mean squares as delineated in ANOVA Table 8.5.2.

Data of a CRH-*pm* (A) with *r* determinations per *ji* - treatment combination can be tabulated as follows:

Table 8.5.1 Tabulation of data

Treat. (A)	Treat. B (A)	Determinations 1	2	.... r	Total for B(A)	Total for (A)
	1	y_{111}	y_{112}	y_{11r}	$y_{11\cdot}$	
	2	y_{121}	y_{122}	y_{12r}	$y_{12\cdot}$	
1	|		|		|	$y_{1\cdot\cdot}$
	m	y_{1m1}	y_{1m2}	y_{1mr}	$y_{1m\cdot}$	
	1	y_{211}	y_{212}	y_{21r}	$y_{21\cdot}$	
	2	y_{221}	y_{222}	y_{22r}	$y_{22\cdot}$	
2	|		|		|	$y_{2\cdot\cdot}$
	m	y_{2m1}	y_{2m2}	y_{2mr}	$y_{2m\cdot}$	
| | |	| | |	| | |			| | |	| | |
	1	y_{p11}	y_{p12}	y_{p1r}	$y_{p1\cdot}$	
	2	y_{p21}	y_{p22}	y_{p2r}	$y_{p2\cdot}$	
p	|		|		|	$y_{p\cdot\cdot}$
	m	y_{pm1}	y_{pm2}	y_{pmr}	$y_{pm\cdot}$	
Total						$y_{\cdot\cdot\cdot}$

Table 8.5.2 ANOVA for CRH-pm (A) with expected mean squares

Source	d.f.	S.S.	M.S.	Expected Model–I A and b fixed	Mean squares Model–II	Model–III
A	$p-1$	SSA	$\dfrac{\text{SSA}}{p-1}$	$\sigma_e^2 + rm\displaystyle\sum_{i=1}^{p}\frac{\alpha_i^2}{p-1}$	$\sigma_e^2 + r\sigma_\beta^2 + rm\sigma_\alpha^2$	$\sigma_e^2 + rm\sigma_\alpha^2$
B(A)	$p(m-1)$	SSB(A)	$\dfrac{\text{SSB(A)}}{p(m-1)}$	$\sigma_e^2 + r\dfrac{\displaystyle\sum_{i=1}^{p}\sum_{j=1}^{m}\beta_{j(i)}^2}{p(m-1)}$	$\sigma_e^2 + r\sigma_\beta^2$	$\sigma_e^2 + r\dfrac{\displaystyle\sum_{i=1}^{p}\sum_{j=1}^{m}\beta_{j(i)}^2}{p(m-1)}$
Error	$pm(r-1)$	SSE	$\dfrac{\text{SSE}}{pm(r-1)} = \sigma_e^2$	σ_e^2	σ_e^2	σ_e^2
Total	$pmr-1$					

Sum of squares for the data given in table (8.5.1) can be calculated as follows:

$$\text{Correction factor (C.F.)} = \frac{y_{...}^2}{rpm}$$

$$\text{Sum of square due to A} = \frac{1}{rm} \sum_{i=1}^{p} y_{i..}^2 - \text{C.F.} = \text{SSA}$$

$$\text{Sum of square due to B (A)} = \frac{1}{r} \sum_{i=1}^{p} \sum_{j=1}^{m} y_{ij.}^2 - \text{C.F.} - \text{SSA} = \text{SSB(A)}$$

$$\text{Total (Determinations) S.S.} = \sum_{i=1}^{p} \sum_{j=1}^{m} \sum_{u=1}^{r} y_{iju}^2 - \text{C.F.} = \text{SSD}$$

Determinations within B(A) = SSD – SSB(A) = SSE

Analysis of variance table for the model (8.5.1) along with expected mean squares for the three types of models have been depicted in Table 8.5.2.

Often one comes across the situation in which the levels of factor B nested in each level of A are not same and also the number of determinations for each combination $j\,(i)$ varies. In such cases, sum of squares for nested design can be calculated as follows:

Suppose for the model (8.5.1),

$i = 1, 2, \ldots\ldots\ldots , p; j = 1, 2, \ldots\ldots\ldots , m_i$ and $u = 1, 2, \ldots\ldots\ldots , r_{j(i)}$. Then,

$$\text{S.S. due to A} = \sum_{i=1}^{p} \frac{y_{i..}^2}{r_{j.}} - \frac{y_{...}^2}{r_{..}} = \text{SSA}$$

$$\text{S.S. due to B (A)} = \sum_{i=1}^{p} \sum_{j=1}^{m_i} \frac{y_{ij.}^2}{r_{j(i)}} - \sum_{i=1}^{p} \frac{y_{i..}^2}{r_{i.}} = \text{SSB(A)}$$

Determinations within B (A), i.e.

$$\text{Error S.S.} = \sum_{i=1}^{p} \sum_{j=1}^{m_i} \sum_{u=1}^{r_{j(i)}} y_{iju}^2 - \sum_{i=1}^{p} \sum_{j=1}^{m_i} \frac{y_{ij.}^2}{r_{j(i)}} = \text{SSE}$$

Where,
$$\sum_{j=1}^{m_i} r_{j(i)} = r_{i.} \text{ and } \sum_{i=1}^{p} \sum_{j=1}^{m_i} r_{j(i)} = r_{..}$$

Rest of the procedure remains same as discussed in case of balanced nested design. Of course, the coefficients in expected mean squares will have to be

Table 8.5.3 ANOVA for CRH–p(A) m(BA) design

Source	d.f.	S.S.	M.S.
A	$p - 1$	SSA	$\dfrac{\text{SSA}}{p-1}$
B(A)	$\displaystyle\sum_i m_i - p$	SSB (A)	$\dfrac{\text{SSB(A)}}{\displaystyle\sum_i m_i - p}$
Error	$r_{..} - \displaystyle\sum_i m_i$	SSE	$\dfrac{\text{SSE}}{r_{..} - \displaystyle\sum_i m_i}$
Total	$r_{..} - 1$		

adjusted in accordance to varying level of factor B and number of determinations. It has been skipped here as the test procedure is not affected by coefficients. This only matters for estimating the parameters.

8.6 THREE-FACTOR NESTED MODEL

If an experiment is conducted taking three factors A, B and C where B is nested in A and C is nested in B (A), then the statistical model is,

$$y_{ijku} = \mu + \alpha_i + \beta_{j\,(i)} + \gamma_{k\,(ij)} + \varepsilon_{u\,(ijk)} \qquad (8.6.1)$$

$i = 1, 2, \ldots\ldots, p; j = 1, 2, \ldots\ldots, m; k = 1, 2, \ldots\ldots, q$ and $u = 1, 2, \ldots\ldots, r$.

Such a model will suit to the problem given in example (v) on page 316 and similar problems. Tabulation and calculation of sum of squares for the factors involved in the model (8.6.1) is trivial. So analysis of variance table along with expected mean squares is given below directly.

Table 8.6.1 ANOVA for three-factor nested model

Source	d.f.	S.S.	M.S.	Expected mean squares (Model-II)
A	$p - 1$	SSA	$\dfrac{\text{SSA}}{p-1}$	$\sigma_e^2 + r\left(1 - \dfrac{q}{Q}\right)\sigma_\gamma^2 + rq\left(1 - \dfrac{m}{M}\right) \times$ $\sigma_\beta^2 + rmq\sigma_\alpha^2$
B(A)	$p\,(m-1)$	SSB(A)	$\dfrac{\text{SSB(A)}}{p(m-1)}$	$\sigma_e^2 + r\left(1 - \dfrac{q}{Q}\right)\sigma_\gamma^2 + rq\sigma_\beta^2$

(Contd.)

Source	d.f.	S.S.	M.S.	Expected mean squares (Model-II)
C(AB)	$pm(q-1)$	SSC(AB)	$\dfrac{\text{SSC(AB)}}{pm(q-1)}$	$\sigma_e^2 + r\sigma_\gamma^2$
Error	$pmq(r-1)$	SSE	$\dfrac{\text{SSE}}{pmq(r-1)}$	σ_e^2
Total	$rpmq-1$	SST	$\displaystyle\sum_{i,j,k,u} y_{ijku}^2 - \text{C.F.}$	

Remarks:

1. In the above ANOVA table, expected mean squares are given only for model-II. In case of model-I, the variances σ_α^2, σ_β^2 and σ_γ^2 should be re-placed by $\dfrac{\sum_i \alpha_i^2}{p-1}$, $\dfrac{\sum_j \beta_j^2}{p(m-1)}$ and $\dfrac{\sum_k \gamma_k^2}{pm(q-1)}$, respectively. For model-III, only those variance(s) are to be replaced by mean deviation sum of squares whose effects are of fixed nature.

2. The expressions $\dfrac{m}{M}$ and $\dfrac{q}{Q}$ are known as the sampling fraction for the populations of levels M and Q of random factors B (A) and C (BA) respectively. In case the factors A and B are fixed, the value of sampling fractions, $\dfrac{m}{M} = 1$ and $\dfrac{q}{Q} = 1$ since, $m = M$ and $q = Q$. If B is random, $\dfrac{m}{M} = 0$ and similarly $\dfrac{q}{Q} = 0$ when C is random.

3. Interestingly, it is noteworthy that there is a correspondence between sum of squares of a nested treatment and sum of squares of the respective treatments in a factorial experiment. For instance,
   SSB(A) = SSB + SSAB.
   SSC(AB) = SSC + SSAC + SSBC + SSABC.

8.7 ANALYSIS OF PARTIALLY NESTED DESIGN

Consider an experiment in which four levels of B(A) are nested in two levels of A and two levels of a third factor treatment C are crossed with A and B(A). Such an experiment can diagrammatically be represented as given below.

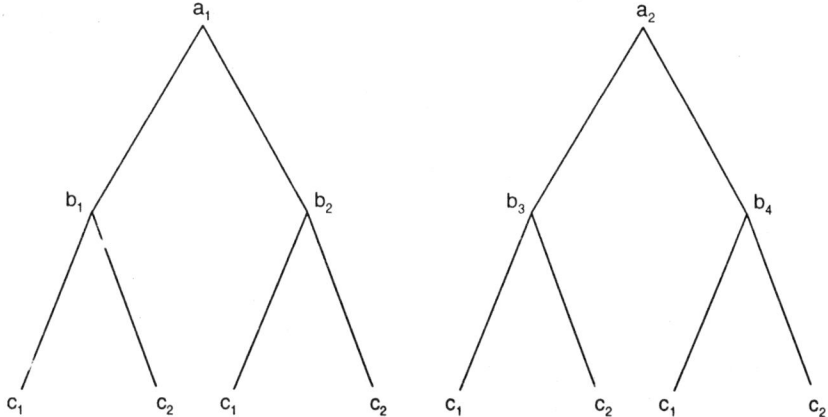

Statistical model for completely randomized design in general is,

$$y_{ijku} = \mu + \alpha_i + \beta_{j\,(i)} + \gamma_k + (\alpha\gamma)_{ik} + (\beta\gamma)_{j\,(i)\,k} + \varepsilon_{ijku} \qquad (8.7.1)$$

$i = 1, 2, \ldots\ldots\ldots , p; \; j = 1, 2, \ldots\ldots\ldots , m; \; k = 1, 2, \ldots\ldots , q$ and $u = 1, 2, \ldots\ldots , r.$

Skeleton analysis of variance table for the model (8.7.1) along with expected mean squares for model-II is presented below.

Table 8.7.1 ANOVA for a partially nested design

Source	d.f.	Expected mean squares
A	$p-1$	$\sigma_e^2 + r\left(1-\dfrac{m}{M}\right)\left(1-\dfrac{q}{Q}\right)\sigma_{\beta\gamma}^2 + rm\left(1-\dfrac{q}{Q}\right)\sigma_{\alpha\gamma}^2$ $+ rq\left(1-\dfrac{m}{M}\right)\sigma_{\beta}^2 + rmq\sigma_{\alpha}^2$
B(A)	$p\,(m-1)$	$\sigma_e^2 + r\left(1-\dfrac{q}{Q}\right)\sigma_{\beta\gamma}^2 + rq\sigma_{\beta}^2$
C	$(q-1)$	$\sigma_e^2 + r\left(1-\dfrac{m}{M}\right)\sigma_{\beta\gamma}^2 + rm\left(1-\dfrac{p}{P}\right)\sigma_{\alpha\gamma}^2 + rpm\sigma_{\gamma}^2$
A × C	$(p\div1)(q-1)$	$\sigma_e^2 + r\left(1-\dfrac{m}{M}\right)\sigma_{\beta\gamma}^2 + rm\sigma_{\alpha\gamma}^2$
B(A) × C	$p\,(m-1) \times (q-1)$	$\sigma_e^2 + r\sigma_{\beta\gamma}^2$
Error	$pmq \times (r-1)$	σ_e^2
Total	$rpmq - 1$	

Remarks:

1. If the effects are fixed (Model-I), then their variances should be replaced by the respective mean deviation sum of square in the column of expected mean squares.

2. The fractions $\frac{p}{P}, \frac{m}{M}$ and $\frac{q}{Q}$ are sampling fractions which take the values 1 when the effects are fixed and 0 when the effects are random.

3. Sum of squares can be calculated in the usual manner. So the expressions are skipped over here.

8.8 HIERARCHICAL CLASSIFICATION IN RANDOMIZED BLOCK DESIGN

Experimental designs using hierarchical classification often require a blocking factor. Hence, the experiment is to be planned in randomized block design. In view of this, one more parameter say, ρ_l will be added in the model (8.5.1). Likewise analysis of variance Table 8.5.2 will have an additional source of variation namely, blocks. Procedure of calculation of sum of squares remains same as in CRH-p (A) m (BA). The details are avoided for the sake of brevity. Following example (8.1) will ifso-facto explain the whole mechanism of analysis of data.

Example 8.1: A training program was launched to update the primary school teachers. Three training methods (M) were applied namely, (*a*) instructions by master trainers (MT) through traditional method, (*b*) instructions by MT with the help of audio-visual aids, (*c*) instructions through computer programs. Four teachers (T) from four schools of three blocks were selected and deputed for training independently. In this way, each teacher represents a school. Every teacher was given five tests of 20 marks each at regular intervals during training. The scores secured by respective teachers are as tabulated below. Totals are also shown alongside.

Teaching methods (M)	Teachers T(M)	Test scores I	II	III	IV	V	Totals for T(M)	Totals for (M)
Method (a)	T_1	12	8	14	12	16	62	
	T_2	9	9	15	10	15	58	
	T_3	11	10	13	12	15	61	239
	T_4	12	8	14	13	11	58	
Method (b)	T_5	19	15	15	19	18	86	
	T_6	14	17	16	17	15	79	
	T_7	15	16	20	17	18	86	341
	T_8	16	19	17	20	18	90	

(Contd.)

Teaching methods (M)	Teachers T(M)	Test scores					Totals for T(M)	Totals for (M)
		I	*II*	*III*	*IV*	*V*		
Method (c)	T_9	13	14	12	11	10	60	
	T_{10}	14	9	7	8	11	49	
	T_{11}	8	9	12	7	6	42	215
	T_{12}	10	13	15	14	12	64	
Grand Total								795

The given experiment follows nested classification and has to be analyzed to test the hypotheses that:

(*i*) Three training methods are equally effective.

(*ii*) There is no variation among teachers (schools) within the training methods.

Taking training methods and teachers as fixed effects, it is evident that both the factors will have to be tested against error.

Calculations

$$C.F. = \frac{(795)^2}{60} = 10533.75$$

$$S.S. \text{ due to (M)} = \frac{1}{20}(239^2 + 341^2 + 215^2) - C.F.$$

$$= 10981.35 - C.F. = 447.65$$

$$S.S. \text{ due to T (M)} = \frac{1}{5}(62^2 + 58^2 + \ldots\ldots + 42^2 + 64^2) - C.F. - S.S. (M)$$

$$= 11057.40 - C.F. - 447.65 = 76.35$$

$$S.S. \text{ due to scores} = (12^2 + 9^2 + \ldots\ldots + 6^2 + 12^2) - C.F.$$

$$= 11293.00 - C.F. = 759.25$$

$$S.S. \text{ due to scores within T (M)} = 759.25 - 76.35 - 447.65 = 235.25$$

ANOVA table

Source	*d.f.*	*S.S.*	*M.S.*	*F-value*
Methods (M)	2	447.65	223.82	45.68*
Teachers nested in methods T(M)	9	76.35	8.48	1.73 NS
Error	48	235.25	4.90	
Total	59	759.25		

Tabulated value: $F_{.05; 2, 48} = 3.19$, $F_{.05; 9, 48} = 2.08$

* Significant at $\alpha = 0.05$.

F-value for the methods is significant which leads to the conclusion that training methods differ significantly and method (*b*) was the best. At the same time, teachers vis-a-vis schools were at par within blocks.

8.9 STRIKING FEATURES OF NESTED OR HIERARCHICAL CLASSIFICATION

Certain points need attention with regard to nested or hierarchical classification. The same are high lighted below.

(*i*) Just like factorial experiments, it is a classification not a design of experiment. The term nested or hierarchical design is used in common parlance.

(*ii*) Process of repeated sampling and sub-sampling in any number of stages leads to nested or hierarchical classification.

(*iii*) The case of several readings or repeated determinations on each experimental unit is also covered under nested or hierarchical classification.

(*iv*) In nested classification, a level of a factor does not refer to the levels of the other factor for the same batch of units. Whereas in cross-classification levels of a factor are referred to the same batch of units of the other factor levels.

(*v*) Nested classification and crossed classification simultaneously be incorporated in multi-factor experiment. Such experiments are called as *nested factorial experiments*.

(*vi*) If the number of determinations or replications per experimental unit is not same, the analysis becomes some what complicated.

For a detailed study of analysis of experimental designs implying nested or hierarchical classification, the readers are referred to the books and research papers given under the heading 'Bibliography'.

QUESTIONS AND EXERCISES

1. What is meant by nested or hierarchical classification?
2. Discuss the following terms,
    (*a*) Hierarchical classification.
    (*b*) Partial hierarchical classification.
    (*c*) Balanced hierarchical classification.
    (*d*) Unbalanced hierarchical classification.
4. Is there any relationship between the sum of squares of nested treatment and the sum of squares of a parallel factorial experiment? If so, justify it through some examples.
5. Discuss the need of hierarchical classification with the help of some applied examples.

6. Give the statistical model for a CR-design involving hierarchical classification and assumptions underlying it.

7. Give the break-up of degrees of a CRH-*pm*(A) *q* C(AB) and also delineate expected mean squares in case when factors A, B and C are random in nature.

8. Usually which of the basic designs are used to conduct an experiment with nested or hierarchical classification?

9. If the same number of determinations on each experimental unit or the same number of replication for all treatments is not taken, then how can the experimental data be analyzed?

10. Explicate diagrammatically the configuration of hierarchical classification by taking three examples from various areas of scientific research.

11. Give the statistical model of randomized block design using hierarchical classification, RBC-*pm* (A) in which *q* levels of treatment B(A) are nested in those of treatment A. Also prepare skeleton analysis of variance table for this design explaining all notations and assumptions.

BIBLIOGRAPHY

1. Dean A, Boss D. *Designs and Analysis of Experiments.* Springer Verlag, New York, 1999.

2. Hicks CR. Fundamentals of Analysis of variance, Part-III –Nested designs in analysis of variance. *Industrial Quality Control*, 1956: 13, Part 4, 13–16.

3. Kirk RE. *Experimental Designs: Procedures for the Behavioral Sciences.* 3rd ed., Brooks/Cole Publishing company, Pacific Grove, USA, 1995.

4. Yandell B. *Practical Data Analysis for designed Experiments.* Chapman and Hall, London, 1997.

9

Transformation in Designed Experimental Data

9.1 REASONING

A transformation is a rational metamorphosis of a set of data usually scores so as to change certain characteristics of the set which meets the requirement of data analysis and still the basic characteristics remain unchanged. For proper analysis of variance, it is necessary that the assumptions for it given in section 1.1 are adhered to, especially three assumptions:

(*a*) Homogeneity of error variance.

(*b*) Normality of error.

(*c*) Additivity of effects.

If the above assumptions do not hold well, there is likelihood that the results that come out to be significant by F or *t*-tests are actually non-significant. Also the estimates of treatment means may be biased. Thus to overcome these lacunae, Bartlett (1947) emphasized on the following requirements for an ideal transformation.

(*i*) The transformed variable should be normally distributed.

(*ii*) After transformation, the variance and means should become independent of each other. In other words, variance should remain unaffected by the changes in mean after transformation.

(*iii*) The transformed scale should be such that the estimated mean is an efficient estimate of population mean.

(*iv*) The variate values should be transformed in such a manner that the effects remain linear and additive.

Above points are further clarified through these examples. In case the response variable follows binomial distribution (n, p), then its mean np and variance npq are related to each other. Again if the criterion variable follows Poisson distribution, then its mean and variance are equal, i.e. they are not independent. In such situations, response variables need proper transformation so as to fulfill the assumptions of analysis of variance.

9.2 BASIS OF TRANSFORMATIONS

One analyzes experimental data assuming that the model is adequate, i.e. ε_{ij}'s are independently and identically distributed normally and in particular its variance is constant. But in reality it is always not so. F-value in ANOVA comes out to be highly significant but under this situation this is always dubious. Because independence of errors earnestly affect type I error as well as power of the test. This compels the analyst to think of some suitable transformation of data.

For an improved analysis, one should think about the nature of variance of inhomogeneity. In an experiment there can be two types of inhomogeneity: (*i*) *Inherent inhomogeneity* – this type of inhomogeneity occur due to assignable causes. For example, the variability in a manufactured product will be more when the operators are inexperienced as compared to experienced operators. In this situation, transformation is not required. (*ii*) *Transformable inhomogeneity* – this kind of inhomogeneity is induced due to inherent nature of the untransformed observations which give rise unnecessarily to complicated statistical model and variance inhomogeneity. Therefore, transformable inhomogeneity needs transformation of data particularly when mean and variance vary simultaneously.

Without entangling into theoretical complexities of transformations, some useful transformations are delineated adequately in the sequel. For a detailed study, the readers are advised to go through the papers: Curtis (1943), Bartlett and Kendall (1946), Box and Fung (1995), Box and Cox (1964), Budescu and appelbaum (1981), Freeman and Tukey (1950), Games and Lucas (1966), Levine and Dunlap (1982, 1983), Games (1983, 1984).

9.3 SQUARE ROOT TRANSFORMATION

As mentioned in section (9.1), if the data follow binomial or Poisson distribution, the mean is proportional to variance. In such a situation, square root transformation is appropriate. In general counts of rare events follow Poisson distribution. For instance, number of defective items produced by various operators in a shift, number of plants lodging per plot in an experiment, number of insects per laef treated under various insecticides, etc.

If variance is a function of mean, i.e. $\sigma^2 = g\ (\mu)$, then an appropriate approximate transformation can be found by evaluating the integral,

$$\int \frac{dy}{\sqrt{g(y)}} = \int \frac{dy}{\sqrt{y}} = 2\sqrt{y} \qquad (9.3.1)$$

2 is a constant multiple, so it has no role in transformation. So a square root transformation is,

$$y' = \sqrt{y} \qquad (9.3.2)$$

Bartlett suggested that if y lies between 0 and 10, especially when some value of y is zero, then he stressed on the use of transformation,

$$y' = \sqrt{y + 0.5} \qquad (9.3.3)$$

Later Freeman and Tukey (1950) propounded a more appropriate transformation as,

$$y' = \sqrt{y} + \sqrt{y + 0.5} \qquad (9.3.4)$$

Following example will further elucidate the theory.

Example 9.1: In an experiment conducted in randomized block design to compare 10 varieties of maize in respect of root lodging of plants. Data are as displayed below:

Varieties	No. of plants lodged (y)			Mean	Var.	Transformed data $y' = \sqrt{y + 0.5}$			Mean	Var.
	R_1	R_2	R_3			R_1	R_2	R_3		
401	4	5	7	5.33	5.44	2.121	2.345	2.739	2.40	0.098
402	1	2	2	1.67	0.34	1.225	1.581	1.581	1.46	0.042
403	2	3	4	3.00	1.00	1.581	1.871	2.121	1.86	0.073
404	1	1	0	0.66	0.34	1.225	1.225	0.707	1.05	0.089
405	5	3	7	5.00	4.00	2.345	1.871	2.739	2.32	0.189
406	2	3	4	3.00	1.00	1.581	1.871	2.121	1.86	0.073
407	1	1	2	1.33	0.34	1.225	1.225	1.581	1.34	0.040
408	4	2	1	2.33	2.33	2.121	0.707	0.707	1.18	0.670
409	1	1	4	2.00	3.00	1.225	1.225	2.121	1.52	0.270
410	4	3	6	4.33	2.33	1.581	1.871	2.550	2.00	0.250

A critical view of the table clearly reveals that there is substantial difference among variance for varieties of untransformed data. After transformation, the variance of the data becomes satisfactorily homogeneous.

9.4 RECIPROCAL TRANSFORMATION

At many occasions, the observations are recorded for occurrence of certain activity per unit time. For instance, number of times a bird flaps its wings per minute, number of eggs laid per month by a hen, number of times a butterfly sits on a flower, and so on. If a graph is plotted between time and number of activity occurred, the shape of the curve is just like a hyperbola. It means the mathematical relation between variable Y and time t is $Yt = C$ (constant) or $Y = C/t$. On putting $1/t = U$, $Y = C U$, which represent a straight line. Hence, for such type of data reciprocal transformation is suitable. For certain life testing data, e.g. life of an animal, a machine part, a computer chip, etc. reciprocal transformation is appropriate. In analysis of variance if the squares of the treatment means are proportional to their respective standard deviations, reciprocal transformation is applied. Let Y be the variable measured, then the transformed variable Y′ is,

$$Y' = \frac{1}{Y} \qquad (9.4.1)$$

If any one or more values of Y are zero, then one must use the transformation formula,

$$Y' = \frac{1}{Y+1} \qquad (9.4.2)$$

9.5 ARCSIN SQUARE ROOT OR ANGULAR TRANSFORMATION

This transformation is suitable in a grouped binomial data. It is an established fact that a binomial response is a dichotomous variable. Often the responses are expressed in percentages. If so they can easily be converted into proportions on dividing each percentage by 100. Responses that can have only two outcomes like success or failure, alive or dead, cured or not cured, etc. are sometimes called *quantal responses*.

The mean and variance of binomial variates are related to each other. But the angular transformation makes them independent. Also it maintains the homogeneity of variances. At this juncture it seems germane to explain the meaning of the term arcsin.

Actually, it is an inverse operation. For instance, arcsin $0.472 = 43.39°$ or $\sin^{-1} 43.39 = 0.472$. In other words, it is that angle θ which is equal to $\sin p$. θ can also be measured in radians. To convert the radian into angle, multiply the radian value by 57.296.

Responses that are proportions 'p' for some treatment doses when plotted for Y and cumulative p on a graph paper, the graph tend to exhibit S-shaped curve. The proportion asymptotically approach to 0 on the left corner tail and 1 on the upper right tail. Arcsin transformation has the property that this stretches the upper and lower tails of S-shaped curve making the relationship between Y and arcsin (p) nearly linear over the border domain of the explanatory variable.

Arcsin square root transformation is,

$$\int \frac{dp}{\sqrt{p(p-1)}} = 2\sin^{-1}\sqrt{p} \qquad (9.5.1)$$

Multiple 2 is immaterial for a distribution. Thus,

$$\theta = \arcsin\sqrt{p} \qquad (9.5.2)$$

$$\sin^{-1}\theta = \sqrt{p} \qquad (9.5.3)$$

The variance of the transformed variable θ is,

$$\sigma_\theta^2 = \frac{180^2}{4\pi^2 n} \qquad (9.5.4)$$

Provided each percentage is based on large number of independent outcomes n.

If θ is measured in radians, then the variance of θ is,

$$\sigma_\theta^2 = \frac{1}{4n} \qquad (9.5.5)$$

Bartlett (1947) suggested that if p is zero, then $1/2n$ or $1/4n$ be substituted for p and $1 - 1/2n$ or $1 - 1/4n$ for p if it is equal to 1. This empirical manipulation improves the equality of variances in angles.

Further this is to emphasize that arcsin transformation is valid when all percentages are based on approximately the same sample sizes n. Also $n \geq 50$ is a desirable group or sample size.

Arcsin transformation has negligible effect on percentages within the range 30 to 70%. So one does not require arcsin transformation if all percentages lie within this range.

Remarks:

1. If any treatment in an experimental design consists of all 0% or 100%, then that treatment should be excluded from the analysis.
2. If all percentages lie between 0–30% or 70–100%, then a square root transformation will be more appropriate.
3. To interpret means after analysis, the arcsin values be retransformed into its original units.

Example 9.2: A trial was planned to see the effect of 12 treatments on shedding of flowers of cotton per plant. Percentages of flowers dropped per plant are displayed below. Design of the trial was randomized block design with three replications. Mean and variance for each treatment are calculated and given alongside.

Data table

Treats.	Percentages of flowers dropped			Mean	Variance
	R_1	R_2	R_3		
T_1	58	59	72	63.00	61.00
T_2	50	48	60	52.67	41.33
T_3	48	36	39	41.00	39.00
T_4	23	33	31	29.00	28.00
T_5	42	55	51	49.33	44.33
T_6	39	54	53	48.67	70.33
T_7	36	24	40	33.33	69.33
T_8	28	34	36	32.67	17.33
T_9	49	68	67	61.33	114.33
T_{10}	61	57	50	56.00	31.00
T_{11}	38	43	26	35.67	76.33
T_{12}	42	56	58	52.00	76.00

Transformed data table $=$ arcsin $\sqrt{p}$

Treats.	Values of θ in angles*			Mean	Variance
	R_1	R_2	R_3		
T_1	49.60	50.18	58.05	52.61	22.23
T_2	45.00	43.85	50.77	46.54	13.75
T_3	43.85	36.87	38.65	39.79	13.15
T_4	28.66	35.06	33.83	32.52	11.53
T_5	40.40	47.87	45.57	44.61	14.64
T_6	38.65	47.29	46.72	44.22	23.35
T_7	36.87	29.33	39.23	35.14	26.74
T_8	31.95	35.67	36.87	34.83	6.58
T_9	44.43	55.55	54.94	51.64	39.08
T_{10}	51.35	49.02	45.00	48.46	10.31
T_{11}	38.06	40.98	30.66	36.57	28.30
T_{12}	40.40	48.45	49.60	46.15	25.13

* Values of arcsin $\sqrt{p}$ are obtained from table B-9.

From the above table, it is evident that the vaiances of untransformed data vary largely as they range from 28.00 to 114.33. On the other hand, the variances of transformed data are virtually homogeneous as they range from 6.58 to 39.08.

9.6 LOGARITHMIC TRANSFORMATION

Problem arises when the data fail to conform to the assumptions specified for analysis of variance. For instance, if the true effects are multip'icative though as per assumption they are required to be additive. To ge over this problem, a transformation is needed which changes the multiplicative model into an additive linear model. Suppose the multiplicative model is,

$$y = \mu \alpha_i \beta_j \varepsilon_{ij} \qquad (9.6.1)$$

Taking logarithm of both side of the model (9.6.1), this single transformation changes the model into additive model.

$$\log y = \log \mu + \log \alpha_i + \log \beta_j + \log \varepsilon_{ij} \qquad (9.6.2)$$

Now it is apparent that the model (9.6.2) is an additive model in terms of logarithm of each component of the model (9.6.1). In routine, hardly any scientist tests whether a model is multiplicative or additive? A transformation is usually chosen by viewing the data critically. Therefore, log transformation is often applied for counts such as number of insects per plant, number of weed plants per plot, number of diseased plants per plot, etc. If one or more counts are zero, then one is added to each count before taking the logarithm. As a general norm, if the standard deviation of the original counts varies directly as the mean, logarithmic transformation stabilizes the variance. Suppose a function $\phi(y) = c^2 y^2$. To find out an approximate transformation, the function $\int \dfrac{dy}{\sqrt{\phi(y)}}$ is to be worked out. Thus,

$$\int \frac{dy}{\sqrt{c^2 y^2}} = \frac{1}{c} \log y. \qquad (9.6.3)$$

As per equation (9.6.3), log transformation is justified.

A rough idea about the type of model can also be had from the graph of data. Anyhow, the normal plot of the transformed data is almost a straight line. This indicates that the assumption of homogeneity of variances shall very likely hold after log transformation. If y is the criterion variable, then the logarithmic transformed variable,

$$y' = \log_{10} y \qquad (9.6.4)$$

In case one or more y's are zero, then use the log transformation as,

$$y' = \log_{10} (y + 1) \qquad (9.6.5)$$

Also when data are positively slewed, this transformation works better.

Example 9.3: A trial was run in randomized block design to ascertain the effect of 9 treatments on the number of weed plants (monocot) in one square metre plot of wheat crop at 60 days after sowing (DAS). Treatments were combinations of CT–conventional tillage, ZT–zero tillage, FIBR–furrow irrigated raised bed, WC–weedy check, HW–hand weeding, herbicide-isoproturon, @ 0.75 k a.i./ha.

Data with regard to weed count at 60 DAS and their logarithmic transformed values are as tabulated below.

Treatments	Original data			Log transformed data		
	R_1	R_2	R_3	R_1	R_2	R_3
CT - WC	28	20	30	1.447	1.301	1.580
CT - HW	12	8	12	1.079	0.903	1.079
CT – Herb	24	16	12	1.380	1.204	1.079
ZT – WC	36	32	20	1.556	1.505	1.301
ZT - HW	8	16	12	0.903	1.206	1.079
ZT – Herb	16	12	20	1.204	1.079	1.301
FIRB - WC	52	40	36	1.716	1.602	1.556
FIRB – HW	12	24	20	1.079	1.380	1.301
FIRB - Herb	20	20	28	1.301	1.301	1.447

The data pertains to the weed count, therefore it becomes essential to transform the original counts into their logarithmic values. Further analysis of data should be carried out only for log transformed values.

9.7 CHOICE BETWEEN LOGARITHMIC AND SQUARE ROOT TRANSFORMATIONS

Logarithmic and square root transformations are applicable when data show direct linear relationship between mean and variance vis-à-vis standard deviation. So the question arises which one to use out of the two in a particular situation. As a rule of thumb, if the ratio of standard deviations to its respective means is constant, a log transformation should be used. This may be checked by plotting a graph of standard deviation against treatment mean. The plotted points will lie approximately in a straight line.

On the other hand, if the variances are proportional to their respective means, then square root transformation should be applied. Mark the difference, for log transformation standard deviation (s) lies in the range $-\infty$ to mean 'm' and for square root transformation s lies between m and ∞.

9.8 LOGIT TRANSFORMATION

Logit transformation is also used for proportional type of responses of scores such as in case of angular transformation. But the main difference between two transformations is that in case of angular transformation, the data are binomial in nature with $p_i = n_i/n$. But when $y_i = p_i/q_i$, then logit transformation

$$\log_e (y_i) = \log_e (p_i/q_i) \qquad (9.8.1)$$

where $\qquad q_i = 1 - p_i$

is more appropriate. The variance of y_i is approximately $1/n_i p_i q_i$. So in logit transformation, y_i's are given the weights $n_i p_i q_i$. Logit transformation is more suitable when $n\, p_i$ and $n\, q_i$ both are 20 or more for all classes. The ratio p_i/q_i is called the *odd ratio* or briefly *odds*. The ratio p_i / q_i can have a minimum value 0. So when p_i / q_i takes the values from less than one to 0, then logit value becomes increasingly negative and if p_i / q_i varies from more than one to ∞, log odd values become increasingly positive.

In logit transformation also S-shaped curve of cumulative distribution is stretched after logit transformation and approaches to linearity.

Note: In case of two-way table, p_i will be changed to p_{ij}, q_i to q_{ij} and n_i to n_{ij}.

Epilogue

An experimenter always tries to analyze data of a trial which gives as valid results as possible. In this endeavor, one tries to fulfill all assumptions on which analysis is based. Though analysis of variance is quite robust, even then the inhomogeneity of variances and non-normality of errors are given much weightage. Through transformations, one tries to especially fulfill these two conditions.

But as envisaged, no transformation ever converts the data to bring about perfect conditions. For instance, S-shaped distribution curve is never stretched exactly to straight line after transformation. Also in angular transformation, there is hardly any situation in which proportions are based on equal number of units. As a matter of fact all transformations are used under some lacunae. Also to decide about a transformation, in some cases, is problematic unless one uses–plot, checks the presence of outliers, tests the homogeneity of variances, etc. Still importance of transformations cannot be ruled out.

In case one is not able to select a suitable transformation, as a recourse he should use non-parametric methods of analysis of data as given in Chapter-3. But unfortunately, non-parametric methods as an alternative to analysis of variance are available only for a few basic designs. Thus, the importance of transformations is very much alive as they can be applied in case of all experimental designs and bring marked improvement.

QUESTIONS AND EXERCISES

1. Write a short note on the importance of transformations in analysis of variance.
2. Does a transformation fulfill all conditions of analysis of variance?

3. Is there any transformation that can be used in all types of experimental data? Justify your answer.

4. What is the criterion which makes one to decide about a particular type of transformation?

5. What cautions have to be taken while transforming a set of data?

6. Should an experimenter use transformation whenever there is inhomogeneity of data?

7. Under what situations a square root transformation is suitable? How to deal with the problem of zero occurring in data set?

8. What types of data attract a scientist to make reciprocal transformation prior to analysis of variance?

9. Arcsin square root transformation is applied to a particular type of data. What is it?

10. In what situations angular transformation is considered proper? Also how to deal with the problem of zeros occurring in data set?

11. Differentiate between logit and angular transformations.

12. What name is given to the data meant for logit transformation?

13. Explain logit transformation.

14. When the model of the design is suspected to be multiplicative, which transformation shall be suitable for analysis of data.

15. Who are rated as pioneer workers in the field of transformations.

16. An experiment was conducted in randomized block design to assess the effect of 5 treatments in respect of damage to mung crop by by pea-pod borer. Ten plants were randomly selected and picked ten pods per plant. Five treatments were:

T_1 - Spray of malathion (0.05%) + endosulfan (0.07%).

T_2 - Three releases of Chrysoperla Carnea @ 25000 neonate larvae/ha at 25, 40 and 55 days after sowing (DAS).

T_3 - Three sprays of neem oil (0.2%) at 25, 40 and 55 DAS.

T_4 - Release of C.Carnea @ 25000 + neem oil (0.2%) spray at 40 and 55 DAS.

T_5 - Release of C.Carnea @ 25000 + neem oil (0.2%) spray at 40 DAS and endosulfan (0.07 %) at 55 DAS.

Table below gives the percentage of damage on third day of application of treatments.

Per cent damage by pea-pod borer

Treatments	Replications		
	R_1	R_2	R_3
T_1	12	13	10
T_2	9	8	8
T_3	11	12	10
T_4	6	5	6
T_5	20	19	20

Analyze the experimental data using arcsin square root transformation.

17. An experiment was laid out in randomized complete block design with 18 treatment combinations of six bioagents (A) and 3 concentrations of neem oil (B). Growth inhibition in per cent was measured for each combination replicated thrice.

Bioagents (A): A_1 - Actenomycet, A_2 - Trichoderm vividea, A_3 - T. Harzanum, A_4 - PGPR- Strain 1, A_5 - PGPR-Strain 2, A_6 - PGPR- Strain 3. Concentrations (B): Three neem oil concentrations B_1, B_2, B_3.

Growth inhibition data in per cent were as tabulated below.

Treatments	% growth inhibition values		
$A_1 B_1$	33.3	27.8	13.3
$A_1 B_2$	27.8	28.9	31.1
$A_1 B_3$	22.2	24.4	21.1
$A_2 B_1$	55.6	50.0	47.8
$A_2 B_2$	33.3	35.6	34.4
$A_2 B_3$	22.2	23.3	25.6
$A_3 B_1$	94.4	95.8	97.8
$A_3 B_2$	88.9	91.1	90.0
$A_3 B_3$	66.7	67.8	68.3
$A_4 B_1$	100.0	100.0	100.0
$A_4 B_2$	88.9	87.8	88.3
$A_4 B_3$	66.7	67.9	67.8
$A_5 B_1$	33.3	35.6	32.8
$A_5 B_2$	22.2	23.3	22.8
$A_5 B_3$	11.1	12.2	12.8
$A_6 B_1$	27.8	30.0	28.9
$A_6 B_2$	22.2	23.3	21.1
$A_6 B_3$	16.7	17.7	17.8

Analyze the experimental data after suitable transformation.

18. The layout of a split plot design with three treatments (ZT, CT, FIRB) in main plots, three treatments (WC, HW, Herbicide) in split plots having three replications is given below. The measurements are the number of tillers/metre^2.

Rep. I			Rep. II			Rep. III		
ZT	CT	FIRB	FIRB	CT	ZT	ZT	FIRB	CT
WC 499	HW 546	Herb 502	HW 552	WC 410	HW 557	HW 518	Herb 530	HW 547
Herb 494	Herb 572	WC 502	WC 489	Herb 536	Herb 521	WC 467	HW 556	WC 474
HW 486	WC 563	HW 547	Herb 592	HW 556	WC 442	Herb 486	WC 553	Herb 507

Analyze the data of split plot design making use of square root transformation.

BIBLIOGRAPHY

1. Bartlett MS. The use of transformation. *Biometrics*, 1947: 3, 39–52.
2. Bartlett MS, Kendall DG. The Statistical analysis of variance–heterogeneity and the logarithmic transformation. *Journal of the Royal Statistical Society*, Suppl., 1946: 128–38.
3. Box GEP, Cox DR. An analysis of transformations. *Journal of Royal Statistical Society*, Ser. B, 1964: 26, 211–52.
4. Box GEP, Fung C. The importance of data transformation in designed experiments for life testing. *Quality Engineering*, 1995: 7, 625–38.
5. Budescu DV, Appelbaum MI. Variance stabilizing transformations and the power of the F-test. *Journal of the Educational Statistics*, 1981: 6, 55–77.
6. Curtis JH. On transformations used in analysis of variance. *Annals of Mathematical Statistics*, 1943: 14, 107–22.
7. Freeman MF, Tukey JW. Transformations related to angular and square root. *Annals of Mathematical Statistics*, 1950: 21, 607–11.
8. Games PA. Curvilinear transformation of the dependent variable. *Psychological Bulletin*, 1983: 93, 382–87.
9. Games PA. Data transformation, power and skew: A rebuttal to Levine and dunlap. *Psychological Bulletin*, 1984: 95, 345–47.
10. Games PA, Lucas PA. Power of the analysis of variance of independent groups on nonnormal and normally transformed data. *Educational and Psychological Measurements*, 1966: 16, 311–27.
11. Levine DW, Dunlap WP. Power of the F-test with skewed data: Should one transform or not? *Psychological Bulletin*, 1982: 92, 272–80.
12. Levine DW, Dunlap WP. Data transformation, power and skew: A rejoinder to Games. *Psychological Bulletin*, 1983: 93, 596–99.

10

Analysis of Covariance

10.1 RATIONALE

It has always been the endeavor of an investigator to reduce experimental error and estimate the treatment effects as precisely as possible in all designs discussed so far and henceforth. The objectives were achieved through blocking factors which were conceivable prior to experimentation. Such a control is termed as *direct control* or *experimental control*. But certain factors are such which affect the criterion variable and are beyond the direct control. For instance, in field experiments one has no control on percentage germination of seeds in plots or the number of weed plants emerged in plots during cropping period. In experiments on educational training methods, the initial IQ of candidates, etc. are such nuisance variables of which the effect cannot be controlled by the investigator. Hence, the effect of such nuisance variable(s) is eliminated through statistical technique known as *statistical control*. Analysis of covariance (ANCOVA) is a device which comes under the category of statistical control.

Analysis of covariance is carried by measuring one or more nuisance variables which are called *concomitant variables* or *covariates* in addition to criterion variable(s). The effect of covariate(s) is removed from the variable of interest through regression technique. Therefore, analysis of covariance is a statistical technique which combines regression analysis with analysis of variance so as to reduce experimental error and estimate treatment effects adjusted for the effect of covariate(s).

Assumptions

Analysis of covariance requires all the assumptions necessitated by analysis of variance. Besides those, some additional assumptions are made which are:

(*i*) The experiment has one or more measurable variables which affect the criterion variable but are nothing to do with the objective of the experiment.

(*ii*) There is no direct control on such nuisance variable(s).

(*iii*) It is assumed that the concomitant variable(s) remains unaffected by the treatments. If this condition is not met, then the adjusted criterion variable will become biased. The theoretical discussion of this aspect is abandoned over here.

(*iv*) The relation between the covariate(s) and criterion variable is linear. In some situations, there may be non-linear relationship between these variables. But it has been kept out of scope of this book. If the pooled within-class correlation between the criterion variable and the covariate is ρ_w and σ_e^2 is the experimental error per unit in ANOVA having n_e degrees of freedom, then the adjusted error in ANCOVA is,

$$\sigma_e^2 \left(1 - \rho_w^2\right)\left(1 + \frac{1}{n_e - 2}\right) \qquad (10.1.1)$$

where ith treatment is replicated n_i times for $i = 1, 2, \ldots\ldots , k$.

(*v*) Fundamental assumption underlying the analysis of covariance is the homogeneity of within-class regression coefficients?

(*vi*) The treatment effects and regression coefficients are additive.

(*vii*) Residuals are distributed normally with mean zero and same variance. Tests to verify the validity of assumptions are available in the literature. But they are not given for the sake of brevity except the test for assumption (*v*). Secondly analysis of covariance is quite robust against many assumptions.

10.2 METHODOLOGY FOR ANALYSIS OF COVARIANCE

Analysis of covariance technique can be applied to any design of experiment. The procedure remains the same except that adjusted treatment means are to be worked out as per the factors involved in the design. For initial works, one may go through the paper of Cochran (1957). Following discussion shall be confined to two designs namely, completely randomized design and randomized block design that too in case of one and two covariates only.

10.3 ANALYSIS OF COVARIANCE IN CRD

Statistical model for analysis of covariance under the assumption that within-class regression coefficients are homogeneous is,

$$y_{ij} = \mu + \tau_i + \beta\, x_{ij} + \varepsilon_{ij} \qquad (10.3.1)$$

$$i = 1, 2, \ldots\ldots ,v \text{ and } j = 1, 2, \ldots\ldots , n_i$$

From the point of view of feasibility, it is preferred to consider the model in terms of deviations from mean for the variable x, i.e.

$$y_{ij} = \mu + \tau_i + \beta\left(x_{ij} - \bar{x}\right) + \varepsilon_{ij}. \qquad (10.3.2)$$

Analysis of covariance procedure can be derived for the model (10.3.1). But the analysis procedure is explicated below without derivation. Paired observations recorded from an experiment in CRD can always be presented in the following tabular form.

Table 10.3.1 Data table

Treatments							
1		*2*	....	*i*	....	*v*	
y_{1j}	x_{1j}	y_{2j}	x_{2j}	y_{ij}	x_{ij}	y_{vj}	x_{vj}
y_{11}	x_{11}	y_{21}	x_{21}	y_{i1}	x_{i1}	y_{v1}	x_{v1}
y_{12}	x_{12}	y_{22}	x_{22}	y_{i2}	x_{i2}	y_{v2}	x_{v2}
|	|	|	|	|	|	|	|
|	|	|	|	|	|	|	|
|	|	|	|	|	|	|	|
y_{1n_1}	x_{1n_1}	y_{2n_2}	x_{2n_2}	y_{in_1}	x_{in_1}	$y_{v_{nk}}$	$x_{v_{nk}}$
$y_1.$	$x_1.$	$y_2.$	$x_2.$	$y_i.$	$x_i.$	$y_v.$	$x_v.$

$$\sum_i y_{i.} = y_{..} \text{ and } \sum_i x_{i.} = x_{..} \text{ Also } \frac{y_{i.}}{n_i} = \overline{y}_i$$

The motif behind analysis of covariance is to find sum of squares and means for all the factors involved in analysis of data adjusted for the concomitant variable. Therefore, the regression sum of square for each factor is subtracted from the sum of square for that factor. In a way, it is the analysis of variance carried out by using only adjusted sum of squares. Adjusted treatment means are obtained through estimated regression equation.

For covariance analysis, following calculations are to be worked out.

Table 10.3.2 Calculation table

Treatments							
1		*2*	....	*i*	....	*v*	
y_{1j}	x_{1i}	y_{2j}	x_{2j}	y_{ij}	x_{ij}	y_{vj}	x_{vj}
$\sum\limits_j y_{1j}^2$	$\sum\limits_j x_{1j}^2$	$\sum\limits_j y_{2j}^2$	$\sum\limits_j x_{2j}^2$	$\sum\limits_j y_{ij}^2$	$\sum\limits_j x_{ij}^2$	$\sum\limits_j y_{vj}^2$	$\sum\limits_j x_{vj}^2$
$S_{yy_1} = \sum\limits_j y_{1j}^2 - \dfrac{y_{1.}^2}{n_1}$		$S_{yy_2} = \sum\limits_j y_{2j}^2 - \dfrac{y_{2.}^2}{n_2}$		$S_{yy_i} = \sum\limits_j y_{ij}^2 - \dfrac{y_{i.}^2}{n_i}$		$S_{yy_v} = \sum\limits_j y_{vj}^2 - \dfrac{y_{v.}^2}{n_v}$	

(Contd.)

				Treatments			
1		2		....	*i*	....	*v*
y_{1j}	x_{1i}	y_{2j}	x_{2j}	y_{ij}	x_{ij}	y_{vj}	x_{vj}

$$S_{xx_1} = \sum_j x_{1j}^2 - \frac{x_{1.}^2}{n_1} \quad S_{xx_2} = \sum_j x_{2j}^2 - \frac{x_{2.}^2}{n_2} \cdots \cdots S_{xx_i} = \sum_j x_{ij}^2 - \frac{x_{i.}^2}{n_i} \cdots \cdots S_{xx_v} = \sum_j x_{vj}^2 - \frac{x_{v.}^2}{n_v}$$

$$S_{y_1 x_1} = \sum_j y_{1j} x_{1j} \quad S_{y_2 x_2} = \sum_j y_{1j} x_{1j} \quad S_{y_2 x_2} = \sum_j y_{2j} x_{2j} \qquad S_{y_i x_i} = \sum_j y_{vj} x_{vj}$$

$$-\frac{y_{1.} x_{1.}}{n_1} \qquad\quad -\frac{y_{2.} x_{2.}}{n_2} \qquad\quad -\frac{y_{i.} x_{i.}}{n_i} \qquad\qquad -\frac{y_{v.} x_{v.}}{n_v}$$

$$\beta_1 = \frac{S_{y_1 x_1}}{S_{xx_1}} \qquad \beta_2 = \frac{S_{y_2 x_2}}{S_{xx_2}} \qquad \beta_i = \frac{S_{y_i x_i}}{S_{xx_i}} \qquad \beta_v = \frac{S_{y_v x_v}}{S_{xx_v}}$$

Further calculate,

Total S.S. for y, x and cross-product (C.P.) xy corrected for mean are:

$$S_{yy} = \sum_{i=1}^{v} \sum_{j=i}^{n_i} y_{ij}^2 - \frac{y_{..}^2}{n} \qquad \text{where,} \sum_i n_i = n$$

$$S_{xx} = \sum_{i=1}^{v} \sum_{j=i}^{n_i} x_{ij}^2 - \frac{x_{..}^2}{n}$$

$$S_{yx} = \sum_{i=1}^{v} \sum_{j=1}^{n_i} y_{ij} x_{ij} - \frac{y_{..} x_{..}}{n}$$

Treatment S.S. for y, x and cross-product (C.P.) xy corrected for mean are:

$$T_{yy} = \sum_{i=1}^{v} \frac{y_{i.}^2}{n_i} - \frac{y_{..}^2}{n}$$

$$T_{xx} = \sum_{i=1}^{v} \frac{x_{i.}^2}{n_i} - \frac{x_{..}^2}{n}$$

$$T_{yx} = \sum_{i=1}^{v} \frac{y_{i.} x_{i.}}{n_i} - \frac{y_{..} x_{..}}{n}$$

Experimental error S.S. for y, x and cross-product (C.P.) xy corrected for mean are:

$$E_{yy} = S_{yy} - T_{yy}$$
$$E_{xx} = S_{xx} - T_{xx}$$
$$E_{yx} = S_{xy} - T_{xy}$$

Total S.S. adjusted for regression due to covariate,

$$S_{T+E} = S_{yy} - \frac{S_{xy}^2}{S_{xx}}$$

Similarly adjusted error S.S. is,

$$S_E = E_{yy} - \frac{E_{xy}^2}{E_{xx}}$$

Therefore, adjusted treatment S.S. is,
$$S_T = S_{T+E} - S_E$$

$$= S_{yy} - E_{yy} - \frac{S_{xy}^2}{S_{xx}} + \frac{E_{xy}^2}{E_{xx}}$$

$$= T_{yy} - \frac{S_{xy}^2}{S_{xx}} + \frac{E_{xy}^2}{E_{xx}}$$

Table 10.3.3 ANCOVA for CRD

Source	d.f.	S.S.	M.S.	F-value
Between treatments	$v - 1$	S_T	$S_T/(v-1)$	$\dfrac{S_T/(v-1)}{S_E \big/ \left(\sum\limits_i n_i - v - 1\right)}$
Within treatments	$\left(\sum\limits_i n_i - v - 1\right)$ $= n_e'$	S_E	S_E / n_e' $= S_E(MS)$	
Total	$n_i - 1$			

F has $\{(v - 1), (\sum n_i - v - 1)\}$ d.f.

In the above table 1, d.f. is reduced in the total d.f. due to regression. This ultimately results into reduction of error d.f. by one.

Calculated value of F is compared with critical value of F for $\{ (v - 1), (\sum n_i - v - 1)\}$ and $\alpha = 0.05$ probability. If it comes out to be significant, then pair wise adjusted treatment means are compared by a suitable post-hoc test like lsd or Duncan's multiple range test, etc. by using within treatments M.S. obtaind in ANCOVA Table 10.3.3.

As per model $(10.3.1), \beta_1, \beta_2, \ldots\ldots\ldots, \beta_v$ are within class regression coefficients. Test of homogeneity of β's can be performed to ensure the validity of the fundamental assumptions and thereby the cogency of tests in analysis of covariance.

The hypothesis $H_0 : \beta_1 = \beta_2 = \ldots\ldots\ldots = \beta_v = \beta$

against $H_1 : \beta_i \neq \beta_{i'}$ for some $i \neq i' = 1, 2, \ldots\ldots\ldots, v,$

can be tested by F-test. Statistic F is,

$$F = \frac{S_2/(v-1)}{S_1 / \left(\sum_i n_i - 2v \right)} \tag{10.3.3}$$

F is distributed with $(v - 1, \sum_i n_i - 2v)$ d.f.

where,

$$S_1 = E_{yy} - \sum_{\substack{\text{Overall} \\ \text{Treats.}}} \left(S_{yx_j}^2 \right) \Big/ S_{xx_j} \tag{10.3.4}$$

$$S_2 = \sum_{\substack{\text{Overall} \\ \text{Treats}}} \left(\frac{S_{yx_j}^2}{S_{xx_j}} \right) - \frac{E_{xy}^2}{E_{xx}} \tag{10.3.5}$$

Compare the calculated value of F with tabulated value of F for $\{(v - 1),$

$(\sum_i n_i - 2v)\}$ d.f. and $\alpha = 0.05$ level of significance. If F is non-significant, then H_0 is accepted. Thus, our assertion is true. This facilitates to pool within class information from all treatment classes to provide a single estimate of the regression parameter β. If F is significant, one cannot be confident about the conclusions drawn from covariance analysis. The formula for estimated value of the pooled within class regression coefficient is,

$$b = \frac{E_{yx}}{E_{xx}} \tag{10.3.6}$$

The hypothesis $H_0: \beta = 0$ can be tested by F-test, where

$$F = \frac{E_{yx}^2}{E_{xx}} \Big/ SE\left(MS_{adj} \right) \tag{10.3.7}$$

Statistic F has 1 and n_e' d.f.

If F comes out to be non-significant, then it leads to the conclusion that the introduction of concomitant variable is not going to improve the analysis of data meaningfully.

Now adjusted ith-treatment mean of the criterion variable can be obtained by the regression equation,

$$\bar{y}_i' = \bar{y}_i - b(\bar{x}_i - \bar{x}) \qquad (10.3.8)$$

S.E. of any apriori contrast $z = c_1\bar{y}_1' + c_2\bar{y}_2' + \ldots\ldots + c_v\bar{y}_v'$ among v treatment means is,

$$S.E.(z) = \sqrt{SE(MS)\left\{\left(\frac{c_1^2}{n_1} + \frac{c_2^2}{n_2} + \ldots\ldots + \frac{c_v^2}{n_v}\right) + \frac{(c_1\bar{x}_1 + c_2\bar{x}_2 + \ldots\ldots + c_v\bar{x}_v)^2}{E_{xx}}\right\}}$$

$$(10.3.9)$$

t-statistic for testing the significance of the contrast z is,

$$t = \frac{z}{S.E.(z)} \qquad (10.3.10)$$

t has $(\sum_i n_i - v - 1)$. Decision about the significance of contrast z is taken in the usual manner.

S. E. of an adjusted treatment mean is,

$$S.E.(\bar{y}_i') = \sqrt{SE(MS)\left\{\frac{1}{n_i} + \frac{(\bar{x}_i - \bar{x})^2}{E_{xx}}\right\}} \qquad (10.3.11)$$

S.E. for any paired contrast of adjusted means is,

$$S.E.(\bar{y}_i' - \bar{y}_{i'}') = \sqrt{SE(MS)\left\{\frac{1}{n_i} + \frac{1}{n_{i'}} + \frac{(\bar{x}_i - \bar{x}_{i'})}{E_{xx}}\right\}} \qquad (10.3.12)$$

For $i \neq i'$

Note: In case all treatments are replicated equal number of times, then $n_i = n$ for $i = 1, 2, \ldots$ $\ldots, v$. The procedure of covariance analysis remains same except that each n_i is replaced by n and $\sum n_i = nv$ and j varies from 1 to n instead of 1 to n_i.

The estimated value r_w of the pooled within-class correlation coefficient ρ_w can be obtained by the formula,

$$r_w = \frac{E_{yx}}{\sqrt{E_{xx}E_{yy}}} \qquad (10.3.13)$$

Value of r_w indicates towards the reduction in error mean square. Greater the value of r_w, more is the reduction in error term. Further by making use of r_w, adjusted error mean square can directly be obtained. Of course, the estimate of σ_e^2 has to be obtained by conducting analysis of variance for the criterion variable.

Example 10.1: Data with regard to scores in statistics (y) of three batches of students taught by three instructors and their scores in a prerequisite mathematics course (x) out of 20 marks are presented below. By experience, it is felt that the performance in mathematics has an impact on the scores in statistics. Assuming that all students are alike in respect of perception, analysis of covariance is carried out for one way classification (CRD) to compare the affectivity of teaching skill of three instructors.

Partial calculations are also displayed under the data table.

Data table

			Instructors				
I		*II*		*III*		*Total*	
y_1	x_1	y_2	x_2	y_3	x_3	y	x
16	14	9	10	15	16		
13	15	14	17	19	12		
9	7	18	15	14	8		
11	10	10	6	10	11		
8	5	12	10	14	9		
		8	10	16	15		
		7	6	8	7		
				18	17		
Total 57	51	78	74	114	95	249	220
Mean 11.40	10.20	11.14	10.57	14.25	11.88	12.45	11.00
S.S. 691	595	958	886	1722	1229	3371	2710
Sum of C.P. 632		900		1418		2950	

Within class S.S. and cross product (C.P.).

$$S_{yy_1} = 691 - \frac{(57)^2}{5} = 41.20, \quad S_{xx_1} = 595 - \frac{(51)^2}{5} = 74.80,$$

$$S_{yx_1} = 632 - \frac{57 \times 51}{5} = 50.60$$

$$b_1 = \frac{50.60}{74.80} = 0.68$$

Similarly,

$$S_{yy_2} = 88.86, \ S_{xx_2} = 103.71, \ S_{yx_2} = 75.43, \ b_2 = 0.73,$$

$$S_{yy_3} = 97.50, \ S_{xx_3} = 100.88, \ S_{yx_3} = 64.25, \ b_3 = 0.64$$

Total S.S. and cross product,

$$S_{yy} = 3371 - \frac{(249)^2}{20} = 270.95, \ S_{xx} = 2710 - \frac{(220)^2}{20} = 290.00,$$

$$S_{yx} = 2950 - \frac{249 \times 220}{20} = 211.00$$

Unadjusted treatment S.S. and C.P.,

$$T_{yy} = \frac{57^2}{5} + \frac{78^2}{7} + \frac{114^2}{8} - \frac{249^2}{20} = 43.39$$

$$T_{xx} = \frac{51^2}{5} + \frac{74^2}{7} + \frac{95^2}{8} - \frac{220^2}{20} = 10.61$$

$$T_{yx} = \frac{57 \times 51}{5} + \frac{78 \times 74}{7} + \frac{114 \times 95}{8} - \frac{249 \times 220}{20} = 20.72$$

Unadjusted error S.S. and C.P.,

$$E_{yy} = 270.95 - 43.39 = 227.56$$
$$E_{xx} = 290.00 - 10.61 = 279.39$$
$$E_{yx} = 211.00 - 20.72 = 190.28$$

Total S.S. adjusted for regression due to covariate,

$$S_{T+E} = 270.95 - \frac{211.00^2}{290.00} = 117.43$$

Adjusted error S.S.,

$$S_E = 227.56 - \frac{190.28^2}{279.39} = 97.97$$

Adjusted treatment S.S.,
$$S_T = 117.43 - 97.97 = 19.46$$

Now the quantities S_1 and S_2 are calculated by the formula (10.3.4) and (10.3.5) respectively so as to perform the test for homogeneity of within-class regression coefficients.

$$S_1 = 227.56 - \left(\frac{50.60^2}{74.80} + \frac{75.43^2}{103.71} + \frac{64.25^2}{100.88} \right)$$

$$= 227.56 - 130.01 = 97.55$$

$$S_2 = 130.01 - \frac{190.28^2}{279.39}$$

$$= 130.01 - 129.59 = 0.42$$

The statistic,

$$F = \frac{0.42/2}{97.55/(20-6)} = 0.03$$

Since the calculated value of F is less than 1, it is concluded that within-class regression coefficients are homogeneous. So the pooled estimate of regression coefficient can safely be used.

ANCOVA table

Source	d.f.	S.S.	M.S.	F-value
Bet. instructors	2	19.46	9.73	9.73/6.12 = 1.59 NS
Error	16	97.97	6.12	
Total	18	117.43		

$F_{.05;\,(2,16)} = 3.63$. Since the calculated value of F = 1.59 is less than the tabulated F-value, it is inferred that there exists no significant difference in the affectivity of three teachers.

Test of H_0: $\beta = 0$ can be performed by the test statistic given in (10.3.7).

$$F = \frac{1}{6.12} \left[\frac{190.28^2}{279.39} \right] = 21.18$$

$F_{.05;\,(1,16)} = 4.49$ is less than the calculated value. Hence, reject H_0. This confirms that the covariate has substantial impact on criterion variable in this case.

Note: Experimental error S.S.(E_{yy}) in ANOVA is 227.56 having 17 d.f.. The estimate of σ_e^2 is 227.56/17 = 13.38.

The estimated value of ρ_w is,

$$r_w = \frac{190.28}{\sqrt{227.56 \times 279.39}} = 0.755$$

or $$r_w^2 = 0.57$$

Error M.S. in ANCOVA can also be obtained by the formula (10.1.1) as given below.

$$S_E^2 \text{(adj.)} = 13.38 \ (1 - 0.57) \left(1 + \frac{1}{17 - 2} \right) = 6.14$$

Error obtained by the formula (10.1.1) is same as in ANCOVA table. Pooled within-class regression coefficient,

$$b = \frac{190.38}{279.39} = 0.68$$

Adjusted mean scores for three instructors by the equation (10.3.8) are,

$$\bar{y}_1' = 11.40 - 0.68(10.20 - 11.00) = 11.94$$
$$\bar{y}_2' = 11.14 - 0.68(10.57 - 11.00) = 11.43$$
$$\bar{y}_3' = 14.25 - 0.68(11.88 - 11.00) = 13.65$$

There is no need of further comparing the means as they have already been found at par through analysis of covariance.

10.4 COVARIANCE ANALYSIS OF RANDOMIZED BLOCK DESIGN

The procedure of analysis of covariance is based on the same assumptions as mentioned earlier. Statistical model for this design is,

$$y_{ij} = \mu + \tau_i + \rho_j + \beta x_{ij} + \varepsilon_{ij} \tag{10.4.1}$$

$$i = 1, 2, \ldots\ldots\ldots, v; j = 1, 2, \ldots\ldots\ldots, k$$

Experimental data for the criterion variable y_{ij} and corresponding covariate x_{ij}-values can always be tabulated in the following manner.

Table 10.4.1 Data table for RBD with covariate values

Blocks	Treatments						Block totals
	1	*2*	....	*i*	....	*v*	
1	y_{11}	y_{21}	....	y_{i1}	....	y_{v1}	$y_{\cdot 1}$
	x_{11}	x_{21}	....	x_{i1}	....	x_{v1}	$x_{\cdot 1}$
2	y_{12}	y_{22}	....	y_{i2}	....	y_{v2}	$y_{\cdot 2}$
	x_{12}	x_{22}	....	x_{i2}	....	x_{v2}	$x_{\cdot 2}$
|							
|							
|							
j	y_{1j}	y_{2j}	....	y_{ij}	....	y_{vj}	$y_{\cdot j}$
|	x_{1j}	x_{2j}	....	x_{ij}	....	x_{vj}	$x_{\cdot j}$

(Contd.)

Blocks	Treatments						Block totals
	1	*2*	....	*i*	....	*v*	
| | *k*	y_{1k} x_{1k}	y_{2k} x_{2k}	....	y_{jk} x_{ik}	....	y_{vk} x_{vk}	| | $y_{\cdot k}$ $x_{\cdot k}$
Treat. totals	$y_1.$ $x_1.$	$y_2.$ $x_2.$	....	$y_i.$ $x_i.$	....	$y_v.$ $x_v.$	$y_{..}$ $x_{..}$
Means	$\bar{y}_1$	$\bar{y}_2$	....	$\bar{y}_i$	....	$\bar{y}_v$	
	$\bar{x}_1$	$\bar{x}_2$	....	$\bar{x}_i$	....	$\bar{x}_v$	

In the above table, $\bar{y}_i = y_i./k$, and $\bar{x}_i = x_i./k$

Follow the same methodology as explicated in section 10.3 and calculate the quantities required for covariance analysis of RBD and prepare ANCOVA table.

Prepare S.S. and C.P. Table 10.4.2 as given on page 352.

Homogeneity of within-class regression coefficients can be tested by F-statistic (10.3.3).

Total S.S. and C.P. are easily calculable by the following formulae.

$$S_{yy} = \sum_{i=1}^{v}\sum_{j=1}^{k} y_{ij}^2 - \frac{y_{..}^2}{vk}$$

$$S_{xx} = \sum_{i=1}^{v}\sum_{j=1}^{k} x_{ij}^2 - \frac{x_{..}^2}{vk}$$

$$S_{yx} = \sum_{i=1}^{v}\sum_{j=1}^{k} y_{ij}x_{ij} - \frac{y_{..}\,x_{..}}{vk}$$

Formulae for treatment S.S and C.P. are:

$$T_{yy} = \sum_{i=1}^{v} \frac{y_{i.}^2}{k} - \frac{y_{..}^2}{vk}$$

$$T_{xx} = \sum_{i=1}^{v} \frac{x_{i.}^2}{k} - \frac{x_{..}^2}{vk}$$

Table 10.4.2 S.S. and C.P. Table

	Treatments				
	1	2	....	i	v
	$\sum_j y_{1j}^2 \quad \sum_j x_{1j}^2$	$\sum_j y_{2j}^2 \quad \sum_j x_{2j}^2$	....	$\sum_j y_{ij}^2 \quad \sum_j x_{ij}^2$	$\sum_j y_{vj}^2 \quad \sum_j x_{vj}^2$
	$S_{yy_1} = \sum_j y_{1j}^2 - \dfrac{y_{1.}^2}{k}$	$S_{yy_2} = \sum_j y_{2j}^2 - \dfrac{y_{2.}^2}{k}$	....	$S_{yy_i} = \sum_j y_{ij}^2 - \dfrac{y_{i.}^2}{k}$	$S_{yy_v} = \sum_j y_{vj}^2 - \dfrac{y_{v.}^2}{k}$
	$S_{xx_1} = \sum_j x_{1j}^2 - \dfrac{x_{1.}^2}{k}$	$S_{xx_2} = \sum_j x_{2j}^2 - \dfrac{x_{2.}^2}{k}$	....	$S_{xx_i} = \sum_j x_{ij}^2 - \dfrac{x_{i.}^2}{k}$	$S_{xx_v} = \sum_j x_{vj}^2 - \dfrac{x_{v.}^2}{k}$
	$S_{yx_1} = \sum_j y_{1j}x_{1j} - \dfrac{y_{1.}x_{1.}}{k}$	$S_{yx_2} = \sum_j y_{2j}x_{2j} - \dfrac{y_{2.}x_{2.}}{k}$	....	$S_{yx_i} = \sum_j y_{ij}x_{ij} - \dfrac{y_{i.}x_{i.}}{k}$	$S_{yx_v} = \sum_j y_{vj}x_{vj} - \dfrac{y_{v.}x_{v.}}{k}$
	$\beta_1 = \dfrac{S_{yx_1}}{S_{xx_1}}$	$\beta_2 = \dfrac{S_{yx_2}}{S_{xx_2}}$		$\beta_i = \dfrac{S_{yx_i}}{S_{xx_i}}$	$\beta_v = \dfrac{S_{yx_v}}{S_{xx_v}}$

$$T_{yx} = \sum_{i=1}^{v} \frac{y_{i.}x_{i.}}{k} - \frac{y_{..}x_{..}}{vk}$$

Similarly S.S. and C.P. terms for replications are:

$$R_{yy} = \sum_{j=1}^{k} \frac{y_{.j}^2}{v} - \frac{y_{..}^2}{vk}$$

$$R_{xx} = \sum_{j=1}^{k} \frac{x_{.j}^2}{v} - \frac{x_{..}^2}{vk}$$

$$R_{yx} = \sum_{j=1}^{k} \frac{y_{.j}x_{.j}}{v} - \frac{y_{..}x_{..}}{vk}$$

S.S. and C.P. for error are obtained by subtraction.

$$E_{yy} = S_{yy} - T_{yy} - R_{yy}$$

$$E_{xx} = S_{xx} - T_{xx} - R_{xx}$$

$$E_{yx} = S_{yx} - T_{yx} - R_{yx}$$

Adjusted error S.S. due to regression is,

$$S_E = E_{yy} - \frac{E_{yx}^2}{E_{xx}}$$

Treatment S.S. adjusted for the effect of the covariate x is obtainable in the following manner. Calculate,

$$A_{yy} = T_{yy} + E_{yy}$$

$$A_{xx} = T_{xx} + E_{xx}$$

$$A_{yx} = T_{yx} + E_{yx}$$

$$S_{T+E} = A_{yy}^2 - \frac{A_{yx}^2}{A_{xx}}$$

Adjusted treatment sum of square,

$$T_{yy}(Adj.) = S_{T+E} - S_E$$

After working out the above quantities, finally the analysis of covariance table 10.4.3 is as given on page 354.

Table 10.4.3 ANCOVA table

Source	d.f.	S.S. and C.P. $\sum y^2$	$\sum xy$	$\sum x^2$	Adj. S.S.	Adj. d.f.	M.S.	F-value
Blocks	$k-1$	R_{yy}	R_{yx}	R_{xx}	$R_{yy} - \dfrac{R_{yx}^2}{R_{xx}} = R'_{yy}$	$(k-1)$	$\dfrac{R'_{yy}}{k-1} = R_{adj}$	
Treatments	$v-1$	T_{yy}	T_{yx}	T_{xx}	$S_{T+E} - S_E = T'_{yy}$	$(v-1)$	$\dfrac{T'_{yy}}{(v-1)} = T_{adj}$	$\dfrac{T_{adj}}{SE'}$
Error	$(k-1)(v-1)$	E_{yy}	E_{yx}	E_{xx}	$E_{yy} - \dfrac{E_{yx}^2}{E_{xx}} = SE$	$(k-1)(v-1)-1 = n'_e$	$\dfrac{SE}{n'_e} = SE'$	
Treat. + Error	$k(v-1)$	A_{yy}	A_{yx}	A_{xx}	$A_{yy} - \dfrac{A_{yx}^2}{A_{xx}} = S_{T+E}$			

F for treatments has d.f. $\{(v-1), n_e'\}$.

On comparing the calculated value of F for $\{(v-1), n_e'\}$ d.f. and α level of significance, the decision about the hypothesis can be taken in the usual manner.

It is always advisable to test the significance of common regression coefficient β because $\beta \neq 0$ ensures that the adjustment due to concomitant variable really improves the analysis and increases the reliability of the tests. H_0: $\beta = 0$ can be tested by the test statistics,

$$F = \frac{E_{yx}^2/E_{xx}}{SE'} \tag{10.4.2}$$

F has d.f. $(1, n_e')$. Decision about H_0 is taken by the routine procedure. To obtain the adjusted treatment means, calculate b, the estimate of β, by the formula,

$$b = \frac{E_{yx}}{E_{xx}} \tag{10.4.3}$$

and the adjusted treatment means by the relation,

$$\bar{y}_i' = \bar{y}_{i'} - b(\bar{x}_i - \bar{x}) \tag{10.4.4}$$

Standard error of an adjusted treatment mean is,

$$\text{S.E.}(\bar{y}_i') = \sqrt{\left[SE'\left\{ \frac{1}{k} + \frac{(\bar{x}_i - \bar{x})^2}{E_{xx}} \right\} \right]} \tag{10.4.5}$$

Standard error of any paired contrast of adjusted treatment means is,

$$\text{S.E.}(\bar{y}_i' - \bar{y}_{i'}) = \sqrt{SE'\left[\left(\frac{2}{k} + \frac{(\hat{x}_i - \bar{x}_{i'})^2}{E_{xx}} \right) \right]} \tag{10.4.6}$$

$\text{S.E.}(\bar{y}_i' - \bar{y}_{i'})$ is used for comparing adjusted pairs of treatment means by any suitable test if F-value for treatments in ANCOVA table comes out to be significant.

Significance of any apriori orthogonal contrast,

$$Z = c_1\bar{y}_1' + c_2\bar{y}_2' + \ldots\ldots\ldots + c_v\bar{y}_v'$$

can be tested by t-test for which the test statistic is,

$$t = \frac{c_1\overline{y}_1' + c_2\overline{y}_2' + \ldots\ldots\ldots + c_v\overline{y}_v'}{\sqrt{\left[SE'\left\{\frac{1}{k}(c_1^2 + c_2^2 + \ldots\ldots\ldots + c_v^2) + \frac{(c_1\overline{x}_1 + c_2\overline{x}_2 + \ldots\ldots\ldots + c_v\overline{x}_v)^2}{E_{xx}}\right\}\right]}}$$

(10.4.7)

Statistic t has n_e' d.f. Contrast Z is significant if $t_{cal} \geq t_{\alpha;n_e'}$, otherwise non-significant.

Example 10.2: An experiment was planned to compare yield potential of seven varieties of pearl millet. The layout of the trial was randomized block design with four replications. It was felt that the number of plants per plot has an impact on yield. Hence the data were collected for yield and number of plants per plot. Plot size was 8×4 m^2. Plot wise yield in kilogram (y) and number of plants per plot (x) were as given below.

Blocks		Varieties						
		HHB-60 V_1	MH-36 V_2	MB-171 V_3	HHB-67 V_4	WCC-75 V_5	CM-46 V_6	MH-179 V_7
1	y	4.47	2.46	3.41	2.26	2.69	3.46	4.05
	x	417	247	288	220	228	304	360
2	y	3.49	3.36	3.46	3.80	2.79	2.79	3.58
	x	278	335	396	336	287	266	305
3	y	3.08	3.80	4.11	3.72	3.81	3.53	4.25
	x	246	358	369	284	377	313	388
4	y	3.74	2.96	3.59	3.03	2.73	3.03	3.93
	x	284	274	352	318	295	262	288

Treatment totals:

$y_1.=14.78, y_2.=12.58, y_3.=14.57, y_4.=12.81, y_5.=12.02, y_6.=12.81, y_7.=15.81$
$x_1.=1225, x_2.=1214, x_3.=1405, x_4.=1158, x_5.=1187, x_6.=1145, x_7.=1341$

Treatment mean:

$$\overline{y}_1 = 3.695, \overline{y}_2 = 3.145, \overline{y}_3 = 3.642, \overline{y}_4 = 3.202,$$

$$\overline{y}_5 = 3.005, \overline{y}_6 = 3.202, \overline{y}_7 = 3.952$$

$$\overline{x}_1 = 306.25, \overline{x}_2 = 303.50, \overline{x}_3 = 351.25, \overline{x}_4 = 289.50,$$

$$\overline{x}_5 = 296.75, \overline{x}_6 = 286.25, \overline{x}_7 = 335.25$$

Block totals:

$$y._1 = 22.80, y._2 = 23.27, y._3 = 26.30, y._4 = 23.01$$
$$x._1 = 2064, x._2 = 2203, x._3 = 2335, x._4 = 2073$$

Grand total	Grand mean
$y.. = 95.38$	$\bar{y} = 3.406$
$x.. = 8675$	$\bar{x} = 309.82$

Analysis of covariance has been carried out taking y as criterion variable and x as covariate. Calculations are run exactly in the same manner as given in section 10.4. Therefore, it needs no further explanation.

Varieties sum of squares, cross-product and regression coefficients:

Variety 1:

$$\sum_j y_{1j}^2 = 4.47^2 + 3.49^2 + 3.08^2 + 3.74^2 = 55.64$$

$$\sum_j x_{1j}^2 = 417^2 + 278^2 + 246^2 + 284^2 = 392345$$

$$S_{yy_1} = 55.64 - \frac{14.78^2}{4} = 1.02$$

$$S_{xx_1} = 392345 - \frac{1225^2}{4} = 17188.75$$

$$S_{yx_1} = 4.47 \times 417 + 3.49 \times 278 + 3.08 \times 246 + 3.74 \times 284 - \frac{14.78 \times 1225}{4}$$

$$= 4654.05 - 4526.38 = 127.67$$

$$\beta_1 = \frac{127.67}{17188.75} = 0.0074$$

Similarly the calculations are made for other varieties and displayed below.

Variety 2:

$$\sum_j y_{2j}^2 = 40.54, \sum_j x_{2j}^2 = 257874, S_{yy_2} = 0.988,$$

$$S_{xx_2} = 8025.00, S_{yx_2} = 3904.66 - 3818.03 = 86.63, \beta_2 = 0.0108$$

Variety 3:

$$\sum_j y_{3j}^2 = 53.38, \sum_j x_{3j}^2 = 499825, S_{yy_3} = 0.3087,$$

$$S_{xx_3} = 6318.75, S_{yx_3} = 5132.51 - 5117.71 = 14.80, \beta_3 = 0.0023$$

Variety 4:

$$\sum_j y_{4j}^2 = 42.57, \sum_j x_{4j}^2 = 343076, S_{yy_4} = 1.54,$$

$$S_{xx_4} = 7835.00, S_{yx_4} = 3794.02 - 3708.50 = 85.52, \beta_4 = 0.0109,$$

Variety 5:

$$\sum_j y_{5j}^2 = 36.99, \sum_j x_{5j}^2 = 363507, S_{yy_5} = 0.87,$$

$$S_{xx_5} = 11264.75, S_{yx_5} = 3655.77 - 3566.94 = 88.83, \beta_5 = 0.0079,$$

Variety 6:

$$\sum_j y_{6j}^2 = 41.40, \sum_j x_{6j}^2 = 329785, S_{yy_6} = 0.37,$$

$$S_{xx_6} = 2028.75, S_{yx_6} = 3692.73 - 3666.86 = 25.87, \beta_6 = 0.0128,$$

Variety 7:

$$\sum_j y_{7j}^2 = 62.73, \sum_j x_{7j}^2 = 456113, S_{yy_7} = 0.24,$$

$$S_{xx_7} = 6542.75, S_{yx_7} = 5330.74 - 5300.30 = 30.44, \beta_7 = 0.0046.$$

Block sum of squares and cross-product.

$$R_{yy} = \frac{1}{7}(22.80^2 + 23.27^2 + 26.30^2 + 23.01^2) - \frac{95.38^2}{28}$$

$$= 326.07 - 324.90 = 1.17$$

$$R_{xx} = \frac{1}{7}(2064^2 + 2203^2 + 2335^2 + 2073^2) - \frac{8675^2}{28}$$

$$= 2694694.14 - 2687700.89 = 6993.25$$

$$R_{yx} = \frac{1}{7}(22.80 \times 2064 + 23.27 \times 2203 +$$

$$26.30 \times 2335 + 23.01 \times 2073) - \frac{95.38 \times 8675}{28}$$

$$= 29633.32 - 29550.77 = 82.55$$

Treatment squares and cross-product.

$$T_{yy} = \frac{1}{4}(14.78^2 + 12.58^2 + \ldots\ldots + 12.81^2 + 15.81^2) - \frac{95.38^2}{28}$$

$$= 327.90 - 324.90 = 3.00$$

$$T_{xx} = \frac{1}{4}(1225^2 + 1214^2 + \ldots\ldots + 1145^2 + 1341^2) - \frac{8675^2}{28}$$

$$= 2701921.25 - 2687700.89 = 14220.36$$

$$T_{yx} = \frac{1}{4}(14.78 \times 1225 + 12.58 \times 1214 + \ldots\ldots$$

$$+ 12.81 \times 1145 + 15.81 \times 1341) - \frac{95.38 \times 8675}{28}$$

$$= 29704.71 - 29550.77 = 153.94$$

Total sum of squares and cross-product.

$$S_{yy} = 4.47^2 + 3.49^2 + \ldots\ldots + 4.25^2 + 3.93^2 - \frac{95.38^2}{28}$$

$$= 333.24 - 324.90 = 8.34$$

$$S_{xx} = 417^2 + 278^2 + \ldots\ldots + 388^2 + 288^2 - \frac{8675^2}{28}$$

$$= 2761125 - 2687700.89 = 73424.11$$

$$S_{yx} = 4.47 \times 417 + 3.49 \times 278 + \ldots\ldots + 4.25 \times 388$$

$$+ 3.93 \times 288 - \frac{95.38 \times 8675}{28}$$

$$= 30164.48 - 29550.77 = 613.71$$

Error sum of squares and cross-product.

$$E_{yy} = 8.34 - 3.00 - 1.17 = 4.17$$

$$E_{xx} = 73424.11 - 14220.36 - 6993.25 = 55210.50$$

$$E_{yx} = 613.71 - 153.94 - 82.55 = 377.22$$

Adjusted error sum of square.

$$S_E = 4.17 - \frac{377.22^2}{55210.50}$$

$$= 4.17 - 2.58 = 1.59$$

Treatment plus error terms,

$$A_{yy} = 3.00 + 4.17 = 7.17$$

$$A_{xx} = 14220.36 + 55210.50 = 69430.86$$
$$A_{yx} = 153.94 + 377.22 = 531.16$$

$$S_{T+E} = 7.17 - \frac{531.16^2}{69430.86}$$

$$= 7.17 - 4.06 = 3.11$$

Adjusted treatment sum of square,

$$T'_{yy} = 3.11 - 1.59 = 1.52$$

Adjusted block sum of square,

$$R'_{yy} = 1.17 - \frac{82.55^2}{6993.25}$$

$$= 1.17 - 0.97 = 0.20$$

ANCOVA table

Source	d.f. (Adj.)	S.S.(Adj.)	M.S.	F-Value
Blocks	3	0.20	0.067	0.71 NS
Treats.	6	1.52	0.253	2.69
Error	17	1.59	0.094	

$F_{.05; 6, 17} = 2.70$ and $F_{.10; 6, 17} = 2.15$

As a matter of fact, the calculated F-value for treatments is slightly less than the tabulated F-value at 5% level of significance. Hence for P = 0.05, hypothesis of equality of treatment effects is accepted. But it will be rejected at little lower level of significance, may it be 0.06 or 0.07. So if one wants to reject H_0 about treatments at a little higher probability of type I error than traditional 5 % probability, he will be quite justified.

The test of homogeneity of within-class regression coefficient can be performed as given below.

S_1 by the formula (10.3.4) is,

$$S_1 = 4.17 - \left(\frac{127.67^2}{17188.75} + \frac{86.63^2}{8025} + \dots\dots + \frac{30.44^2}{6542.75} \right)$$

$$= 4.17 - (0.9483 + 0.9352 + 0.0347 + 0.9335 + 0.7005 + 0.3299 + 0.1416)$$

$$= 4.17 - 4.03 = 0.14$$

Similarly by the formula (10.3.5),

$$S_2 = 4.03 - \frac{377.22^2}{55210.50}$$

$$= 4.03 - 2.58 = 1.45$$

Statistic by the formula (10.3.3),

$$F = \frac{1.45/6}{0.14/14} = \frac{0.24}{0.01} = 24.00$$

On comparing the calculated F with $F_{.05;\,1,17} = 4.45$, it is inferred that within-class regression coefficients are not homogeneous. Hence, the pooled estimate of regression coefficients shall not be a correct proposition. Thus, the adjusted treatment means using b as a single within-class of regression coefficients will not be as accurate as expected.

Estimate of pooled β is,

$$b = \frac{377.22}{55210.50} = 0.0068$$

Test of hypothesis, H_0: $\beta = 0$, by the formula (10.3.7) is,

$$F = \frac{(377.22)^2/55210.50}{0.094} = 27.45$$

$F_{.05;\,1,17} = 4.45$ which is less than 27.45. Hence, pooled regression coefficient is significantly greater than zero. This ensures that the introduction of number of plants per plot as covariate has meaningfully improved the process of analysis of data for comparing the varieties.

Adjusted treatment means are:

$$\bar{y}_1' = 3.695 - 0.0068(306.25 - 309.82) = 3.672$$

$$\bar{y}_2' = 3.145 - 0.0068(303.50 - 309.82) = 3.188$$

Similarly,

$$\bar{y}_3' = 3.360,\ \bar{y}_4' = 3.340,\ \bar{y}_5' = 3.094,\ \bar{y}_6' = 3.362,\ \bar{y}_7' = 3.779$$

Standard error of an adjusted treatment mean by the formula (10.4.5) is,

$$S.E.(\bar{y}_1') = \sqrt{0.094\left\{\frac{1}{4} + \frac{(306.25 - 309.82)^2}{55210.50}\right\}}$$

$$= 0.153$$

S.E. for any other adjusted treatment mean can be worked out in the like manner.

Since the hypothesis of equality of means has been accepted, there is no need of pairwise comparisons.

Note: Solved example (10.2) has been given with two purposes, (*i*) it fully explains the method of analysis of covariance of a randomized block design, (*ii*) it also reflects on the complexities which an analyst may be confronted with while carrying out the analysis of covariance.

10.5 TACKLING MISSING PLOTS THROUGH COVARIANCE ANALYSIS

Problem of missing plots has already been discussed in sections 2.3 and 2.9. Also, analysis of covariance as a remedy for analysis of data having missing value(s) in an orthogonal design has been mentioned in Chapter 2. If there is/ are one or more missing values in an experiment, then according to Bartlett (1937), analysis of covariance makes one to estimate the missing value(s) and analyze the data for any design precisely.

Consider one missing value in the experiment and a concomitant variable x associated with the criterion variable y. x takes on the value 0 corresponding to all available observations on y and –1 for the missing value. Also the missing value y is taken to be zero. Now carry on the analysis of covariance as discussed in the previous section.

$$\hat{y} = y_0 - bx_0 \tag{10.5.1}$$

Since, $$y_0 = 0, \, x_0 = -1, \hat{y} = b = \frac{E_{yx}}{E_{xx}} \tag{10.5.2}$$

If there are two or more missing values, one has to take as many pseudo-variables as the number of missing values and then carry out the analysis of covariance. But the analysis becomes more and more complicated as the number of missing values increases. Hence, it is not advisable to recommend this method when the number of missing values is large.

Example 10.3: Data pertaining to five treatments tried in randomized block design with four replications with regard to chlorophyll content in fresh leaves of pea in mg/g are presented below. Any how one observation was lost during chemical analysis.

Estimate the missing value and analyze the experimental data by the method of analysis of covariance.

Treatments were as named here.

S_0 - control, S_1 - elemental sulphur (250 kg/ha), S_2 - gypsum (250 kg/ha), S_3 – H_2SO_4 (0.1 % foliar spray), S_4 – F_e- EDDHA (0.2 % foliar spray).

Treatment and block totals, means along with the values for pseudo-variate have also been displayed in the data table given below for the sake of brevity.

Treats.	Replications					Mean
	I	*II*	*III*	*IV*	*Total*	
S_0	0.679	0.852	0.513	0.507	2.551	0.6378
	(0)	(0)	(0)	(0)	(0)	(0)
S_1	0.952	1.002	0(NA)	0.621	2.575	0.6438
	(0)	(0)	(−1)	(0)	(−1)	(−0.25)
S_2	0.899	0.919	0.718	0.679	3.215	0.8038
	(0)	(0)	(0)	(0)	(0)	(0)
S_3	0.986	0.949	0.845	0.780	3.560	0.8900
	(0)	(0)	(0)	(0)	(0)	(0)
S_4	0.911	0.922	0.668	0.746	3.247	0.8118
	(0)	(0)	(0)	(0)	(0)	(0)
Total	4.427	4.644	2.744	3.333	15.148	
	(0)	(0)	(−1)	(0)	(−1)	

Analysis of data using covariance analysis technique has been delineated below.

$$S_{yy} = 0.679^2 + 0.952^2 + \ldots\ldots + 0.780^2 + 0.746^2 - \frac{15.148^2}{20}$$

$$= 12.510 - 11.473 = 1.037$$

$$S_{xx} = 1 - \frac{(-1)^2}{20} = 1 - 0.05 = 0.95$$

$$S_{yx} = 0 - \frac{15.148 \times (-1)}{20} = 0.7574$$

$$T_{yy} = \frac{1}{4}(2.551^2 + 2.575^2 + 3.215^2 + 3.560^2 + 3.247^2) - \frac{15.148^2}{20}$$

$$= 11.6728 - 11.4731 = 0.1997$$

$$T_{xx} = \frac{(-1)^2}{4} - \frac{(-1)^2}{20} = 0.2500 - 0.0500 = 0.2000$$

$$T_{yx} = \frac{-2.575}{4} - \frac{15.148 \times (-1)}{20} = -0.6438 + 0.7574 = 0.1134$$

$$R_{yy} = \frac{1}{5}(4.427^2 + 4.644^2 + 2.744^2 + 3.333^2) - \frac{15.148^2}{20}$$

$$= 11.9607 - 11.4731 = 0.4876$$

$$R_{xx} = \frac{(-1)^2}{5} - \frac{(-1)^2}{20} = 0.20 - 0.05 = 0.15$$

$$R_{yx} = \frac{1}{5}\{2.744 \times (-1)\} - \frac{15.148 \times (-1)}{20} = -0.5488 + 0.7574 = 0.2086$$

$$E_{yy} = 1.0369 - 0.1997 - 0.4876 = 0.3496$$

$$E_{xx} = 0.95 - 0.20 - 0.15 = 0.60$$

$$E_{yx} = 0.7574 - 0.1136 - 0.2086 = 0.4352$$

The estimate of the missing value is,

$$b = \frac{0.4352}{0.60} = 0.7253$$

Adjusted error sum of square,

$$S_E = 0.3496 - \frac{0.4352^2}{0.60} = 0.3496 - 0.3157 = 0.0339$$

$$A_{yy} = 0.1997 + 0.3496 = 0.5493$$

$$A_{xx} = 0.20 + 0.60 = 0.80$$

$$A_{yx} = 0.1134 + 0.4352 = 0.5486$$

$$S_{T+E} = 0.5493 - \frac{0.5486^2}{0.80} = 0.5493 - 0.3762 = 0.1731$$

Adjusted treatment S.S. $= 0.1731 - 0.0339 = 0.1392$

Adjusted replication S.S. $= 0.4876 - \frac{0.2086^2}{0.15} = 0.4876 - 0.2901 = 0.1975$

ANCOVA table

Source	Adj. d.f.	Adj. S.S.	M.S.	F-value
Replications	3	0.1975	0.0658	21.22*
Treatments	4	0.1392	0.0348	11.22*
Error	11	0.0339	0.0031	

$F_{0.05;\,3,\,11} = 3.59$, $F_{0.05;\,4,11} = 3.36$

Comparison of F-values for replications and treatments with the corresponding tabulated values (given below the table) reveals that both are significant at 5% level of significance.

Values of adjusted treatment means by the equation $\bar{y}_i' = \bar{y}_i - b(\hat{x}_i - \bar{x})$ are,

$$\bar{y}_0' = \frac{2.551}{4} - 0.7253(0 + 0.05) = 0.6015$$

$$\bar{y}_1' = \frac{2.575}{4} - 0.7253(-0.25 + 0.05) = 0.7888$$

$$\bar{y}_2' = \frac{3.215}{4} - 0.7253(0 + 0.05) = 0.7675$$

$$\bar{y}_3' = \frac{3.560}{4} - 0.7253(0 + 0.05) = 0.8537$$

$$\bar{y}_4' = \frac{3.247}{4} - 0.7253(0 + 0.05) = 0.7755$$

Since the treatment effects differ significantly, it is necessitated to compare all pairs of treatment means to identify which of them diverge significantly. Such a comparison is accomplished in the following manner.

First find the standard errors of the paired comparisons of adjusted treatment means by the formula (10.4.6). There shall be two standard errors, one for comparing treatment means not having any missing value, the other for comparing a treatment mean with the treatment mean having a missing value.

$$S.E.(\bar{y}_i' - \bar{y}_{i'}') = \sqrt{0.0031 \times \frac{2}{4}} = 0.0394$$

for $i \neq i'$ and $\bar{x}_i = \bar{x}_{i'} = 0$

Also,

$$S.E.(\bar{y}_i' - \bar{y}_{i'}') = \sqrt{0.0031 \times \left\{ \frac{2}{4} + \frac{(0 + 0.25)^2}{0.6} \right\}} = 0.0434$$

where either of $\bar{x}_i$ or $\bar{x}_{i'} = 0.25$

For 11 d.f. and P = 0.05, tabulated $t = 2.201$.

On multiplying the standard errors by 2.201, the required values of critical differences for comparing pairs of adjusted treatment means are obtained as follows:

$$\text{C.D}_1 = 0.0394 \times 2.201 = 0.0867$$
$$\text{C.D}_2 = 0.0434 \times 2.201 = 0.0955$$

Table of absolute differences of all pairs of adjusted treatment means is as displayed below.

Treatments	S_1	S_2	S_3	S_4
S_0	0.1878*	0.1660*	0.2522*	0.1740*
S_1		0.0213	0.0649	0.0133
S_2			0.0862	0.0080
S_3				0.0782

*significant difference

Differences in the preceding table in a column or row of S_1 are compared with CD_2 and others with CD_1. Comparison reveals that all treatments are superior than control and at par with each other.

10.6 ANALYSIS OF COVARIANCE WITH MULTIPLE COVARIATES

It has already been clarified that analysis of covariance technique is applicable to any number of covariates and to any experimental design. But the author never came across a situation where more than two covariates have ever been taken into consideration. Hence, analysis of covariance with two covariates for randomized complete block design is explained which can be extended to any other design. Let there be an experiment in RBD - v, k with measurements on a criterion variable Y and two concomitant variables X and Z. The observations y_{ij} for ith treatment (i =1, 2, ..., v) and jth block (j = 1, 2, ..., k) can be tabulated as follows:

Table 10.6.1 Data table

Blocks	Treatments						Totals
	1	*2*	...	*i*	...	*v*	
1	$y_{11}\ x_{11}\ z_{11}$	$y_{21}\ x_{21}\ z_{21}$	...	$y_{i1}\ x_{i1}\ z_{i1}$	...	$y_{v1}\ x_{v1}\ z_{v1}$	$y_{\cdot 1}\ x_{\cdot 1}\ z_{\cdot 1}$
2	$y_{12}\ x_{12}\ z_{12}$	$y_{22}\ x_{22}\ z_{22}$	...	$y_{i2}\ x_{i2}\ z_{i2}$	...	$y_{v2}\ x_{v2}\ z_{v2}$	$y_{\cdot 2}\ x_{\cdot 2}\ z_{\cdot 2}$
:				:	:		:
:				:	:		:
:				:	:		:
k	$y_{1k}\ x_{1k}\ z_{1k}$	$y_{2k}\ x_{2k}\ z_{2k}$	...	$y_{ik}\ x_{ik}\ z_{ik}$	...	$y_{vk}\ x_{vk}\ z_{vk}$	$y_{\cdot k}\ x_{\cdot k}\ z_{\cdot k}$
Totals	$y_{1\cdot}\ x_{1\cdot}\ z_{1\cdot}$	$y_{2\cdot}\ x_{2\cdot}\ z_{2\cdot}$	...	$y_{i\cdot}\ x_{i\cdot}\ z_{i\cdot}$	...	$y_{v\cdot}\ x_{v\cdot}\ z_{v\cdot}$	$y_{\cdot\cdot}\ x_{\cdot\cdot}\ z_{\cdot\cdot}$
Means	$\bar{y}_1\ \bar{x}_1\ \bar{z}_1$	$\bar{y}_2\ \bar{x}_2\ \bar{z}_2$	...	$\bar{y}_i\ \bar{x}_i\ \bar{z}_i$	...	$\bar{y}_v\ \bar{x}_v\ \bar{z}_v$	$\bar{y}\ \bar{x}\ \bar{z}$

The analysis of covariance is carried out under the fundamental assumption that within-class regression coefficients are homogeneous. Hence, estimates of pooled within-class regression coefficients b_{yx} and b_{yz} are computed. Multiple regression equation is,

$$\hat{y}_{ij} = b_{yx}(x_{ij} - \bar{x}_i) + b_{yz}(z_{ij} - \bar{z}_i) + \bar{y}_i \qquad (10.6.1)$$

Now the sum of squares and cross-products shall be worked out to obtain the values of b_{yx} and b_{yz} and also sum of squares for ANCOVA.

$S_{yy}, S_{xx}, S_{yx}, T_{yy}, T_{xx}, T_{yx}, R_{yy}, R_{xx}, R_{yx}$ are calculable by the expressions given in section (10.4). Additional terms which are needed in this case can be obtained by the following expressions.

$$S_{zz} = \sum_{i=1}^{v} \sum_{j=1}^{k} z_{ij}^2 - \frac{z_{..}^2}{vk}$$

$$S_{yz} = \sum_{i=1}^{v} \sum_{j=1}^{k} y_{ij} z_{ij} - \frac{y_{..}z_{..}}{vk}$$

$$S_{xz} = \sum_{i=1}^{v} \sum_{j=1}^{k} x_{ij} z_{ij} - \frac{x_{..}z_{..}}{vk}$$

Treatment S.S. and C.P.,

$$T_{zz} = \sum_{i=1}^{v} \frac{z_{i.}^2}{k} - \frac{z_{..}^2}{vk}$$

$$T_{yz} = \sum_{i=1}^{v} \frac{y_{i.}z_{i.}}{k} - \frac{y_{..}z_{..}}{vk}$$

$$T_{xz} = \sum_{i=1}^{v} \frac{x_{i.}z_{i.}}{k} - \frac{x_{..}z_{..}}{vk}$$

Similarly S.S. and C.P. terms for replications are:

$$R_{zz} = \sum_{j=1}^{k} \frac{z_{.j}^2}{v} - \frac{z_{..}^2}{vk}$$

$$R_{yz} = \sum_{j=1}^{k} \frac{y_{.j}z_{.j}}{v} - \frac{y_{..}z_{..}}{vk}$$

$$R_{xz} = \sum_{j=1}^{k} \frac{x_{.j}z_{.j}}{v} - \frac{x_{..}z_{..}}{vk}$$

Making use of the above expressions, the values of pooled within-class regression coefficients can be estimated by the following formulae.

E_{yy}, E_{xx}, and E_{yx} are same as in section (10.4) and additional error terms are obtained as follows:

$$E_{zz} = S_{zz} - T_{zz} - R_{zz}$$
$$E_{yz} = S_{yz} - T_{yz} - R_{yz}$$
$$E_{xz} = S_{xz} - T_{xz} - R_{xz}$$

$$b_{yx} = \frac{E_{zz}E_{yx} - E_{yz}E_{zx}}{\Delta} \tag{10.6.2}$$

$$b_{yz} = \frac{E_{xx}E_{yz} - E_{zx}E_{yx}}{\Delta} \tag{10.6.3}$$

where, $\qquad \Delta = E_{xx}E_{zz} - E_{zx}^2$

Now the sum of square due to error is,

$$S_E' = E_{yy} - b_{yx}E_{yx} - b_{yz}E_{yz} \tag{10.6.4}$$

Combined S.S. and C.P. terms are as given below.

$$A_{yy} = T_{yy} + E_{yy} \; ; \; A_{xx} = T_{xx} + E_{xx} \; ; \; A_{yx} = T_{yx} + E_{yx} \; ;$$
$$A_{zz} = T_{zz} + E_{zz} \; ; \; A_{yz} = T_{yz} + E_{yz} \; ; \; A_{zx} = T_{zx} + E_{xz} \; ;$$

Let the regression equation associated with the combined data be,

$$y_{ij}' = b_{y.x}'(x_{ij} - \bar{x}) + b_{y.z}'(z_{ij} - \bar{z}) + \bar{y} \tag{10.6.5}$$

Thus for the above regression equation, the partial regression coefficients are obtained by the formulae,

$$b_{y.x}' = \frac{A_{zz}A_{yx} - A_{yz}A_{zx}}{\Delta'} \tag{10.6.6}$$

$$b_{y.z}' = \frac{A_{xx}A_{yz} - A_{zx}A_{yx}}{\Delta'} \tag{10.6.7}$$

where, $\qquad \Delta' = A_{xx}A_{zz} - A_{zx}^2$

Adjusted (Treatment + Error) S.S.,

$$S_{T+E}' = A_{yy} - b_{y.x}'A_{yx} - b_{y.z}'A_{yz} \tag{10.6.8}$$

Therefore,

Adjusted treatment $S.S. = S_{T+E}' - S_E' = T_{yy}' \tag{10.6.9}$

Since the testing of block differences is of secondary interest and estimation of error sum of squares is free from block sum of square, a rough estimate of adjusted block sum of square can be obtained by the formula given below.

Adjusted block $S.S. = R_{yy} - \dfrac{R_{yx}^2}{R_{xx}} - \dfrac{R_{yz}^2}{R_{zz}} = R'_{yy}$ (10.6.10)

Further, since two concomitant variables are involved, two degrees of freedom will be deducted from total degrees of freedom. This would result into reduction of error degrees of freedom by 2. Hence, the ANCOVA table will be as follows:

Table 10.6.2 ANCOVA table

Source	Adj. d.f.	Adj. S.S.	Adj. M.S.	F-value
Block	$k - 1$	R'_{yy}	$R'_{yy}/(k - 1) = RMS$	RMS/SE(MS) = F_1
Treatment	$v - 1$	T'_{yy}	$T'_{yy}/(v - 1) = TMS$	TMS/SE(MS) = F_2
Error	$(k - 1)(v - 1)$ $- 2 = n'_e$	S'_E	$S'_E/n'_e = SE(MS)$	
Combined	$kv - 3$			

Comparing the calculated values of F_1 and F_2 with the tabulated values of F for corresponding degrees of freedom, it is trivial to decide about the null hypotheses for the factors under consideration.

The adjusted mean for treatment i is,

$$\bar{y}'_i = \bar{y}_i - b_{yx}(\bar{x}_i - \bar{x}) - b_{yz}(\bar{z}_i - \bar{z}) \qquad (10.6.11)$$

Average effective error per experimental unit of an adjusted treatment mean is,

$$S'^2_{\bar{y}} = SE(MS) \left\{ 1 + \frac{T_{xx}E_{zz} - 2T_{zx}E_{zx} + T_{zz}E_{xx}}{(v-1)\Delta} \right\} \qquad (10.6.12)$$

Average effective error variance of the difference between any two adjusted treatment means is,

$$S^2_{\bar{y}_i - \bar{y}_{i'}} = \frac{2S'^2_{\bar{y}}}{k} \text{ for } i \neq i' \qquad (10.6.13)$$

For any apriori paired contrast $(\bar{y}_i - \bar{y}_u)$ for $i \neq u$, the effective error variance is,

$$S'^2_y = SE(MS) \left[\frac{2}{k} + \frac{(\bar{x}_i - \bar{x}_u)^2 E_{zz} - 2(\bar{x}_i - \bar{x}_u)(\bar{z}_i - \bar{z}_u)E_{zx} + (\bar{z}_i - \bar{z}_u)^2 E_{xx}}{\Delta} \right]$$

$$(10.6.14)$$

Formula (10.6.14) is more exact than (10.6.13) for comparing any two adjusted treatment means. On multiplying error variance for a paired difference of adjusted treatment means by t-value for α level of significance and n_e' d.f., the value of critical difference (C.D.) is obtained. If C.D. $\geq$ the difference between two treatment means, then the treatments differ significantly, otherwise not.

10.7 ANALYSIS OF COVARIANCE FOR TWO MISSING VALUES EXEMPLIFIED

Briefing

Problems of missing observations arise only in balanced block designs. The same has already been discussed and dealt within section (2.4). But analysis of covariance is also a technique to deal with this problem. In this method, two covariates x and z are introduced in a data set. The missing values for the criterion variable y are coded as zero. For one missing y-value, $x = 1$, $z = 0$ and for the other $x = 0$, $z = 1$. Rest all x and z are coded as zero. Analysis of covariance of data is carried exactly in the same manner for RBD as discussed in section (10.6). Missing values are estimated by the equation (10.6.11). Another amendment in the procedure is with regard to standard errors as delineated below.

The standard error of the difference between two treatment means having no missing value is,

$$\text{S.E.}_{d1} = \sqrt{\frac{2SE(MS)}{k}} \tag{10.7.1}$$

Standard error of the difference between two treatment means, of which only one treatment has a missing value, is,

$$\text{S.E.}_{d2} = \sqrt{SE(MS)\left\{\frac{2}{k} + \frac{E_{zz}}{k^2\Delta}\right\}} \tag{10.7.2}$$

Standard error of the difference between two treatment means, of which both treatments have a missing value, is,

$$\text{S.E.}_{d3} = \sqrt{SE(MS)\left\{\frac{2}{k} + \frac{1}{k^2\Delta}\left(E_{xx} + 2E_{zx} + E_{xx}\right)\right\}} \tag{10.7.3}$$

Entire process of analysis of covariance of data with two missing values is expatiated through the example (10.4).

Example 10.4: For the sake of comparability the data of example (2.4) is analyzed again. The data have been presented in the same pattern as in data table (10.6.1).

Blocks	Varieties														
	V_1			V_2			V_3			V_4			V_5		
	y_1	x_1	z_1	y_2	x_2	z_2	y_3	x_3	z_3	y_4	x_4	z_4	y_5	x_5	z_5
B_1	20.2	0	0	18.5	0	0	0	−1	0	27.7	0	0	32.0	0	0
B_2	41.4	0	0	28.8	0	0	25.5	0	0	30.8	0	0	34.7	0	0
B_3	29.0	0	0	15.0	0	0	11.5	0	0	22.5	0	0	29.2	0	0
B_4	25.6	0	0	24.8	0	0	13.5	0	0	31.3	0	0	30.5	0	0
Totals	116.2	0	0	87.1	0	0	50.5	−1	0	112.3	0	0	126.4	0	0
Mean	29.05	0	0	21.78	0	0	12.62	−.25	0	28.08	0	0	31.60	0	0

Blocks	Varieties												Total		
	V_6			V_7			V_8			V_9					
	y_6	x_6	z_6	y_7	x_7	z_7	y_8	x_8	z_8	y_9	x_9	z_9	y	x	z
	31.3	0	0	25.5	0	0	39.2	0	0	20.6	0	0	215.0	−1	0
	35.2	0	0	43.7	0	0	46.3	0	0	31.8	0	0	318.2	0	0
	16.3	0	0	0	0	−1	20.1	0	0	19.5	0	0	163.1	0	−1
	18.7	0	0	18.1	0	0	20.2	0	0	26.4	0	0	209.1	0	0
Totals	101.5	0	0	87.3	0	−1	125.8	0	0	98.3	0	0	905.4	−1	−1
Mean	25.38	0	0	21.82	0	−.25	31.45	0	0	24.58	0	0			

Over all means,

$$\bar{y} = \frac{905.4}{36} = 25.15, \bar{x} = \frac{-1.0}{36} = -0.0278, \bar{z} = \frac{-1.0}{36} = -0.0278$$

Sum of squares and cross-products are calculated below without giving any explanation as per expressions given in earlier sections.

$$S_{yy} = 20.2^2 + 18.5^2 + \ldots\ldots\ldots + 20.2^2 + 26.4^2 - \frac{905.4^2}{36}$$

$$= 26534.54 - 22770.81 = 3763.73$$

$$S_{xx} = (-1)^2 - \frac{(-1)^2}{36}$$

$$= 1 - 0.0278 = 0.9722$$

Also, $S_{zz} = 0.9722$

$$S_{yz} = 0 - \frac{(905.4)(-1)}{36} = 25.15$$

Similarly, $$S_{zx} = 0 - \frac{(-1)(-1)}{36} = -0.0278$$

$$T_{yy} = \frac{1}{4}(116.2^2 + 87.1^2 + \ldots\ldots + 125.8^2 + 98.3^2) - \frac{905.4^2}{36}$$

$$= 23909.8550 - 22770.81 = 1139.0450$$

$$T_{xx} = \frac{(-1)^2}{4} - \frac{(-1)^2}{36}$$

$$= 0.25 - 0.0278 = 0.2222$$

Also, $$T_{zz} = 0.2222$$

$$T_{yz} = \frac{(87.3)(-1)}{4} - \frac{(905.4)(-1)}{36}$$

$$= -21.825 + 25.15 = 3.3250$$

$$T_{yx} = \frac{(50.5)(-1)}{4} - \frac{(905.4)(-1)}{36}$$

$$= -12.625 + 25.15 = 12.5250$$

$$T_{xx} = 0 - \frac{(-1)(-1)}{36} = -0.0278$$

$$R_{yy} = \frac{1}{9}(215.0^2 + 318.2^2 + 163.1^2 + 209.1^2) - \frac{905.4^2}{36}$$

$$= 24200.0733 - 22770.81 = 1429.2633$$

$$R_{xx} = \frac{(-1)^2}{9} - \frac{(-1)^2}{36}$$

$$= 0.1111 - 0.0278 = 0.0833$$

Similarly, $$R_{zz} = 0.0833$$

$$R_{yz} = \frac{(163.1)(-1)}{9} - \frac{(905.4)(-1)}{36}$$

$$= -18.1222 + 25.15 = 7.0278$$

$$R_{yx} = \frac{(215.0)(-1)}{9} - \frac{(905.4)(-1)}{36}$$

$$= -23.889 + 25.15 = 1.2611$$

$$R_{zx} = 0 - \frac{(-1)(-1)}{36} = -0.0278$$

$E_{yy} = 3763.73 - 1139.0450 - 1429.2633 = 1195.4217$

$E_{xx} = 0.9722 - 0.2222 - 0.0833 = 0.6667$

$E_{zs} = -0.0278 + 0.0278 + 0.0278 = 0.0278$

$E_{zz} = 0.9722 - 0.2222 - 0.0833 = 0.6667$

$E_{yz} = 25.15 - 3.3250 - 7.0278 = 14.7972$

$E_{yx} = 25.15 - 12.5250 - 1.2611 = 11.3639$

Now the values of b_{yx} and b_{yz} by the formulae (10.6.2) and (10.6.3) are,

$$b_{yx} = \frac{0.6667 \times 11.3639 - 14.7972 \times 0.0278}{0.6667 \times 0.6667 - (0.0278)^2}$$

$$= \frac{7.1649}{0.4437} = 16.1482$$

$$b_{yz} = \frac{0.6667 \times 14.7972 - 0.0278 \times 11.3639}{0.4437}$$

$$= \frac{9.5494}{0.4437}$$

$$= 21.5222$$

$SE' = 1195.4217 - 16.1482 \times 11.3639 - 21.5222 \times 14.7972$

$\qquad = 693.45$

$A_{yy} = 1139.0450 + 1195.4217 = 2334.4667$

$A_{xx} = 0.2222 + 0.6667 = 0.8889$

$A_{zz} = 0.2222 + 0.6667 = 0.8889$

$A_{yz} = 3.3250 + 14.7972 = 18.1222$

$A_{yx} = 12.5250 + 11.3639 = 23.8889$

$A_{zx} = -0.0278 + 0.0278 = 0$

Now by the formulae (10.6.6) and (10.6.7),

$$b'_{yx} = \frac{0.8889 \times 23.8889 - 0 \times 18.1222}{0.8889 \times 0.8889 - 0} = \frac{21.2348}{0.7901}$$

$$= 26.8761$$

$$b'_{yz} = \frac{0.8889 \times 18.1222 - 0 \times 23.8889}{0.8889 \times 0.8889 - 0} = \frac{16.1088}{0.7901}$$

$$= 20.3883$$

By the formula (10.6.8),

$$S'_{T+E} = 2334.4667 - 26.8761 \times 23.8889 - 20.3883 \times 18.1222$$
$$= 1322.94$$

Hence, the adjusted treatment sum of square is,

$$T'_{SS} = 1322.94 - 693.45 = 629.49$$

Since the experiment is conducted to test the null hypothesis of equality of treatment effects, abridged analysis of variance table is as follows:

ANOVA table

Source	d.f	S.S.	M.S.	F-value
Blocks	3	629.49	78.69	2.496*
Treatments	8	693.45	31.52	
Error	22			
Total	33	1322.94		

Tabulated value of $F_{.05; 8, 22} = 2.40$

Adjusted treatment means by the equation,

$$\overline{y_i}' = \overline{y_i} - b'_{yx}(\overline{x_i} - \overline{x}) - b'_{yz}(\overline{z_i} - \overline{z})$$

are,

$$\overline{y_1} = 29.05 - 26.8761 \times \{0 - (-0.0278)\} - 20.3883 \times \{0 - (-0.0278)\}$$
$$= 29.05 - 0.7472 - 0.5668 = 27.7360$$
$$y'_2 = 21.78 - 0.7472 - 0.5668 = 20.4660$$
$$\overline{y_3} = 12.62 - 26.8761 \times (-0.25 + 0.0278) - 0.5668 = 18.0250$$
$$\overline{y_4} = 28.08 - 0.7472 - 0.5668 = 26.7660$$
$$y'_5 = 31.60 - 0.7472 - 0.5668 = 30.2860$$
$$y'_6 = 25.38 - 0.7472 - 0.5668 = 24.0660$$
$$y'_7 = 21.82 - 0.7472 - 20.3883 \times (-0.25 + 0.0278) = 25.6031$$
$$y'_8 = 31.45 - 0.7472 - 0.5668 = 30.1360$$
$$y'_9 = 24.58 - 0.7472 - 0.5668 = 23.2660$$

F-value in ANOVA reveals that the treatments differ significantly. Thus, pairwise contrasts of treatment means are to be tested. Standard errors for three types of pairwise contrasts through (10.7.1) – (10.7.3) and their critical differences on multiplying them by t-value at P = 0.05 and 22 d.f., i.e. $t = 2.074$, are,

$$SE_{d1} = \sqrt{\frac{2 \times 31.52}{4}} = 3.9699$$

$$C.D_1 = 3.9699 \times 2.074 = 8.2336$$

$$SE_{d2} = \sqrt{31.52 \times \left(\frac{2}{4} + \frac{0.6667}{16 \times 0.4437}\right)} = 4.3267$$

$$C.D_2 = 4.3267 \times 2.074 = 8.9736$$

$$SE_{d3} = \sqrt{31.52 \times \left\{\frac{2}{4} + \frac{1}{16 \times 0.4437}(0.6667 + 2 \times 0.0278 + 0.6667)\right\}}$$

$$= 4.6826$$

$$C.D_3 = 4.6826 \times 2.074 = 9.7117$$

To compare any two varieties except V_3 or V_7, use $C.D_1$. While comparing either V_3 or V_7 with any other variety, use $C.D_2$. Whereas, the significance of difference between V_3 and V_7 means is decided by comparing with $C.D_3$.

Significant and non-significant differences between variety means are precisely shown in the following difference table.

Table of absolute mean differences

Adj. mean		20.47 V_2	18.02 V_3	26.77 V_4	30.29 V_5	24.07 V_6	25.60 V_7	30.14 V_8	23.27 V_9
27.74	V_1	7.27	9.72*	0.97	2.55	3.67	2.14	2.40	4.47
20.73	V_2		2.45	6.30	9.82*	3.60	5.13	9.67*	2.80
18.02	V_3			8.75	12.27*	6.05	7.58	12.12*	5.25
26.77	V_4				3.52	2.70	1.17	3.37	3.50
30.29	V_5					6.22	4.69	0.15	7.02
24.07	V_6						1.53	6.07	0.80
25.60	V_7							4.54	2.33
30.14	V_8								6.87

On comparing the absolute mean differences with the respective critical difference, the values marked with asterisk (*) show that the corresponding varieties differ significantly at $P = 0.05$ whereas other varieties do not.

CONCLUDING REMARK

In many experiments, additional quantitative information is available or can be gathered regarding concomitant variables or covariates. These covariates are those which are likely to affect the response variable directly or indirectly and are not affected by the treatments. Further these covariates are beyond the control of the experimenter. For instance, the percentage germination of seeds or in other words plant population, the age of patients in a drug trial, initial weight of animals in trials on effect of feeds, number of insects per unit area in an experiment to test the efficacy of insecticides, etc.

Analysis of covariance is an analytical device to minimize the effect of covariates so as to estimate the actual effect of treatments mean and getting more exact results by incorporating the regression information in ANOVA model or so called through ANCOVA model. This technique is applicable in all designs but its use is mostly seen in CRD, RBD and LSD designs. Estimation of missing values is another important feature of this technique.

This chapter has equipped the readers with concepts of ANCOVA procedures and methods of their application in a variety of situations. They can utilize this knowledge in any design of experiment and in any branch of research.

QUESTIONS AND EXERCISES

1. What do you understand by covariates (auxiliary variable)? Define and discuss.
2. In what respects analysis of covariance improves the procedure of analysis of variance?
3. What is the basic principle on which analysis of covariance is based?
4. Can analysis of covariance be carried out in case of incomplete data? If so, what procedure is there for it?
5. Can analysis of covariance involve any number of covariates? Are there any limitations too?
6. What are the assumptions underlying analysis of covariance?
7. Can the procedure of analysis of covariance be applied to all designs? Justify your answer.
8. Explain the method of analysis of covariance in randomized block design when it is considered appropriate to involve two auxiliary variables.
9. Name the pioneer statisticians who propounded the theory of analysis of covariance and their contribution.
10. Explain the methodology of estimating the missing values through analysis of covariance and thereby testing the hypotheses about treatment means.
11. In an intercropping experiment on sorghum, twenty treatment combinations were taken consisting of four intercrops and five weeding methods. The experimental design was randomized block design with three replications.

Intercrops:	Weeding methods:
C_1 – Sole sorghum	W_1 – Weedy check
C_2 – Sorghum + ground nut	W_2 – Hand weeding
C_3 – Sorghum + black gram	W_3 – Alachlor
C_4 – Sorghum + soybean	W_4 – Pedimethalin
	W_5 – Fluchloralin

The data with regard to yield (gram/m^2) and weed density at 60 days after sowing (DAS) were recorded as given below.

Treatments			Blocks	
		B_1	B_2	B_3
C_1W_1	y	228.0	188.0	262.0
	x	86	52	46
C_1W_2	y	433.0	341.0	397.3
	x	20	28	28
C_1W_3	y	342.0	377.7	357.9
	x	52	62	38
C_1W_4	y	315.5	359.7	381.4
	x	46	20	26
C_1W_5	y	378.7	341.2	315.8
	x	66	92	78
C_2W_1	y	150.0	243.0	204.2
	x	60	24	60
C_2W_2	y	261.0	259.2	257.3
	x	26	24	26
C_2W_3	y	213.0	276.0	233.4
	x	36	36	52
C_2W_4	y	249.0	205.0	234.0
	x	34	30	28
C_2W_5	y	200.0	264.0	232.3
	x	64	96	96
C_3W_1	y	144.0	190.3	229.0
	x	78	50	56
C_3W_2	y	283.0	212.0	251.2
	x	20	28	32
C_3W_3	y	231.2	233.3	241.6
	x	48	60	48
C_3W_4	y	219.8	212.0	217.3
	x	56	20	20

(Contd.)

Treatments		Blocks		
		B_1	B_2	B_3
C_3W_5	y	220.9	222.2	214.2
	x	68	72	64
C_4W_1	y	203.0	194.0	195.2
	x	90	52	62
C_4W_2	y	265.3	234.0	275.0
	x	26	36	34
C_4W_3	y	271.0	227.0	239.3
	x	28	56	52
C_4W_4	y	241.2	216.0	247.0
	x	30	28	30
C_4W_5	y	230.3	258.0	215.0
	x	70	90	76

Analyze the experimental data to compare treatment means with regard to crop yield making use of the auxiliary information of weed density per meter square.

BIBLIOGRAPHY

1. Bartlett MS. Some examples of statistical methods of research in agriculture and biology. *Suppl., Journal of Royal Statistical Society*, 1937: 4, 137–70.
2. Cochran WG. Analysis of covariance. Its nature and uses. *Biometrics*, 1957: 13, 261–81.
3. Coons I. The analysis of covariance as a missing plot technique. *Biometrics*, 1957: 13, 387–405.
4. Das MN. Analysis of covariance in two way classification with disproportionate cell frequencies. *Journal of the Indian Society of Agricultural Statistics*, 1953: 5, 161–178.
5. Delbury DB. The analysis of covariance. *Biometrics*, 1948: 4, 153–70.
6. Nair KR. The application of the technique of analysis of covariance to field experiments with several missing values or mixed up plots. *Sankhyá*, 1940: 4, 581–88.

11

Incomplete Block Designs

11.1 GENERAL DISCUSSION

The foremost and prime requirement in the formation of blocks is that they should be homogeneous in respect of the variable(s) which are likely to influence the treatment effects. This condition is often not met if the set of treatments is large or the units in a block are too many and are of heterogeneous nature. Such a situation generally arises in field, animal, meteorological experiments and many others. In these situations, one has no option but to go for incomplete block designs. As the name implies the blocks are incomplete in the sense that the blocks do not contain full set of treatments in each and every block. Incomplete blocks were dealt under confounding. But confounding is possible only in factorial experiments. Also the partial information is lost about the treatment confounded with the blocks which is often not desirable.

11.2 THEORIZATION

The treatments in incomplete blocks are assigned in such a manner that each pair of treatments occurs together in blocks equal number of times. This enables the estimation and comparison of treatment effects with equal precision. Such a design is termed as *balanced incomplete block design* (BIBD). If the pairs of treatments do not occur together in the blocks of a design equal number of times, they are categorized as *partially balanced incomplete block design* (PBIBD). Following layouts will give a clear cut idea of incomplete block designs.

Balanced incomplete block designs with seven treatments in blocks of 3 units are:

Layout 11.2.1 (*a*) BIBD with seven treatments

			Blocks			
I	*II*	*III*	*IV*	*V*	*VI*	*VII*
1	2	3	4	5	6	7
2	3	4	5	6	7	1
4	5	6	7	1	2	3

It can be verified that every pair of treatments occurs once in one of the seven blocks. In other words, seven blocks constitute a balanced design. Also separate replications cannot be formed because 7 is not divisible by 3, the block size.

There is another way also in which seven treatments in blocks of size 3 can be arranged eliminating two-way heterogeneity. Such a layout was given by Youden (1937, 1940) in the form of incomplete Latin squares. This design is named after him as *Youden square*.

Layout 11.2.1 (*b*) BIBD with 7 treatments controlling
two-way heterogeneity

				Columns			
Rows	*I*	*II*	*III*	*IV*	*V*	*VI*	*VII*
(*i*)	7	1	2	3	4	5	6
(*ii*)	1	2	3	4	5	6	7
(*iii*)	3	4	5	6	7	1	2

In the layout (11.2.1-b), each pair of treatments occurs once in a row and once in a column.

Youden square as a balanced incomplete design with 7 treatments in blocks of size four is given below.

Layout 11.2.2 BIBD with 7 treatments

				Columns			
Rows	*I*	*II*	*III*	*IV*	*V*	*VI*	*VII*
(*i*)	6	7	1	2	3	4	5
(*ii*)	4	5	6	7	1	2	3
(*iii*)	2	3	4	5	6	7	1
(*iv*)	5	6	7	1	2	3	4

In the above design, each pair of treatments occurs twice in each block.
Layout of a BIB design with 9 treatments in blocks of size 3 units can be

given as displayed below. Since nine is a multiple of 3, the blocks can be divided into four replications.

Layout 11.2.3 BIB the design with 4 replications

Blocks	*Rep. I*			*Blocks*	*Rep. II*			*Blocks*	*Rep. III*			*Blocks*	*Rep. IV*		
(1)	1	2	3	(4)	1	4	7	(7)	1	5	9	(10)	1	8	6
(2)	4	5	6	(5)	2	5	8	(8)	7	2	6	(11)	4	2	9
(3)	7	8	9	(6)	3	6	9	(9)	4	8	3	(12)	7	5	3

It is easily verifiable that all pairs of treatments occur once in any of the twelve blocks. Hence, the design is balanced.

Layout displayed below corroborate the definition of partially balance incomplete block design.

An incomplete Latin square with six treatments and blocks of size 4 given by Youden (1940) leads to partially balanced incomplete block design.

Layout 11.2.4 PBIB design

Rows	*Blocks*					
	I	*II*	*III*	*IV*	*V*	*VI*
(*i*)	1	2	3	4	5	6
(*ii*)	2	3	4	5	6	1
(*iii*)	5	6	1	2	3	4
(*iv*)	4	5	6	1	2	3

Layout (11.2.4) holds the property that either a pair of treatment occurs four times in a block or twice. For instance, treatments 1 and 4, 3 and 6, 2 and 5, etc. occur together four times in the same block, whereas treatments 1 and 2, 2 and 3, 1 and 6 etc. occur together twice in the same block. Thus, the design is partially balanced.

For different values of v, the number of treatments and k, the block sizes, plans are listed by Youden (1937) and Cochran and Cox in his book *Experimental Designs* (1957), pp. 522-544.

Construction of BIB or PBIB designs will not at all be discussed in this chapter. Most of the incomplete block designs have already been constructed through various techniques such as euclidean geometry (EG), projective geometry (PG), combinatorics, arrays, mutually orthogonal latin squares (MOLS), graphics, etc. This chapter will confine to the concepts and analysis of incomplete block designs.

Incomplete block designs are usually specified by certain notations and characterized by the relationship among them which they hold and vice-versa.

These notations are called parameters of the design in common parlance. Same set of parameters shall be followed throughout this chapter. Consider a design having v treatments, arranged in b blocks ($b \geq v$) of size k, r replications and each pair of treatments occur together in λ blocks. The symbols (v, b, r, k, λ) are known as the parameters of the design. If λ is same for all pairs of treatments, the design is balanced. Since λ has to be an integer, some more relations hold good, which are,

$$b \times v = b \times k \tag{11.2.1}$$
$$r(k-1) = \lambda(v-1) \tag{11.2.2}$$
$$b \geq v \tag{11.2.3}$$

Further for a block design there exist a matrix N of order ($v \times b$) with its elements n_{ij}, where n_{ij} is the number of times the ith treatment occurs in jth block. If n_{ij} takes only two values 0 and 1, then it is known as *binary design*. Also the matrix N is called *incidence matrix* of the design. In general, if n_{ij} takes n-values, then it is known as *n-ary design*. For details of n-ary designs, the readers are referred to Sharma and Agarwal (1976).

11.3 SIMPLE LATTICE

A balance two-dimensional (double) design with k^2 treatments having one restriction is called a *simple lattice*. Also in this design, the treatments should be assigned to each block in such a manner that λ is same for all pairs of treatments. It has been proved that for a design to be balanced, minimum required number of blocks is $k(k+1)$. Thus, at least ($k+1$) replications are needed for a k^2 simple lattice. This property of having separate replications for BIBD holds only when v is a multiple of k and especially for lattice square design. Most of the BIB designs do not hold this property. In general, in an m-dimensional balanced lattice design, the number of treatments is k^m, where k is a prime number or prime power.

Following layout with 9 treatments in blocks of size 3 arranged in 12 columns and 12 rows divided into 4 replications depicts a lattice square design.

Layout 11.3.1 Lattice square design

Cols.	Rep. I			Cols.	Rep. II			Cols.	Rep. III			Cols.	Rep. IV		
(1)	1	2	3	(4)	1	4	7	(7)	1	6	8	(10)	1	9	5
(2)	4	5	6	(5)	2	5	8	(8)	9	2	4	(11)	6	2	7
(3)	7	8	9	(6)	3	6	9	(9)	5	7	3	(12)	8	4	3

It may be confirmed that every pair of treatments occurs once in a row and once in a column. Hence it is a balanced design.

11.4 QUASI-FACTORIAL DESIGN

Here it seems germane to point out that lattice square design is a sort of partially confounded symmetrical factorial design involving two factors. Some of the treatment combinations occur once in a block while others do not. Likewise there may be cubic lattice as well. Because of the analogue between factorial designs and lattice designs, they are also called *quasi-factorial designs*. The analogy will be shown in example (11.1).

11.5 INTRABLOCK AND INTERBLOCK INFORMATION

From the above discussion, it is apparent that an incomplete block design consists of b blocks each of size k ($k < b$). k treatments contained in a block are randomly allotted to experimental units. Since the sets of k treatments are allotted to blocks randomly, the blocks may be taken as random effects. At the same time, the blocks consisting of k units have their own effect. In this way, the information about treatment effects for their estimation and testing comes from two sources, (i) the information from the differences between pairs of treatments occurring in the same block, i.e. within block information and is known as *intrablock information* and the error associated with estimates from intrablock information is termed as *intrablock error* and is denoted by E_e. (ii) The information from block differences containing ith treatment say T_i. Such information is called *interblock information* and the error associated with it is called *interblock error* and is denoted by E_b, i.e. the mean square error for blocks adjusted for treatments within replications. Yates' (1939, 1940) was first to propound the idea of intrablock and interblock analysis. Usually E_b is larger than E_e.

11.6 ANALYSIS OF A BIB DESIGN

From the above discussion, it is clear that incomplete block designs are non-orthogonal. Hence, the data have to be analyzed by the method already discussed in section (2.7). The analysis is based on the usual statistical model,

$$y_{iju} = \mu + \tau_i + \beta_{j(i)} + \pi_u + e_{iju} \qquad (11.6.1)$$

where, y_{iju} is the observation for ith treatment in jth block of the uth replication. Further, μ, τ_i, $\beta_{j(i)}$ and π_u represent the effects of the mean, ith treatment, jth block containing treatment i and uth replication respectively. e_{iju}'s are intrablock errors assumed to be normally and independently distributed with mean zero and variance σ_e^2. Intrablock analysis will be carried first and then inter block analysis will follow.

Intrablock analysis

There may be two approaches for intrablock analysis. (i) With the help of incidence matrix, (ii) from the formulae directly. Simple lattice is a binary design. So it is simple to prepare the following incidence matrix table.

Suppose,

T_i – ith treatment total for $i = 1, 2, \ldots\ldots\ldots, v$

β_j – Total of jth block for $j = 1, 2, \ldots\ldots\ldots, b$

$n_{i.} = r$ as each treatment is replicated r times.

$n_{.j} = k$, as each block contains k treatments

Q_i – treatment total adjusted for blocks.

Table 11.6.1 Incidence matrix table

Blocks	Treatments					Block size $n_{.j}$	Block totals
	1	*2*	......	*i*	...... *v*		
1	1	1	......	0	 1	k	B_1
2	0	1	......	1	 0	k	B_2
⋮	⋮	⋮	⋮	⋮	⋮	⋮	⋮
⋮	⋮	⋮	⋮	⋮	⋮	⋮	⋮
⋮	⋮	⋮	⋮	⋮	⋮	⋮	⋮
j	0	0	......	1	 1	k	B_j
⋮	⋮	⋮	⋮	⋮	⋮	⋮	⋮
⋮	⋮	⋮	⋮	⋮	⋮	⋮	⋮
⋮	⋮	⋮	⋮	⋮	⋮	⋮	⋮
b	0	1		0	1	k	B_b
$n_{i.}$	r	r		r	r		
Treat. total	T_1	T_2	......	T_i	...... T_v		

Where, $n_{ij} = 1$, if ith treatment occurs in jth block.

　　　 $= 0$, otherwise.

Formula for ith treatment adjusted for blocks,

$$Q_i = T_i - \frac{1}{k}\sum_j n_{ij}B_j \tag{11.6.2}$$

where, $n_{ij} = 1$, if treatment i occurs in jth block.

　　　 $= 0$, otherwise.

Sum of squares due to treatments adjusted for blocks,

$$= \frac{k}{\lambda(v-1)}\sum_i Q_i^2 \tag{11.6.3}$$

This approach has not been pursued for analysis of simple lattice. Rather a simple method of calculating the sum of square due to blocks adjusted for treatment effects is elucidated. For intrablock analysis, one would require the block effects adjusted for treatments within replications. Let W_i denotes a function of the sum of blocks in which treatment i occurs. W_i can be estimated by the formula,

$$W_i = k\,T_i - (k + 1)\,B_{(i)} + G \qquad (11.6.4)$$

In the above formula, G is the grand total and $B_{(i)}$ is the sum of block totals in which the treatment i appears.

Check that, $\displaystyle\sum_i B_{(i)} = kG$ and $\displaystyle\sum_i W_i = 0$.

Sum of square due to blocks within replications adjusted for treatment effects is,

$$= \frac{\displaystyle\sum_i W_i^2}{k^3(k+1)} \qquad (11.6.5)$$

Analysis of variance using intrablock information will be as given in Table 11.5.2. for a simple lattice, $v = k^2$, $b = k\,(k + 1)$, $r = k + 1$, $\lambda = 1$.

Table 11.6.2. Intrablock analysis ANOVA table

Source	d.f.	S.S.	M.S.
Replications	k	$\dfrac{1}{k}\sum_u R_u^2 - C.F. = R_{yy}$	$R_{yy}/k = R_y$
Treats.(Unadj.)	$k^2 - 1$	$\dfrac{1}{k+1}\sum_i T_i^2 - C.F. = T_{yy}$	$T_{yy}/(k^2 - 1) = T_y$
Blocks (Adj.) within Reps.	$k^2 - 1$	$\dfrac{\displaystyle\sum_i W_i^2}{k^3(k+1)} = B_{yy}$	$B_{yy}/(k^2 - 1) = E_b$
Intrablock error	$(k-1)$ $\times (k^2 - 1)$	By difference $= E_{yy}$	$E_{yy}/(k-1)(k^2 - 1) = E_e$
Total	$k^2(k+1) - 1$	$\displaystyle\sum_i\sum_j\sum_u y_{iju}^2 - \dfrac{G^2}{k^2(k+1)}$ $= S_{yy}$	

Where,

R_u – uth replication total for $u = 1, 2, \ldots\ldots , (k + 1)$.

G – total of all observations.

$$E_{yy} = S_{yy} - R_{yy} - T_{yy} - B_{yy}$$

Analysis with recovery of interblock information

The general principle of combining two estimates may be utilized in pooling intrablock and interblock information. In case of lattice square design, the best linear unbiased estimate of the sum of the effects due to treatment i is given by the relation,

$$T_i' = T_i + \mu W_i \tag{11.6.6}$$

where,

$$\mu = \frac{E_b - E_e}{k^2 E_e} \tag{11.6.6.1}$$

$$\left[\text{Check: } \sum_i T_i' = \sum_i T_i \right]$$

If per chance, $E_b < E_e$, becomes negative. But in this situation μ is taken as zero.

Also the effective error variance associated with pooled information is,

$$E_e' = E_e (1 + k\mu) \tag{11.6.7}$$

Adjusted treatment sum of square,

$$\text{S.S. (Treats. adj.)} = \frac{\sum_i T_i'^2}{k+1} - \frac{G^2}{k^2(k+1)} \tag{11.6.8}$$

Mean S.S. due to treatments (Adj.) $= \dfrac{\text{S.S.Treats.}(Adj.)}{k^2 - 1}$ \hfill (11.6.9)

Test statistic,

$$F = \frac{\text{MS Treat } (Adj.)}{E_e'} \tag{11.6.10}$$

F has $\{(k^2 - 1), (k - 1)(k^2 - 1)\}$ d.f. By comparing calculated F value with the critical value of F for pre-decided level of significance, α and $\{(k^2 - 1), (k - 1)(k^2 - 1)\}$ d.f., the decision can be taken in the usual manner. If F is significant, one would definitely be interested to test pairwise treatment means. To perform post-hoc test using lsd, the variance of the difference between two adjusted treatment means is obtained by the formula,

$$\text{Var. } \left(\overline{y}_i - \overline{y}_{i'} \right) = \frac{2E_e'}{r} \tag{11.6.11}$$

Thus, least significant difference,

$$\text{lsd for} \left(\overline{y}_i - \overline{y}_{i'} \right) = \sqrt{\frac{2E_e'}{r}} x t_{0.05;(k-1)(k^2-1)} \tag{11.6.12}$$

Further, one may use any other appropriate post-hoc test for comparison of paired treatment effects using E_e'.

Example 11.1: An experiment was conducted in simple lattice to compare the quality of nine genotypes of maize in terms of hundred seed weight. Data with respect to 100 seed weight in grams have been recorded for genotypes V_1, V_2, ..., V_9 along with the layout. Also an analogy is shown between a partially confounded 3^2 factorial experiment with treatments A and B and a simple lattice to support the theory.

Simple lattice with parameters (9, 12, 4, 3, 1)
Figures show 100 seed weight (gm.)

Blocks	*Replication I A*			*Confounded*	*Block totals*
1	(00)	(01)	(02)	$x_1 = 0 \bmod 3$	69.8
	V_1 23.9	V_2 25.9	V_3 20.0		
2	(10)	(11)	(12)	$x_1 = 1 \bmod 3$	67.2
	V_4 21.4	V_5 21.8	V_6 24.0		
3	(20)	(21)	(22)	$x_1 = 2 \bmod 3$	62.8
	V_7 18.9	V_8 26.9	V_9 17.0		
Total					199.8

Blocks	*Replication II B*			*Confounded*	*Block totals*
4	(00)	(10)	(20)	$x_2 = 0 \bmod 3$	64.1
	V_1 23.0	V_4 21.0	V_7 20.1		
5	(01)	(11)	(21)	$x_2 = 1 \bmod 3$	73.1
	V_2 25.0	V_5 22.3	V_8 25.8		
6	(02)	(12)	(22)	$x_2 = 2 \bmod 3$	64.7
	V_3 21.8	V_6 24.8	V_9 18.1		
Total					201.9

Blocks	*Replication III AB*			*Confounded*	*Block totals*
7	(00)	(12)	(21)	$x_1 + x_2 = 0 \bmod 3$	74.7
	V_1 24.3	V_6 23.3	V_8 27.1		
8	(01)	(10)	(22)	$x_1 + x_2 = 1 \bmod 3$	65.0
	V_2 26.4	V_4 21.0	V_9 17.6		
9	(02)	(11)	(20)	$x_1 + x_2 = 2 \bmod 3$	61.1
	V_3 20.2	V_5 21.2	V_7 19.7		
Total					200.8

(Contd.)

Blocks	Replication IV AB^2			Confounded	Block totals
10	(00)	(11)	(22)	$x_1 + 2x_2 = 0 \bmod 3$	63.1
	V_1 22.8	V_5 22.0	V_9 18.3		
11	(02)	(10)	(21)	$x_1 + 2x_2 = 1 \bmod 3$	69.7
	V_3 20.6	V_4 21.5	V_8 27.6		
12	(01)	(12)	(20)	$x_1 + 2x_2 = 2 \bmod 3$	72.6
	V_2 27.2	V_6 25.1	V_7 20.3		
Total					205.4

[**Hint:** $V_1, V_2, V_3, V_4, V_5, V_6, V_7, V_8, V_9 \equiv$ 00, 01, 02, 10, 11, 12, 20, 21, 22.]

Nota bene:

(*i*) Through the layout equivalence is shown between a partially confounded 3^2 factorial and simple lattice with nine non-factorial treatments.

(*ii*) x_1 shows the levels of factor A and x_2, the levels of factor B

(*iii*) This design is also a nested design as the blocks are nested within replications.

(*iv*) Block and replication totals have been worked out and placed along with the experimental data.

Calculations

Grand total, G = 807.9

Correction factor, C.F. $= \dfrac{807.9^2}{36} = 18131.35$

Total S.S. $= 23.9^2 + 25.9^2 + \ldots\ldots + 25.1^2 + 20.3^2 - $ C.F.

$= 18430.99 - 18131.35 = 299.64$

Rep.S.S. $= \dfrac{1}{9}(199.8^2 + 201.9^2 + \ldots\ldots + 200.8^2 + 205.4^2) - $ C.F.

$= 18132.60 - 18131.35 = 1.25$

Now various quantities required for analysis are calculated by the formulae given in the theory portion.

Table-I

Treats. T_i	Treat. totals	Block totals $B_{(i)}$	W_i	Adj. treat. T_i^2	Adj. treat. mean $\bar{T}_i'$
V_1	94.0	271.7	3.1	94.2356	23.5589
V_2	104.5	280.5	-0.6	104.4544	26.1136

(Contd.)

Treats. T_i	Treat. totals	Block totals $B_{(i)}$	W_i	Adj. treat. T_i^2	Adj. treat. mean $\bar{T}_i'$
V_3	82.6	265.3	−5.5	82.1820	20.5455
V_4	84.9	266.0	−1.4	84.7936	21.1984
V_5	87.3	264.5	11.8	88.1968	22.0492
V_6	97.2	279.2	−17.3	95.8852	23.9713
V_7	79.0	260.6	2.5	79.1900	19.7975
V_8	107.4	280.3	8.9	108.0764	27.0191
V_9	71.0	255.6	−1.5	70.8860	17.7215
Total	807.9 = G	2423.7 = 3G	00	807.9000	

In the above table,

$$T_1 = 23.9 + 23.0 + 24.3 + 22.8 = 94.0$$
$$B_{(1)} = 69.8 + 64.1 + 74.7 + 63.1 = 271.7$$
$$W_1 = 3 \times 94.0 - 4 \times 271.7 + 807.9 = 3.1$$
$$T_1' = 94.0 + 0.076 \times 3.1 = 94.2356$$
$$\bar{T}_1' = \frac{94.2356}{4} = 23.5589$$

All other entries in the above table are calculated in the same manner.

Treat. S.S. (unadj.) $= \dfrac{1}{4}(94.0^2 + 104.5^2 + \ldots\ldots + 107.4^2 + 71.0^2) - \text{C.F.}$

$$= 18418.23 - 18131.35 = 286.88$$

Blocks (adj.) within rep. S.S.

$$= \frac{1}{108}\left\{(3.1)^2 + (-0.6)^2 + \ldots\ldots\ldots + (8.9)^2 + (-1.5)^2\right\} = \frac{568.42}{108} = 5.26$$

Table-II ANOVA for interblock analysis as per table (11.5.2)

Source	d.f.	S.S.	M.S.
Replications	3	1.25	0.42
Treats. (unadj.)	8	286.88	35.86
Blocks (adj.) within reps.	8	5.26	$0.6575 = E_b$
Intrablock error	16	6.25	$0.3906 = E_e$
Total	35	299.64	

Now analyze the data using interblock information since $E_b > E_e$.

From (11.6.6.1),

$$\mu = \frac{0.6575 - 0.3906}{9 \times 0.3906} = 0.076$$

Adjusted treatment total are calculated by the formula (11.6.6) and placed in the former table. In the next column, adjusted treatment means are given which are obtained as $T_i'/4$.

The effective error M.S. by the form by the formula (10.6.7) is,

$$E_e' = 0.3906 \ (1 + 3 \times 0.076)$$
$$= 0.4796$$

Adjusted Treatment sum of square by (11.5.8) is,
Treat. (adj.) S.S.

$$\frac{1}{4}(94.2356^2 + 104.4544^2 + \ldots\ldots\ldots + 108.0764^2 + 70.8860^2) - \text{C.F.}$$

$$= 18420.98 - 18131.34 = 289.64$$

Treat. (adj.) M.S. $= \dfrac{289.64}{8} = 36.20$

Thus, $\qquad F = \dfrac{36.20}{0.4796} = 75.48$

Tabulated value of F at $= 0.05$ and $(8, 16)$ d.f. is 2.59, which is less than the calculated F. Hence, there is significant difference among genotypes with regard to 100 seed weight.

The least significant difference between any two genotypes means by the formula (11.6.12) is,

$$lsd = \sqrt{\frac{2 \times 0.4796}{4}} \times 2.12 = 1.0382$$

Adjusted treatment means are arranged in ascending order and non-significant differences are underlined as displayed in the following diagram.

V_9	V_7	V_3	V_4	V_5	V_1	V_6	V_2	V_8
17.7215	19.7975	20.5455	21.1984	22.0492	23.5589	23.9713	26.1136	27.0191

The conclusions are self-explanatory and hence need no further interpretation.

11.7 ANALYSIS OF VARIANCE OF A BALANCE YOUDEN SQUARE DESIGN

As already stated, a Youden square design is an incomplete Latin square. So the estimates of treatments in a Youden square are not independent of the row-block effects. Let us consider a Youden square with v treatments, b row-blocks of size v units and c column-blocks of size b units. Thus, the statistical model for row-column block design is,

$$y_{ijh} = \mu + \tau_i + \theta_j + \Phi_h + e_{ijh} \qquad (11.7.1)$$

$i = 1, 2, \ldots, v; j = 1,2, \ldots, b; h = 1, 2, \ldots, k$

Total number of treatments $= b\ c = v\ r$.

where, r is the number of times each treatment occurs.

y_{ijh} - response due to ith treatment in jth row-block and hth column-block.

τ_i - ith treatment effect.

θ_j - jth row-block effect

Φ_h - hth column-block effect

e_{ijh} - error terms independently and identically distributed as N $(0, \sigma_e^2)$.

It is also assumed that there is no interaction between row and column blocking factors.

Introduce a notation n_{ijh} defined as $n_{ijh} = 1$, if the ith treatment occurs in jth row-block and hth column-block. Otherwise, $n_{ijh} = 0$. As a matter of fact, n_{ijh} are the elements of the binary incidence matrix. Let B_j be the sum of the values in jth row-block.

Intrablock analysis of variance table for balanced Youden square design is given below.

Table 15.7.1 ANOVA for Youden square

Source	d.f.	S.S.	M.S.	F-value
Row-blocks (unadj)	$b - 1$	SSθ	MSθ	$\dfrac{MS\theta}{MSE}$, $= F_\theta$
Col-blocks (unadj.)	$c - 1$	SSΦ	MSΦ	$\dfrac{MS\theta}{MSE}$, $= F_\varphi$
Treatments (adj.)	$v - 1$	SST$_{Adj}$	MST$_{Adj}$	$\dfrac{MST_{Adj.}}{MSE}$, $= F_T$
Error	$bc - b - c - v + 2$	SSE	SSE	
Total	$bc - 1$	SST		

In the above table, row-block, column-block and total sum of square are calculated in the usual manner as in case of a complete Latin square.

Adjusted Treatment S.S.

$$\text{SST}_{\text{Adj.}} = \sum_{i=1}^{v} \hat{Q}_i \hat{T}_i \qquad (11.7.2)$$

where
$$\hat{T}_i = \frac{c}{\lambda v} Q_i \qquad (11.7.2.1)$$

$$Q_i = \hat{T}_i - \frac{1}{c} \sum_{j=1}^{b} n_{j.i} B_{i.h} \qquad (11.7.3)$$

In the above expression, $\frac{1}{c}\sum_j n_{j.i} B_{i.h}$, $B_{i.h}$ is the sum of row-block totals in which the ith treatment occur.

$$SSE' = SST - SS\theta - SS\phi - SST_{Adj.} \qquad (11.7.4)$$

As a whole, the procedure is almost same as given for analysis of balanced lattice square design. Null hypothesis about no difference between treatment effects is tested in the usual manner by finding the F-value as the ratio of $SST_{Adj.}/SSE'$. F has d.f. { $(v-1)$, $(bc-v-b-c+2)$)}.

Example 11.2: A trial was planned to see the impact of seven price discount and prize schemes on the weekly sales of products marketed at seven stores of a company for consecutive four weeks of a month. The trial was carried out in a Youden square design. Figures in parentheses show the sales in thousands of rupees per week.

Row-blocks	Column-blocks				
	(i)	*(ii)*	*(iii)*	*(iv)*	*Total*
I	6 (35)	4 (28)	2 (51)	5 (41)	155
II	7 (28)	5 (53)	3 (25)	6 (42)	148
III	1 (56)	6 (38)	4 (26)	7 (21)	141
IV	2 (63)	7 (36)	5 (46)	1 (38)	183
V	3 (27)	1 (49)	6 (40)	2 (65)	181
VI	4 (22)	2 (58)	7 (32)	3 (28)	140
VII	5 (48)	3 (34)	1 (42)	4 (32)	156
Total	279	296	262	267	1104

Analysis of data to test the equality of seven schemes is carried as per procedure explained in section (11.7). In this example, $v = 7$, $b = 7$, $r = c = 4$, $\lambda = 2$. To calculate various sum of squares and treatment means, following table is prepared.

Treats. (Schemes)	Treat. total	Row-block total ($B_{i.h}$)	Q_i	$\hat{T}_i$ $2/7\, Q_i$	$Q_i \hat{T}_i$	Treat. mean
1	185	661	19.75	5.643	111.449	46.25
2	237	659	72.25	20.643	1491.457	59.25
3	114	625	−42.25	−12.071	510.000	28.50
4	108	592	−40.00	−11.428	457.120	27.00
5	188	642	27.50	7.857	216.068	47.00
6	155	625	−1.25	−0.357	0.446	38.75
7	117	612	−36.00	−10.286	370.296	29.25

$$\text{C.F.} = \frac{1104^2}{28} = 43599.143$$

$$\text{SST}_{\text{Adj}} = 111.449 + 1491.457 + \ldots\ldots + 0.446 + 370.296 = 3156.836$$

$$\text{Row-block S.S.} = \frac{1}{4}(155^2 + 148^2 + \ldots\ldots + 140^2 + 156^2) - \text{C.F.}$$

$$= 43999.000 - 43599.143 = 399.857$$

$$\text{Col.-block S.S.} = \frac{1}{7}(279^2 + 296^2 + 262^2 + 267^2) - \text{C.F.}$$

$$= 43627.143 - 43599.143 = 28.00$$

$$\text{Total S.S.} = 35^2 + 28^2 + \ldots\ldots + 42^2 + 32^2 - \text{C.F}$$

$$= 47654.000 - 43599.143 = 4054.875$$

$$\text{Error S.S.} = 4054.857 - 399.857 - 28.00 - 3156.877 = 470.123$$

ANOVA table

Source	d.f.	S.S.	M.S.	F-value
Row-block (unadj.)	6	399.857	66.643	1.07 NS
Col.-block (unadj.)	3	28.00	9.333	< 1 NS
Treatments (adj.)	6	3156.836	531.139	13.56*
Error	12	470.123	39.177	
Total	27			

Critical values of F at = 0.05 and d.f. (3,12), (6,12) are,

$$F_{.05;\,(3,\,12)} = 3.49 \text{ and } F_{.05;\,(6,\,12)} = 3.00$$

On comparing the calculated F-values with the corresponding critical values of F it is apparent that there is a significant difference among various schemes.

For pairwise comparison of various schemes, any suitable post-hoc test can be applied as discussed in Chapter 1.

11.8 OTHER BALANCED INCOMPLETE BLOCK DESIGNS

As discussed in section (11.1), if a design is such that all blocks are not having the full set of treatments and each pair of treatment occurs together in the same block an equal number of times in the whole design (λ is constant for all pairs), then it is said to be a balanced design. Further a BIB design with parameters v, b, r, k, λ fulfils the conditions given from (11.2.1) through (11.2.3). In the category of incomplete block designs, lattice square, lattice cubic or rectangular lattice are prominent ones. But they do not exist for all combinations of v, b, k. Also balancing in a number of situations requires such a large number of replications which deters the use of lattice designs. Construction is also a tough task. Therefore, an experimenter is to think of other kinds of BIB designs.

Another difficulty with BIB designs is that many times it is not possible to group the blocks into distinct replications. So BIB designs are constructed consisting of blocks alone. Once we have decided about v and k, immediately the question arises what minimum number of replications for treatments be taken which leads to an integral number of blocks and makes it feasible to construct a BIB design. Consider an experiment with $v = 5$ and $k = 3$. By the relation (11.2.1),

$$b \times k = v \times r$$
$$\therefore \quad b \times 3 = 5 \times r$$
$$b = (5 \times r)/3$$

For $r = 3$, $b = 5$ and from (11.2.2), $3 \times 2 = \lambda (5 - 1) = 4\lambda$ or $\lambda = 3/2$, which is not possible. So one has to look for larger number of replications which result into integral values for b and λ. For $r = 6$, $b = 10$ and $\lambda = 3$, a design with parameters (5, 10, 6, 3, 3) is attainable as delineated below.

BIB design with parameters (5, 10, 6, 3, 3)

Blocks	Treatments			Blocks	Treatments		
(1)	1	2	3	(6)	1	2	4
(2)	1	2	5	(7)	1	3	4
(3)	1	4	5	(8)	1	3	5
(4)	2	3	4	(9)	2	3	5
(5)	3	4	5	(10)	2	4	5

In this way, it is simple to construct a design as there are only $5C_3$ blocks each having 3 treatments from 1 to 5 in order. Of course, the blocks and entries within blocks can be randomized independently.

It has been worked out that the number of effective replication for within treatment information is $r\,E$, where E is the efficiency factor given by the formula,

$$E = \frac{v(k-1)}{k(v-1)} = \frac{1-\frac{1}{k}}{1-\frac{1}{v}} = \frac{v\lambda}{rk} \qquad (11.8.1)$$

For the previous design with $v = 5$, $r = 6$, $k = 3$ and $\lambda = 3$, E = 5/6 = 0.83 is quite high for a BIB design.

. Another design with parameters (7, 7, 3, 3, 1) is illustrated below. In this plan also, blocks are not grouped into replications.

Blocks	Treatments			Blocks	Treatments		
(1)	1	2	4	(5)	5	6	1
(2)	2	3	5	(6)	6	7	2.
(3)	3	4	6	(7)	7	1	3
(4)	4	5	7				

One can easily verify that for this design, the efficiency $E = \dfrac{7 \times 1}{3 \times 3} = 0.78$.

A list of various plans of BIB designs is provided by W.G. Cochran and G.M. Cox in his book 'Experimental Designs' (1957), Asia Publishing House, from pages 471 through 482.

11.9 ANALYSIS OF VARIANCE OF BIB DESIGN WITH NO DISTINCT REPLICATIONS

Analysis of variance for BIB designs in which no distinct replications are formed can easily be performed. As a matter of fact, the procedure is almost the same in this situation as in case of lattice square design.

Statistical model for the BIB designs having no distinct replications is same as (11.6.1) except that the factor π_u does not appear in the model. Secondly, a linear model is assumed in which the block effects are definitely additive. Statistical model is given without any explanation as this contains the same parameters as in model (11.6.1).

$$y_{iju} = \mu + \tau_i + \beta_{j(i)} + e_{iju} \qquad (11.9.1)$$

Obviously, the sum T_i of treatment i over r observations in all blocks provide an estimate of the right hand side parameters in (11.9.1) which can be expressed as,

$$T_i = r\mu + r\tau_i + \sum_i \beta_{j(i)} \qquad (11.9.2)$$

Without implicating into derivation, direct steps for analysis of variance are presented here. For a BIB design with parameters v, b, r, k, λ.

T_i - sum of observations for treatment i in all blocks.

$B_{j(i)}$ - the aggregate of the totals of all blocks in which the treatment i appears.

A quantity Q_i, the effect of treatment i adjusted for blocks is defined as,

$$Q_i = k\,T_i - B_{j(i)} = k\,r\,E\,\tau_i \qquad (11.9.3)$$

where E is given by (11.8.1). In case, there is no confounding with block effects, $E = 1$, Otherwise, $E \leq 1$. An expedient meaning of E is that it provides a measure of relative amount of within block treatment information per replication. Further a BIB design is more efficient than a randomized block design if,

$$\sigma_e^2 < E\sigma^2 \qquad (11.9.4)$$

where σ_e^2 and σ^2 are the error variances of a BIB and randomized block designs respectively having equal number of replications.

Analysis of variance table for intrablock analysis shall be as follows:

Table 11.8.1 ANOVA for intrablock analysis

Source	d.f.	S.S.	M.S.	F-value
Treats. (unadj)	$v - 1$	T_{yy}	$T_{yy}/(v-1) = T_y$	T_y/E_e
Blocks (adj.)	$b - 1$	B_{yy}	$B_{yy}/(b-1) = E_b$	E_b/E_e
Intrablock error	$rv - v - b + 1 = n_e$	S_{ee}	$S_{ee}/n_e = E_e$	
Total	$rv - 1$	S_{yy}		

where,

$$\text{S.S. due to blocks (unadj.)} = \sum_j B_j^2 - \frac{G^2}{rv} \qquad (11.9.5)$$

In the above expression, B_j = Total of jth block.

$$\text{S.S. due to treats. (adj. for blocks)} = \frac{\sum\limits_i Q_i^2}{k^2 r v} \qquad (11.9.6)$$

Suppose T_i' is the estimate of the ith treatment total which contain intrablock and interblock information.

T_i' can be expressed as,

$$T_i' = T_i + \mu W_i \qquad (11.9.7)$$

where

$$\mu = \frac{(b-1)(E_b - E_e)}{v(k-1)(b-1)E_b + (v-k)(b-v)E_e} \qquad (11.9.8)$$

and
$$W_i = (v - k) T_i - (v - 1) B_{j\,(i)} + (k - 1) G \qquad (11.9.9)$$

The effective error sum of square for adjusted treatment totals T_i' is,
$$E_e' = E_e\{1 + (v - k)\,\mu\} \qquad (11.9.10)$$

Sum of square for adjusted treatment effects is,

$$\text{S.S. due to treats. (adj.)} = \frac{\sum_i (T_i')^2}{r} - \frac{G^2}{rv} \qquad (11.9.11)$$

and
$$F = \frac{\text{S.S.Treat (adj.)}}{E_e'} \qquad (11.9.12)$$

Standard error for the difference between two adjusted treatment means is,

$$\text{S.E.} = \sqrt{\frac{2E_e'}{r}} \qquad (11.9.13)$$

If F for adjusted treatment effects is significant, then for multiple comparisons of paired means any method like lsd, Duncan's multiple range test or any other post-hoc test can be applied. To compare any two treatment totals T_i and $T_{i'}$ for $i \neq i'$, one can use F-test where the statistic,

$$F = \frac{(T_i - T_{i'})^2}{2rE_e'} \qquad (11.9.14)$$

and for any two treatment means,

$$F = \frac{(\bar{T}_i - \bar{T}_{i'})^2}{2E_e'/r} \qquad (11.9.15)$$

F has $\{1, (rv - v - b + 1)\}$ d.f. The test can be performed at any predecided level of significance α.

Special case: For a BIB design with $k < 5$ and $b < 10$ simultaneously, usually interblock information may be disregarded. In this situation, the adjusted treatment total,

$$T_i^0 = \frac{Q_i}{kE} + r\,\bar{G} \qquad (11.9.16)$$

And adjusted treatment mean,

$$\bar{T}_i^0 = \frac{Q_i}{rkE} + \bar{G} \qquad (11.9.17)$$

The difference between two treatment totals T_i^0 and $T_{i'}^0$ based on only intrablock information can be tested by F-statistic,

$$F = \frac{(T_i^0 - T_{i'}^0)^2 E}{2rE_e} \qquad (11.9.18)$$

where F has $\{1, (r\,v - v - b + 1)\}$ d.f. E is the efficiency factor.

Note: A numerical example in this case is avoided as the procedure is exactly the same as for lattice designs. The only change required in the procedure is to make use of the formulae for Q_i, and T_i etc. Also there is nothing like variation due to replications.

11.10 PARTIALLY BALANCE INCOMPLETE BLOCK DESIGNS

In the series of balanced incomplete block designs, it is evident that balancing requires an inordinately large number of blocks or distinct replications since λ is to be same and an integral value for all pairs of treatments occurring together in the same block. But such a requirement is often not met in practice for various reasons as enunciated below.

1. Availability of material is often limited and it is not possible to take a large number of replications. So one has to restrict to a limited number of replications which leads to unbalanced designs.

2. In many situations, an investigator may desire more precision on some of the pairs of treatments and less on the others. So a design is required which is not fully balance.

3. At a time more material may be available for some treatments and less for others. So one can choose a design which can utilize the material properly by choosing different number of replication for various treatments. This will obviously lead to an unbalanced design.

4. For an *m*-dimensional lattice design, one of the foremost condition is that the number of treatments should be k^m where k is a prime number or a prime power. Such a restriction is often difficult to meet. Thus, number of treatments is a factor more appertained to the objectives of the experiment rather than the requirement of the design.

5. Balanced incomplete block designs are not available for all number of treatments and block sizes. In such situations, one has to resort to some other kind of incomplete block designs, more likely the partially balanced incomplete block designs.

To get over the above mentioned difficulties, Bose and Nair (1939) evolved a new series of balanced incomplete block designs, in which λ is not same for all pairs of treatments, which they called partially balanced incomplete block (PBIB) designs. They further defined *m*-associate classes. Consider a PBIB design with parameters $v, b, r, k, \lambda_1, \lambda_2, \ldots, \lambda_m$. In PBIB design following relations hold:

(*i*) Each treatment occurs in r blocks.

(*ii*) Various pairs of treatments occur together in $\lambda_1, \lambda_2, \ldots, \lambda_m$ blocks which means that any treatment i and rest of the treatments are divided into m groups which occur together with i in m sets of blocks.

(*iii*) Given any treatment θ, each of the n_i treatments occur together with θ in λ_i blocks such that $\sum_i n_i = v - 1, \sum_i n_i \lambda_i = r(k-1)$ for $i = 1, 2, \ldots,$

m. These n_i treatments are called *ith-associates* of θ.

(*iv*) Given any two treatments say, θ and Φ which are *i*th associates among themselves, the number of treatments which are common between *j*th associate of θ and *k*th-associate of Φ is denoted by p^i_{jk}.

(*v*) Further p^i_{jk} is constant for any two treatments named as θ and Φ provided they are *i*th associates.

(*vi*) $p^i_{jk} = p^i_{kj}$, for example if $i = 2, j = 1, k = 2$, then $p^2_{12} = p^2_{21}$.

(*vii*) $n_i p^i_{jk} = n_j p^j_{ki} = n_k p^k_{ji}$.

(*viii*) $\sum_i p^i_{jk} = \begin{cases} n_j - 1 \text{ for } i = j \\ n_j \quad \text{ for } i \neq j \end{cases}$

Bose and Nair called the parameters $v, b, r, k, \lambda_1, \lambda_2, \ldots, \lambda_m$ and $n_1, n_2,$

$\ldots, n_m$ as parameters of *first kind* and p^i_{jk} as the parameters of *second kind*.

In the sequel, the discussion of PBIB design will be confined to two associate classes only. Hence forth, the parameters $v, b, r, k, \lambda_1, \lambda_2$ and $p^1_{11}, p^1_{12}, p^1_{21}, p^1_{22}, p^2_{11}, p^2_{12}, p^2_{21}, p^2_{22}$ shall only be the part of subject matter. As regards second kind of parameters p^i_{jk}, there are as many symmetrical matrices as the number of associate classes. The matrices are,

$$p^1_{jk} = \begin{pmatrix} p^1_{11} & p^1_{12} \\ p^1_{21} & p^1_{22} \end{pmatrix} \text{ and } p^2_{jk} = \begin{pmatrix} p^2_{11} & p^2_{12} \\ p^2_{21} & p^2_{22} \end{pmatrix}$$

Thus for PBIB designs with two associate classes, the definitions can be specified as follows:

(*i*) Pairs of treatments i and j that occur together in the same block are called *first associates* and those which do not are termed as *second associates*.

(*ii*) Every pair of treatments occurs together eiu..i in λ_1 blocks or λ_2 blocks.

(*iii*) Pairs of treatments which occur together in λ_1 blocks are known as *first associates* and in λ_2 blocks are called *second associates*.

The concept of two associate classes is expatiated through an example.

Example 11.3: Consider a partially balanced incomplete block design with parameters $v = b = 10$, $r = k = 4$, $\lambda_1 = 1$, $\lambda_2 = 2$.

Ten treatments are denoted by the numerals 1 to 10.

Blocks	Treatments			
I	1	8	4	7
II	2	7	8	3
III	3	9	6	1
IV	4	6	9	2
V	5	2	1	10
VI	6	5	4	8
VII	7	10	2	6
VIII	8	1	10	9
IX	9	3	7	5
X	10	4	3	5

For the above design, first and second associates of each treatment 1 through 10 are listed and then it is shown that the properties from (*iii*) to (*viii*) hold true.

It seems germane to point out that all those treatments which occur together in one block only shall be listed as first associates and those which occur together in two blocks are classified as second associates.

Table of association scheme

Treatments	First associates						Second associates		
1	2	3	4	5	6	7	8	9	10
2	1	3	4	5	8	9	6	7	10
3	1	2	4	6	8	10	5	7	9
4	1	2	3	7	9	10	5	6	8
5	1	2	6	7	8	9	3	4	10
6	1	3	5	7	8	10	2	4	9
7	1	4	5	6	9	10	2	3	8
8	2	3	5	6	9	10	1	4	7
9	2	4	5	7	8	10	1	3	6
10	3	4	6	7	8	9	1	2	5

For this design $n_1 = 6$, $n_2 = 3$, $r = 4$, $k = 4$, $\lambda_1 = 1$, $\lambda_2 = 2$.

For the relation (iii), $\sum_i n_i \lambda_i = 6 \times 1 + 3 \times 2 = 12$

$$r(k-1) = 4(4-1) = 12$$

Therefore, relation (iii) holds.

From the above table, it can easily be seen that treatments common to first associates of treat. -1 and treat. -2 are 3, 4 and 5. Therefore, $p_{11}^1 = 3$. Again treatments 6 and 7 are common to the first associates of treat. -1 and second associates of treat. -2. Hence, $p_{12}^1 = 2$. Similarly, $p_{21}^1 = 2$ and $p_{22}^1 = 1$.

Consider again two treatments say, 1 and 8 which are second associates. It is easy to verify that treatments 2, 3, 5 and 6 are common to first associates of treat. -1 and treat. -8. Therefore, $p_{11}^2 = 4$. Also common treatments which are first associate to treat. -1 and second associates to treat. -8 are 4 and 7. Hence, $p_{12}^2 = 2$. Similarly, $p_{21}^2 = 2$ and $p_{22}^2 = 0$, as no treatment is common between 8, 9, 10 and 1, 4, 7. Preceding values of p_{jk}^i for $i, j = 1, 2$ confirm the relations (iv) to (vi).

For relation (vii), $n_1 = 6$, $n_2 = 3$, $p_{12}^1 = 2$ and $p_{11}^2 = 4$, $n_1 p_{21}^1 = 6 \times 2$ and $n_2 p_{11}^2 = 3 \times 4 = 12$. Hence, relation (vii) is verfied.

For relation $(viii)$, $p_{11}^1 + p_{12}^1 = 3 + 2 = 5 = (n_1 - 1)$.

Similarly, $\qquad p_{21}^1 + p_{22}^2 = 2 + 0 = 2 = (n_2 - 1)$.

$$p_{21}^1 + p_{22}^1 = 2 + 1 = 3 = n_2 \text{ for } j = 2.$$

$$p_{11}^2 + p_{12}^2 = 4 + 2 = 6 = n_1 \text{ for } j = 1.$$

This substantiates $(viii)$.

Example 11.4: Consider another famous example of Youden square which is a PBIB design with parameters $v = b = 6$, $r = k = 4$, $\lambda_1 = 4$, $\lambda_2 = 2$. Treatments are Arabic numbers 1, 2, 3, 4, 5, 6.

Blocks	Treatments			
I	1	4	2	5
II	2	5	3	6
III	3	6	1	4
IV	4	1	5	2
V	5	2	6	3
VI	6	3	4	1

Treatments which are first and second associate to each of the treatments 1 to 6 are tabulated below.

Treatments	First associates	Second associates			
1	4	2	3	5	6
2	5	1	3	4	6
3	6	1	2	4	5
4	1	2	3	5	6
5	2	1	3	4	6
6	3	1	2	4	5

In the above association scheme, it is to note that all those treatments which occur with a treatment i ($i = 1, 2, 3, 4, 5, 6$) four times in the same block are listed as first associates and in two blocks only as second associates.

In the present example, $n_1 = 1$, $n_2 = 4$.

It is trivial to verify the properties (*iii*) to (*viii*) hold good in the same way as shown in example (11.3).

11.11 ANALYSIS OF VARIANCE OF A PBIB DESIGN

From the discussion so far, it is amply clear that the analysis of data of an incomplete block design utilizes intrablock and interblock information. Block effects are considered to be random variables and are another source of error. Hence, block totals contain information about treatment effects. Such estimates are called *interblock estimates*. Yates (1940) obtained interblock estimates alone in terms of block totals. But in PBIB designs, often this is not possible since the number of blocks can be less than the number of treatments. But general method of analysis of incomplete block designs given by C.R. Rao (1947) resolved this difficulty. He obtained the combined estimates using intrablock and interblock information. The same is being followed over here.

For intrablock analysis, the model of a PBIB design having no distinct replications is given as,

$$y_{iju} = \mu + \tau_i + \beta_{j(i)} + e_{iju} \qquad (11.11.1)$$

All terms involved in this model are same as in model (11.9.1). On the basis of this model, analysis of data is carried out in the following manner.

Consider a PBIB design with parameters (v, b, r, k, λ_1, λ_2). Also some terms are decoded below.

n_1 – Number of treatments which are first associates to a treatment i.

n_2 – Number of treatments which are second associates to a treatment i.

$k\, Q_i = k\, T_i - B_{j(i)}$, where $B_{j(i)}$ is the sum of those block totals which contain the treatment i.

$S_1(Q_i)$ – The sum of Q_i's for the treatments which are first associates to the treatment i.

$S_2(Q_i)$ – The sum of Q_i's for the treatments which are second associates to the treatment i.

Relation that holds between Q_i, $S_1(Q_i)$ and $S_2(Q_i)$ is,

$$Q_i + S_1(Q_i) + S_2(Q_i) = 0 \qquad (11.11.2)$$

Interblock estimates are obtained by minimizing the expression,

$$\sum_j \left(B_j - k\mu - \sum_i n_{ij}t_i \right)^2 \qquad (11.11.3)$$

with respect to μ and t_i.

Interblock estimates obtained by the reduced normal equation are,

$$rt_i + \sum_{i \neq i'} \lambda_{ii'}t_{i'} = B_{j(i)} - \frac{r}{b}G \qquad (11.11.4)$$

$$= P_i \text{ (say)}$$

where,

$$\lambda_{ii'} = \sum_j n_{ij}n_{i'j}$$

Intrablock estimates are obtained by minimizing the quantity,

$$\sum_j \left(y_{ij} - \mu - t_i - b_j \right)^2$$

The resulting normal equation yielding the estimates is,

$$r(k-1)t_i - \sum_{i \neq i'} \lambda_{ii'}t_{i'} = kQ_i \qquad (11.11.5)$$

$$i, i^2 = 1, 2, \ldots\ldots, v$$

Then the combined estimate of the ith treatment can be obtained by linearly combining the equations (11.11.4) and (11.11.5) with weights w'/k and w.

The weighted equation is,

$$\frac{r}{k}\left\{ w(k-1) + w' \right\}t_i - \frac{w - w'}{k} \sum_{i \neq i'} \lambda_{ii'}t_{i'} = wQ_i + \frac{w'}{k}P_i \qquad (11.11.6)$$

Weights w and w' are estimated as,

$$w = \frac{1}{E_e} \text{ and } \frac{1}{w'} = \frac{k(b-1)E_b' - (v-k)E_e}{v(r-1)} \qquad (11.11.7)$$

E_e and E_b' are intrablock error mean square and adjusted block mean square, respectively.

Roy and Shah (1962) obtained the individual maximum likelihood estimates of σ_e^2 and σ_b^2 under normality assumption and thereby the estimates of σ_e^2 and $(\sigma_e^2 + k\sigma_b^2)$ are,

$$\hat{\sigma}_e^2 = E_e \qquad (11.11.8)$$

and $$Est\left(\sigma_e^2 + k\sigma_b^2\right) = \frac{E_e}{b-1} \sum_{m=1}^{2} \frac{1}{c_m} + \frac{1}{b-1}\left(\frac{1}{k}B_{(i)}^2 - \frac{G^2}{bk}\right) \qquad (11.11.9)$$

where $$B_{(i)} = B_{j(i)} - \sum_{j,u} \theta_{iju} t_i$$

$\theta_{iju} = 1$ if ith treatment occurs in the uth cell of the jth block,
$\quad\quad\quad = 0$ otherwise.

$C_m = (r\,k - \Phi_m)/\Phi_m$ whereas Φ_m's are the positive latent roots of the matrix NN'. C_m is known as *design constant*. To overcome this tedium, Bose et al (1954) have given the tables of PBIB designs with two associate classes. Obviously there shall be two C_m's, i.e. C_1 and C_2.

Intrablock estimate of the ith treatment effect in general is obtained by the relation,

$$r\,k\,(k-1)\,t_i = (k - C_2)\,Q_i + (C_1 - C_2)\,S_1\,(Q_i) \qquad (11.11.10)$$

For the design given in example (11.3), $C_1 = 4/15$ and $C_2 = 8/15$.
Alternative: The author has preferred the analysis without entangling C_i's.

Many times the values of C_1 and C_2 are not available. In that situation, intrablock estimate of the ith treatment is obtained by the equation (when $n_2 > n_1$),

$$t_i = \frac{k}{\Delta^*}\,[\,B_{22}\,Q_i - B_{12}\,S_1(Q_i)] \qquad (11.11.11)$$

where,

$$B_{12} = (\lambda_2 - \lambda_1); \; B_{22} = r(k-1) + \lambda_2 + (\lambda_2 - \lambda_1)\left(p_{11}^1 - p_{11}^2\right)$$

$$A_{12} = r(k-1) + \lambda_2; \; A_{22} = (\lambda_2 - \lambda_1)\,p_{12}^2$$

and $$\Delta^* = A_{12}B_{22} - A_{22}B_{12}$$

Alternatively, if $n_2 < n_1$, t_i is estimated by the following equation.

$$\hat{t}_i = \frac{k}{\Delta^*}\left[B_{21}Q_i - B_{11}S_2(Q_i)\right] \qquad (11.11.12)$$

where,

$$B_{11} = (\lambda_1 - \lambda_2); \; B_{21} = r(k-1) + \lambda_1 + (\lambda_1 - \lambda_2)\left(p_{22}^2 - p_{22}^1\right)$$

$$A_{11} = r(k-1) + \lambda_1; \; A_{21} = (\lambda_1 - \lambda_2)\,p_{12}^1$$

$$\Delta = A_{11}B_{21} - A_{21}B_{11}$$

Note: It is to verify that the value of Δ and Δ^* in both the situations remain the same.

$$\text{Treat. S.S. (adj. for blocks)} = \sum_i t_i Q_i = T' \qquad (11.11.13)$$

Analysis of variance table for intrablock analysis is as given below.

Table 11.11.1 ANOVA with intrablock information

Source	d.f.	S.S.	M.S.
Blocks (unadj.)	$b-1$	$\dfrac{1}{k}\sum_j y_{.j}^2 - \dfrac{G^2}{rv} = B$	$\dfrac{B}{b-1} = E_b$
Treats. (adj. for blocks)	$v-1$	T'	$\dfrac{T'}{v-1} = T''$
Intrablock error	$rv - b - v$ $+ 1 = n_e$	$T_{ss} - B - T' + 1 = E_{ss}$	$\dfrac{E_{ss}}{n_e} = E_e$
Total	$rv-1$	$\sum y_{ij}^2 - \dfrac{G^2}{rv} = T_{ss}$	

The variance of the difference between two intrablock estimates of the treatment i and i' (for $i \neq i'$) obtained by the formula (11.11.11), when i and i' are first associates, is

$$= \frac{2k}{\Delta}(B_{12} + B_{22})E_e \qquad (11.11.15)$$

and when i and i' are second associates, the variance of the difference is,

$$= \frac{2k}{\Delta}B_{22} \times E_e \qquad (11.11.16)$$

If t_i's are estimated by the formula (11.11.12), then the variance of the difference between t_i and $t_{i'}$, when t_i and $t_{i'}$ are first associates, is

$$= \frac{2k}{\Delta}B_{21} \times E_e \qquad (11.11.17)$$

and when t_i and $i_{i'}$ are second associates, the variance of the difference is,

$$= \frac{2k}{\Delta}(B_{11} + B_{21})E_e \qquad (11.11.18)$$

The average variance of any two treatment differences using only intra-block information is,

$$= \frac{2k}{(v-1)\Delta}\left\{(t-1)B_{22} + n_1 B_{12}\right\}E_e \quad (11.11.19)$$

Efficiency factor of the design is,

$$\frac{(v-1)\Delta}{rk\left\{(v-1)B_{22} + n_1 B_{12}\right\}} \quad (11.11.20)$$

In case of estimation through (11.11.12), replace B_{12} by B_{11} and B_{22} by B_{21} and Δ by Δ^* in the last two expressions for variance and efficiency factors.

Now to extract and make use of interblock information, proceed as follows:

Calculate block sum of square eliminating treatment effects by the relation,

$$\text{Block S.S.(adj.)} = \text{Blocks S.S.(unadj)} + \text{Treat.S.S.(adj.)}$$
$$- \text{Treat.S.S.(unadj.)} = B' \quad (11.11.21)$$

Block and treatment sum of square (unadjusted) are calculated in the usual way.

Analysis of variance table is prepared as follows:

Table 11.10.2 Simple ANOVA

Source	d.f.	S.S.	M.S.
Blocks (adj.)	$b-1$	B'	$\dfrac{B'}{b-1} = E'_b$
Treats. (unadj.)	$v-1$	T	$\dfrac{T}{v-1} = T_{MS}$
Error	$rv-b-v+1$	E_{SS}	$\dfrac{E_{SS}}{rv-b-v+1} = E'_e$
Total	$rv-1$		

Find the value of the weights w and w' by the formulae given in (11.11.7). Under interblock analysis E_b in the formula for w^2, E_b^2 has to be put in from the foregoing table.

Now some more terms are to be worked out for further calculations as presented below.

$$R = r\left(w + \frac{w'}{k-1}\right)$$

$$\Lambda_1 = \lambda_1(w - w')$$

$$\Lambda_2 = \lambda_2(w - w')$$

$$A'_{12} = R(k-1) + \Lambda_2$$

$$A'_{22} = (\Lambda_2 - \Lambda_1)p_{12}^2$$

$$B'_{12} = \Lambda_2 - \Lambda_1$$

$$B'_{22} = A'_{12} + B'_{12}(p_{11}^1 - p_{11}^2)$$

$$\Delta = A'_{12}B'_{22} - A'_{22}B'_{12}$$

$$A'_{11} = R(k-1) + \Lambda_1$$

$$A'_{21} = (\Lambda_1 - \Lambda_2)p_{22}^1$$

$$B'_{11} = \Lambda_1 - \Lambda_2$$

$$B'_{21} = A'_{11} + B'_{11}(p_{22}^2 - p_{22}^1)$$

$$\Delta' = A'_{11}B'_{21} - A'_{21}B'_{11}$$

Estimate of the ith treatment effect mean utilizing interblock information,

$$\overline{t_i}' = \overline{t} + \frac{1}{\Delta'}\{B'_{21}P_i - B'_{11}S_2(P_i)\} \qquad (11.11.22)$$

where,

$$P_i = \frac{w}{k}(kQ_i) + \frac{w'}{k}B_{j(i)} - r\,w'\overline{G} \qquad (11.11.22.1)$$

In the above formula for t'_i, $S_2(P_i)$ is the sum of those P_i's which are second associates to treatment i and $\overline{G}$ in (11.11.22.1) is the grand mean.

$$\left[\text{Check: } \sum_i \overline{t_i}' = v\overline{t}\right]$$

Variance of the difference between two treatments utilizing both intra-block and interblock information in cases:

(*i*) When the treatments are first associates to treatment i is,

$$\text{var}(t_i - t_{i'}) = \frac{2k}{\Delta'}B'_{21} \quad \text{for} \quad i \neq i' \qquad (11.11.23)$$

(*ii*) When the treatments are second associates to treatment i is,

$$\text{var}(t_i - t_{i'}) = \frac{2k}{\Delta}B'_{22} \quad \text{for} \quad i \neq i' \qquad (11.11.23.1)$$

Rather than obtaining the variances for first and second associates separately, it is convenient to calculate the average variance of the differences between any two treatment estimates obtained by making use of intrablock and interblock information. The formula for average variance is,

$$\frac{2k}{(v-1)\Delta'}\left\{(v-1)B'_{22} + n_1 B'_{12}\right\} \qquad (11.11.24)$$

Average variance can be used for comparing pairwise treatments without identifying whether they are first associates and second associates.

Experimental data of any partially balanced incomplete block design can be analyzed easily by studying the analysis procedure described in this section. Application of the same is delineated through the following example.

Example 11.5: An experiment was conducted with nine treatments as various lengths of cold storage to compare their effect on the tenderness of beef roasts. The plan of experiment was a partially balanced incomplete block design with blocks of size three units. Each block may represent a source of variation which an investigator desired to control.

The layout with treatment number in parentheses and observations in round off values are given below parentheses. Block totals are also displayed in the last column.

Layout with data

Blocks	Layout and observations			Block totals
I	(3)	(8)	(4)	
	26	27	25	78
II	(2)	(7)	(4)	
	17	27	34	78
III	(1)	(7)	(5)	
	7	32	33	72
IV	(7)	(8)	(9)	
	25	27	30	82
V	(4)	(5)	(6)	
	25	40	37	102
VI	(3)	(9)	(5)	
	24	32	32	88
VII	(1)	(8)	(6)	
	10	21	26	57
VIII	(2)	(9)	(6)	
	28	38	30	96
IX	(1)	(2)	(3)	
	17	23	29	69

Association scheme of the above design is prepared and tabulated below.

Treat. no.	1st associates						2nd associates	
1	2	3	5	6	7	8	4	9
2	1	3	4	6	7	9	5	8
3	1	2	4	5	8	9	6	7
4	2	3	5	6	7	8	1	9
5	1	3	4	6	7	9	2	8
6	1	2	4	5	8	9	3	7
7	1	2	4	5	8	9	3	6
8	1	3	4	6	7	9	2	5
9	2	3	5	6	7	8	1	4

First kind of parameters of the design are:

$V = 9$, $b = 9$, $r = 3$, $k = 3$, $\lambda_1 = 1$, $\lambda_2 = 0$, $n_1 = 6$, $n_2 = 2$.

Second kind of the parameters are,

$$p_{11}^1 = 3,\ p_{12}^1 = 2,\quad p_{21}^1 = 2,\ p_{22}^1 = 0$$

$$p_{11}^2 = 6,\ p_{12}^2 = 0,\quad p_{21}^2 = 0,\ p_{22}^2 = 1$$

In the matrix form,

$$p_{ij}^1 = \begin{pmatrix} 3 & 2 \\ 2 & 0 \end{pmatrix} \text{ and } p_{ij}^2 = \begin{pmatrix} 6 & 0 \\ 1 & 0 \end{pmatrix};\ i,\ j = 1,\ 2$$

For the analysis of data with recovery of intrablock and interblock information, proceed step by step as follows.

Prepare a table of calculations as given below.

Treat. No.	T_i	$B_{j\,(i)}$	$K Q_i = k T_i - B_{j\,(i)}$	Q_i	$S_2 Q_i$	Intrablock estimate $\hat{t}_i$
1	34	198	−96	−32.00	9.33	−14.74
2	68	243	−39	−13.00	20.33	−6.91
3	79	235	2	0.67	14.67	−0.52
4	84	258	−6	−2.00	−20.67	0.26
5	105	262	53	17.67	−10.34	8.43
6	93	255	24	8.00	7.34	3.15
7	84	232	20	6.67	8.67	2.48
8	75	217	8	2.67	4.67	0.93
9	100	266	34	11.33	−34.00	6.92
Total	722	2166	0	0	0	0

Treatment grand mean = $\bar{t}$ = 26.7407

$$\left[\text{Check: } \sum_i B_{j(i)} = 3T_i, \sum_i Q_i = 0, \sum_i S_2(Q_i) = 0 \right]$$

Obviously, $\quad G = 722, \bar{t} = 26.7407, \text{C.F.} = \dfrac{722^2}{27} = 19306.81$

$$\text{Total S.S.} = 26^2 + 27^2 + \dots\dots + 23^2 + 29^2 - \text{C.F.}$$
$$= 20842.00 - 19306.81 = 1535.19$$

$$\text{Treat.S.S. (unadj.)} = \frac{1}{3}(34^2 + 68^2 + \dots\dots + 75^2 + 100^2) - \text{C.F.}$$
$$= 20477.33 - 19306.81 = 1170.52$$

$$\text{Blocks S.S. (unadj.)} = \frac{1}{3}(78^2 + 78^2 + \dots\dots + 96^2 + 69^2) - \text{C.F.}$$
$$= 19816.64 - 19306.81 = 509.63$$

In the present example, n_2 is less than n_1 as $n_1 = 6$ and $n_2 = 2$. Thus, calculate the following quantities and thereby intrablock estimates by the formula (11.11.12).

$$A_{11} = 3 \times 2 + 1 = 7$$
$$A_{21} = (1 - 0) \times 2 = 2$$
$$B_{11} = 1 - 0 = 1$$
$$B_{21} = 3 \times 2 + 1 + (1 - 0)(1 - 0) = 8$$
$$\Delta = 7 \times 8 - 2 \times 1 = 54$$

$$\hat{t}_i = \frac{3}{54} \{(8Q_i - S_2(Q_i)\}$$

$$= \frac{1}{18} \{(8Q_i - S_2(Q_i))\}$$

Putting the values of Q_i and $S_2(Q_i)$ for $i = 1, 2, \dots\dots, 9$, the treatment estimates are obtained as follows: $t_1 = -14.74, t_2 = -6.91, t_3 = -0.52, t_4 = 0.26, t_5 = 8.43, t_6 = 3.15, t_7 = 2.48, t_8 = 0.93, t_9 = 6.92$.

$$\left[\text{Check: } \sum_i t_i = 0 \right]$$

Treat. S.S. (adj. for blocks) = $Q_i t_i$ = 832.22

Blocks S.S. (adj. for treat. effects)

= Block (unadj.) S.S. + Treats. (adj.) S.S. – Treats.(unadj.) S.S.

= 509.86 + 832.22 – 1170.52 = 171.56

ANOVA utilizing intrablock information

Source	d.f.	S.S.	M.S.
Blocks (unadj.)	8	509.86	$63.73 = E_b$
Treats.(adj.)	8	832.22	$104.03 = T''$
Intrablock error	10	193.11	$19.31 = E_e$
Total	26	1535.19	

ANOVA utilizing interblock information

Source	d.f.	S.S.	M.S.
Blocks (adj.)	8	171.56	$21.44 = E_b'$
Treats.(unadj.)	8	1170.52	$146.32 = T_{MS}$
Error	10	193.11	$19.31 = E_e$
Total	26	1535.19	

Estimate of the variance of the difference between intrablock estimates which are first associates from the formula (11.11.17) is,

$$= \frac{2 \times 3}{54} \times 8 \times 19.31 = 17.16$$

and those which are second associates by the formula (11.11.18) is,

$$= \frac{2 \times 3}{54} \times (1 + 8) \times 19.31 = 19.31$$

Efficiency factor of the design using (11.11.20) is,

$$= \frac{(9 - 1) \times 54}{3 \times 3 \times [(9 - 1) \times 9 + 6 \times (-1)]} = 0.727 = 72.7\%$$

where, $\qquad B_{22} = r(k - 1) + \lambda_2 + (\lambda_2 - {}_1)(p_{11}^1 - p_{11}^2)$

$$= 3 \times 2 + 0 + (0 - 1)(3 - 6) = 9$$

and $\qquad B_{12} = \lambda_2 - \lambda_1 = -1$

Obtain the value of the weights w and w' by the formula (11.11.7).

$$w = \frac{1}{E_e} = \frac{1}{19.31} = 0.05$$

$$w' = \frac{r(v - 1)}{k(b - 1)E_b' - (v - k)E_e}$$

$$= \frac{9 \times 2}{3 \times 8 \times 21.44 - (9-3) \times 19.31} = \frac{18}{398.7} = 0.045$$

To utilize interblock information and to find combined estimates of the treatments, following quantities are first worked out.

$$R = 3 \ (0.052 + 0.045/2) = 0.2235$$

$$\Lambda_1 = 1 \times (0.052 - 0.045) = 0.007$$

$$\Lambda_2 = 0 \times (0.052 - 0.045) = 0$$

$$A'_{12} = 0.2235 \ (3 - 1) + 0 = 0.4470$$

$$A'_{22} = (0 - 0.007) \times 0 = 0$$

$$B'_{12} = -0.007$$

$$B'_{22} = 0.447 - 0.007 \ (3 - 6) = 0.468$$

$$\Delta' = 0.447 \times 0.468 - 0 \times (-0.007) = 0.209$$

$$A'_{11} = 0.2235 \times (3 - 1) + 0.007 = 0.454$$

$$A'_{21} = (0.007 - 0) \times 0 = 0$$

$$B'_{11} = 0.007 - 0 = 0.007$$

$$B'_{21} = 0.454 + 0.007 \ (1 - 0) = 0.461$$

$$\Delta' = 0.454 \times 0.461 - 0 \times 0.007 = 0.209$$

Now calculate the values of P_i by the formula (11.11.22.1) and subsequently $S_2(P_i)$. Once P_i's and respective values of $S_2(P_i)$ are obtained, it is trivial to obtain the estimated values of t_i's from the formula (11.11.22). The same have been calculated and tabulated below.

$$P_i = \frac{0.052}{3}(kQ_i) + \frac{0.045}{3}B_{j(i)} - 3 \times 0.045 \times 26.74$$

$$= 0.017 \times k \ Q_i + 0.015 \ B_{j(i)} - 3.6099$$

$$P_1 = 0.017 \times (-96) + 0.015 \times 198 - 3.6099 = -2.2719$$

$$S_2(P_1) = 0.1581 + 0.9581 = 1.1162$$

Estimated values of the treatment means utilizing both intrablock and interblock information are computed by the formula (11.11.22).

$$\overline{t}'_i = 26.7407 + \frac{1}{0.209}\left\{0.461 \times P_i - 0.007 \times S_2(P_i)\right\}$$

$$= 26.7407 + 4.7847\left\{0.461 \times P_i - 0.007 \times S_2(P_i)\right\}$$

$$= 26.7407 + 2.2057 \times P_i - 0.0335 \times S_2(P_i)$$

$$\overline{t}'_1 = 26.7407 + 2.2057 \times (-2.2719) - 0.0335 \times 1.1162$$

$$= 21.6921$$

In the same manner, other values of S_2 (P_i) and t_i's are calculated and placed in the following table.

Treat. no.	P_i	$S_2(P_i)$	t_i'
1	−2.2719	1.1162	21.6921
2	−0.6279	1.0022	25.3222
3	−0.0509	0.8332	26.6005
4	0.1581	−1.3138	27.1334
5	1.2211	−0.8468	29.4652
6	0.6231	0.1592	28.1098
7	0.2101	0.5722	27.1850
8	−0.2189	0.5932	26.2380
9	0.9581	−2.1138	28.9248
Total	0.0009	0.0018	240.671

Check: $\sum_i \bar{t_i}' = v\bar{t} = 9 \times 26.7407 = 240.666$

Totals of the middle two columns are nearly 0 due to rounding of figures.

Variance of the difference between two treatments utilizing both intrablock and interblock information when the treatments are first associates by the formula (11.11.23) is,

$$V_1 = \frac{2 \times 3}{0.209} \times 0.461 = 13.2344$$

and when the treatments are second associates is,

$$V_2 = \frac{2 \times 3}{0.209} \times 0.468 = 13.4354$$

Nota bene: The variance for the difference between any two treatment estimates based on only intrablock information for first associates is 17.16 whereas utilizing intrablock and interblock information is 13.2344.

Also in case of second associates, the variance for the difference between any two treatment estimates based on only intrablock information is 19.31 whereas utilizing intrablock and interblock information it is 13.4354. This substantiates that the use of interblock information has resulted into meaningful reduction of variances.

Average variance for comparing pairwise treatment means without identifying whether they are first associates or second associates by the formula (11.11.24) is,

$$\frac{2 \times 3}{8 \times 0.209} \{8 \times 0.468 + 6 \times (-0.007)\} = 13.2847$$

For pairwise comparison of treatments, least significant difference is found out. If one wants, he can apply any other post-hoc test. Bonferroni's approach can also be a good option as the number of treatments is nine. At present least significant difference,

$$\text{lsd} = \sqrt{\left(\frac{2}{3} \times 13.2847\right)} \times 2.228 = 6.6305$$

For pairwise comparison estimated $\bar{\tau}'s$ are arranged in ascending order as given below.

T_1	T_2	T_8	T_3	T_4	T_7	T_6
21.6921	25.3222	26.2380	26.6005	27.1334	27.1850	28.1098
					T_9	T_5
					28.9248	29.4652

Pairwise comparison using least significant difference method reveals that treatment T_1 is significantly inferior than T_9 and T_5. All remaining pairs of treatments are at par with each other.

Concluding remark

There are a large number of incomplete block designs and the variations in them. Hence all of them cannot be covered in volumes of books what to say of a chapter. Thus, the matter is ended at this stage. The author feels that enough has been given in this chapter and also more than enough has been left out. Still one can be confident that the designs and analysis covered in this chapter will fulfill the requirement of students, teachers and research scientists. For details, one should go through the references given at the end of this chapter.

QUESTIONS AND EXERCISES

1. Write an essay on the importance of incomplete block designs.
2. In what sense, incomplete block designs differ from confounded designs?
3. In what manner, quasi-factorial design is analogous to lattice design?
4. When an incomplete block design is said to be balanced? Also give some examples of balanced incomplete block designs.
5. What are the parameters of a balanced incomplete block design? What relations do they hold?
6. Is it possible to construct balanced lattice square for any number of treatments? Justify your answer.
7. Define and discuss intrablock and interblock information and their role in analysis of variance.

8. Explain the procedure of intrablock analysis of data of a balanced incomplete block designs.
9. How can the analysis of a balanced lattice be carried out with recovery of interblock information?
10. Give the layout of two balanced incomplete block designs for 9 and 7 treatments.
11. Explain the procedure of analyzing data of a balanced incomplete block design having no distinct replications.
12. What are the merits and demerits of partially balanced incomplete block designs?
13. Give a Youden's square design which represents a partially balanced incomplete block design.
14. Clarify the concept of associate classes in partially balanced incomplete block designs through proper examples.
15. What do you understand by parameters of first and second kind in case of incomplete block designs?
16. Explicate the concept of design contrasts.
17. Grain yields of maize per plant (grams) with layout of a 4 × 4 lattice with parameters 16, 20, 5, 4, 1 are given below.

	Rep.I					Rep.-II		
Bls.					Bls.			
(1)	1-21.0	2-19.0	3-18.5	4-27.5	(5) 1-30.0	5-30.5	9-20.5	13-22.6
(2)	5-31.0	6-31.5	7-25.3	8-39.0	(6) 2-19.0	6-23.0	10-24.0	14-19.0
(3)	9-39.0	10-26.0	11-21.0	12-42.0	(7) 3-18.2	7-27.4	11-21.0	15-35.0
(4)	13-40.0	14-19.0	15-31.5	16-25.5	(8) 4-25.2	8-32.5	12-39.5	16-22.0

	Rep.III					Rep.-IV		
Bls.					Bls.			
(9)	1-20.0	6-32.0	11-24.0	16-24.0	(13) 1-20.0	14-18.0	7-28.6	12-32.5
(10)	5-39.0	2-17.0	15-31.0	12-42.0	(14) 13-18.0	2-25.0	11-20.0	8-38.5
(11)	9-40.0	14-21.0	3-26.0	8-32.6	(15) 5-30.5	10-43.0	3-18.0	16-29.0
(12)	13-42.6	10-25.0	7-26.0	4-25.0	(16) 9-40.0	6-32.0	15-29.5	4-32.0

	Rep.-V			
Bls.				
(17)	1-30.6	10-20.6	15-34.4	8-31.5
(18)	9-38.5	2-19.4	7-23.8	16-27.5
(19)	13-38.6	6-17.6	3-20.4	12-33.4
(20)	5-30.6	14-23.0	11-20.8	4-24.0

Analyze the experimental data and draw conclusions.

18. A trial in Youden square design with parameters $v = 7$, $b = 7$, $r = 3$, $k = 3$, $\lambda = 1$, $E = 0.78$ was planned. Observations with regard to 100-grain weight of maize are presented below.

Row-blocks	Col.-blocks		
(1)	7 - 17.8	1 - 23.3	3 - 18.1
(2)	1 - 22.5	2 - 18.3	4 - 18.2
(3)	2 - 15.6	3 - 17.0	5 - 18.8
(4)	3 - 17.1	4 - 17.8	6 - 20.5
(5)	4 - 13.4	5 - 18.0	7 - 21.0
(6)	5 - 19.9	6 - 20.4	1 - 22.3
(7)	6 - 19.4	7 - 20.6	2 - 17.5

Analyze the data of the Youden square experimental design and interpret the results.

19. Analyze the data of a balanced incomplete block design with parameters $v = 5$, $b = 10$, $r = 6$, $k = 3$, $\lambda = 3$ for the variable average number of seeds per siliqua in mustard.

Bls. Treats. with obs.   Bls. Treats. with obs.
(1) 1-7.5 2-6.4 3-7.6   (6) 1-7.6 2-6.9 4-6.8
(2) 1-6.6 2-6.0 5-7.8   (7) 1-6.5 3-8.2 4-7.6
(3) 1-5.4 4-7.8 5-8.7   (8) 1-8.0 3-7.7 5-7.2
(4) 2-6.6 3-7.0 4-7.3   (9) 2-7.2 3-8.4 5-8.3
(5) 3-7.5 4-8.7 5-7.3   (10) 2-8.3 4-7.0 5-8.7

20. The layout of a partially balanced incomplete block design with parameters $v = b = 9$, $r = k = 4$, $n_1 = 4$, $n_2 = 4$, $\lambda_1 = 1$, $\lambda_2 = 2$ along with hypothetical data is given below. Figures in table before hyphen (-) show the treatment number and figures after hyphen show the average number of grain rows per ear.

Incomplete

Blocks	Treatments with data			
(1)	1-13.8	2-17.0	5-15.0	9-13.2
(2)	1-13.0	3-12.8	5-11.0	6-10.8
(3)	1-12.0	6-12.2	7-12.4	8-11.8
(4)	1-14.0	4-14.2	7-14.0	9-13.8
(5)	4-12.8	6-12.2	8-13.0	9-12.0
(6)	3-14.0	5-12.0	7-12.8	8-11.6
(7)	2-16.0	3-13.8	4-13.0	8-10.8
(8)	2-15.0	3-13.6	6-10.2	9-14.8
(9)	2-17.0	4-15.8	5-15.0	7-14.8

Do the complete analysis of variance of data of incomplete block design and draw conclusions.

BIBLIOGRAPHY

1. Bose RC, Nair RK. Partially balanced incomplete block designs. *Sankhyā* , 1939: 4, 337–72.
2. Bose RC, Clatworthy WH, Shrikhande SS. Tables of partially balanced design with associate classes. *North Carolina. Agricultural experiment station, Technical Bulletin*, 1954: 107.
3. Cochran WJ, Cox GM. *Experimental designs*, 2nd ed. John Wiley, New York, 1957.
4. Dean A, Daniel boss. *Designs and Analysis of Experiments*. Springer-Verlag, New York, 1999.
5. John PWM. An application of a balanced incomplete block design. *Technometrics*, 1961: 3, 51–4.
6. Rao CR. General methods of analysis of incomplete block designs. *Journal of the American Statistical Association*, 1947: 42, 541–61.
7. Roy J, Shah KR. Recovery of inter-block information. *Sankhyā* , 1962: (A), 24, 269–80.
8. Shadish WR. *Quasi-experimental designs*. In International Encyclopedia of the Social and Behavioral Sciences (N.J. Smelser and P.B. Baltes, eds.). Elsevier, New York, 2001.
9. Sharma SD, Agarwal BL. Some aspects of construction of balanced n-ary designs. *Sankhyā* , 1976: 199–201.
10. Trochim WMK. Advances in Quasi Experimental Design and Analysis. *Hoboken*, 31, New Jersey, 1986.
11. Yates F. Incomplete Latin squares. *Journal of Agricultural Sciences*, 1936: 26, 301–15.
12. Yates F. The recovery of inter-block information in variety trials arranged in three dimensional lattice. *Annals of eugenics*, 1939: 9, 136–56.
13. Yates F. The recovery of inter-block information in balanced incomplete block dsigns. *Annals of eugenics*, 1940: 10, 317–25.
14. Youden WJ. Use of incomplete block replications in estimating tobacco-mosaic virus. *Contributions from Boyce Thompson Institute*, 1937: 9(1), 41–48, (Reprinted in Jan., 1972 issue of the *Journal of Quality Technology*.).
15. Youden WJ. Experimental designs to increase accuracy of greenhouse studies. *Biometrics*, 1940: 7, 124.
16. Youden WJ. Randomization and experimentation. *Technometrics*, 1972: 14, 13–22.

12

Multivariate Analysis of Variance

12.1 PREAMBLE

Analysis of variance (ANOVA) is most frequently used technique with which the readers are well familiar and is exorbitantly exploited in the preceding chapters. Multivariate analysis of variance abbreviated as MANOVA is an extension of univariate ANOVA to multivariate analysis that makes possible to compare k group means (centroids) based on more than one say, p dependent variables simultaneously. To be clearer, a centroid is a p-dimensional vector of population means of a group. MANOVA technique is applicable for experimental as well as survey data. A researcher should be wary of interpreting the results of MANOVA since they are somewhat complicated as compared to ANOVA.

In this chapter, the author shall focus on three points, (*i*) conceptual understanding of MANOVA, (*ii*) methodology of MANOVA, (*iii*) interpretation of results obtained from MANOVA.

Also a brief description of causal models underlying MANOVA and their relationship with other techniques is succinctly covered in this chapter.

12.2 CONCEPTUAL UNDERSTANDING ABOUT MANOVA

In a number of research studies, investigators do not confine to a single measure on a response variable to evaluate the group or treatment differences. But their interest lies in a number of measures or components (variates) that might be affected by the treatment(s) and/or criteria that are useful to measure group differences. For instance, one may be interested to test the efficacy of a drug on persons of groups to alleviate hypertension and heart disorder by comparing it with a control group who have been administered a placebo. A sociologist may like to study the differences between male and female groups with regard to opinion about inflation in city and rural dwellers.

If two or more variables are measured and simple ANOVA is run for each individual variable, one has to ensure that there are zero correlations between these variables or otherwise correlations are of no interest. But such a situation rarely exists. Therefore, MANOVA has distinct advantage over ANOVA because the tests involved in MANOVA considering that correlation exists between the variables.

Cook and Campbell (1979) emphasized that even if a researcher's main interest lies in assessing the effects on only one construct, it is better to multi-operationalize the construct especially in social sciences in which the measurements are somewhat vague.

Cole et al (1981) accentuated that in social sciences, multi-operational construct and multivariate analysis always provide more reliable results in assessing the effects of treatments as compared to mono-operationalize construct.

12.3 REASONS FOR USING MANOVA

There are many reasons for using MANOVA instead of ANOVA which are delineated below.

1. Researchers are often interested in testing equality of mean differences on several variables simultaneously.
2. MANOVA controls the overall desired α-level of significance than separate univariate analysis of each individual criterion variable.
3. A researcher wants to compare the means simultaneously in consideration of the relationship among criterion variables rather than assessing them in isolation.
4. MANOVA forms the basis for other multivariate techniques such as:
   (a) An investigator may look at the relationship among p-criterion variables for the group comparisons.
   (b) One may like to reduce the number of p-criterion variables to some less number of variables. This is done through factor analysis.
   (c) An analyst may require to select variables to form groups that are almost alike in contributing to groups. This kind of analysis is categorized as discriminant analysis.
   (d) MANOVA enhances the scope of vivid interpretation of results due to simultaneous consideration of criterion variables.

12.4 ASSUMPTIONS UNDERLYING MANOVA

MANOVA is also based on cetain assumptions almost similar to ANOVA. Assumptions for MANOVA are:

1. Experimental units like persons, families, plots, etc. are selected randomly from the target populations.
2. Observations are statistically independent to one another.

3. Every individual criterion variable follows normal distribution and jointly within each group they follow multivariate normal distribution.
4. k-groups have a commom population covariance matrix. This is possible when two assumptions are met:
   (*i*) For each individual dependent variable, the condition of homogeneity be met.
   (*ii*) The correlation between any two dependent variables be same in all k groups.

12.5 HYPOTHESES TESTING IN ANOVA AND MANOVA

Consider k populations and let their means be μ_1, μ_2, ..., μ_k, respectively. Then in case of univariate populations for single variable X in ANOVA, the null hypothesis H_0 under test is that means of normal distribution having a common variance versus alternative hypothesis H_1 that at least two of them differ significantly. Notationally,

H_0: $\mu_1 = \mu_2 = ... = \mu_k$ is tested against H_1: $\mu_j \neq \mu_{j'}$ for some $j \neq j'$.
$$j \text{ and } j' = 1, 2, ..., k$$

In consideration of MANOVA, let us confine to p-dimensional space as comparison of k centroids of k ellipsoids with equal dimensions but may have different centers. Denoting the mean vector of a group by $\underset{\sim}{\mu}_{ij}$ having p criterion variables and also suppose that ith variable has n_i observations for $i = 1$, $2, ..., p$ and $j = 1, 2, ..., k$. The hypothesis under test is,

H_0: $\underset{\sim}{\mu}_1 = \underset{\sim}{\mu}_2 = ... = \underset{\sim}{\mu}_k$ vs. H_1: at least two of them are not equal.

$$\underset{\sim}{\mu}_1 = \begin{pmatrix} \mu_{11} \\ \mu_{21} \\ \vdots \\ \vdots \\ \mu_{p1} \end{pmatrix}, \underset{\sim}{\mu}_2 = \begin{pmatrix} \mu_{12} \\ \mu_{22} \\ \vdots \\ \vdots \\ \mu_{p2} \end{pmatrix} .... \underset{\sim}{\mu}_k = \begin{pmatrix} \mu_{1k} \\ \mu_{2k} \\ \vdots \\ \vdots \\ \mu_{pk} \end{pmatrix}$$

and common covariance matrix for all groups is,

$$= \begin{bmatrix} \sigma_{11} & \sigma_{12} & .... & \sigma_{1p} \\ \sigma_{21} & \sigma_{22} & .... & \sigma_{p2} \\ \vdots & \vdots & & \vdots \\ \vdots & \vdots & & \vdots \\ \sigma_{p1} & \sigma_{p2} & .... & \sigma_{pp} \end{bmatrix}$$

and grand mean vector over all groups is,

$$\mu_2 = \begin{pmatrix} \mu_1 \\ \mu_2 \\ \vdots \\ \vdots \\ \mu_k \end{pmatrix}$$

Nearness of normal curves or overlapping in figures (*a*) and (*b*) depends on how close are the population means of three groups.

For multivariate test of means in MANOVA, consider $k = 3$ and $p = 2$ (two-dimensional space). Diagrammatic presentation will be as shown below.

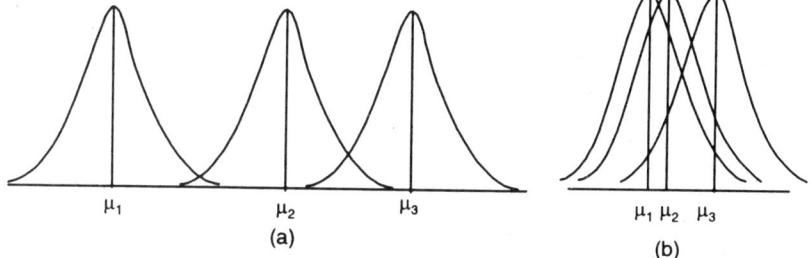

Fig. 12.5.1 Comparison of three means in case of univariate normal distribution with common variance σ^2.

Nearness of normal curves or overlapping in figures (*a*) and (*b*) depends how close the population means of three groups are.

For multivariate test of means in MANOVA, consider $k = 3$ and $p = 2$ (two-dimensional space).

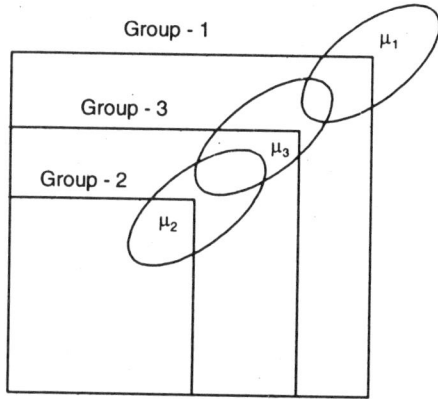

Fig. 12.5.2 Comparison of three centroids of ellipsoids with equal dimensions in two-dimensional spaces.

Graphical depiction of curves in more than two-dimensional spaces becomes too complicated. Hence, the author has confined to three groups and two variables only for clear understanding. In Figure 12.5.2, three centroids are located on parallel lines vis-a-vis three ellipsoids are parallel which affirm that the three groups have a common covariance matrix.

12.6 TESTING OF HYPOTHESES IN MANOVA

Similar to ANOVA, tests of hypothesis in MANOVA are conducted in two stages. As a first step, an omnibus test is performed about mean vectors to test $H_0 : \underset{\sim}{\mu}_1 = \underset{\sim}{\mu}_2 = \ldots\ldots\ldots = \underset{\sim}{\mu}_k$ as given in the last section. If it is rejected, then as a second step some other test is accomplished to compare pairwise group differences just like post-hoc tests in ANOVA. For MANOVA, no exact tests are available. But some other suitable multivariate test is used depending on the number of groups and purpose of study. Further there is hardly any absolute test for comparing group differences when an over all test is found significant. There is lot of controversy over this issue. Discussion of the controversy has been kept out of scope of this chapter. Anyhow, if some wants to go through this topic, he may read the research papers by Leary and Altmaier (1980), Bray and Maxwell (1982), Strahan (1982). To test H_0 in MANOVA, some commonly used tests are discussed over here.

Consider a trial in which k groups, where jth group consists of n_k respondents, are studied for comparison in respect of p variables or items simultaneously. The data of the trial can be tabulated as given in table (12.6.1).

Observation $y_{iuj} \rightarrow i$th observation for uth variable and jth group.

$i = 1, 2, \ldots\ldots , n_j ; u = 1, 2, \ldots\ldots , p; j = 1, 2, \ldots\ldots , k$

Table 12.6.1 Data table

Groups (G)	Variables					
	y_1	y_2	....	y_u	....	y_p
1	y_{111}	y_{121}	....	y_{1u_1}	....	y_{1p_1}
2	y_{211}	y_{221}	....	y_{2u_1}	....	y_{2p_1}
G_1 $\vdots$				$\vdots$		
i	y_{i11}	y_{i21}	....	y_{iu_1}	....	y_{ip_1}
$\vdots$			$\vdots$			
n_1	y_{n_111}	y_{n12}	....	$y_{n_1u_1}$	....	$y_{n_1p_1}$
Mean	$\bar{y}_{.11}$	$\bar{y}_{.21}$	....	$\bar{y}_{.u_1}$	....	$\bar{y}_{.p_1}$

(Contd.)

Groups (G)	y_1	y_2	Variables $\cdots$	y_u	$\cdots$	y_p
1	y_{112}	y_{122}	$\cdots$	y_{1u_2}	$\cdots$	y_{1p_2}
2	y_{212}	y_{222}	$\cdots$	y_{2u_2}	$\cdots$	y_{2p_2}
G_2 $\vdots$			$\vdots$			
i	y_{i12}	y_{i22}	$\cdots$	y_{iu_2}	$\cdots$	y_{ip_2}
$\vdots$				$\vdots$		
n_2	y_{n_212}	y_{n_222}	$\cdots$	$y_{n_2u_2}$	$\cdots$	$y_{n_2p_2}$
Mean	$\bar{y}_{.12}$	$\bar{y}_{.22}$	$\cdots$	$\bar{y}_{.u_2}$	$\cdots$	$\bar{y}_{.p_2}$
$\vdots$				$\vdots$		
1	y_{11k}	y_{12k}	$\cdots$	y_{1u_k}	$\cdots$	y_{1p_k}
2	y_{21k}	y_{22k}	$\cdots$	y_{2u_k}	$\cdots$	y_{2p_k}
G_k $\vdots$			$\vdots$			
i	y_{i1k}	y_{i2k}	$\cdots$	y_{iu_k}	$\cdots$	y_{ip_k}
$\vdots$						
n_k	y_{n_k1k}	y_{n_k2k}	$\cdots$	$y_{n_ku_k}$	$\cdots$	$y_{n_kp_k}$
Mean	$\bar{y}_{.1k}$	$\bar{y}_{.2k}$	$\cdots$	$\bar{y}_{.uk}$	$\cdots$	$\bar{y}_{.p_k}$

From the Table 12.6.1, it is simple to denote sample means as given below. Sample mean vector for jth group is,

$$\bar{y}._{j} = \begin{pmatrix} \bar{y}_{.1j} \\ \bar{y}_{.2j} \\ | \\ \bar{y}_{.uj} \end{pmatrix} \text{ where } \bar{y}_{.uj} = \sum_{i=1}^{n_j} \frac{y_{iu_j}}{n_j} \qquad j = 1, 2, \ldots, k$$

Sample grand mean vector over all k groups is given by,

$$\bar{y}. = \begin{pmatrix} \bar{y}_{.1.} \\ \bar{y}_{.2.} \\ | \\ \bar{y}_{.p.} \end{pmatrix} \text{ where } \bar{y}_{.j.} = \sum_{j=1}^{k}\sum_{i=1}^{n_k} y_{iu_j} \Big/ \sum_{j=1}^{k} n_j$$

or $\quad \bar{y}_{.j.} = \sum_{j=1}^{k} n_j \bar{y}_{.uj} \Big/ n$

where n represents total number of observations.

In the discussion henceforth one suffix u for p variables has been dropped so that the formula and expression can better be understood. It is to be considered himself as if it is there when it is required.

Let group parameters be (μ_j, Σ) for $j = 1, 2, \ldots\ldots, k$.

Statistical model for one way variation is,

$$y_{ij} = \mu + \alpha_j + \varepsilon_{ij} \tag{12.6.1}$$

where, y_{ij}, μ, α_j and ε_{ij} are $(p \times 1)$ vectors when p variables are considered simultaneously.

Thus, testing $H_{0p}: \mu_1 = \mu_2 = \ldots\ldots\ldots = \mu_k$ is equivalent to testing, $H_{0p}: \alpha_1 = \alpha_2 = \ldots\ldots = \alpha_k$ parallel to univariate situation in ANOVA. Sum of squares shall be calculated resulting into matrix.

Within group sum of squares and cross-product matrix is obtained as,

$$W = \sum_{j=1}^{k}\sum_{i=1}^{n_j} (y_{ij} - \bar{y}_{.j})(y_{ij} - \bar{y}_{.j})' \tag{12.6.2}$$

for p variables.

Matrix W can also be obtained as sum of individual matrices of sum of squares and cross-products, i.e.

$$W = S_1 + S_2 + \ldots\ldots + S_k \tag{12.6.2.1}$$

Among group sum of squares matrix is given as,

$$G = \sum_{j=1}^{k}\sum_{i=1}^{n_j} (\bar{y}_{.j} - \bar{y}.)(\bar{y}_{.j} - \bar{y}.)' \tag{12.6.3}$$

$$= \sum_{j=1}^{k} n_j (\bar{y}_{.j} - \bar{y}.)(\bar{y}_{.j} - \bar{y}.)' \tag{12.6.3.1}$$

for p variables. Where $\bar{y}.$ is the grand mean.

Total sum of squares matrix is obtained by the formula,

$$T = \sum_{j=1}^{k}\sum_{i=1}^{n_j} (y_{ij} - \bar{y}.)(y_{ij} - \bar{y}.)' \tag{12.6.4}$$

Also $\qquad\qquad T = W + G \tag{12.6.5}$

To test H_{0p}, *Wilk's lambda* likelihood ratio test can be applied. The test statistic is,

$$\Lambda = \frac{|W|}{|W+G|} \qquad (12.6.6)$$

Λ has p parameters whereas number of variables $(k-1)$ and $(n-k)$ refer to the dimension of y, where $n = \sum_{j=1}^{k} n_j$. Also degrees of freedom for G and W are $(k-1)$ and $(n-k)$, respectively.

For large samples, the test statistic under H_{0p} is,

$$F = \frac{1-y}{y}\frac{m_2}{m_1} \qquad (12.6.7)$$

F is distributed with m_1 and m_2 d.f., whereas $y = \Lambda^{1/s}$

for
$$s = \sqrt{\frac{p^2(k-1)^2 - 4}{p^2 + (k-1)^2 - 5}} \qquad (12.6.8)$$

Also, $m_1 = p(k-1)$, and $m_2 = s\left[n - \frac{p+k+2}{2}\right] - \frac{p(k-1)}{2} + 1$

Alternative test:

Λ can approximately be distributed as Bartlett's χ^2. For large sample under H_{0p}, the test statistic is,

$$\chi^2 = -\left[(n-1) - \frac{p+k}{2}\right]\log_e \Lambda \qquad (12.6.9)$$

χ^2 has $p(k-1)$ d.f.

Note: When k is small, statistic F is preferred over χ^2.

12.7 ANALYSIS IN CASE OF TWO GROUPS

When only two groups are involved, administering a battery of questions to two groups of respondents and creating response profiles can be compared across groups. Then the parallelism, i.e. profiles are similar to each other can be tested by Hotteling-T^2. For more than two groups, analysis is carried through multivariate analysis of variance.

For $k = 2$, MANOVA procedure is extremely simplified. Suppose for group 1, $y \sim N(\mu_1, \Sigma)$ and for group 2, $z \sim N(\mu_2, \Sigma)$, where p is the number of queries in the battery. μ_1 and μ_2 are the vectors of means for groups 1 and

2 respectively and Σ, the covariance matrix for both the groups is same. Also the difference of estimators of two group means, i.e. $(\bar{y}_{\cdot 1} - \bar{y}_{\cdot 2})$ is distributed normally as $N_p \left\{ (\mu_1 - \mu_2), \left(\dfrac{1}{n_1} + \dfrac{1}{n_2} \right) \Sigma \right\}$, where n_1 and n_2 are the sample sizes of two groups.

The hypothesis under test is,

$$H_{0p} : \mu_1 = \mu_2 \quad \text{vs.} \quad H_{1p} : \mu_1 \neq \mu_2$$

Hypothesis H_{0p} can be tested by Hotelling's T^2.

Formula for statistic T^2 is,

$$T^2 = (\bar{y}_{\cdot 1} - \bar{y}_{\cdot 2})' \bar{S}^{-1} (\bar{y}_{\cdot 1} - \bar{y}_{\cdot 2}) \left(\frac{n_1 n_2}{n_1 + n_2} \right) \qquad (12.7.1)$$

Where $\bar{S}$ is the estimator of the covariance matrix obtained as,

$$\bar{S} = \frac{(n_1 - 1)S_1 + (n_2 - 1)S_2}{n_1 + n_2 - 2} \qquad (12.7.2)$$

If $n_1 = n_2$, $\qquad \bar{S} = \frac{1}{2}(S_1 + S_2) \qquad (12.7.2.1)$

In the formula (12.7.2), S_1 and S_2 are covariance matrices of groups 1 and 2, respectively.

Under H_{0p} the statistic, $\qquad \dfrac{(n_1 + n_2 - p - 1)T^2}{(n_1 + n_2 - 2)p} \qquad (12.7.3)$

is distributed as F with p and $(n_1 + n_2 - p - 1)$ d.f.

Statistic T^2 is also related to Mahalanobis distance D^2.

True $D^2 = (\mu_1 - \mu_2)' \Sigma^{-1} (\mu_1 - \mu_2)$ is estimated by $(\bar{y}_{\cdot 1} - \bar{y}_{\cdot 2})' S^{-1} (\bar{y}_{\cdot 1} - \bar{y}_{\cdot 2})$.

The relation between Mahalanobis D_2^2 for two groups and Hotelling's T^2 is,

$$D_2^2 = \left(\frac{1}{n_1} + \frac{1}{n_2} \right) T^2 \qquad (12.7.4)$$

The distribution of D_2^2 is same as given in (12.7.3).

12.8 CONFIDENCE INTERVAL

Confidence interval of the individual ith component of $(\mu_1 - \mu_2)$ at $(1 - \alpha)$ % level of confidence can be obtained by the expression,

$$\left(\bar{y}_{.u1} - \bar{y}_{.u2}\right) \pm c\sqrt{\bar{s}_{uu}^2\left(\frac{1}{n_1} + \frac{1}{n_2}\right)} \qquad (12.8.1)$$

$$u = 1, 2, \ldots\ldots, p$$

where,
$$c = \frac{(n_1 + n_2 - 2)p}{(n_1 + n_2 - p - 1)} F_{\alpha;\{p,(n_1+n_2-p-1)\}}$$

and $\bar{s}_{uu}^2$ is the (u, u)th diagonal element of $\bar{S}$. Substituting the values for the notations appearing in (12.8.1), one can easily find the limits of the confidence interval.

Bonferroni's approach

As given in section (1.19) under Bonferroni's test, the expression (1.19.2) can be extended to multivariate analysis. For two groups, confidence interval for jth individual component using Bonferroni's approximation is,

$$(\bar{y}_{.u1} - \bar{y}_{.u2}) \pm \sqrt{\bar{s}_{uu}^2\left(\frac{1}{n_1} + \frac{1}{n_2}\right)} \times t_{\frac{\alpha}{2p};(n_1+n_2-p-1)} \qquad (12.8.2)$$

$$u = 1, 2, \ldots\ldots, p$$

All notations in (12.8.2) are same except that instead of c, critical value of t is used that too at $\alpha/2p$ level of significance and $(n_1 + n_2 - p - 1)$ d.f.

12.9 POST-HOC TEST

Once the null hypothesis H_{0p} is rejected, there is always a need to test which of the pairwise groups differ significantly from one another. Comparison of any two vector means can be carried through *Hotelling's* T^2. For comparing r and s groups vector means, i.e. to test $H_0: \mu_r = \mu_s$ based on sample sizes n_1 and n_2 for all $r, s, = 1, 2, \ldots\ldots, k$ and $r \neq s$, Hotelling T^2 statistic can be used. The test statistic is,

$$T^2 = (\bar{y}_{.r} - \bar{y}_{.s})'\bar{S}^{-1}(\bar{y}_{.r} - \bar{y}_{.s})\left(\frac{n_r n_s}{n_r + n_s}\right)\left\{\frac{(n_r + n_s - p - 1)}{(n_r + n_s - 2)p}\right\} \qquad (12.9.1)$$

where,
$$\bar{S} = \frac{(n_r - 1)S_r + (n_s - 1)S_s}{(n_r + n_s - 2)} \qquad (12.9.1.1)$$

If $n_r = n_s$,

$$\bar{S} = \frac{1}{2}(S_r + S_s) \qquad (12.9.1.1)$$

Statistic T^2 is compared with $F_{\alpha; \left[p, (n_r + n_s - p - 1) \right]}$ since under H_0 statistic

$$\left\{ \frac{(n_r + n_s - p - 1)}{(n_r + n_s - 2)p} \right\} T^2 \text{ follows F-distribution with } p \text{ and } (n_1 + n_2 - p - 1) \text{ d.f.}$$

For a predecided level of significance α and using Bonferroni's approximate critical level of significance $= 2\alpha/k \ (k - 1)p$.

Again once a pair of two group means is declared significant, a researcher shall further be curious to test that which of the component means of the two groups differ significantly. This is equivalent to conduct univariate multiple comparison under ANOVA parallel to post-hoc tests. Using simple t-statistic for uth variable in rth and sth groups utilizing Bonferroni's approximation, confidence interval can be obtained by the formula,

$$(\overline{y}_{\cdot ur} - \overline{y}_{\cdot us}) \pm \sqrt{\overline{s}_{uu}^2 \left(\frac{1}{n_r} + \frac{1}{n_s} \right)} \times t_{\alpha' : (n-k)} \qquad (12.9.10)$$

where, $\alpha' = 2\alpha/k(k - 1)$, and $\overline{s}_{uu}^2$ is the (u, u)th diagonal element of the matrix $\overline{S}$, obtained by the relation,

$$\overline{S} = \frac{(n_1 - 1)S_1 + (n_2 - 1)S_2 + \ldots\ldots + (n_k - 1)S_k}{n - k} \qquad (12.9.10.1)$$

Test under Bonferroni's approximation is quite satisfactory.

12.10 REPEATED MEASURES

The topic *repeated measures* has already been touched in section 1.2 along with replications. Here it is discussed again to distinguish multivariate observations from repeated observations. In multivariate data, various characters are measured on the same subject for the response variable. On the other hand, in repeated measures same character is measured more than once on the same unit at regular intervals or through various ways (instruments or persons). In both the situations, these measurements shall be correlated. In case of repeated measures, MANOVA is not applicable, instead repeated measure ANOVA model is used. For this topic, readers are referred to Everitt (1995), Crowder and Hand (1990).

Example 12.1: A survey was conducted on four species of surface grass hopper in a taxonomical study. Samples of fifteen male insects were collected and their length up to genetalia, body length up to wing tip and width were measured. Four sample groups consisted of species, S_1-*chrotogonus traehyptenus*, S_2-*Hieroghyphus nigrorephetus*, S_3-*Oedaleus abruptus*, S_4-*Oedaleus senegalensis*. Specieswise measurements were as presented below. In the following example, only y_{uj} has been used in place of y_{iuj} as the number of insects in all groups is same, i.e. $n_j = 15$.

Spe-length Ge	*Length Wi*	*Width*	*Spe-length Ge*	*Length Wi*	*Width*
y_{11}	y_{12}	y_{13}	y_{12}	y_{22}	y_{32}
Cies mm	mm	mm	*cies* mm	mm	mm
S_1 12.51	13.20	4.42	S_2 37.8	38.0	7.62
S_1 13.40	14.30	4.13	S_2 38.7	39.1	7.44
S_1 15.60	16.50	3.82	S_2 41.4	41.6	7.60
S_1 12.40	13.50	4.11	S_2 36.5	36.8	7.47
S_1 12.42	12.20	4.10	S_2 30.3	31.4	7.40
S_1 11.30	13.10	3.82	S_2 32.8	31.5	7.38
S_1 12.60	13.50	4.24	S_2 33.2	33.7	7.55
S_1 12.52	14.20	4.32	S_2 36.6	36.0	6.89
S_1 13.20	13.90	4.37	S_2 34.4	34.6	7.50
S_1 13.51	12.90	4.54	S_2 38.5	37.7	7.40
S_1 12.70	13.40	4.20	S_2 38.2	38.5	7.35
S_1 12.40	13.90	4.28	S_2 42.6	42.9	7.44
S_1 13.40	13.40	3.84	S_2 35.1	35.4	7.47
S_1 14.20	13.90	3.97	S_2 31.0	31.3	7.62
S_1 13.80	14.40	2.89	S_2 30.4	30.6	7.60
Total 195.96	206.30	61.05	Total 537.5	539.10	111.73
Mean 13.064	13.753	4.070	Mean 35.833	35.940	7.449

Spe-length Ge	*Length Wi*	*Width*	*Spe-length Ge*	*Length Wi*	*Width*
y_{13}	y_{23}	y_{33}	y_{14}	y_{24}	y_{34}
Cies mm	mm	mm	*cies* mm	mm	mm
S_3 19.1	18.9	6.0	S_4 23.0	23.3	7.02
S_3 17.1	16.7	5.0	S_4 23.0	23.4	7.02
S_3 19.1	18.8	6.0	S_4 24.7	24.9	7.40
S_3 19.1	19.0	6.0	S_4 25.0	25.2	7.30
S_3 19.1	18.8	6.0	S_4 23.0	23.3	8.00
S_3 19.1	18.9	6.0	S_4 24.0	24.4	6.00
S_3 19.1	18.7	6.0	S_4 24.7	25.0	7.30
S_3 18.5	18.2	6.1	S_4 24.4	24.8	7.50
S_3 18.5	18.4	6.3	S_4 26.1	26.4	6.90
S_3 18.2	18.1	6.1	S_4 29.0	29.2	8.10
S_3 18.1	17.9	6.2	S_4 25.0	25.2	7.30
S_3 20.0	19.7	6.0	S_4 24.3	24.5	7.20
S_3 19.0	18.8	5.2	S_4 25.0	25.3	7.20
S_3 19.4	19.1	6.0	S_4 26.0	26.2	7.08
S_3 16.3	16.0	6.1	S_4 24.0	24.1	7.00
Total 279.70	276.00	89.00	Total 371.20	375.20	108.32
Mean 18.647	18.400	5.933	Mean 24.746	25.014	7.221

In the above table y_{uj} is denoted as uth variable in jth group-suffix, i for observation number is kept silent. Here $i = 1, 2, \ldots\ldots, 15$ as $n_1 = n_2 = n_3 = n_4 = 15$. Though it appears in the calculations.

Calculation follow as per theory.

Grand total: $y_{.1.} = 1384.36$, $y_{.2.} = 1396.60$, $y_{.3.} = 370.10$

Grand mean $\bar{y}_{.1.} = 23.070$, $\bar{y}_{.2.} = 23.277$, $\bar{y}_{.3.} = 6.168$

Groupwise sum of squares for each variable corrected for mean,

$$\sum_i y_{11}^2 = 2574.07 - \frac{195.6^2}{15} = 2574.07 - 2560.02 = 14.05$$

$$\sum_i y_{21}^2 = 2850.09 - \frac{206.3^2}{15} = 2850.09 - 2837.31 = 12.78$$

$$\sum_i y_{31}^2 = 250.65 - \frac{61.05^2}{15} = 250.65 - 248.47 = 2.18$$

Similarly,

$$\sum_i y_{12}^2 = 19465.25 - \frac{537.5^2}{15} = 204.83$$

$$\sum_i y_{22}^2 = 19577.83 - \frac{539.1^2}{15} = 202.58$$

$$\sum_i y_{32}^2 = 832.60 - \frac{111.73^2}{15} = 0.36$$

$$\sum_i y_{13}^2 = 5227.67 - \frac{279.7^2}{15} = 12.20$$

$$\sum_i y_{23}^2 = 5091.04 - \frac{276.0^2}{15} = 12.64$$

$$\sum_i y_{33}^2 = 529.80 - \frac{89.0^2}{15} = 1.73$$

$$\sum_i y_{14}^2 = 9218.24 - \frac{371.2^2}{15} = 32.28$$

$$\sum_i y_{24}^2 = 9416.02 - \frac{375.2^2}{15} = 31.02$$

$$\sum_i y_{34}^2 = 785.47 - \frac{108.32^2}{15} = 1.19$$

$$\sum_i y_{11}y_{21} = 2704.95 - \frac{195.96 \times 206.30}{15} = 9.85$$

$$\sum_i y_{11}y_{31} = 795.93 - \frac{195.96 \times 61.05}{15} = -1.63$$

$$\sum_i y_{21}y_{31} = 837.90 - \frac{206.3 \times 61.05}{15} = -1.74$$

$$\sum_i y_{12}y_{22} = 19519.08 - \frac{537.5 \times 539.1}{15} = 201.33$$

$$\sum_i y_{12}y_{32} = 4002.45 - \frac{537.5 \times 111.73}{15} = -1.20$$

$$\sum_i y_{22}y_{32} = 4014.96 - \frac{539.1 \times 111.73}{15} = -0.62$$

$$\sum_i y_{13}y_{23} = 5158.83 - \frac{279.7 \times 276.0}{15} = 12.35$$

$$\sum_i y_{13}y_{33} = 1660.37 - \frac{279.7 \times 89.0}{15} = 0.82$$

$$\sum_i y_{23}y_{33} = 1638.59 - \frac{276.0 \times 89.0}{15} = 0.99$$

$$\sum_i y_{14}y_{24} = 9316.54 - \frac{371.2 \times 375.2}{15} = 31.59$$

$$\sum_i y_{14}y_{34} = 2684.04 - \frac{371.2 \times 108.32}{15} = 3.46$$

$$\sum_i y_{24}y_{34} = 2712.75 - \frac{375.2 \times 108.32}{15} = 3.31$$

Between groups sum of squares and cross-products,

$$\sum_j y_{\cdot 1}^2 = \frac{1}{15}(195.96^2 + 537.5^2 + 279.7^2 + 371.2^2) - \frac{1384.36^2}{60}$$

$$= 36221.87 - 31940.88 = 4280.99$$

$$\sum_j y_{.2}^2 = \frac{1}{15}(206.3^2 + 539.1^2 + 276.0^2 + 375.2^2) - \frac{1396.6^2}{60}$$

$$= 36675.97 - 32508.19 = 4167.78$$

$$\sum_j y_{.3}^2 = \frac{1}{15}(61.05^2 + 111.73^2 + 89.0^2 + 108.32^2) - \frac{370.1^2}{60}$$

$$= 2390.99 - 2282.90 = 108.09$$

$$\sum_j y_{.1}y_{.2} = \frac{1}{15}(195.96 \times 206.3 + 537.5 \times 539.1$$

$$+ 279.7 \times 276.0 + 371.2 \times 375.2) - \frac{1384.36 \times 1396.6}{60}$$

$$= 36444.28 - 32223.29 = 4220.99$$

$$\sum_j y_{.1}y_{.3} = \frac{1}{15}(195.96 \times 61.05 + 537.5 \times 111.73$$

$$+ 279.7 \times 89.0 + 371.2 \times 108.32) - \frac{1384.36 \times 370.1}{60}$$

$$= 9141.33 - 8539.19 = 602.14$$

$$\sum_j y_{.2}y_{.3} = \frac{1}{15}(206.3 \times 61.05 + 539.1 \times 111.73$$

$$+ 276.0 \times 89.0 + 375.2 \times 108.32) - \frac{1396.6 \times 370.1}{60}$$

$$= 9202.26 - 8614.69 = 587.57$$

In the above calculations, $y.u$ is the sum over 15 observation for uth variable, $u = 1, 2, 3$. $\sum_j$ is for four groups

Between groups sum of squares and cross-products matrix,

$$G = \begin{array}{c} \\ y_1 \\ y_2 \\ y_3 \end{array} \begin{array}{ccc} y_1 & y_2 & y_3 \\ \left[\begin{array}{ccc} 4280.99 & 4220.99 & 602.14 \\ 4220.99 & 4167.78 & 587.57 \\ 602.14 & 587.57 & 108.09 \end{array}\right] \end{array}$$

Within groups S.S. and C.P.,

$$\sum_i y_{1.}^2 = 14.05 + 204.83 + 12.20 + 32.28 = 263.36$$

$$\sum_i y_{2.}^2 = 12.78 + 202.58 + 12.64 + 31.02 = 259.02$$

$$\sum_i y_{3.}^2 = 2.18 + 0.36 + 1.73 + 1.19 = 5.46$$

$$\sum_i y_{1.}y_{2.} = 9.85 + 201.33 + 12.35 + 31.59 = 255.12$$

$$\sum_i y_{1.}y_{3.} = -1.63 - 1.20 + 0.82 + 3.46 = 1.45$$

$$\sum_i y_{2.}y_{3.} = -1.74 - 0.62 + 0.99 + 3.31 = 1.94$$

Within groups S.S. and C.P. matrix,

$$W = \begin{array}{c} y_1 \\ y_2 \\ y_3 \end{array} \begin{matrix} y_1 & y_2 & y_3 \\ \begin{bmatrix} 263.36 & 255.12 & 1.45 \\ 255.12 & 259.02 & 1.94 \\ 1.45 & 1.94 & 5.46 \end{bmatrix} \end{matrix}$$

Total groups S.S. and C.P. matrix,

$$T = W + G = \begin{array}{c} y_1 \\ y_2 \\ y_3 \end{array} \begin{matrix} y_1 & y_2 & y_3 \\ \begin{bmatrix} 4544.35 & 4476.11 & 603.59 \\ 4476.11 & 4426.80 & 589.51 \\ 603.59 & 589.51 & 113.55 \end{bmatrix} \end{matrix}$$

Correlation matrix based on total groups S.S. and C.P.,

$$R = \begin{array}{c} y_1 \\ y_2 \\ y_3 \end{array} \begin{matrix} y_1 & y_2 & y_3 \\ \begin{bmatrix} 1.000 & 0.998 & 0.840 \\ 0.998 & 1.000 & 0.831 \\ 0.840 & 0.831 & 1.000 \end{bmatrix} \end{matrix}$$

Correlation matrix clearly reveals that there is high degree of correlation between the variables y_1, y_2 and y_3.

Determinants of matrices W and T,

$$|W| = 263.36(259.02 \times 5.46 - 1.94 \times 1.94) - 255.12 \times$$
$$(255.12 \times 5.46 - 1.94 \times 1.45)$$
$$+ 1.45(255.12 \times 1.94 - 259.02 \times 1.45)$$
$$= 371465.4876 - 354653.0781 + 173.0630 = 16985.4725$$

$|T| = 4544.35 \, (4426.80 \times 113.55 - 589.51 \times 589.51) - 4476.11 \times$
$\quad (4476.11 \times 113.55 - 589.51 \times 603.59) + 603.59 \times$
$\quad (4476.11 \times 589.51 - 4426.80 \times 603.59)$
$\quad = 705015457.30 - 682337982.80 - 20075769.12 = 2601705.38$

By the formula (12.6.6),

$$\Lambda = \frac{16985.4725}{2601705.38} = 0.006528$$

For the given problem, $p = 3$, $k = 4$. By the formula (12.6.8),

$$s = \sqrt{\frac{9 \times 9 - 4}{9 + 9 - 5}} = 2.43$$

$$y = \Lambda^{\frac{1}{2.43}} = 0.1261$$

$$m_1 = 3 \times 3 = 9, m_2 = 2.43 \left(60 - \frac{3 + 4 + 2}{2} \right) - \frac{3 \times 3}{2} + 1 = 131.36$$

By the formula (12.6.8),

$$F = \frac{1 - 0.1261}{0.1261} \times \frac{131.36}{9} = 101.150$$

F has d.f. (9, 131.36).

Tabulated value of F for $= 0.05$ and (9, 131) d.f. is almost 1.95. Since the calculated value of F $= 101.150$ is greater than 1.95, null hypothesis is rejected. It means that four species of grass hopper are different from one another.

Covariance matrices for the four species (groups) are:

$$S_1 = \begin{bmatrix} 14.05 & 9.85 & -1.63 \\ 9.85 & 12.78 & -1.74 \\ -1.63 & -1.74 & 2.18 \end{bmatrix}$$

$$S_2 = \begin{bmatrix} 204.83 & 201.33 & -1.20 \\ 201.33 & 202.58 & -0.62 \\ -1.20 & -0.62 & 0.36 \end{bmatrix}$$

$$S_3 = \begin{bmatrix} 12.20 & 12.35 & 0.82 \\ 12.35 & 12.64 & 0.99 \\ 0.82 & 0.99 & 1.73 \end{bmatrix}$$

$$S_4 = \begin{bmatrix} 32.28 & 31.59 & 3.46 \\ 31.59 & 31.02 & 3.31 \\ 3.46 & 3.31 & 1.19 \end{bmatrix}$$

From the above matrices, it can be verified that,

$$W = S_1 + S_2 + S_3 + S_4$$

This confirms the relation (12.6.2.1).

Since all groups have same number of observation for all variables, matrix

$$\bar{S} = \frac{1}{2}(S_1 + S_2) = \begin{bmatrix} 109.44 & 105.59 & -1.41 \\ 105.59 & 107.68 & -1.18 \\ -1.41 & -1.18 & 1.27 \end{bmatrix}$$

Now the confidence interval by the formula (12.8.1) is,

$$(\bar{y}_{\cdot 11} - \bar{y}_{\cdot 12}) \pm c \sqrt{s_{11}^2 \left(\frac{1}{n_1} + \frac{1}{n_2} \right)}$$

$$\bar{y}_{\cdot 11} = 13.064, \bar{y}_{\cdot 12} = 13.753, n_1 = n_2, s_{11}^2 = 109.44$$

$$c = \frac{(15+15-2) \times 3}{(15+15-3-1)} F_{.05,(3,26)} = \frac{28 \times 3}{26} \times 2.98 = 9.63$$

Confidence interval for $(\mu_{11} - \mu_{12})$ is,

$$\text{C.I.} = (13.064 - 13.753) \pm 9.63 \sqrt{109.44 \left(\frac{1}{15} + \frac{1}{15} \right)}$$

$$= -0.689 \pm 9.63 \times 3.82$$

$$= -0.689 \pm 36.79$$

Thus, the confidence limits are,

$$L = -37.48 \text{ and } U = 36.101$$

Similarly, the confidence intervals for all other pairs of means can be obtained.

Since statistic F is significant, there is need to compare group mean vectors.

For testing equality of a pair of mean vectors say, μ_1 and μ_3 make use of formula (12.9.1). To perform the test, the value of Hotelling's T^2 is to be calculated. It will require the value of $\bar{S}^{-1}$, where

$$\bar{S} = \frac{S_1 + S_3}{2} = \begin{bmatrix} 13.125 & 11.100 & -0.405 \\ 11.100 & 12.71 & -0.375 \\ -0.405 & -0.375 & 1.955 \end{bmatrix}$$

$\bar{S}^{-1}$ can either be found by pivotal condensation method or directly by the formula (A – 2 – 8) given appendix-A in Agarwal's (2009) book. Since the matrix $\bar{S}$ is of order (3 × 3) in all pairwise post-hoc tests, it is advised to use direct method. Thus, the inverse of $\bar{S}$ is,

$$|\bar{S}| = 13.125 \times 24.7074 - 11.1 \times 21.5486 - 0.405 \times 0.9850 = 84.6910$$

$$\bar{S}^{-1} = \begin{bmatrix} 0.29173 & -0.25444 & 0.01163 \\ -0.25444 & 0.30104 & 0.00503 \\ 0.01163 & 0.00503 & 0.51008 \end{bmatrix}$$

Check: It should always be verified that $\bar{S}\,\bar{S}^{-1} = I$.

$$T^2 = (5.583 \quad 4.647 \quad 1.863) \begin{bmatrix} 0.29173 & -0.25444 & 0.01163 \\ -0.25444 & 0.30104 & 0.00503 \\ 0.01163 & 0.00503 & 0.51008 \end{bmatrix} \begin{bmatrix} 5.583 \\ 4.647 \\ 1.863 \end{bmatrix} \frac{15 \times 15}{15 + 15}$$

$$= (0.4680 \quad -0.0123 \quad 0.9087) \begin{pmatrix} 5.583 \\ 4.647 \\ 1.863 \end{pmatrix} \times \frac{15 \times 15}{15 + 15}$$

$$= 4.25 \times 7.5 = 31.875$$

$$F = \frac{(n_r + n_s - p - 1)}{p(n_r + n_s - 2)} T^2 = \frac{15 + 15 - 3 - 1}{3(15 + 15 - 2)} \times 31.875 = 9.866$$

Bonferroni's approximate level of significance for $\alpha = 0.05 = \dfrac{0.05 \times 2}{4 \times 3}$

$$= 0.0083$$

Tabulated value of F for = 0.0083 and (3,26) d.f. is slightly more than 5.41. Since calculated value of F is greater than the tabulated value, it is concluded that species S_1 and S_3 differ significantly in respect of their body length and breath. Similarly other Pairs of species can be tested. This is left to the readers as an exercise.

12.11 MULTIVARIATE DATA ANALYSIS

Multivariate data analysis techniques are in vogue for two centuries. Bivariate data analysis became popular due to ease of understanding and computation. But the studies for large number of dependent variables measured simultaneously on individuals of a sample remain confined due to its complexities. But the invention of computers prompted the researchers working in different areas to use them in a real sense. Some of the multivariate data analysis methods are popularly applied in marketing, administrative, social, industrial, agricultural research and so on. Multivariate data analysis of variance (MANOVA) is one of them which has been covered in this chapter succinctly. Other methods of multivariate data analysis do not form the part of this chapter. So they are discussed in brief just to acquaint the readers with their purpose and utility.

12.12 DISCRIMINANT ANALYSIS

If the null hypothesis of mean vectors, $H_0: \underset{\sim}{\mu}_1 = \underset{\sim}{\mu}_2 = \ldots\ldots = \underset{\sim}{\mu}_k$ is rejected in MANOVA, then a researcher will be interested to characterize the groups of means that show the equality among the mean vectors. Discriminant analysis can also help in classifying the observations whose identity of group membership is unknown. For example, the data from samples of voters divided into six categories namely, illiterate, primary, secondary, graduates, post-graduates and technical degree holders regarding fairness of elections are collected in a big city. A battery of ten questions is canvassed to each individual and their opinion is taken about each question on a scale from 1 to 10. Now the groups of various levels of education can be categorized according to their resemblance with regard to mean vectors. Secondly, the individuals can be classified on the basis of their answers of whom the identity is not known.

12.13 PRINCIPAL COMPONENTS ANALYSIS

Principal component analysis is concerned with dimension reduction of data covariance matrix say, X of dimension $p \times p$ based on n observations on each of the p characteristic variables under consideration. While planning a research project, one is unaware of the fact that how many variables are sufficient to account for most of the variation present in p original variables. So to reduce p original variables to k affective variables for $k < p$, principal component analysis is an analytical tool which helps to achieve this goal.

The reduction can be carried through principal components of the data matrix X. Principal components are the eigen-vectors (latent roots) associated with orthogonal decomposition of the $p \times p$ sample covariance matrix X. Largest eigen values account for most of the portion of the total variability. For instance, in the problem given with discriminant analysis, one likes to reduce the number of questions from ten to a lesser number. In that situation, principal component analysis shall be applicable. Selection of variables in multiple regression equation is provided by the subset obtained by principal component analysis out of all explanatory variables initially included in the model. Another advantage is that the components so obtained are uncorrelated. Hence, the presence of one component does not affect the regression coefficient of the other component.

12.14 FACTOR ANALYSIS

Purpose of factor analysis is to explain the variation among many variables in terms of some underlying unobservable random variables known as factors. Model for factor analysis specifies a small number of common factors. Statistical regression model for factor analysis, so called *common factor analysis model* consists of three kinds of variables as given below.

$$y - \mu = A f + \varepsilon \qquad (12.14.1)$$

where, μ is $(p \times 1)$ vector of means, $(y - \mu)$ is a vector of order $(p \times 1)$ of elements $(y_i - \mu_i)$ for $i = 1, 2, \ldots\ldots, p$. f is a $(p \times 1)$ vector of linearly independent common factors f_j, $j = 1, 2, \ldots\ldots, k$ for $k < p$. A is the $(p \times k)$ matrix of unknown factor loadings a_{ij}. ε is a $(p \times 1)$ vector of unique factors ε_i. Any portion of the variance not explained by the common factors is assigned to error terms which are called *unique factors*.

Assumptions underlying model (12.14.1) are:
1. f is distributed independently with mean 0 and variance 1.
2. ε's are identically and independently distributed with mean 0 and variance σ_ε^2.

Under these assumptions, the covariance matrix is of the form,

$$\Sigma = E[(y - \mu)(y - \mu)'] = AA' + \psi$$

$E(f) = E(\varepsilon) = 0$

$E(ff') = I$, a $(k \times k)$ identity matrix.

$E(\varepsilon\varepsilon') = \Psi$, a $(p \times p)$ diagonal matrix with diagonal elements σ_ε^2.

$E(\varepsilon f') = 0$, i.e. the unique factors and common factors are independent.

This regression type model is used for grouping variables into subsets in such a way that the variables within each set are highly correlated whereas the variables belonging to different subsets are uncorrelated. Main difference between principal component analysis and factor analysis can be explained as former one is concerned with explaining the variance of the variables and the later is concerned with explaining the covariance of the variables.

12.15 CLUSTER ANALYSIS

Purpose of cluster sampling is to classify the objects into groups, so called *clusters*, in such a manner that profile of objects in the same cluster are similar whereas in two different clusters they are distinct. There are two approaches which are generally used for doing cluster analysis: (*i*) hierarchical clustering, (*ii*) inter-cluster distance approach. Details of these methods do not form the part of this chapter.

12.16 CONJOINT ANALYSIS

Conjoint analysis is a statistical technique mostly used in marketing research to assess the preferences of buyers towards product features while purchasing an article or services. It was first time used in 1970 and mainly developed from 1980 onwards. The objective of conjoint analysis is achieved by canvassing a number of combinations of product features as a profile considered jointly. From the results of survey, one would like to determine the desirability of the combination. For instance, what combination of features of television set are preferred by the purchaser in general out of many like, its outer look, screen size, color contrast, sound system, cost, operational ease, etc.

A number of algorithms may be applied to estimate *utility function*. A utility function indicates the perceived value of features and preferences. The choice of algorithm is task based and some of techniques like maximum likelihood, logistic regression, linear programming, hierarchical Bayesian procedure, etc. are often used. The data analysis is carried on computers by using appropriate statistical software especially prepared for conjoint analysis.

Concluding remark

Analysis of variance (ANOVA) is an integral part of experimental designs. In ANOVA model, there is one dependent or criterion variable and one or more explanatory variables which are linearly related. In this way, ANOVA is a special case of multiple regression. But the researches or studies in various fields or studies like psychology, education, social sciences, marketing, organizational administration, agriculture, etc. do not confine to one criterion variable but are based on a battery of related factors. These factors may be quantitative or categorical in nature. The multivariate regression model using categorical dependent variables is known as *qualitative response model*. Due to availability of computer packages for various multivariate data analysis techniques, discussed in the text of this chapter, have become popular. The reason being their utility in advance research or quick reliable results for practical purposes.

The motive of Chapter 12 is to familiarize the readers with the technique of multivariate analysis of variance (MANOVA) as an extension of ANOVA. Each and every aspect of MANOVA is fully explained. Not only this, each step of multivariate data analysis is expatiated clearly so that an inquisitive mind does not face any difficulty in analyzing data himself. The same is substantiated by an example. Good understanding of analysis of data further facilitates to interpret the results in the correct perspective.

Besides MANOVA, other multivariate techniques are also mentioned objectively so that one can make proper use of any of them. Multivariate techniques as a whole require one full volume of a book. So except MANOVA details of other multivariate methods are not included as they are not directly related to the approach of this book.

QUESTIONS AND EXERCISES

1. In what situations, multivariate analysis of variance is to be carried out?
2. Differentiate between ANOVA model and MANOVA model.
3. What are the assumptions underlying MANOVA?
4. What hypothesis is generally tested in case of MANOVA?
5. What test statistics is used in MANOVA?
6. If there are only two groups, give the test statistic for the test of hypothsis.
7. Give the relation between Mahalanobis D^2 and Hotelling T^2.
8. How can the post-hoc test be applied in case of MANOVA?
9. Give the formula for confidence interval of the difference of two mean vectors.
10. Distinguish between repeated measures and multivariate response variables.
11. Describe the situations in which the following multivariate data techniques are suitable:
    (*a*) Discriminant analysis.
    (*b*) Cluster analysis.
    (*c*) Factor analysis.
    (*d*) Principal component analysis.
    (*e*) Conjoint analysis.
12. In what respects, multivariate data analysis methods given in question – 11 differ from MANOVA?
13. What do you understand by covariance matrix and what is its role in multivariate data analysis.
14. How Bonferroni's approach is useful in multivariate data analysis?
15. A survey was conducted to know the opinion of three groups of news paper subscribers about four supplements of news paper supplied weekly with the main paper. The supplements are:
    (*a*) Fashion and films supplement – X_1,
    (*b*) Supplement for children – X_2
    (*c*) Supplement for women – X_3
    (*d*) Economic and investment supplement – X_4.

Respondents were divided into three categories namely, (*i*) students – 1, (*ii*) house wives – 2 (*iii*) business man – 3. Only 10 persons were selected randomly from each of the category for survey. They were requested to rank these supplements from 1 to 4 as, excellent – 1, very good – 2, good – 3, ordinary – 4. Their responses in codes are displayed below.

Category	Variables			
	X_1	X_2	X_3	X_4
1	2	3	4	3
1	1	2	3	2
1	2	3	2	2
1	1	2	3	4
1	2	2	4	3
1	1	3	3	1
1	3	2	3	4
1	2	4	1	3
1	1	2	3	4
1	2	2	4	2
2	3	2	1	4
2	2	2	2	3
2	3	3	2	3
2	2	2	1	4
2	3	4	2	3
2	2	3	1	3
2	1	3	2	2
2	2	2	1	3
2	2	3	1	3
2	3	2	2	4
3	3	3	3	2
3	2	3	3	1
3	2	2	3	2
3	2	3	3	2
3	3	3	3	1
3	2	4	4	2
3	1	2	3	2
3	2	3	4	1
3	3	3	4	2
3	1	3	3	1

* Hypothetical example

Perform the complete multivariate analysis of variance to compare the means of four variables over the three categories of paper subscribers.

BIBLIOGRAPHY

1. Agarwal BL. *Basic Statistics*, 5th ed. New Age International Publishers, New Delhi, 2009.
2. Anderson TW. *An Introduction to Multivariate Statistical Analysis*, 2nd ed., John Wiley, New York, 1984.
3. Bray JH, Maxwell SE. Analyzing and interpreting significant MANOVA. *Review of Educational Research*, 1980: 52, 340–67.
4. Cliff N. *Analyzing Multivariate Data*. Harcourt Brace Jovanovich, Orlando, PL, 1987.
5. Cole DA, Howard GS, Maxwell SE. Effects if mono-operationalization in construct validation efforts. *Journal of Consulting and Clinical Psychology*, 1981: 49, 395–405.
6. Cook TD, Campbell DT. *Quasi-Experimentation: Design and Analysis Issues*. Houghton Miffin Company, Boston, 1979.
7. Crowder M, Hand D. *Analysis of Repeated Measures*. Chapman and Hall, New York, 1990.
8. Everitt B. The analysis of repeated measures: A practical review with examples. *The Statistician*, 1995: 44, 113–35.
9. Finn D. *A General Model for Multivariate Analysis*. Rinehols Winston, Fort Worth, 1974.
10. Gnanadesikan R. *Methods for Statistical Data Analysis of Multivariate Observations*. John Wiley, New York, 1997.
11. Green PE, Carroll JD. *Mathematical Tools for Applied Multivatiate Analysis*. Academic Press, New York, 1976.
12. Jobson JD. *Applied Multivariate Data Analysis: Regression and Experimental Designs*, Vol. 1. Springer-Verlag, New York, 1991.
13. Jobson JD. *Applied Multivariate Data Analysis: Categorical and Multivariate Methods*, Vol. 2. Springer-Verlag, New York, 1992.
14. Johnson RA, Wichem, DW. *Applied Multivariate Statistical Analysis*, 2nd ed. Englewood Cliffs, Princeton Hall, New Jersey, 1988.
15. Lachenbruch PA. *Discriminant analysis*. Hafuer Press, New York, 1975.
16. Leary MR, Altamaier EM. Type I error in consulting: A plea for multivariate analysis. *Journal of Counseling Psychology*, 1980: 27, 611–15.
17. Seber G. *Multivariate Observations*. Johm Wiley, New York, 1984.
18. Strahan RI. Multivariate analysis and problem of type I error. *Journal of Counseling Psychology*, 1982: 29, 175–79.
19. Tatsuoka M. Multivariate Analysis: *Techniques for Educational and Psychological Research*, 2nd ed. Macmillan Publishing Co., New York, 1988.

Introductory Matrix Algebra

DEFINITIONS

Matrix: A matrix is a two-dimensional array of elements (real or complex numbers) having m rows and n columns. It is denoted as $A = ((a_{ij}))$ for $i = 1$, $2, \ldots, m$ and $j = 1, 2, \ldots, n$. This is called the matrix of order $(m \times n)$. Complex numbers are not considered in the foregoing discussion. In full form, it is written as below.

$$A = \begin{bmatrix} a_{11} & a_{12} & ----- & a_{1j} & ---- & a_{1n} \\ a_{21} & a_{22} & \ldots & a_{2j} & \ldots & a_{2n} \\ \vdots & & & & & \\ a_{i1} & a_{i2} & \ldots & a_{ij} & \ldots & a_{in} \\ \vdots & & & & & \\ a_{m1} & a_{m2} & \ldots & a_{mj} & \ldots & a_{mn} \end{bmatrix}$$

Usually a_{ij}'s take on any real values. For example, a matrix of order (3×4) in real values is of the type,

$$A = \begin{bmatrix} 13.5 & 4.6 & 2.1 & 5.0 \\ 6.8 & 12.0 & 11.6 & 9.8 \\ 7.7 & 6.3 & 8.7 & 7.9 \end{bmatrix}$$

Square matrix: If the number of rows and columns of a matrix A are equal, i.e. $m = n$, then A is said to be a square matrix of order m.

Column vector: A single column of a matrix with m row elements is called a *column vector* and is denoted by $\underset{\sim}{a}$ of order $(m \times 1)$. Each column of the above matrix is a column vector. In particular,

$$\underset{\sim}{a}_1 = \begin{bmatrix} 13.5 \\ 6.8 \\ 7.7 \end{bmatrix}, \underset{\sim}{a}_2 = \begin{bmatrix} 4.6 \\ 12.0 \\ 6.3 \end{bmatrix}, \underset{\sim}{a}_3 = \begin{bmatrix} 2.1 \\ 11.6 \\ 8.7 \end{bmatrix}, \underset{\sim}{a}_4 = \begin{bmatrix} 5.0 \\ 9.8 \\ 7.9 \end{bmatrix}$$

Therefore, a matrix **A** is a row vector of the column vectors $\underline{a}_1, \underline{a}_2, \underline{a}_3$ and $\underline{a}_4$.

Row vector: A single row of a matrix with n column elements is called a *row vector* and is denoted by $\underline{a}'$. Thus, for the above matrix, the row vectors are,

$$\underline{a}_1' = \begin{bmatrix} 13.5 & 4.6 & 2.1 & 5.0 \end{bmatrix}, \underline{a}_2' = \begin{bmatrix} 6.8 & 12.0 & 11.6 & 9.8 \end{bmatrix}, \underline{a}_3' = \begin{bmatrix} 7.7 & 6.3 & 8.7 & 7.9 \end{bmatrix}$$

Therefore, a matrix **A** is a column vector of the row vectors:

$$\underline{a}_1', \quad \underline{a}_2', \quad \underline{a}_3'.$$

Transpose of a matrix: If the rows of a matrix **A** are taken as columns and columns as rows, then the resulting matrix is said to be the *transpose* of **A** and is denoted as **A'**. Its order is also reversed. For instance,

If
$$\mathbf{A} = \begin{bmatrix} a_{11} & a_{12} & a_{13} & a_{14} \\ a_{21} & a_{22} & a_{23} & a_{24} \\ a_{31} & a_{32} & a_{33} & a_{34} \end{bmatrix}, \text{ then } \mathbf{A}' = \begin{bmatrix} a_{11} & a_{21} & a_{31} \\ a_{12} & a_{22} & a_{32} \\ a_{13} & a_{23} & a_{33} \\ a_{14} & a_{24} & a_{34} \end{bmatrix}$$
$$(3 \times 4) \qquad\qquad (4 \times 3)$$

Submatrix: When some row(s) and/or some column(s) of a matrix are deleted, the remaining matrix is called a *submatrix*. Suppose in matrix **A** of order (3 × 4), first row and first two columns are deleted, then the remaining matrix,

$$\mathbf{A}^\circ = \begin{bmatrix} a_{23} & a_{24} \\ a_{33} & a_{34} \end{bmatrix}$$

is called a submatrix of order (2 × 2).

Let **A** and **B** be the two matrices of the same order, then the following entities are always true.

$(\mathbf{A} + \mathbf{B})' = \mathbf{A}' + \mathbf{B}'$

$(\mathbf{A} - \mathbf{B})' = \mathbf{A}' - \mathbf{B}'$

$(\mathbf{A} \times \mathbf{A})' = \mathbf{B}' \times \mathbf{A}'$

$(\mathbf{c} \times \mathbf{B})' = \mathbf{c} \times \mathbf{A}' = \mathbf{A}' \times \mathbf{c}$ where c is a scalar quantity.

Symmetric matrix: A square matrix of order m is called a *symmetric matrix* if the elements $a_{ij} = a_{ji}$. Following two matrices are symmetric matrices.

$$A = \begin{bmatrix} a & b & c \\ b & d & e \\ c & e & f \end{bmatrix} \text{ or in numerals, } A = \begin{bmatrix} 3 & 5 & 8 \\ 5 & 6 & 7 \\ 8 & 7 & 4 \end{bmatrix}$$

Diagonal matrix: A matrix is known as *diagonal matrix* if only diagonal elements exist and off-diagonal elements are zero, e.g. a matrix A of order (3 × 3) is of the type given below.

$$A = \begin{bmatrix} a & 0 & 0 \\ 0 & d & 0 \\ 0 & 0 & f \end{bmatrix} \text{ or in numerals, } A = \begin{bmatrix} 3 & 0 & 0 \\ 0 & 6 & 0 \\ 0 & 0 & 4 \end{bmatrix}$$

Identity matrix: A square matrix of order m with all its diagonal elements unity (1) and off-diagonal elements zero (0) is an *identity matrix* and is denoted by I. For $m = 3$,

$$I = \begin{bmatrix} 1 & 0 & 0 \\ 0 & 1 & 0 \\ 0 & 0 & 1 \end{bmatrix}$$

Null matrix: A matrix whose all elements are zero is called a *null matrix*. It is denoted by **0**. A null matrix of order (3 × 4) is,

$$0 = \begin{bmatrix} 0 & 0 & 0 & 0 \\ 0 & 0 & 0 & 0 \\ 0 & 0 & 0 & 0 \end{bmatrix}$$

Trace of a matrix: The sum of diagonal elements of a square matrix A is called the *trace* of A and is written as tr (A). For a matrix of order m with its

Elements a_{ij} for i, $j = 1, 2, \ldots\ldots , m$, the trace of A is,

$$\text{tr } (A) = \sum_{i=1}^{m} a_{ii} = a_{11} + a_{22} + \ldots\ldots + a_{mm}$$

Equality of matrices: Two matrices $A = ((a_{ij}))$ and $B = ((b_{ij}))$ of the same order are said to equal if each and every element of A is equal to the corresponding element of B, i.e. $a_{ij} = b_{ij}$ for all i and j.

Additive inverse of matrices: If each element of a matrix A is equal but opposite in sign to each corresponding element of another matrix B, then $[A + B]$ is a null matrix. In such a case, matrix A is called additive inverse of B and vice-versa. Thus, $A + B = B + A = 0$. For instance, if

$$\mathbf{A} = \begin{bmatrix} 3 & 2 \\ 5 & -4 \\ -6 & 1 \end{bmatrix} \text{ and } \mathbf{B} = \begin{bmatrix} -3 & -2 \\ -5 & 4 \\ 6 & -1 \end{bmatrix}$$

Then,
$$[\mathbf{A} + \mathbf{B}] = \begin{bmatrix} 0 & 0 \\ 0 & 0 \\ 0 & 0 \end{bmatrix} = \mathbf{0}$$

Operations on Matrices

Addition of matrices: For matrices $\mathbf{A} = ((a_{ij}))$ and $\mathbf{B} = ((b_{ij}))$ of the same order, the additive matrix $((\mathbf{A} + \mathbf{B})) = ((a_{ij} + b_{ij}))$. For example,

If
$$\mathbf{A} = \begin{bmatrix} 3 & 2 \\ 5 & 4 \\ 1 & 6 \end{bmatrix} \text{ and } \mathbf{B} = \begin{bmatrix} 7 & -3 \\ -8 & 2 \\ 5 & -4 \end{bmatrix}, \text{ then } \mathbf{A} + \mathbf{B} = \begin{bmatrix} 10 & -1 \\ -3 & 6 \\ 6 & 2 \end{bmatrix}$$

Also, $\mathbf{A} + \mathbf{B} = \mathbf{B} + \mathbf{A}$ (Cumulative law).

This rule can be extended to any number of matrices.

Scalar multiplication of a matrix: If a matrix $\mathbf{A}$ is multiplied by a constant value c, then each element of the matrix $\mathbf{A}$ is multiplied by c.

If
$$\mathbf{A} = \begin{bmatrix} a_{11} & a_{21} & a_{31} \\ a_{12} & a_{22} & a_{32} \\ a_{13} & a_{23} & a_{33} \\ a_{14} & a_{24} & a_{34} \end{bmatrix} \text{ then } c\mathbf{A} = \begin{bmatrix} ca_{11} & ca_{21} & ca_{31} \\ ca_{12} & ca_{22} & ca_{32} \\ ca_{13} & ca_{23} & ca_{33} \\ ca_{14} & ca_{24} & ca_{34} \end{bmatrix}$$

Also, $c.\mathbf{A} = \mathbf{A}.c$

Multiplication of two matrices: Multiplication of two matrices $\mathbf{A}$ and $\mathbf{B}$ denoted as $\mathbf{AB}$ is possible only if number of columns in $\mathbf{A}$ is equal to the number of rows in $\mathbf{B}$. The product of matrix $\mathbf{A}$ $(m \times n)$ with $\mathbf{B}$ $(n \times p)$, i.e.,

$\mathbf{AB} = \mathbf{C}$ (say) will be of order $(m \times p)$. The elements of the product matrix $((c_{ij}))$ shall be,

$$C_{iu} = \sum_{j=1}^{n} a_{ij} b_{ju}$$

$$i = 1, 2, \ldots\ldots, m; u = 1, 2, \ldots\ldots, p.$$

For instance,

If
$$\mathbf{A} = \begin{bmatrix} a_{11} & a_{12} \\ a_{21} & a_{22} \\ a_{31} & a_{32} \end{bmatrix} \text{ and } \mathbf{B} = \begin{bmatrix} b_{11} & b_{12} & b_{13} \\ b_{21} & b_{22} & b_{23} \end{bmatrix},$$

$$3 \times 2 \qquad\qquad\qquad 2 \times 3$$

$$\text{then,} \quad C = \begin{bmatrix} a_{11}b_{11} + a_{12}b_{21} & a_{11}b_{12} + a_{12}b_{22} & a_{11}b_{13} + a_{12}b_{23} \\ a_{21}b_{11} + a_{22}b_{21} & a_{21}b_{12} + a_{22}b_{22} & a_{21}b_{13} + a_{22}b_{23} \\ a_{31}b_{11} + a_{32}b_{21} & a_{31}b_{12} + a_{32}b_{22} & a_{31}b_{13} + a_{32}b_{23} \end{bmatrix}$$

$$3 \times 3$$

Distributive law of matrices: If **A**, **B**, **C**, etc. are the matrices of the same order and a, b, c, etc., are some constant, then the following relations always hold between matrices.

$$(AB)' = B' \times A'$$
$$A (B + C) = A \times B + A \times C$$
$$A (B - C) = A \times B - A \times C$$
$$a (B + C) = a \times B + a \times C = (B + C) a$$
$$a (B - C) = a \times B - a \times C = (B - C) a$$

These relations can be extended to any number of matrices.

Kronecker product of matrices: Let there be any two matrices, $A = ((a_{ij}))$ of order $(m \times n)$ and $B = ((b_{ij}))$ of order $(p \times q)$. Kronecker product of **A** and **B** is denoted by $A \otimes B$. The product matrix **C** (say) will contain each element of **A** multiplied by the matrix **B**. The order of **C** will be (mp, nq). Notationally,

$$C = A \otimes B = \begin{bmatrix} a_{11}B & a_{12}B & \cdots & a_{1j}B & \cdots & a_{1n}B \\ a_{21}B & a_{22}B & \cdots & a_{2j}B & \cdots & a_{2n}B \\ & & & & & \\ a_{i1}B & a_{i2}B & \cdots & a_{ij}B & \cdots & a_{in}B \\ & & & & & \\ a_{m1}B & a_{m2}B & \cdots & a_{mj}B & \cdots & a_{mn}B \end{bmatrix}$$

For example,

$$\text{If} \qquad A = \begin{bmatrix} 3 & -1 \\ 2 & 5 \\ -4 & 7 \end{bmatrix}; \ B = \begin{bmatrix} 6 & 1 \\ 3 & 2 \end{bmatrix}.$$

$$\text{Then} \qquad C = A \otimes B = \begin{bmatrix} 3\begin{pmatrix} 6 & 1 \\ 3 & 2 \end{pmatrix} & -1\begin{pmatrix} 6 & 1 \\ 3 & 2 \end{pmatrix} \\ 2\begin{pmatrix} 6 & 1 \\ 3 & 2 \end{pmatrix} & 5\begin{pmatrix} 6 & 1 \\ 3 & 2 \end{pmatrix} \\ -4\begin{pmatrix} 6 & 1 \\ 3 & 2 \end{pmatrix} & 7\begin{pmatrix} 6 & 1 \\ 3 & 2 \end{pmatrix} \end{bmatrix}$$

$$
=\begin{bmatrix}
18 & 3 & -6 & -1 \\
9 & 6 & -3 & -2 \\
12 & 2 & 30 & 5 \\
6 & 4 & 15 & 10 \\
-24 & -4 & 42 & 7 \\
-12 & -8 & 21 & 14
\end{bmatrix}
$$

$$6 \times 4$$

Verify that, $m \times p = 3 \times 2 = 6, n \times q = 2 \times 2 = 4$.

Here it is to emphasize that each element of Kronecker product matrix **C** is obtained by using *horizontal product operation* on rows of **B**.

Properties of Kronecker products,

$$\mathbf{A} \otimes \mathbf{B} \neq \mathbf{B} \otimes \mathbf{A} \text{ except in some special cases.}$$

$$(\mathbf{A} + \mathbf{B}) \otimes \mathbf{C} = (\mathbf{A} \otimes \mathbf{C}) + (\mathbf{B} \otimes \mathbf{C})$$

$$(\mathbf{A} \otimes \mathbf{B}) \times (\mathbf{C} \otimes \mathbf{D}) = \mathbf{AC} \otimes \mathbf{CD}$$

$$(\mathbf{A} \otimes \mathbf{B})' = \mathbf{A}' \otimes \mathbf{B}'$$

$$\mathrm{tr}(\mathbf{A} \otimes \mathbf{B}) = \mathrm{tr}(\mathbf{A}).\mathrm{tr}(\mathbf{B})$$

Determinant

Definition: A determinant of a square matrix **A** is a scalar quantity and is denoted by $|A|$.

Properties

(*i*) Number of rows and columns in a determinant are always equal.

(*ii*) If all elements of any row or column are zero, then the value of the determinant is zero.

(*iii*) The sign of a determinant changes (+ to – or – to +) if any two rows (columns) are interchanged amongst themselves.

(*iv*) If **A** is a diagonal matrix, then its determinant $|A|$ is equal to the product of the diagonal elements e.g.,

$$
|A| = \begin{vmatrix}
d_1 & 0 & 0 \\
0 & d_2 & 0 \\
0 & 0 & d_3
\end{vmatrix} = d_1 \times d_2 \times d_3
$$

(*v*) If for any element a_{ij} of a determinant, *i*th row and *j*th column are deleted, then the determinant with the remaining elements is known as the *minor* of a_{ij}.

(*vi*) A minor of a_{ij} with its proper sign which is determined by $(-1)^{i+j}$ is called the cofactor of a_{ij}.

(*vii*) The value of determinant **A** of order (2×2) is obtained as,

$$|A| = \begin{vmatrix} a_{11} & a_{12} \\ a_{21} & a_{22} \end{vmatrix} = a_{11}\, a_{22} - a_{12}\, a_{21}$$

(*viii*) Value of a determinant of any order is worked out as the sum of each element of a row or column multiplied by the value of their respective cofactor. For example,

$$|A| = \begin{vmatrix} 3 & 4 & 5 & 4 \\ 5 & 6 & 2 & 3 \\ 2 & 3 & 1 & 2 \\ 1 & 7 & 1 & 8 \end{vmatrix}$$

$$= 3\begin{vmatrix} 6 & 2 & 3 \\ 3 & 1 & 2 \\ 7 & 1 & 8 \end{vmatrix} - 5\begin{vmatrix} 4 & 5 & 4 \\ 3 & 1 & 2 \\ 7 & 1 & 8 \end{vmatrix} + 2\begin{vmatrix} 4 & 5 & 4 \\ 6 & 2 & 3 \\ 7 & 1 & 8 \end{vmatrix} - 1\begin{vmatrix} 4 & 5 & 6 \\ 6 & 2 & 3 \\ 3 & 1 & 2 \end{vmatrix}$$

$$= 3\left\{ 6\begin{vmatrix} 1 & 2 \\ 1 & 8 \end{vmatrix} - 3\begin{vmatrix} 2 & 3 \\ 1 & 8 \end{vmatrix} + 7\begin{vmatrix} 2 & 3 \\ 1 & 2 \end{vmatrix} \right\} - 5\left\{ 4\begin{vmatrix} 1 & 2 \\ 1 & 8 \end{vmatrix} - 3\begin{vmatrix} 5 & 4 \\ 1 & 8 \end{vmatrix} + 7\begin{vmatrix} 5 & 4 \\ 1 & 2 \end{vmatrix} \right\}$$

$$+ 2\left\{ 4\begin{vmatrix} 2 & 3 \\ 1 & 8 \end{vmatrix} - 6\begin{vmatrix} 5 & 4 \\ 1 & 8 \end{vmatrix} + 7\begin{vmatrix} 5 & 4 \\ 2 & 3 \end{vmatrix} \right\} - 1\left\{ 4\begin{vmatrix} 2 & 3 \\ 1 & 2 \end{vmatrix} - 6\begin{vmatrix} 5 & 6 \\ 1 & 2 \end{vmatrix} + 3\begin{vmatrix} 5 & 6 \\ 2 & 3 \end{vmatrix} \right\}$$

$$= 3\{6 \times 6 - 3 \times 13 + 7 \times 1\} - 5\{4 \times 6 - 3 \times 36 + 7 \times 6\}$$
$$+ 2\{4 \times 13 - 6 \times 36 + 7 \times 7\} - 1\{ 4 \times 1 - 6 \times 4 + 3 \times 3\}$$
$$= 3 \{36 - 39 + 7\} - 5\{24 - 108 + 42\} + 2\{52 - 216 + 49\}$$
$$- 1\{ 4 - 24 + 9\}$$
$$= 3 \times 4 + 5 \times 42 - 2 \times 115 + 1 \times 11$$
$$= 12 + 210 - 230 + 11 = 3$$

(*ix*) If any two rows or columns of a determinant are identical or a row (column) is a scalar multiple of any other row (column), then the value of the determinant is zero.

(*x*) If for any two square matrices **A** and **B**, **AB = C**, then $|A| \times |B| = |C|$

Back to matrices

Rank of a matrix: Rank of a matrix is the maximum number of independent rows or columns in that matrix. For a matrix of order $(m \times n)$, the matrix is said to have *full rank,* i.e. rank $(A) = \min (m , n)$. Also for any matrix **A**, row rank is equal to column rank.

For any two matrices **A** and **B**,

$$\text{Rank} (A + B) \leq \text{rank } (A) + \text{rank } (B)$$
$$\text{Rank} (A \times B) \leq \min \{\text{rank } (A), \text{rank } (B)\}$$

Singular matrix: If the determinant of a matrix **A** is zero, i.e. $|A| = 0$, then matrix **A** is known as singular matrix. In other words, if the rank of a matrix is not full, then it is a *singular matrix.*

Non-singular matrix: If $|A| \neq 0$, then matrix **A** is said to be *non-singular matrix.* In this case, the rank of a matrix is full.

Adjoint matrix: A matrix consisting of cofactors of all elements of $|A|$ is called its *adjoint matrix* and is usually denoted by A*.

Inverse of matrix: If for any non-singular matrix **A** of order $(m \times n)$, there exists a matrix **B** such that $AB = I$, then **B** is said to be the inverse of **A** and is denoted as A^{-1} or $B = A^{-1}$. Thus, $A \ A^{-1} = I$.

Formula for obtaining an inverse matrix of **A** is,

$$A^{-1} = \frac{(A*)'}{|A|}$$

where $(A*)'$ is the transpose of the cofactors matrix **A***.

In other words, elements c_{ij} of an inverse matrix A^{-1} of A $(m \times n)$ are,

$$c_{ij} = \frac{Cofactors(a_{ji})}{|A|}$$

$$i, j = 1, 2, \ldots\ldots\ldots , m.$$

Determinants and cofactors are evaluated in the manner they are worked out under the heading *Determinants.*

Properties

(*i*) It is worth pointing out that the inverse of a symmetric matrix is always a symmetric matrix.

(*ii*) Inverse of a diagonal matrix is a diagonal matrix with the reciprocals of the respective diagonal elements. For example,

$$\text{If } A = \begin{bmatrix} d_1 & 0 & 0 \\ 0 & d_2 & 0 \\ 0 & 0 & d_3 \end{bmatrix}, \text{ then } A^{-1} = \begin{bmatrix} 1/d_1 & 0 & 0 \\ 0 & 1/d_2 & 0 \\ 0 & 0 & 1/d_3 \end{bmatrix}$$

(*iii*) $A^{-1} = (A^{-1})' = (A')^{-1}$

Example: Consider a matrix **A** of order (3 × 3),

$$A = \begin{bmatrix} 3 & 4 & 2 \\ 5 & 7 & 4 \\ 6 & 8 & 9 \end{bmatrix}$$

$$|A| = 3 \times (63 - 32) - 5 \times (36 - 16) + 6 \times (16 - 14)$$
$$= 3 \times 31 - 5 \times 20 + 6 \times 2$$
$$= 93 - 100 + 12 = 5$$

Cofactor matrix,

$$A^* = \begin{bmatrix} (7 \times 9 - 4 \times 8) & -(5 \times 9 - 4 \times 6) & (5 \times 8 - 7 \times 6) \\ -(4 \times 9 - 2 \times 8) & (3 \times 9 - 2 \times 6) & -(3 \times 8 - 4 \times 6) \\ (4 \times 4 - 2 \times 7) & -(3 \times 4 - 2 \times 5) & (3 \times 7 - 4 \times 5) \end{bmatrix}$$

$$= \begin{bmatrix} 31 & -21 & -2 \\ -20 & 15 & 0 \\ 2 & -2 & 1 \end{bmatrix}$$

$$\left(A^*\right)' = \begin{bmatrix} 31 & -20 & 2 \\ -21 & 15 & -2 \\ -2 & 0 & 1 \end{bmatrix}$$

$$A^{-1} = \frac{1}{5} \begin{bmatrix} 31 & -20 & 2 \\ -21 & 15 & -2 \\ -2 & 0 & 1 \end{bmatrix} = \begin{bmatrix} 6.2 & -4.0 & 0.4 \\ -4.2 & 3.0 & -0.4 \\ -0.4 & 0 & 0.2 \end{bmatrix}$$

It is easy to verify that $A A^{-1} = I$.

$$AA^{-1} = \begin{bmatrix} 3 & 4 & 2 \\ 5 & 7 & 4 \\ 6 & 8 & 9 \end{bmatrix} \begin{bmatrix} 6.2 & -4 & 0.4 \\ -4.2 & 3 & -0.4 \\ -0.4 & 0 & 0.2 \end{bmatrix} = \begin{bmatrix} 1 & 0 & 0 \\ 0 & 1 & 0 \\ 0 & 0 & 1 \end{bmatrix} = I$$

Note: Finding inverse directly by evaluating the cofactor matrix and the value of the determinants is not cumbersome so long as the order of the matrix is not more than three. As soon as the order of the matrix is large, it becomes tedious to find inverse by the direct method. So another method known as *pivotal condensation method* is mostly used. Computer packages are also available to find the inverse of non-singular matrix by pivotal condensation method.

Example: Inverse of the matrix **A** given in the previous example has again been found out by pivotal condensation method for three reasons:

(*i*) It will be conspicuous to understand pivotal condensation method.

(*ii*) Two methods will be easily comparable.

(*iii*) For the sake of brevity

First the inverse of matrix is found out by pivotal condensation method. Below algorithm step by step explanation is provided.

Row no.		A			B		
(*i*)	3	4	2	1	0	0	
(*ii*)	5	7	4	0	1	0	
(*iii*)	6	8	9	0	0	1	
(*iv*)	1	4/3	2/3	1/3	0	0	1st pivotal row
(*v*)	0	1/3	2/3	−5/3	1	0	
(*vi*)	0	0	5	−2	0	1	
(*vii*)		1	2	−5	3	0	2nd pivotal row
(*viii*)		0	5	−2	0	1	
(*ix*)			1	−2/5	0	1/5	3rd pivotal row
(*x*)	1	4/3	2/3	1/3	0	0	
(*xi*)	0	1	2	−5	3	0	
(*xii*)	0	0	1	−2/5	0	1/5	
(*xiii*)	1	0	−2	7	−4	0	
(*xiv*)	0	1	0	−21/5	3	−2/5	
(*xv*)	0	0	1	−2/5	0	1/5	
(*xvi*)	1	0	0	31/5	−4	2/5	
(*xvii*)	0	1	0	−21/5	3	−2/5	
(*xviii*)	0	0	1	−2/5	0	1/5	
		I			A^{-1}		

Operations

1. Put **A** = I, left and right hand side of vertical line.
2. Divide first row elements by 3 on both sides of vertical line to obtain first element as 1. The row so obtained is known as 1st pivotal row.
3. Multiply first pivotal row element by 5 and subtract each element from second row corresponding elements to reduce the 1st column element to zero. 1st element, 5 − 5 = 0, 2nd row ele., 7− 20/3 =1/3, 3rd ele.,

4 − 10/ 3 = 2/3, 4th ele., 0 − 5/3 = −5/3, 5th ele., 1 − 0 = 1, 6th ele., 0 − 0 = 0

4. Multiply 1st pivotal row by 6 and repeat the operations as in step 3 to reduce the 3rd element of first column to zero. 3rd row, 2nd ele. = 8 − 24/3 = 0, 3rd row, 3rd ele. = 9 − 12/3 = 5, right side elements are 0 − 6/3 = − 2 and 0, 1.

5. Now ignore row (*iv*) and first column of second set.

6. Now in the remaining entries, divide row five elements by 1/3 to bring 1 in the (2,2)th position.

7. Since per chance (3,2)th element is zero, there is no need to do the operation on 2nd row to bring zero at this position. Otherwise it should have been converted to zero.

8. Delete again 1st row and 1st column of the third set.

9. Divide by 5 to get (3,3)th element 1.

10. Rewrite again all three pivotal rows with lower triangle elements zero.

11. At this stage, the operations are performed so as to reduce the upper triangle elements zero.

12. Multiply row (*xi*) by 4/3 and subtract row (*x*) element by element. The elements come out to be, 2/3 − 8/3 = − 2, 1/3 +20/3 = 7, 0 − 4/3 × 3 = − 4 and next element remains zero.

13. Multiply row (*xi*) by 1/3 and subtract from row (*x*) elements respectively so as to obtain (2,3)th element zero. Other values are written in the like manner without giving explanation.

14. Multiply row (*xv*) by 2 and add element by element to row (*xiii*) yielding the row (*xvi*)

15. Copy rows (*xiv*) and (*xv*) as shown in rows (*xvii*) and (*xviii*). In this way left side matrix is reduced to **I** and right hand matrix to $\mathbf{A}^{-1}$. It can be checked that this is the same matrix as found out by direct method.

Note: Pivotal condensation method discussed above can be extended to any higher order matrix without any complicacies.

Quadratic form in *n* variables

If $y_1, y_2, \dots\dots, y_n$ are *n* variables and *a*'s are some constants, then an equation of all possible paired multiples of *y*'s is of the form,

$$a_{11}y_1^2 + a_{12}y_1y_2 + a_{13}y_1y_3 + \dots\dots\dots + a_{1n}y_1y_n$$
$$a_{21}y_2y_1 + a_{22}y_2^2 + a_{23}y_2y_3 + \dots\dots\dots + a_{2n}y_2y_n$$
$$\vdots$$
$$a_{n1}y_ny_1 + a_{n2}y_ny_2 + a_{n3}y_ny_3 + \dots\dots\dots + a_{nn}y_n^2$$

In short above quadratic form can be written as $\sum_{i=1}^{n} \sum_{j=1}^{n} a_{ij} y_i y_j$

for $i, j = 1, 2, \ldots, n$.

In matrix notation, above quadratic form can be written as,

$$
\begin{pmatrix} y_1 & y_2 & y_3 & \cdots & y_n \end{pmatrix}
\begin{bmatrix}
a_{11} & a_{12} & a_{13} & \cdots & a_{1n} \\
a_{21} & a_{22} & a_{23} & \cdots & a_{2n} \\
a_{31} & a_{32} & a_{33} & \cdots & a_{3n} \\
\vdots & \vdots & \vdots & & \vdots \\
\vdots & \vdots & \vdots & & \vdots \\
\vdots & \vdots & \vdots & & \vdots \\
a_{n1} & a_{n2} & a_{n3} & \cdots & a_{nn}
\end{bmatrix}
\begin{pmatrix} y_1 \\ y_2 \\ y_3 \\ \vdots \\ \vdots \\ \vdots \\ y_n \end{pmatrix}
$$

$$
\begin{array}{ccc}
y' & \mathbf{A} & y \\
1 \times n & n \times n & n \times 1
\end{array}
$$

Where, y is a column vector and $\mathbf{A}$ is a symmetric matrix of order $(n \times n)$.

$\mathbf{A}$ is called the matrix of quadratic form and its determinant $|\mathbf{A}|$ is called the *Discriminant* of the quadratic form. Obviously, this implies that $\mathbf{A}$ is a square matrix. The *rank* of $y'\mathbf{A}y$ is same as the rank of $\mathbf{A}$.

Matrix $\mathbf{A}$ is *positive definite* if $y'\mathbf{A}y > 0$ for all y and *positive semidefinite* if $y'\mathbf{A}y \geq 0$ for all y and $y'\mathbf{A}y = 0$ for at least one y. Similarly, $\mathbf{A}$ is *negative definite* if $y'\mathbf{A}y < 0$ for all y and *negative semidefinite* if $y'\mathbf{A}y \leq 0$ for all y and $y'\mathbf{A}y = 0$ for at least one y.

Eigenvalues

$y'\mathbf{A}y$ is a quadratic form in n variables where $\mathbf{A}$ is necessarily a square symmetric matrix. Let us find a vector y, not equal to zero, which maximizes $y'\mathbf{A}y$ subject to the condition $y' y = 1$. Utilizing Lagrangian multiplier for maximization, vector y is obtained by differentiating partially with respect to y the expression, $y'\mathbf{A}y - (y' y - 1)$ and equating to zero.

$$\frac{\partial}{\partial y}\left\{ y'\mathbf{A}y - \lambda(y' y - 1) \right\} = 0$$

$$2y\mathbf{A} - 2\lambda y\, \mathbf{I} = 0$$

or

$$y\,\mathbf{A} - \lambda y\, \mathbf{I} = 0$$

$$y\,|\mathbf{A} - \lambda\,\mathbf{I}| = 0$$

Since $y \neq 0$, λ is to be found out which satisfies the equation $|\mathbf{A} - \lambda\mathbf{I}| = 0$. This equation is called the *characteristic equation* of $\mathbf{A}$ $(n \times n)$ which is an nth degree polynomial of λ. Any value of λ which satisfies the characteristic equation is called an *eigenvalue* or *characteristic root* or *latent root*. The word eigen in German means characteristic. If **A** is a singular matrix, i.e. $|A| = 0$ implies that at least one eigenvalue is equal to zero.

Example: Consider a 2×2 matrix $\mathbf{A} = \begin{bmatrix} 4 & 3 \\ 1 & 2 \end{bmatrix}$. The characteristic equation,

$$|\mathbf{A} - \lambda \mathbf{I}| = \left| \begin{bmatrix} 4 & 3 \\ 1 & 2 \end{bmatrix} - \lambda \begin{bmatrix} 1 & 0 \\ 0 & 1 \end{bmatrix} \right| = 0$$

$$\begin{vmatrix} 4 - \lambda & 3 \\ 1 & 2 - \lambda \end{vmatrix} = 0$$

$$(4 - \lambda)(2 - \lambda) - 3 = 0$$
$$\lambda^2 - 6 + 5 = 0$$

or
$$(\lambda - 5)(\lambda - 1) = 0$$
$$\lambda = 5, 1$$

This equation gives two eigenvalues since the rank of $|A| \neq 0$.

Example: Now take a situation when $|A| = 0$. Let $|A| = \begin{vmatrix} 4 & 6 \\ 2 & 3 \end{vmatrix} = 0$.

Similarly as above,

$$(4 - \lambda)(3 - \lambda) - 12 = 0$$
or
$$\lambda^2 - 7\lambda = 0$$
$$\lambda(\lambda - 7) = 0$$

so $\lambda = 0$ and 7. This shows that if **A** is not of full rank, at least one λ is zero. In the present example, one eigenvalue is zero and the other one is 7.

Remark: Matrix is a fully developed branch of mathematics. Here it is given to acquaint the readers with necessary portion so that they can understand multivariate analysis without difficulty.

Statistical Tables

Table B-1 Logarithms

	0	1	2	3	4	5	6	7	8	9	1	2	3	4	5	6	7	8	9
10	0000	0043	0086	0128	0170	0212	0253	0294	0334	0374	5	9	13	17	21	26	30	34	38
											4	8	12	16	20	24	28	32	36
11	0414	0453	0492	0531	0569	0607	0645	0682	0719	0755	4	8	12	16	20	23	27	31	35
											4	7	11	15	18	22	26	29	33
12	0792	0828	0864	0899	0934	0969	1004	1038	1072	1106	3	7	11	14	18	21	25	28	32
											3	7	10	14	17	20	24	27	31
13	1139	1173	1206	1239	1271	1303	1335	1367	1399	1430	3	6	10	13	16	19	23	26	29
											3	7	10	13	16	19	22	25	29
14	1461	1492	1523	1553	1584	1614	1644	1673	1703	1732	3	6	9	12	15	19	22	25	28
											3	6	9	12	14	17	20	23	26

(Contd.)

	0	1	2	3	4	5	6	7	8	9	1	2	3	4	5	6	7	8	9
15	1761	1790	1818	1847	1875	1903	1931	1959	1987	2014	3	6	9	11	14	17	20	23	26
											3	6	8	11	14	17	19	22	25
16	2041	2068	2095	2122	2148	2175	2201	2227	2253	2279	3	6	8	11	14	16	19	22	24
											3	5	8	10	13	16	18	21	23
17	2304	2330	2355	2380	2405	2430	2455	2480	2504	2529	3	5	8	10	13	15	18	20	23
											3	5	8	10	12	15	17	20	22
18	2553	2577	2601	2625	2648	2672	2695	2718	2742	2765	2	5	7	9	12	14	17	19	21
											2	4	7	9	11	14	16	18	21
19	2788	2810	2833	2856	2878	2900	2923	2945	2967	2989	2	4	7	9	11	13	16	18	20
											2	4	6	8	11	13	15	17	19
20	3010	3032	3054	3075	3096	3118	3139	3160	3181	3201	2	4	6	8	11	13	15	17	19
21	3222	3243	3263	3284	3304	3324	3345	3365	3385	3404	2	4	6	8	10	12	14	16	18
22	3434	3444	3464	3483	3502	3522	3541	3560	3579	3598	2	4	6	8	10	12	14	15	17
23	3617	3636	3655	3674	3692	3711	3729	3747	3766	3784	2	4	6	7	9	11	13	15	17
24	3802	3820	3838	3856	3874	3892	3909	3927	3945	3962	2	4	5	7	9	11	12	14	16
25	3979	3997	4014	4031	4048	4065	4082	4099	4116	4133	2	3	5	7	9	10	12	14	15
26	4150	4166	4183	4200	4216	4232	4249	4265	4281	4298	2	3	5	7	8	10	11	13	15
27	4314	4330	4346	4362	4378	4393	4409	4425	4440	4456	2	3	5	6	8	9	11	13	14

(Contd.)

	0	1	2	3	4	5	6	7	8	9	1	2	3	4	5	6	7	8	9
28	4472	4487	4502	4518	4533	4548	4564	4579	4594	4609	2	3	5	6	8	9	11	12	14
29	4624	4639	4654	4669	4683	4698	4713	4728	4742	4757	1	3	4	6	7	9	10	12	13
30	4771	4786	4800	4814	4829	4843	4857	4871	4886	4900	1	3	4	6	7	9	10	11	13
31	4914	4928	4942	4955	4969	4983	4997	5011	5024	5038	1	3	4	6	7	8	10	11	12
32	5051	5065	5079	5092	5105	5119	5132	5145	5159	5172	1	3	4	5	7	8	9	11	12
33	5185	5198	5211	5224	5237	5250	5263	5276	5289	5302	1	3	4	5	6	8	9	10	12
34	5315	5328	5340	5353	5366	5378	5391	5403	5416	5428	1	3	4	5	6	8	9	10	11
35	5441	5453	5465	5478	5490	5502	5514	5527	5539	5551	1	2	4	5	6	7	9	10	11
36	5563	5575	5587	5599	5611	5623	5635	5647	5658	5670	1	2	4	5	6	7	8	10	11
37	5682	5694	5705	5717	5729	5740	5752	5763	5775	5786	1	2	3	5	6	7	8	9	10
38	5798	5809	5821	5832	5843	5855	5866	5877	5888	5899	1	2	3	5	6	7	8	9	10
39	5911	5922	5933	5944	5955	5966	5977	5988	5999	6010	1	2	3	4	5	7	8	9	10
40	6021	6031	6042	6053	6064	6075	6085	6096	6107	6117	1	2	3	4	5	6	8	9	10
41	6128	6138	6149	6160	6170	6180	6191	6201	6212	6222	1	2	3	4	5	6	7	8	9
42	6232	6243	6253	6263	6274	6284	6294	6304	6314	6325	1	2	3	4	5	6	7	8	9
43	6335	6345	6355	6365	6375	6385	6395	6405	6415	6425	1	2	3	4	5	6	7	8	9
44	6435	6444	6454	6464	6474	6484	6493	6503	6513	6522	1	2	3	4	5	6	7	8	9
45	6532	6542	6551	6561	6571	6580	6590	6599	6609	6618	1	2	3	4	5	6	7	8	9
46	6628	6637	6646	6656	6665	6675	6684	6693	6702	6712	1	2	3	4	5	5	7	8	8
47	6721	6730	6739	6749	6758	6767	6776	6785	6794	6803	1	2	3	4	5	5	6	7	8
48	6812	6821	6830	6839	6848	6857	6866	6875	6884	6893	1	2	3	4	4	5	6	7	8

(Contd.)

	0	1	2	3	4	5	6	7	8	9	1	2	3	4	5	6	7	8	9
49	6902	6911	6920	6928	6937	6946	6955	6964	6972	6981	1	2	3	4	4	5	6	7	8
50	6990	6998	7007	7016	7024	7033	7042	7050	7059	7067	1	2	3	4	4	5	6	7	8
51	7076	7084	7093	7101	7110	7118	7126	7135	7143	7152	1	2	3	3	4	5	6	7	8
52	7160	7168	7177	7185	7193	7202	7210	7218	7226	7235	1	2	2	3	4	5	6	7	8
53	7243	7251	7259	7267	7275	7284	7292,	7300	7308	7316	1	2	2	3	4	5	6	6	7
54	7324	7332	7340	7348	7356	7364	7372	7380	7388	7396	1	2	2	3	4	5	6	6	7
55	7404	7412	7419	7427	7435	7443	7451	7459	7466	7474	1	2	2	3	4	5	5	6	7
56	7482	7490	7497	7505	7513	7520	7528	7536	7543	7551	1	2	2	3	4	5	5	6	7
57	7559	7566	7574	7582	7589	7597	7604	7612	7619	7627	1	2	2	3	4	5	5	6	7
58	7634	7642	7649	7657	7664	7672	7679	7686	7694	7701	1	1	2	3	4	4	5	6	7
59	7709	7716	7723	7731	7738	7745	7752	7760	7767	7774	1	1	2	3	4	4	5	6	7
60	7782	7789	7796	7803	7810	7818	7825	7832	7839	7846	1	1	2	3	4	4	5	6	7
61	7853	7860	7868	7875	7882	7889	7896	7903	7910	7917	1	1	2	3	3	4	5	6	6
62	7924	7931	7938	7945	7952	7959	7966	7973	7980	7987	1	1	2	3	4	4	5	6	6
63	7993	8000	8007	8014	8021	8028	8035	8041	8048	8055	1	1	2	3	3	4	5	5	6
64	8062	8069	8075	8082	8089	8096	8102	8109	8116	8122	1	1	2	3	3	4	5	5	6
65	8129	8136	8142	8149	8156	8162	8169	8176	8182	8189	1	1	2	3	3	4	5	5	6
66	8195	8202	8209	8215	8222	8228	8235	8241	8248	8254	1	1	2	3	3	4	5	5	6
67	8261	8267	8274	8280	8287	8293	8299	8306	8312	8319	1	1	2	3	3	4	5	5	6
68	8325	8331	8338	8344	8351	8357	8363	8370	8376	8382	1	1	2	3	3	4	4	5	6
69	8388	8395	8401	8407	8414	8420	8426	8432	8439	8445	1	1	2	2	3	4	4	5	6

(Contd.)

Logarithms (contd.)

N	0	1	2	3	4	5	6	7	8	9	1	2	3	4	5	6	7	8	9
70	8451	8457	8463	8470	8476	8482	8488	8494	8500	8506	1	1	2	2	3	4	4	5	6
71	8513	8519	8525	8531	8537	8543	8549	8555	8561	8567	1	1	2	2	3	4	4	5	5
72	8573	8579	8585	8591	8597	8603	8609	8615	8621	8627	1	1	2	2	3	4	4	5	5
73	8633	8639	8645	8651	8657	8663	8669	8675	8681	8686	1	1	2	2	3	4	4	5	5
74	8692	8698	8704	8710	8716	8722	8727	8733	8739	8745	1	1	2	2	3	4	4	5	5
75	8751	8756	8762	8768	8774	8779	8785	8791	8797	8802	1	1	2	2	3	3	4	5	5
76	8808	8814	8820	8825	8831	8837	8842	8848	8854	8859	1	1	2	2	3	3	4	5	5
77	8865	8871	8876	8882	8887	8893	8899	8904	8910	8915	1	1	2	2	3	3	4	4	5
78	8921	8927	8932	8938	8943	8949	8954	8960	8965	8971	1	1	2	2	3	3	4	4	5
79	8976	8982	8987	8993	8998	9004	9009	9015	9020	9025	1	1	2	2	3	3	4	4	5
80	9031	9036	9042	9047	9053	9058	9063	9069	9074	9079	1	1	2	2	3	3	4	4	5
81	9085	9090	9096	9101	9106	9112	9117	9122	9128	9133	1	1	2	2	3	3	4	4	5
82	9138	9143	9149	9154	9159	9165	9170	9175	9180	9186	1	1	2	2	3	3	4	4	5
83	9191	9196	9201	9206	9212	9217	9222	9227	9232	9238	1	1	2	2	3	3	4	4	5
84	9243	9248	9253	9258	9263	9269	9274	9279	9284	9289	1	1	2	2	3	3	4	4	5
85	9294	9299	9304	9309	9315	9320	9325	9330	9335	9340	1	1	2	2	3	3	4	4	5
86	9345	9350	9355	9360	9365	9370	9375	9380	9385	9390	1	1	2	2	3	3	4	4	5
87	9395	9400	9405	9410	9415	9420	9425	9430	9435	9440	1	1	2	2	3	3	4	4	5
88	9445	9450	9455	9460	9465	9469	9474	4479	9484	9489	0	1	1	2	2	3	3	4	4
89	9494	9499	9504	9509	9513	9518	9523	9528	9533	9538	0	1	1	2	2	3	3	4	4
90	9542	9547	9552	9557	9562	9566	9571	9576	9581	9586	0	1	1	2	2	3	3	4	4

(Contd.)

	0	1	2	3	4	5	6	7	8	9	1	2	3	4	5	6	7	8	9
91	9590	9595	9600	9605	9609	9614	9619	9624	9628	9633	0	1	1	2	2	3	3	4	4
92	9638	9643	9647	9652	9657	9661	9666	9671	9675	9680	0	1	1	2	2	3	3	4	4
93	9685	9689	9694	9699	9703	9708	9713	9717	9722	9727	0	1	1	2	2	3	3	4	4
94	9731	9736	9741	9745	9750	9754	9759	9763	9768	9773	0	1	1	2	2	3	3	4	4
95	9777	9782	9786	9791	9795	9800	9805	9809	9814	9818	0	1	1	2	2	3	3	4	4
96	9823	9827	9832	9836	9841	9845	9850	9854	9859	9863	0	1	1	2	2	3	3	4	4
97	9868	9872	9877	9881	9886	9890	9894	9899	9903	9908	0	1	1	2	2	3	3	4	4
98	9912	9917	9921	9926	9930	9934	9939	9943	9948	9952	0	1	1	2	2	3	3	4	4
99	9956	9961	9965	9969	9974	9978	9983	9987	9991	9996	0	1	1	2	2	3	3	3	4

Table B-2 Antilogarithms

	0	1	2	3	4	5	6	7	8	9	1	2	3	4	5	6	7	8	9
.00	1000	1002	1005	1007	1009	1012	1014	1016	1019	1021	0	0	1	1	1	1	2	2	2
.01	1023	1026	1028	1030	1033	1035	1038	1040	1042	1045	0	0	1	1	1	1	2	2	2
.02	1047	1050	1052	1054	1057	1059	1062	1064	1067	1069	0	0	1	1	1	1	2	2	2
.03	1072	1074	1076	1079	1081	1084	1086	1089	1091	1094	0	0	1	1	1	1	2	2	2
.04	1096	1099	1102	1104	1107	1109	1112	1114	1117	1119	0	1	1	1	1	2	2	2	2
.05	1122	1125	1127	1130	1132	1135	1138	1140	1143	1146	0	1	1	1	1	2	2	2	2
.06	1148	1151	1153	1156	1159	1161	1164	1167	1169	1172	0	1	1	1	1	2	2	2	2
.07	1175	1178	1180	1183	1186	1189	1191	1194	1197	1199	0	1	1	1	1	2	2	2	3
.08	1202	1205	1208	1211	1213	1216	1219	1222	1225	1227	0	1	1	1	1	2	2	2	3
.09	1230	1233	1236	1239	1242	1245	1247	1250	1253	1256	0	1	1	1	1	2	2	2	3
.10	1259	1262	1265	1268	1271	1274	1276	1279	1282	1285	0	1	1	1	1	2	2	2	3
.11	1288	1291	1294	1297	1300	1303	1306	1309	1312	1315	0	1	1	1	2	2	2	2	3
.12	1318	1321	1324	1327	1330	1334	1337	1340	1343	1346	0	1	1	1	2	2	2	2	3
.13	1349	1352	1355	1358	1361	1365	1368	1371	1374	1377	0	1	1	1	2	2	2	3	3
.14	1380	1384	1387	1390	1393	1396	1400	1403	1406	1409	0	1	1	1	2	2	2	3	3
.15	1413	1416	1419	1422	1426	1429	1432	1435	1439	1442	0	1	1	1	2	2	2	3	3
.16	1445	1449	1452	1455	1459	1462	1466	1469	1472	1476	0	1	1	1	2	2	2	3	3
.17	1479	1483	1486	1489	1493	1496	1500	1503	1507	1510	0	1	1	1	2	2	2	3	3

(Contd.)

	0	1	2	3	4	5	6	7	8	9	1	2	3	4	5	6	7	8	9
.18	1514	1517	1521	1524	1528	1531	1535	1538	1542	1545	0	1	1	1	2	2	2	3	3
.19	1549	1552	1556	1560	1563	1567	1570	1574	1578	1581	0	1	1	1	2	2	3	3	3
.20	1585	1589	1592	1596	1600	1603	1607	1611	1614	1618	0	1	1	1	2	2	3	3	3
.21	1622	1626	1629	1633	1637	1641	1644	1648	1652	1656	0	1	1	2	2	2	3	3	3
.22	1660	1663	1667	1671	1675	1679	1683	1687	1690	1694	0	1	1	2	2	2	3	3	3
.23	1698	1702	1706	1710	1714	1718	1722	1726	1730	1734	0	1	1	2	2	2	3	3	4
.24	1738	1742	1746	1750	1754	1758	1762	1766	1770	1774	0	1	1	2	2	2	3	3	4
.25	1778	1782	1786	1791	1795	1799	1803	1807	1811	1816	0	1	1	2	2	3	3	3	4
.26	1820	1824	1828	1832	1837	1841	1845	1849	1854	1858	0	1	1	2	2	3	3	3	4
.27	1862	1866	1871	1875	1879	1884	1888	1892	1897	1901	0	1	1	2	2	3	3	3	4
.28	1905	1910	1914	1919	1923	1928	1932	1936	1941	1945	0	1	1	2	2	3	3	4	4
.29	1950	1954	1959	4963	1968	1972	1977	1982	1986	1991	0	1	1	2	2	3	3	4	4
.30	1995	2000	2004	2009	2014	2018	2023	2028	2032	2037	0	1	1	2	2	3	3	4	4
.31	2042	2046	2051	2056	2061	2065	2070	2075	2080	2084	0	1	1	2	2	3	3	4	4
.32	2089	2094	2099	2104	2109	2113	2118	2123	2128	2133	0	1	1	2	2	3	3	4	4
.33	2138	2143	2148	2153	2158	2163	2168	2173	2178	2183	0	1	1	2	3	3	3	4	4
.34	2188	2193	2198	2203	2208	2213	2218	2223	2228	2234	1	1	2	2	3	3	4	4	5
.35	2239	2244	2249	2254	2259	2265	2270	2275	2280	2286	1	1	2	2	3	3	4	4	5
.36	2291	2296	2301	2307	2312	2317	2323	2328	2333	2339	1	1	2	2	3	3	4	4	5
.37	2344	2350	2355	2360	2366	2371	2377	2382	2388	2393	1	1	2	2	3	3	4	4	5
.38	2399	2404	2410	2415	2421	2427	2432	2438	2443	2449	1	1	2	2	3	3	4	4	5

(Contd.)

	0	1	2	3	4	5	6	7	8	9	1	2	3	4	5	6	7	8	9
.39	2455	2460	2466	2472	2477	2483	2489	2495	2500	2506	1	1	2	2	3	3	4	5	5
.40	2512	2518	2523	2529	2535	2541	2547	2553	2559	2564	1	1	2	2	3	4	4	5	5
.41	2570	2576	2582	2588	2594	2600	2606	2612	2618	2624	1	1	2	2	3	4	4	5	5
.42	2630	2636	2642	2649	2655	2661	2667	2673	2679	2685	1	1	2	2	3	4	4	5	6
.43	2692	2698	2704	2710	2716	2723	2729	2735	2742	2748	1	1	2	3	3	4	4	5	6
.44	2754	2761	2767	2773	2780	2786	2793	2799	2805	2812	1	1	2	3	3	4	5	5	6
.45	2818	2825	2831	2838	2844	2851	2858	2864	2871	2877	1	1	2	3	3	4	5	5	6
.46	2884	2891	2897	2904	2911	2917	2924	2931	2938	2944	1	1	2	3	3	4	5	5	6
.47	2951	2958	2965	2972	2979	2985	2992	2999	3006	3013	1	1	2	3	3	4	5	5	6
.48	3020	3027	3034	3041	3048	3055	3062	3069	3076	3083	1	1	2	3	4	4	5	6	6
.49	3090	3097	3105	3112	3119	3126	3133	3141	3148	3155	1	1	2	3	4	4	5	6	6
.50	3162	3170	3177	3184	3192	3199	3206	3214	3221	3228	1	1	2	3	4	5	5	6	7
.51	3236	3243	3251	3258	3266	3273	3281	3289	3296	3304	1	1	2	3	4	5	5	6	7
.52	3311	3319	3327	3334	3342	3350	3357	3365	3373	3381	1	2	2	3	4	5	5	6	7
.53	3388	3396	3404	3412	3420	3428	3436	3443	3451	3459	1	2	2	3	4	5	6	6	7
.54	3467	3475	3483	3491	3499	3508	3516	3524	3532	3540	1	2	2	3	4	5	6	6	7
.55	3548	3556	3565	3573	3581	3589	3597	3606	3614	3622	1	2	2	3	4	5	6	7	8
.56	3631	3639	3648	3656	3664	3673	3681	3690	3698	3707	1	2	2	3	4	5	6	7	8
.57	3715	3724	3733	3741	3750	3758	3767	3776	3784	3793	1	2	3	3	4	5	6	7	8
.58	3802	3811	3819	3828	3837	3846	3855	3864	3873	3882	1	2	3	4	4	5	6	7	8
.59	3890	3899	3908	3917	3926	3936	3945	3954	3963	3972	1	2	3	4	5	5	6	7	8

(Contd.)

	0	1	2	3	4	5	6	7	8	9	1	2	3	4	5	6	7	8	9
.60	3981	3990	3999	4009	4018	4027	4036	4046	4055	4064	1	2	3	4	5	6	6	7	8
.61	4074	4083	4093	4102	4111	4121	4130	4140	4150	4159	1	2	3	4	5	6	7	8	9
.62	4169	4178	4188	4198	4207	4217	4227	4236	4246	4256	1	2	3	4	5	6	7	8	9
.63	4266	4276	4285	4295	4305	4315	4325	4335	4345	4355	1	2	3	4	5	6	7	8	9
.64	4365	4375	4385	4395	4406	4416	4426	4436	4446	4457	1	2	3	4	5	6	7	8	9
.65	4467	4477	4487	4498	4508	4519	4529	4539	4550	4560	1	2	3	4	5	6	7	8	9
.66	4571	4581	4592	4603	4613	4624	4634	4645	4656	4667	1	2	3	4	5	6	7	9	10
.67	4677	4688	4699	4710	4721	4732	4742	4753	4764	4775	1	2	3	4	5	7	8	9	10
.68	4786	4797	4808	4819	4831	4842	4853	4864	4875	4887	1	2	3	4	5	7	8	9	10
.69	4898	4909	4920	4932	4943	4955	4966	4977	4989	5000	1	2	3	5	6	7	8	9	10
.70	5012	5023	5035	5047	5058	5070	5082	5093	5105	5117	1	2	4	5	6	7	8	9	11
.71	5129	5140	5152	5164	5176	5188	5200	5212	5224	5236	1	2	4	5	6	7	8	10	11
.72	5248	5260	5272	5284	5297	5309	5321	5333	5346	5358	1	3	4	5	6	8	9	10	11
.73	5370	5383	5395	5408	5420	5433	5445	5458	5470	5483	1	3	4	5	6	8	9	10	11
.74	5495	5508	5521	5534	5546	5559	5572	5585	5598	5610	1	3	4	5	6	8	9	10	12
.75	5623	5636	5649	5662	5675	5689	5702	5715	5728	5741	1	3	4	5	7	8	9	10	12
.76	5754	5768	5781	5794	5808	5821	5834	5848	5861	5875	1	3	4	5	7	8	9	11	12
.77	5888	5902	5916	5929	5943	5957	5970	5984	5998	6012	1	3	4	5	7	8	10	11	12
.78	6026	6039	6053	6067	6081	6095	6109	6124	6138	6152	1	3	4	6	7	8	10	11	13
.79	6166	6180	6194	6209	6223	6237	6252	6266	6281	6295	1	3	4	6	7	9	10	11	13
.80	6310	6324	6339	6353	6368	6383	6397	6412	6427	6442	1	3	4	6	7	9	10	12	13

(Contd.)

	0	1	2	3	4	5	6	7	8	9	1	2	3	4	5	6	7	8	9
.81	6457	6471	6486	6501	6516	6531	6546	6561	6577	6592	2	3	5	6	8	9	11	12	14
.82	6607	6622	6637	6653	6668	6683	6699	6714	6730	6745	2	3	5	6	8	9	11	12	14
.83	6761	6776	6792	6808	6823	6839	6853	6871	6887	6902	2	3	5	6	8	9	11	13	14
.84	6918	6934	6950	6966	6982	6998	7015	7031	7047	7063	2	3	5	6	8	10	11	13	15
.85	7079	7096	7112	7129	7145	7161	7178	7194	7211	7228	2	3	5	7	8	10	12	13	15
.86	7244	7261	7278	7295	7311	7328	7345	7362	7379	7396	2	3	5	7	8	10	12	13	15
.87	7413	7430	7447	7464	7482	7499	7516	7534	7551	7568	2	3	5	7	9	10	12	14	16
.88	7586	7603	7621	7638	7656	7674	7691	7709	7727	7745	2	4	5	7	9	11	12	14	16
.89	7762	7780	7798	7816	7834	7852	7870	7889	7907	7925	2	4	5	7	9	11	13	14	16
.90	7943	7962	7980	7998	8017	8035	8054	8072	8091	8110	2	4	6	7	9	11	13	15	17
.91	8128	8147	8166	8185	8204	8222	8241	8260	8279	8299	2	4	6	8	9	11	13	15	17
.92	8318	8337	8356	8375	8395	8414	8433	8453	8472	8492	2	4	6	8	10	12	14	15	17
.93	8511	8531	8551	8570	8590	8610	8630	8650	8670	8690	2	4	6	8	10	12	14	16	18
.94	8710	8730	8750	8770	8790	8810	8831	8851	8872	8892	2	4	6	8	10	12	14	16	18
.95	8913	8933	8954	8974	8995	9016	9036	9057	9078	9099	2	4	6	8	11	12	15	17	19
.96	9120	9141	9162	9183	9204	9226	9247	9268	9290	9314	2	4	6	9	11	13	15	17	19
.97	9333	9354	9376	9397	9419	9441	9462	9484	9506	9528	2	4	7	9	11	13	15	17	20
.98	9550	9572	9594	9616	9638	9661	9683	9705	9727	9750	2	4	7	9	11	13	16	18	20
.99	9772	9795	9817	9840	9863	9886	9908	9931	9954	9977	2	5	7	9	11	14	16	18	20

Table B-3 One digit random numbers

9	8	0	9	5	9	8	1	5	7	1	3	7	3	2	7	1	2	1	9
9	3	8	6	1	4	4	4	5	9	2	9	7	5	8	3	9	4	1	0
9	8	4	4	4	0	2	7	2	0	6	1	4	6	3	0	9	1	4	0
3	4	1	2	1	1	5	4	8	5	7	3	2	3	6	6	9	2	3	0
3	7	5	1	7	6	7	6	7	0	1	5	8	3	0	8	7	6	2	1
0	7	1	9	2	2	3	3	3	9	5	4	1	2	9	7	7	6	1	2
8	4	1	9	2	2	3	3	9	5	9	2	6	1	4	0	0	1	0	1
1	0	9	0	2	6	9	9	2	9	5	7	3	2	8	8	3	5	7	0
0	7	1	3	1	3	1	3	5	8	6	7	9	2	6	1	4	0	0	1
2	4	6	7	2	3	9	0	1	1	7	5	1	7	0	4	9	0	8	6
5	2	7	7	0	5	1	2	0	2	6	9	1	5	9	1	1	6	4	1
9	2	5	7	4	4	2	8	8	9	1	5	3	2	3	3	3	6	9	2
1	3	4	1	5	2	9	9	0	0	9	9	3	9	6	0	5	3	4	1
2	3	7	8	2	7	2	3	8	6	7	1	8	1	4	9	6	1	6	9
7	6	9	8	3	4	3	4	1	3	7	9	0	8	7	0	8	6	7	7
1	7	7	9	9	4	2	0	9	3	6	9	8	2	1	1	2	2	6	3
9	6	8	4	7	2	8	2	2	9	2	6	2	0	9	3	1	3	1	3
3	6	1	2	4	1	9	6	1	3	1	9	3	5	3	4	2	0	4	9
5	1	6	1	3	4	4	7	8	8	5	3	7	5	9	7	7	5	3	6
6	6	0	4	7	1	2	6	2	3	2	5	5	3	8	8	1	4	9	4
9	9	2	7	0	7	4	2	5	0	0	5	9	8	5	7	9	7	2	2
2	6	6	1	2	1	0	3	0	0	3	2	2	5	2	4	0	3	6	1
1	7	7	1	4	6	6	6	5	4	1	2	1	2	2	6	5	1	4	1
1	3	5	0	6	2	1	1	1	1	9	3	7	1	1	4	4	1	1	2
8	8	0	1	6	3	2	0	3	2	6	1	2	6	4	2	5	6	1	7
6	0	3	4	9	6	8	4	3	0	4	8	4	6	9	4	4	5	7	4
6	9	1	4	0	6	0	8	8	3	4	1	1	8	9	8	1	7	8	4
1	9	9	5	1	3	9	0	6	0	9	6	7	9	0	9	5	5	1	6
9	8	5	1	0	6	2	0	5	3	7	6	3	7	2	3	3	2	0	7
8	3	9	1	5	1	7	5	9	3	8	6	2	7	2	4	8	4	4	6

Table B-4 Two-digit random numbers

```
13 31 76 54 08 18 96 35 03 32 73 43 64 16 63 58 00 91 79 97
10 30 01 58 94 10 92 69 19 12 02 91 08 50 33 69 21 79 40 05
20 00 43 15 72 46 23 56 84 19 81 98 29 62 32 10 53 30 15 40
22 43 76 59 00 00 27 53 22 07 32 26 40 30 04 98 17 08 95 74
51 36 38 46 83 70 28 20 98 36 20 34 55 23 02 40 54 30 39 29

97 36 48 20 88 81 84 79 53 02 79 63 51 36 19 99 45 23 53 06
73 33 18 79 84 41 89 19 08 64 26 94 07 75 42 22 95 19 73 40
12 09 40 49 12 80 69 94 62 68 48 63 99 02 98 80 52 61 20 61
76 02 18 21 87 07 42 86 71 40 55 05 45 78 80 32 35 04 04 40
57 26 29 61 70 97 20 14 76 23 56 04 48 17 21 57 48 90 50 78

64 15 38 92 44 63 08 86 25 85 64 39 55 37 90 16 75 38 78 56
18 91 82 36 73 72 37 75 60 17 98 27 54 04 68 15 44 28 42 52
61 31 58 66 32 63 01 05 16 15 38 24 99 10 89 87 83 26 68 14
74 78 49 83 14 04 17 41 71 07 10 01 58 08 30 59 44 03 09 66
00 87 70 95 70 17 14 51 10 17 38 09 85 47 53 33 84 17 04 54

20 54 82 81 84 83 42 49 78 31 01 81 68 41 25 77 29 05 10 12
31 05 62 61 49 14 49 35 38 27 05 76 92 38 67 50 11 50 44 44
18 85 10 42 09 48 33 02 41 73 04 92 80 22 39 67 85 18 40 62
28 32 06 52 05 05 13 47 27 53 16 66 61 45 93 94 15 99 53 66
92 68 77 64 97 51 63 68 06 74 20 50 79 00 17 90 57 01 62 25

55 21 39 44 03 42 41 48 39 44 90 19 66 92 33 11 20 42 95 15
09 83 91 34 27 04 11 23 16 92 44 03 04 94 72 31 65 47 85 04
29 66 20 14 34 97 10 44 95 07 93 03 89 29 20 54 15 61 83 39
89 43 93 48 97 60 02 85 04 58 74 29 44 81 89 58 55 04 64 53
72 56 65 12 16 12 15 81 87 96 76 15 65 84 76 19 19 33 28 35

97 12 69 71 63 29 20 88 95 93 00 37 44 31 10 05 28 58 58 16
03 41 05 12 94 34 02 99 84 78 56 90 81 58 07 12 22 88 93 86
52 85 72 13 48 99 27 80 48 98 61 48 69 80 07 12 78 97 48 46
35 58 53 42 27 69 16 25 80 74 67 67 16 52 08 88 86 51 55 95
61 16 57 20 11 85 07 94 13 55 68 71 75 13 97 90 02 99 49 82

03 41 05 12 94 34 02 99 84 78 56 90 81 58 07 12 22 88 93 86
52 85 72 13 48 99 27 80 48 98 61 48 69 80 07 12 78 97 02 46
```

(Contd.)

B-4 (Contd.)

```
35 58 52 42 27 69 17 29 80 74 67 16 52 08 88 86 51 55 95 27
61 16 57 20 11 85 07 94 13 55 68 71 75 13 97 90 02 99 49 32
38 61 18 96 87 33 18 05 71 26 61 34 03 73 59 62 58 86 58 61

83 18 46 07 16 64 02 80 39 64 45 74 74 45 64 94 25 77 90 36
93 05 71 05 24 54 38 76 92 74 85 64 77 73 49 34 69 60 75 80
54 80 77 49 22 86 43 41 12 42 46 08 43 76 58 66 07 27 68 29
65 35 64 56 95 98 62 35 48 52 27 24 54 71 65 69 67 19 03 09
13 40 26 86 73 42 46 61 19 54 71 65 86 58 35 86 45 39 10 52

10 52 72 00 14 99 92 92 70 67 69 00 22 27 08 27 06 22 19 10
65 43 36 81 09 52 22 99 18 03 50 35 21 28 91 24 12 42 66 13
90 20 10 63 14 95 48 76 86 96 47 38 04 13 67 73 89 01 49 97
27 38 16 13 54 85 09 40 65 59 91 27 18 80 90 98 35 54 93 85
52 85 69 14 81 30 69 18 35 19 79 19 12 52 35 54 91 40 56 06

00 06 72 53 99 41 59 14 89 61 77 53 12 56 22 23 08 74 22 71
92 55 56 07 12 40 59 86 57 04 11 37 63 57 32 69 70 86 42 63
42 19 56 67 41 51 10 59 66 20 05 27 75 87 06 69 58 94 97 51
38 00 43 90 75 26 51 59 94 54 35 93 20 22 51 66 83 45 69 74
25 79 15 53 39 99 61 37 19 49 13 35 31 04 75 13 03 49 02 54

03 99 15 77 82 31 03 56 64 11 47 08 60 80 60 08 56 43 05 77
78 93 10 24 14 11 71 95 25 80 67 56 03 79 36 63 26 87 27 88
75 98 65 88 89 26 17 51 25 62 42 72 35 81 02 51 31 98 39 22
67 64 77 59 35 44 12 27 76 90 60 77 31 52 85 17 46 00 58 84
13 36 70 15 79 55 15 69 11 96 02 66 85 20 19 15 93 86 90 77

35 75 73 08 77 52 00 51 08 87 85 48 21 80 25 66 06 81 09 94
80 59 74 35 85 79 24 31 03 74 30 78 68 54 34 55 45 59 70 03
16 28 74 27 69 99 39 82 91 70 01 70 48 62 19 56 88 78 89 52
00 66 54 47 69 26 20 04 67 51 45 32 47 48 46 37 47 53 13 01
28 50 03 74 95 31 52 77 49 07 01 75 60 53 57 70 06 14 35 83
```

Table B-5 Abridged *t*-table and *Z*-values

	Levels of significance			
Two-tailed test	0.20	0.10	0.05	0.02
One-tailed test	0.10	0.05	0.025	0.01
d.f. = υ				
1	3.078	6.314	12.706	31.821
2	1.886	2.920	4.303	6.965
3	1.638	2.353	3.182	4.541
4	1.533	2.132	2.776	3.747
5	1.476	2.015	2.571	3.365
6	1.440	1.943	2.447	3.143
7	1.415	1.895	2.365	2.998
8	1.397	1.860	2.306	2.896
9	1.383	1.833	2.262	2.821
10	1.372	1.812	2.228	2.764
11	1.363	1.796	2.201	2.718
12	1.356	1.782	2.179	2.681
13	1.350	1.771	2.160	2.650
14	1.345	1.761	2.145	2.624
15	1.341	1.753	2.131	2.602
16	1.337	1.746	2.120	2.583
17	1.333	1.740	2.110	2.567
18	1.330	1.734	2.101	2.552
19	1.328	1.729	2.093	2.539
20	1.325	1.725	2.086	2.528
21	1.323	1.721	2.080	2.518
22	1.321	1.717	2.074	2.508
23	1.319	1.714	2.069	2.500
24	1.318	1.711	2.064	2.492
25	1.316	1.708	2.060	2.485
26	1.315	1.706	2.056	2.479
27	1.314	1.703	2.052	2.473
28	1.313	1.701	2.048	2.467
29	1.311	1.699	2.045	2.462
30	1.310	1.697	2.042	2.457

(Contd.)

Table B-5 (Contd.)

Levels of significance				
Two-tailed test	0.20	0.10	0.05	0.02
One-tailed test	0.10	0.05	0.025	0.01
d.f. = υ				
31	1.309	1.696	2.040	2.453
32	1.309	1.694	2.037	2.449
33	1.308	1.692	2.035	2.445
34	1.307	1.691	2.032	2.441
35	1.306	1.690	2.030	2.438
36	1.306	1.688	2.028	2.434
37	1.305	1.687	2.026	2.431
38	1.304	1.686	2.024	2.429
39	1.304	1.685	2.023	2.426
40	1.303	1.684	2.021	2.423
45	1.301	1.679	2.014	2.412
50	1.299	1.676	2.009	2.403
55	1.297	1.673	2.004	2.396
60	1.296	1.671	2.000	2.390
65	1.295	1.669	1.997	2.385
70	1.294	1.667	1.994	2.381
75	1.293	1.665	1.992	2.377
80	1.292	1.664	1.988	2.374
85	1.292	1.663	1.990	2.371
90	1.291	1.662	1.987	2.368
95	1.291	1.661	1.985	2.366
100	1.290	1.660	1.984	2.364
Z	1.282	1.645	1.960	2.326

Source: Website and values interpolated by the author.

Table B-6 Abridged chi-square table

d.f.	$\chi^2_{.100}$	$\chi^2_{.050}$	$\chi^2_{.025}$	$\chi^2_{.010}$
1	2.706	3.841	5.024	7.879
2	4.605	5.991	7.378	9.210
3	6.251	7.815	9.348	11.345
4	7.779	8.488	11.143	13.277
5	9.236	11.070	12.833	15.086
6	10.645	12.592	14.449	16.812
7	12.017	14.067	16.013	18.475
8	13.362	15.507	17.535	20.090
9	14.684	16.919	19.023	21.666
10	15.987	18.307	20.483	23.209
11	17.275	19.675	21.920	24.725
12	18.549	21.026	23.337	26.217
13	19.812	22.362	24.736	27.688
14	21.064	23.685	26.119	29.141
15	22.307	24.996	27.488	30.578
16	23.542	26.296	28.845	32.000
17	24.769	27.587	30.191	33.409
18	25.989	28.869	31.526	34.805
19	27.204	30.144	32.852	36.191
20	28.412	31.410	34.170	37.566
21	29.615	32.671	35.479	38.932
22	30.813	33.924	36.781	40.289
23	32.007	35.172	38.076	41.638
24	33.196	36.415	39.364	42.980
25	34.382	37.652	40.646	44.314
26	35.563	38.885	41.923	45.642
27	36.741	40.113	43.194	46.963
28	37.916	41.337	44.461	48.278
29	39.088	42.557	45.722	49.588
30	40.256	43.773	46.979	50.892
31	41.422	42.557	46.979	52.191
32	42.585	46.194	49.480	53.486

Critical values of χ^2_α for α level of significance

(*Contd.*)

Table B-6 (Contd.)

	Critical values of χ_α^2 for α level of significance			
d.f.	$\chi^2_{.100}$	$\chi^2_{.050}$	$\chi^2_{.025}$	$\chi^2_{.010}$
33	43.745	47.400	50.725	54.776
34	44.903	48.602	51.966	56.061
35	46.059	49.802	53.203	57.342
36	47.212	50.998	54.437	58.619
37	48.363	52.192	55.668	59.893
38	49.513	53.384	56.896	61.162
39	50.660	54.572	58.120	62.428
40	51.805	55.758	59.342	63.691
45	57.505	61.656	65.410	69.957
50	63.167	67.505	71.420	76.154
55	68.796	73.311	77.380	82.292
60	74.397	79.082	83.298	88.379
65	79.973	84.821	89.177	94.422
70	85.527	90.531	95.023	100.425
75`	91.061	96.217	100.839	106.393
80	96.578	101.879	106.629	112.329
85	102.079	107.522	112.393	118.236
90	107.565	113.145	118.136	124.116
95	113.038	118.752	123.858	129.973
100	118.498	124.342	129.561	135.807

Source: Website and values interpolated by the author.

Table B-7 Upper critical values of the F-distribution

Degrees of freedom of the variance in numerator = υ_1

Degrees of freedom of the variance in denominator = υ_2

Level of significance = α

Table is provided for three most commonly used
levels of significance, i.e. 10%, 5% and 1%.

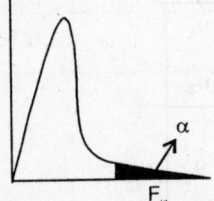

υ_1 υ_2	α	1	2	3	4	5	6	7	8	9	10
1	0.10	39.86	49.50	53.59	55.83	57.24	58.20	58.91	59.44	59.86	60.20
	0.05	161.45	199.50	215.71	224.58	230.16	233.99	236.77	238.90	240.5	241.90
	0.01	4052	5000	5403	5625	5764	5859	5928	5982	6022	6056
2	0.10	8.53	9.00	9.16	9.24	9.29	9.33	9.35	9.37	9.38	9.39
	0.05	18.51	19.00	19.16	19.25	19.30	19.33	19.35	19.37	19.38	19.40
	0.01	98.50	99.00	99.17	99.25	99.30	99.33	99.35	99.37	99.39	99.40
3	0.10	5.54	5.46	5.39	5.34	5.31	5.28	5.27	5.25	5.24	5.23
	0.05	10.13	9.55	9.28	9.12	9.01	8.94	8.89	8.85	8.81	8.79
	0.01	34.12	30.82	29.46	28.71	28.24	27.91	27.67	27.49	27.35	27.23
4	0.10	4.54	4.32	4.19	4.11	4.05	4.01	3.98	3.96	3.94	3.92
	0.05	7.71	6.94	6.59	6.39	6.26	6.16	6.09	6.04	6.00	5.96
	0.01	21.20	18.00	16.69	15.98	15.52	15.21	14.98	14.80	14.66	14.55
5	0.10	4.06	3.78	3.62	3.52	3.45	3.40	3.37	3.34	3.32	3.30
	0.05	6.61	5.79	5.41	5.19	5.05	4.95	4.88	4.82	4.77	4.74
	0.01	16.26	13.27	12.06	11.39	10.97	10.67	10.46	10.29	10.16	10.05
6	0.10	3.78	3.46	3.29	3.18	3.11	3.06	3.01	2.98	2.96	2.94
	0.05	5.99	5.14	4.76	4.53	4.39	4.28	4.21	4.15	4.10	4.06
	0.01	13.75	10.92	9.78	9.15	8.75	8.47	8.26	8.10	7.98	7.87
7	0.10	3.59	3.26	3.07	2.96	2.88	2.83	2.78	2.75	2.72	2.70
	0.05	5.59	4.74	4.35	4.12	3.97	3.87	3.79	3.73	3.68	3.64
	0.01	12.25	9.55	8.45	7.85	7.46	7.19	6.99	6.84	6.72	6.62
8	0.10	3.46	3.11	2.92	2.81	2.73	2.67	2.62	2.59	2.56	2.54
	0.05	5.32	4.46	4.07	3.84	3.69	3.58	3.50	3.44	3.89	3.35
	0.01	11.26	8.65	7.59	7.01	6.63	6.37	6.18	6.03	5.91	5.81

(Contd.)

Table B-7 (Contd.)

υ_1 α υ_2		1	2	3	4	5	6	7	8	9	10
9	0.10	3.36	3.01	2.81	2.69	2.61	2.55	2.50	2.47	2.44	2.42
	0.05	5.12	4.26	3.86	3.63	3.48	3.37	3.29	3.23	3.18	3.14
	0.01	10.56	8.02	6.99	6.42	6.06	5.80	5.61	5.47	5.35	5.26
10	0.10	3.29	2.92	2.73	2.60	2.52	2.46	2.41	2.38	2.35	2.32
	0.05	4.96	4.10	3.71	3.48	3.33	3.22	3.14	3.07	3.02	2.98
	0.01	10.04	7.56	6.55	5.99	5.64	5.39	5.20	5.06	4.94	4.85
11	0.10	3.22	2.86	2.66	2.54	2.45	2.39	2.34	2.30	2.27	2.25
	0.05	4.84	3.98	3.59	3.36	3.20	3.10	3.01	2.95	2.90	2.85
	0.01	9.65	7.21	6.22	5.67	5.32	5.07	4.89	4.74	4.63	4.54
12	0.10	3.18	2.81	2.61	2.48	2.39	2.33	2.28	2.24	2.21	2.19
	0.05	4.75	3.88	3.49	3.26	3.11	3.00	2.91	2.85	2.80	2.75
	0.01	9.33	6.93	5.95	5.41	5.06	4.82	4.64	4.50	4.39	4.30
13	0.10	3.14	2.76	2.56	2.43	2.35	2.28	2.23	2.20	2.16	2.14
	0.05	4.67	3.81	3.41	3.18	3.02	2.92	2.83	2.77	2.71	2.67
	0.01	9.07	6.70	5.74	5.20	4.86	4.62	4.44	4.30	4.19	4.10
14	0.10	3.10	2.73	2.52	2.39	2.31	2.24	2.19	2.15	2.12	2.10
	0.05	4.60	3.74	3.34	3.11	2.96	2.85	2.76	2.70	2.65	2.60
	0.01	8.86	6.51	5.56	5.04	4.70	4.46	4.28	4.14	4.03	3.94
15	0.10	3.07	2.70	2.49	2.36	2.27	2.21	2.16	2.12	2.09	2.06
	0.05	4.54	3.68	3.29	3.06	2.90	2.79	2.71	2.64	2.59	2.54
	0.01	8.68	6.36	5.42	4.89	4.56	4.32	4.14	4.00	3.90	3.80
16	0.10	3.05	2.67	2.46	2.33	2.24	2.18	2.13	2.09	2.06	2.03
	0.05	4.49	3.63	3.24	3.01	2.85	2.74	2.66	2.59	2.54	2.49
	0.01	8.53	6.23	5.29	4.77	4.44	4.20	4.03	3.89	3.78	3.69
17	0.10	3.03	2.64	2.44	2.31	2.22	2.15	2.10	2.06	2.03	2.00
	0.05	4.45	3.59	3.20	2.96	2.81	2.70	2.61	2.55	2.49	2.45
	0.01	8.40	6.11	5.18	4.67	4.34	4.10	3.93	3.79	3.68	3.59
18	0.10	3.01	2.62	2.42	2.29	2.20	2.13	2.08	2.04	2.00	1.98
	0.05	4.41	3.56	3.16	2.93	2.77	2.66	2.58	2.51	2.46	2.41
	0.01	8.28	6.01	5.09	4.58	4.25	4.02	3.84	3.70	3.60	3.51

(Contd.)

Table B-7 (Contd.)

v_1 v_2	α	1	2	3	4	5	6	7	8	9	10
19	0.10	2.99	2.61	2.40	2.27	2.18	2.11	2.06	2.02	1.98	1.96
	0.05	4.38	3.52	3.13	2.90	2.74	2.63	2.54	2.48	2.42	2.38
	0.01	8.18	5.93	5.01	4.50	4.17	3.94	3.76	3.63	3.52	3.43
20	0.10	2.98	2.59	2.38	2.25	2.16	2.09	2.04	2.00	1.96	1.94
	0.05	4.35	3.49	3.10	2.87	2.71	2.60	2.51	2.45	2.39	2.35
	0.01	8.10	5.85	4.94	4.43	4.10	3.87	3.70	3.56	3.46	3.37
21	0.10	2.96	2.58	2.37	2.23	2.14	2.08	2.02	1.98	1.95	1.92
	0.05	4.32	3.47	3.07	2.84	2.68	2.57	2.49	2.42	2.37	2.32
	0.01	8.02	5.78	4.87	4.37	4.04	3.81	3.64	3.51	3.40	3.31
22	0.10	2.95	2.56	2.35	2.22	2.13	2.06	2.01	1.97	1.93	1.90
	0.05	4.30	3.44	3.05	2.82	2.66	2.55	2.46	2.40	2.34	2.30
	0.01	7.94	5.72	4.82	4.31	3.99	3.76	3.59	3.45	3.35	3.26
23	0.10	2.94	2.55	2.34	2.21	2.12	2.05	2.00	1.95	1.92	1.89
	0.05	4.28	3.42	3.03	2.80	2.64	2.53	2.44	2.38	2.32	2.28
	0.01	7.88	5.66	4.77	4.26	3.94	3.71	3.54	3.41	3.30	3.21
24	0.10	2.93	2.54	2.33	2.20	2.10	2.04	1.98	1.94	1.91	1.88
	0.05	4.26	3.40	3.01	2.78	2.62	2.51	2.42	2.36	2.30	2.26
	0.01	7.82	5.61	4.72	4.22	3.90	3.67	3.50	3.36	3.26	3.17
25	0.10	2.92	2.53	2.32	2.18	2.09	2.02	1.97	1.93	1.90	1.87
	0.05	4.24	3.38	2.99	2.76	2.60	2.49	2.40	2.34	2.28	2.24
	0.01	7.77	5.57	4.68	4.18	3.86	3.63	3.46	3.32	3.22	3.13
30	0.10	2.88	2.49	2.28	2.14	2.05	1.98	1.93	1.88	1.85	1.82
	0.05	4.17	3.32	2.92	2.69	2.53	2.42	2.33	2.27	2.21	2.16
	0.01	7.56	5.39	4.51	4.02	3.70	3.47	3.30	3.17	3.07	2.98
35	0.10	2.86	2.46	2.25	2.11	2.02	1.95	1.90	1.85	1.82	1.79
	0.05	4.12	3.27	2.87	2.64	2.48	2.37	2.28	2.22	2.16	2.11
	0.01	7.42	5.27	4.40	3.91	3.59	3.37	3.20	3.07	2.96	2.88
40	0.10	2.84	2.44	2.23	2.09	2.00	1.93	1.87	1.83	1.79	1.76
	0.05	4.08	3.23	2.84	2.61	2.45	2.34	2.25	2.18	2.12	2.08
	0.01	7.31	5.18	4.31	3.83	3.51	3.29	3.12	2.99	2.89	2.80

(Contd.)

Table B-7 (Contd.)

υ_1 / α / υ_2	1	2	3	4	5	6	7	8	9	10
45 0.10	2.82	2.42	2.21	2.07	1.98	1.91	1.86	1.81	1.77	1.74
45 0.05	4.06	3.20	2.81	2.58	2.42	2.31	2.22	2.15	2.10	2.05
45 0.01	7.23	5.11	4.25	3.77	3.45	3.23	3.07	2.94	2.83	2.74
50 0.10	2.81	2.41	2.20	2.06	1.97	1.90	1.84	1.80	1.76	1.73
50 0.05	4.03	3.18	2.79	2.56	2.40	2.29	2.20	2.13	2.07	2.03
50 0.01	7.17	5.06	4.20	3.72	3.41	3.19	3.02	2.89	2.78	2.70
55 0.10	2.80	2.40	2.19	2.05	1.96	1.88	1.83	1.78	1.75	1.72
55 0.05	4.02	3.16	2.77	2.54	2.38	2.27	2.18	2.11	2.06	2.01
55 0.01	7.12	5.01	4.16	3.68	3.37	3.15	2.98	2.85	2.75	2.66
60 0.10	2.79	2.39	2.18	2.04	1.95	1.88	1.82	1.78	1.74	1.71
60 0.05	4.00	3.15	2.76	2.52	2.37	2.25	2.17	2.10	2.04	1.99
60 0.01	7.08	4.98	4.13	3.65	3.34	3.12	2.95	2.82	2.72	2.63
70 0.10	2.78	2.38	2.16	2.03	1.93	1.86	1.80	1.76	1.72	1.69
70 0.05	3.98	3.13	2.74	2.50	2.35	2.23	2.14	2.07	2.02	1.97
70 0.01	7.01	4.92	4.07	3.60	3.29	3.07	2.91	2.78	2.67	2.58
80 0.10	2.77	2.37	2.15	2.02	1.92	1.85	1.79	1.75	1.71	1.68
80 0.05	3.96	3.11	2.72	2.49	2.33	2.21	2.13	2.06	2.00	1.95
80 0.01	6.96	4.88	4.04	3.56	3.26	3.04	2.87	2.74	2.64	2.55
90 0.10	2.76	2.36	2.15	2.01	1.91	1.84	1.78	1.74	1.70	1.67
90 0.05	3.95	3.10	2.71	2.47	2.31	2.20	2.11	2.04	1.98	1.94
90 0.01	6.92	4.85	4.01	3.54	3.23	3.01	2.84	2.72	2.61	2.52
100 0.10	2.76	2.36	2.14	2.00	1.91	1.83	1.78	1.73	1.70	1.66
100 0.05	3.94	3.09	2.70	2.46	2.30	2.19	2.10	2.03	1.98	1.93
100 0.01	6.90	4.82	3.98	3.51	3.21	2.99	2.82	2.69	2.59	2.50
120 0.10	2.75	2.35	2.13	1.99	1.90	1.82	1.77	1.72	1.68	1.65
120 0.05	3.92	3.07	2.68	2.45	2.29	2.17	2.09	2.02	1.96	1.91
120 0.01	6.85	4.79	3.95	3.48	3.17	2.96	2.79	2.66	2.56	2.47

Table B-7 F-table

υ_1 / α / υ_2	11	12	13	14	16	18	20	40	60	120
1 0.10	60.47	60.71	60.93	61.07	61.35	61.57	61.74	62.53	62.79	63.06
0.05	243.0	243.9	244.7	245.4	246.5	247.3	248.0	251.1	252.2	253.3
0.01	6083	6106	6126	6143	6170	6192	6209	6257	6313	6339
2 0.10	9.40	9.41	9.42	9.42	9.43	9.44	9.44	9.47	9.47	9.48
0.05	19.40	19.41	19.42	19.42	19.43	19.44	19.45	19.47	19.48	19.49
0.01	99.41	99.42	99.42	99.43	99.44	99.44	99.45	99.47	99.48	99.49
3 0.10	5.22	5.22	5.21	5.20	5.20	5.19	5.18	5.16	5.15	5.14
0.05	8.76	8.74	8.73	8.72	8.69	8.68	8.66	8.59	8.57	8.55
0.01	27.13	27.05	26.98	26.92	26.83	26.75	26.69	26.41	26.32	26.22
4 0.10	3.91	3.90	3.89	3.88	3.86	3.85	3.84	3.80	3.79	3.78
0.05	5.94	5.91	5.89	5.87	5.84	5.82	5.80	5.72	5.69	5.66
0.01	14.45	14.37	14.31	14.25	14.15	14.08	14.02	13.75	13.65	13.56
5 0.10	3.28	3.27	3.26	3.25	3.23	3.22	3.21	3.16	3.14	3.12
0.05	4.70	4.68	4.66	4.64	4.60	4.58	4.56	4.46	4.43	4.40
0.01	9.96	9.89	9.82	9.77	9.68	9.61	9.55	9.29	9.20	9.11
6 0.10	2.92	2.91	2.89	2.88	2.86	2.85	2.84	2.78	2.76	2.74
0.05	4.03	4.00	3.98	3.96	3.92	3.90	3.87	3.77	3.74	3.70
0.01	7.79	7.72	7.66	7.60	7.52	7.45	7.40	7.14	7.06	6.97
7 0.10	2.68	2.67	2.65	2.64	2.62	2.61	2.60	2.54	2.51	2.49
0.05	3.60	3.58	3.55	3.53	3.49	3.47	3.44	3.34	3.30	3.27
0.01	6.54	6.47	6.41	6.36	6.28	6.21	6.16	5.91	5.82	5.74
8 0.10	2.52	2.50	2.49	2.48	2.46	2.44	2.42	2.36	2.34	2.32
0.05	3.31	3.28	3.26	3.24	3.20	3.17	3.15	3.04	3.01	2.97
0.01	5.73	5.67	5.61	5.56	5.48	5.41	5.36	5.12	5.03	4.95
9 0.10	2.40	2.38	2.36	2.35	2.33	2.31	2.30	2.23	2.21	2.18
0.05	3.10	3.07	3.05	3.02	2.99	2.96	2.94	2.83	2.79	2.75
0.01	5.18	5.11	5.06	5.00	4.92	4.86	4.81	4.57	4.48	4.40

(Contd.)

Table B-7 (Contd.)

υ_1	α	11	12	13	14	16	18	20	40	60	120
υ_2											
10	0.10	2.30	2.28	2.27	2.26	2.23	2.22	2.20	2.13	2.11	2.08
	0.05	2.94	2.91	2.89	2.86	2.83	2.80	2.77	2.66	2.62	2.58
	0.01	4.77	4.71	4.65	4.60	4.52	4.46	4.41	4.17	4.08	4.00
11	0.10	2.23	2.21	2.19	2.18	2.16	2.14	2.12	2.05	2.03	2.00
	0.05	2.82	2.79	2.76	2.74	2.70	2.67	2.65	2.53	2.49	2.45
	0.01	4.46	4.40	4.34	4.29	4.21	4.15	4.10	3.86	3.78	3.69
12	0.10	2.17	2.15	2.13	2.12	2.09	2.08	2.06	1.99	1.96	1.93
	0.05	2.72	2.69	2.66	2.64	2.60	2.57	2.54	2.43	2.38	2.34
	0.01	4.22	4.16	4.10	4.05	3.97	3.91	3.86	3.62	3.54	3.45
13	0.10	2.12	2.10	2.08	2.07	2.04	2.02	2.01	1.93	1.90	1.88
	0.05	2.64	2.60	2.58	2.55	2.52	2.48	2.46	2.34	2.30	2.25
	0.01	4.02	3.96	3.90	3.86	3.78	3.72	3.66	3.43	3.34	3.25
14	0.10	2.07	2.05	2.04	2.02	2.00	1.98	1.96	1.89	1.86	1.83
	0.05	2.56	2.53	2.51	2.48	2.44	2.41	2.39	2.27	2.22	2.18
	0.01	3.86	3.80	3.74	3.70	3.62	3.56	3.51	3.27	3.18	3.09
15	0.10	2.04	2.02	2.00	1.98	1.96	1.94	1.92	1.85	1.82	1.79
	0.05	2.51	2.46	2.45	2.42	2.38	2.35	2.33	2.20	2.16	2.11
	0.01	3.73	3.67	3.61	3.56	3.48	3.42	3.37	3.13	3.05	2.96
16	0.10	2.00	1.98	1.97	1.95	1.93	1.91	1.89	1.81	1.78	1.75
	0.05	2.46	2.42	2.40	2.37	2.33	2.30	2.28	2.15	2.11	2.06
	0.01	3.62	3.55	3.50	3.45	3.37	3.31	3.26	3.02	2.93	2.84
17	0.10	1.98	1.96	1.94	1.92	1.90	1.88	1.86	1.81	1.78	1.75
	0.05	2.41	2.38	2.35	2.33	2.29	2.26	2.23	2.10	2.06	2.01
	0.01	3.52	3.46	3.40	3.35	3.28	3.21	3.16	2.92	2.83	2.75
18	0.10	1.95	1.93	1.92	1.90	1.88	1.85	1.84	1.75	1.72	1.69
	0.05	2.37	2.34	2.31	2.29	2.25	2.22	2.19	2.06	2.02	1.97
	0.01	3.43	3.37	3.32	3.27	3.19	3.13	3.08	2.84	2.75	2.66
19	0.10	1.93	1.91	1.89	1.88	1.85	1.83	1.81	1.73	1.70	1.67
	0.05	2.34	2.31	2.28	2.26	2.22	2.18	2.16	2.03	1.98	1.93
	0.01	3.36	3.30	3.24	3.20	3.12	3.05	3.00	2.76	2.67	2.58

(Contd.)

Table B-7 (Contd.)

υ_1 υ_2	α	11	12	13	14	16	18	20	40	60	120
20	0.10	1.91	1.89	1.88	1.86	1.83	1.81	1.79	1.71	1.68	1.64
	0.05	2.31	2.28	2.25	2.22	2.18	2.15	2.12	1.99	1.95	1.90
	0.01	3.29	3.23	3.18	3.13	3.05	2.99	2.94	2.69	2.61	2.52
21	0.10	1.90	1.88	1.86	1.84	1.82	1.79	1.78	1.69	1.66	1.62
	0.05	2.28	2.25	2.22	2.20	2.16	2.12	2.10	1.96	1.92	1.87
	0.01	3.24	3.17	3.12	3.07	2.99	2.93	2.88	2.64	2.55	2.46
22	0.10	1.88	1.86	1.84	1.82	1.80	1.78	1.76	1.67	1.64	1.60
	0.05	2.26	2.23	2.20	2.17	2.13	2.10	2.07	1.94	1.89	1.84
	0.01	3.18	3.12	3.07	3.20	2.94	2.88	2.83	2.58	2.50	2.40
23	0.10	1.87	1.84	1.83	1.81	1.78	1.76	1.74	1.66	1.62	1.59
	0.05	2.24	2.20	2.18	2.15	2.11	2.08	2.05	1.91	1.86	1.81
	0.01	3.14	3.07	3.02	2.97	2.89	2.83	2.78	2.54	2.45	2.35
24	0.10	1.85	1.83	1.81	1.78	1.77	1.75	1.73	1.64	1.61	1.57
	0.05	2.22	2.18	2.16	2.13	2.09	2.05	2.03	1.89	1.84	1.79
	0.01	3.09	3.03	2.98	2.93	2.85	2.80	2.74	2.49	2.40	2.31
25	0.10	1.84	1.82	1.80	1.78	1.76	1.74	1.72	1.63	1.59	1.56
	0.05	2.20	2.16	2.14	2.11	2.07	2.04	2.01	1.87	1.82	1.77
	0.01	3.06	2.99	2.94	2.89	2.81	2.75	2.70	2.45	2.36	2.27
30	0.10	1.79	1.77	1.75	1.74	1.71	1.69	1.67	1.57	1.54	1.50
	0.05	2.13	2.09	2.06	2.04	2.00	1.96	1.93	1.79	1.74	1.68
	0.01	2.91	2.84	2.79	2.74	2.66	2.60	2.55	2.30	2.21	2.11
35	0.10	1.76	1.74	1.72	1.70	1.67	1.65	1.63	1.53	1.50	1.46
	0.05	2.08	2.04	2.01	1.99	1.94	1.91	1.88	1.74	1.68	1.62
	0.01	2.80	2.74	2.69	2.64	2.56	2.50	2.44	2.19	2.10	2.00
40	0.10	1.74	1.72	1.70	1.69	1.65	1.62	1.60	1.51	1.47	1.42
	0.05	2.04	2.00	1.97	1.95	1.90	1.87	1.84	1.69	1.64	1.58
	0.01	2.73	2.66	2.61	2.56	2.48	2.42	2.37	2.11	2.02	1.92
45	0.10	1.72	1.70	1.68	1.66	1.63	1.60	1.58	1.49	1.45	1.40
	0.05	2.01	1.97	1.94	1.92	1.87	1.84	1.81	1.66	1.60	1.54
	0.01	2.67	2.61	2.55	2.51	2.43	2.36	2.31	2.05	1.96	1.86

(Contd.)

Table B-7 (Contd.)

υ_1 α υ_2		11	12	13	14	16	18	20	40	60	120
50	0.10	1.70	1.68	1.66	1.64	1.61	1.59	1.57	1.47	1.44	1.38
	0.05	1.99	1.95	1.92	1.90	1.85	1.81	1.78	1.63	1.57	1.51
	0.01	2.62	2.56	2.51	2.46	2.38	2.32	2.26	2.01	1.95	1.81
55	0.10	1.69	1.67	1.65	1.63	1.60	1.58	1.56	1.45	1.42	1.36
	0.05	1.97	1.93	1.90	1.88	1.83	1.80	1.76	1.61	1.55	1.49
	0.01	2.59	2.53	2.47	2.42	2.34	2.28	2.23	1.97	1.87	1.76
60	0.10	1.68	1.66	1.64	1.62	1.59	1.56	1.54	1.44	1.40	1.35
	0.05	1.95	1.92	1.89	1.86	1.82	1.78	1.75	1.59	1.53	1.47
	0.01	2.56	2.50	2.44	2.39	2.32	2.25	2.20	1.94	1.84	1.73
70	0.10	1.66	1.64	1.62	1.60	1.57	1.55	1.53	1.42	1.38	1.33
	0.05	1.93	1.89	1.86	1.84	1.79	1.75	1.72	1.56	1.50	1.44
	0.01	2.51	2.45	2.40	2.35	2.27	2.20	2.15	1.89	1.79	1.68
80	0.10	1.65	1.63	1.61	1.59	1.56	1.53	1.51	1.40	1.36	1.31
	0.05	1.91	1.88	1.84	1.82	1.77	1.73	1.70	1.54	1.48	1.41
	0.01	2.48	2.42	2.36	2.31	2.23	2.17	2.12	1.85	1.75	1.63
90	0.10	1.64	1.62	1.60	1.58	1.55	1.52	1.50	1.39	1.35	1.30
	0.05	1.90	1.86	1.83	1.80	1.76	1.72	1.69	1.53	1.46	1.40
	0.01	2.45	2.39	2.33	2.29	2.21	2.14	2.09	1.82	1.72	1.60
100	0.10	1.64	1.61	1.59	1.57	1.54	1.52	1.49	1.38	1.34	1.28
	0.05	1.89	1.85	1.82	1.79	1.75	1.71	1.68	1.52	1.45	1.38
	0.01	2.43	2.37	2.31	2.26	2.19	2.12	2.07	1.80	1.69	1.57
120	0.10	1.62	1.60	1.58	1.56	1.53	1.50	1.48	1.37	1.32	1.26
	0.05	1.87	1.83	1.80	1.77	1.73	1.69	1.66	1.50	1.43	1.35
	0.01	2.40	2.34	2.27	2.20	2.15	2.08	2.03	1.76	1.66	1.53

Source: Website and values interpolated by the author.

Table B-8 Critical values for Duncan's multiple range test

Critical values q' (p, df; 0.05) for Duncan's multiple range tests

df p>	2	3	4	5	6	7	8	9	10	11	12	13	14	15	16	17	18	19	20
1	17.969	17.969	17.969	17.969	17.969	17.969	17.969	17.969	17.969	17.969	17.969	17.969	17.969	17.969	17.969	17.969	17.969	17.969	17.969
2	6.085	6.085	6.085	6.085	6.085	6.085	6.085	6.085	6.085	6.085	6.085	6.085	6.085	6.085	6.085	6.085	6.085	6.085	6.085
3	4.501	4.516	4.516	4.516	4.516	4.516	4.516	4.516	4.516	4.516	4.516	4.516	4.516	4.516	4.516	4.516	4.516	4.516	4.516
4	3.926	4.013	4.033	4.033	4.033	4.033	4.033	4.033	4.033	4.033	4.033	4.033	4.033	4.033	4.033	4.033	4.033	4.033	4.033
5	3.635	3.749	3.796	3.814	3.814	3.814	3.814	3.814	3.814	3.814	3.814	3.814	3.814	3.814	3.814	3.814	3.814	3.814	3.814
6	3.46	3.586	3.649	3.68	3.694	3.697	3.697	3.697	3.697	3.697	3.697	3.697	3.697	3.697	3.697	3.697	3.697	3.697	3.697
7	3.344	3.477	3.548	3.588	3.611	3.622	3.625	3.625	3.625	3.625	3.625	3.625	3.625	3.625	3.625	3.625	3.625	3.625	3.625
8	3.261	3.398	3.475	3.521	3.549	3.566	3.575	3.579	3.579	3.579	3.579	3.579	3.579	3.579	3.579	3.579	3.579	3.579	3.579
9	3.199	3.339	3.42	3.47	3.502	3.523	3.536	3.544	3.547	3.547	3.547	3.547	3.547	3.547	3.547	3.547	3.547	3.547	3.547
10	3.151	3.293	3.376	3.43	3.465	3.489	3.505	3.516	3.522	3.525	3.525	3.525	3.525	3.525	3.525	3.525	3.525	3.525	3.525
11	3.113	3.256	3.341	3.397	3.435	3.462	3.48	3.493	3.501	3.506	3.509	3.51	3.51	3.51	3.51	3.51	3.51	3.51	3.51
12	3.081	3.225	3.312	3.37	3.41	3.439	3.459	3.474	3.484	3.491	3.495	3.498	3.498	3.498	3.498	3.498	3.498	3.498	3.498
13	3.055	3.2	3.288	3.348	3.389	3.419	3.441	3.458	3.47	3.478	3.484	3.488	3.49	3.49	3.49	3.49	3.49	3.49	3.49
14	3.033	3.178	3.268	3.328	3.371	3.403	3.426	3.444	3.457	3.467	3.474	3.479	3.482	3.484	3.484	3.484	3.484	3.484	3.484
15	3.014	3.16	3.25	3.312	3.356	3.389	3.413	3.432	3.446	3.457	3.465	3.471	3.476	3.478	3.48	3.48	3.48	3.48	3.48
16	2.998	3.144	3.235	3.297	3.343	3.376	3.402	3.422	3.437	3.449	3.458	3.465	3.47	3.473	3.476	3.477	3.477	3.477	3.477
17	2.984	3.13	3.222	3.285	3.331	3.365	3.392	3.412	3.429	3.441	3.451	3.459	3.465	3.469	3.472	3.474	3.475	3.475	3.475
18	2.971	3.117	3.21	3.274	3.32	3.356	3.383	3.404	3.421	3.435	3.445	3.454	3.46	3.465	3.469	3.472	3.473	3.474	3.474
19	2.96	3.106	3.199	3.264	3.311	3.347	3.375	3.397	3.415	3.429	3.44	3.449	3.456	3.462	3.466	3.469	3.472	3.473	3.474
20	2.95	3.097	3.19	3.255	3.303	3.339	3.368	3.39	3.409	3.423	3.435	3.445	3.452	3.459	3.463	3.467	3.47	3.472	3.473
21	2.941	3.088	3.181	3.247	3.295	3.332	3.361	3.385	3.403	3.418	3.431	3.441	3.449	3.456	3.461	3.465	3.469	3.471	3.473
22	2.933	3.08	3.173	3.239	3.288	3.326	3.355	3.379	3.398	3.414	3.427	3.437	3.446	3.453	3.459	3.464	3.467	3.47	3.472

(Contd.)

df p->	2	3	4	5	6	7	8	9	10	11	12	13	14	15	16	17	18	19	20
23	2.926	3.072	3.166	3.233	3.282	3.32	3.35	3.374	3.394	3.41	3.423	3.434	3.443	3.451	3.457	3.462	3.466	3.469	3.472
24	2.919	3.066	3.16	3.226	3.276	3.315	3.345	3.37	3.39	3.406	3.42	3.431	3.441	3.449	3.455	3.461	3.465	3.469	3.472
25	2.913	3.059	3.154	3.221	3.271	3.31	3.341	3.366	3.386	3.403	3.417	3.429	3.439	3.447	3.454	3.459	3.464	3.468	3.471
26	2.907	3.054	3.149	3.216	3.266	3.305	3.336	3.362	3.382	3.4	3.414	3.426	3.436	3.445	3.452	3.458	3.463	3.468	3.471
27	2.902	3.049	3.144	3.211	3.262	3.301	3.332	3.358	3.379	3.397	3.412	3.424	3.434	3.443	3.451	3.457	3.463	3.467	3.471
28	2.897	3.044	3.139	3.206	3.257	3.297	3.329	3.355	3.376	3.394	3.409	3.422	3.433	3.442	3.45	3.456	3.462	3.467	3.47
29	2.892	3.039	3.135	3.202	3.253	3.293	3.326	3.352	3.373	3.392	3.407	3.42	3.431	3.44	3.448	3.455	3.461	3.466	3.47
30	2.888	3.035	3.131	3.199	3.25	3.29	3.322	3.349	3.371	3.389	3.405	3.418	3.429	3.439	3.447	3.454	3.46	3.466	3.47
31	2.884	3.031	3.127	3.195	3.246	3.287	3.319	3.346	3.368	3.387	3.403	3.416	3.428	3.438	3.446	3.454	3.46	3.465	3.47
32	2.881	3.028	3.123	3.192	3.243	3.284	3.317	3.344	3.366	3.385	3.401	3.415	3.426	3.436	3.445	3.453	3.459	3.465	3.47
33	2.877	3.024	3.12	3.188	3.24	3.281	3.314	3.341	3.364	3.383	3.399	3.413	3.425	3.435	3.444	3.452	3.459	3.465	3.47
34	2.874	3.021	3.117	3.185	3.238	3.279	3.312	3.339	3.362	3.381	3.398	3.412	3.424	3.434	3.443	3.451	3.458	3.464	3.469
35	2.871	3.018	3.114	3.183	3.235	3.276	3.309	3.337	3.36	3.379	3.396	3.41	3.423	3.433	3.443	3.451	3.458	3.464	3.469
36	2.868	3.015	3.111	3.18	3.232	3.274	3.307	3.335	3.358	3.378	3.395	3.409	3.421	3.432	3.442	3.45	3.457	3.464	3.469
37	2.865	3.013	3.109	3.178	3.23	3.272	3.305	3.333	3.356	3.376	3.393	3.408	3.42	3.431	3.441	3.449	3.457	3.463	3.469
38	2.863	3.01	3.106	3.175	3.228	3.27	3.303	3.331	3.355	3.375	3.392	3.407	3.419	3.431	3.44	3.449	3.457	3.463	3.469
39	2.861	3.008	3.104	3.173	3.226	3.268	3.301	3.33	3.353	3.373	3.391	3.406	3.418	3.43	3.44	3.448	3.456	3.463	3.469
40	2.858	3.005	3.102	3.171	3.224	3.266	3.3	3.328	3.352	3.372	3.389	3.404	3.418	3.429	3.439	3.448	3.456	3.463	3.469
48	2.843	2.991	3.087	3.157	3.211	3.253	3.288	3.318	3.342	3.363	3.382	3.398	3.412	3.424	3.435	3.445	3.453	3.461	3.468
60	2.829	2.976	3.073	3.143	3.198	3.241	3.277	3.307	3.333	3.355	3.374	3.391	3.406	3.419	3.431	3.441	3.451	3.46	3.468
80	2.814	2.961	3.059	3.13	3.185	3.229	3.266	3.297	3.323	3.346	3.366	3.384	3.4	3.414	3.427	3.438	3.449	3.458	3.467
120	2.8	2.947	3.045	3.116	3.172	3.217	3.254	3.286	3.313	3.337	3.358	3.377	3.394	3.409	3.423	3.435	3.446	3.457	3.466
240	2.786	2.933	3.031	3.103	3.159	3.205	3.243	3.276	3.304	3.329	3.35	3.37	3.388	3.404	3.418	3.432	3.444	3.455	3.466
Inf	2.772	2.918	3.017	3.089	3.146	3.193	3.232	3.265	3.294	3.32	3.343	3.363	3.382	3.399	3.414	3.428	3.442	3.454	3.466

(Contd.)

Critical values q' (p, df, 0.01) for Duncan's multiple range tests

df p->	2	3	4	5	6	7	8	9	10	11	12	13	14	15	16	17	18	19	20
1	90.024	90.024	90.024	90.024	90.024	90.024	90.024	90.024	90.024	90.024	90.024	90.024	90.024	90.02	90.024	90.024	90.024	90.024	90.024
2	14.036	14.036	14.036	14.036	14.036	14.036	14.036	14.036	14.036	14.036	14.036	14.036	14.036	14.036	14.036	14.036	14.036	14.036	14.036
3	8.26	8.321	8.321	8.321	8.321	8.321	8.321	8.321	8.321	8.321	8.321	8.321	8.321	8.321	8.321	8.321	8.321	8.321	8.321
4	6.511	6.677	6.74	6.755	6.755	6.755	6.755	6.755	6.755	6.755	6.755	6.755	6.755	6.755	6.755	6.755	6.755	6.755	6.755
5	5.702	5.893	5.989	6.04	6.065	6.074	6.074	6.074	6.074	6.074	6.074	6.074	6.074	6.074	6.074	6.074	6.074	6.074	6.074
6	5.243	5.439	5.549	5.614	5.655	5.68	5.694	5.701	5.703	5.703	5.703	5.703	5.703	5.703	5.703	5.703	5.703	5.703	5.703
7	4.949	5.145	5.26	5.333	5.383	5.416	5.439	5.454	5.464	5.47	5.472	5.472	5.472	5.472	5.472	5.472	5.472	5.472	5.472
8	4.745	4.939	5.056	5.134	5.189	5.227	5.256	5.276	5.291	5.302	5.309	5.313	5.316	5.317	5.317	5.317	5.317	5.317	5.317
9	4.596	4.787	4.906	4.986	5.043	5.086	5.117	5.142	5.16	5.174	5.185	5.193	5.199	5.202	5.205	5.206	5.206	5.206	5.206
10	4.482	4.671	4.789	4.871	4.931	4.975	5.01	5.036	5.058	5.074	5.087	5.098	5.106	5.112	5.117	5.12	5.122	5.123	5.124
11	4.392	4.579	4.697	4.78	4.841	4.887	4.923	4.952	4.975	4.994	5.009	5.021	5.031	5.039	5.045	5.05	5.054	5.057	5.059
12	4.32	4.504	4.622	4.705	4.767	4.815	4.852	4.882	4.907	4.927	4.944	4.957	4.969	4.978	4.986	4.993	4.998	5.002	5.005
13	4.26	4.442	4.56	4.643	4.706	4.754	4.793	4.824	4.85	4.871	4.889	4.904	4.917	4.927	4.936	4.944	4.95	4.955	4.96
14	4.21	4.391	4.508	4.591	4.654	4.703	4.743	4.775	4.802	4.824	4.843	4.859	4.872	4.884	4.894	4.902	4.909	4.916	4.921
15	4.167	4.346	4.463	4.547	4.61	4.66	4.7	4.733	4.76	4.783	4.803	4.82	4.834	4.846	4.857	4.866	4.874	4.881	4.887
16	4.131	4.308	4.425	4.508	4.572	4.622	4.662	4.696	4.724	4.748	4.768	4.785	4.8	4.813	4.825	4.835	4.843	4.851	4.858
17	4.099	4.275	4.391	4.474	4.538	4.589	4.63	4.664	4.692	4.717	4.737	4.755	4.771	4.785	4.797	4.807	4.816	4.824	4.832
18	4.071	4.246	4.361	4.445	4.509	4.559	4.601	4.635	4.664	4.689	4.71	4.729	4.745	4.759	4.771	4.782	4.792	4.801	4.808
19	4.046	4.22	4.335	4.418	4.483	4.533	4.575	4.61	4.639	4.664	4.686	4.705	4.722	4.736	4.749	4.76	4.771	4.78	4.788
20	4.024	4.197	4.312	4.395	4.459	4.51	4.552	4.587	4.617	4.642	4.664	4.684	4.701	4.716	4.729	4.741	4.751	4.761	4.769
21	4.004	4.177	4.291	4.374	4.438	4.489	4.531	4.567	4.597	4.622	4.645	4.664	4.682	4.697	4.711	4.723	4.734	4.743	4.752
22	3.986	4.158	4.272	4.355	4.419	4.47	4.513	4.548	4.578	4.604	4.627	4.647	4.664	4.68	4.694	4.706	4.718	4.728	4.737
23	3.97	4.141	4.254	4.337	4.402	4.453	4.496	4.531	4.562	4.588	4.611	4.631	4.649	4.665	4.679	4.692	4.703	4.713	4.723
24	3.955	4.126	4.239	4.322	4.386	4.437	4.48	4.516	4.546	4.573	4.596	4.616	4.634	4.651	4.665	4.678	4.69	4.7	4.71

(Contd.)

df p->	2	3	4	5	6	7	8	9	10	11	12	13	14	15	16	17	18	19	20
25	3.942	4.112	4.224	4.307	4.371	4.423	4.466	4.502	4.532	4.559	4.582	4.603	4.621	4.638	4.652	4.665	4.677	4.688	4.698
26	3.93	4.099	4.211	4.294	4.358	4.41	4.452	4.489	4.52	4.546	4.57	4.591	4.609	4.626	4.64	4.654	4.666	4.677	4.687
27	3.918	4.087	4.199	4.282	4.346	4.397	4.44	4.477	4.508	4.535	4.558	4.579	4.598	4.615	4.63	4.643	4.655	4.667	4.677
28	3.908	4.076	4.188	4.27	4.334	4.386	4.429	4.465	4.497	4.524	4.548	4.569	4.587	4.604	4.619	4.633	4.646	4.657	4.667
29	3.898	4.065	4.177	4.26	4.324	4.376	4.419	4.455	4.486	4.514	4.538	4.559	4.578	4.595	4.61	4.624	4.637	4.648	4.659
30	3.889	4.056	4.168	4.25	4.314	4.366	4.409	4.445	4.477	4.504	4.528	4.55	4.569	4.586	4.601	4.615	4.628	4.64	4.65
31	3.881	4.047	4.159	4.241	4.305	4.357	4.4	4.436	4.468	4.495	4.519	4.541	4.56	4.577	4.593	4.607	4.62	4.632	4.643
32	3.873	4.039	4.15	4.232	4.296	4.348	4.391	4.428	4.459	4.487	4.511	4.533	4.552	4.57	4.585	4.6	4.613	4.625	4.635
33	3.865	4.031	4.142	4.224	4.288	4.34	4.383	4.42	4.452	4.479	4.504	4.525	4.545	4.562	4.578	4.592	4.606	4.618	4.629
34	3.859	4.024	4.135	4.217	4.281	4.333	4.376	4.413	4.444	4.472	4.496	4.518	4.538	4.555	4.571	4.586	4.599	4.611	4.622
35	3.852	4.017	4.128	4.21	4.273	4.325	4.369	4.406	4.437	4.465	4.49	4.511	4.531	4.549	4.565	4.579	4.593	4.605	4.616
36	3.846	4.011	4.121	4.203	4.267	4.319	4.362	4.399	4.431	4.459	4.483	4.505	4.525	4.543	4.559	4.573	4.587	4.599	4.611
37	3.84	4.005	4.115	4.197	4.26	4.312	4.356	4.393	4.425	4.452	4.477	4.499	4.519	4.537	4.553	4.568	4.581	4.594	4.605
38	3.835	3.999	4.109	4.191	4.254	4.306	4.35	4.387	4.419	4.447	4.471	4.493	4.513	4.531	4.548	4.562	4.576	4.589	4.6
39	3.83	3.993	4.103	4.185	4.249	4.301	4.344	4.381	4.413	4.441	4.466	4.488	4.508	4.526	4.542	4.557	4.571	4.584	4.595
40	3.825	3.988	4.098	4.18	4.243	4.295	4.339	4.376	4.408	4.436	4.461	4.483	4.503	4.521	4.537	4.552	4.566	4.579	4.591
48	3.793	3.955	4.064	4.145	4.209	4.261	4.304	4.341	4.374	4.402	4.427	4.45	4.47	4.489	4.506	4.521	4.535	4.548	4.561
60	3.762	3.922	4.03	4.111	4.174	4.226	4.27	4.307	4.34	4.368	4.394	4.417	4.437	4.456	4.474	4.489	4.504	4.518	4.53
80	3.732	3.89	3.997	4.077	4.14	4.192	4.236	4.273	4.306	4.335	4.36	4.384	4.405	4.424	4.442	4.458	4.473	4.487	4.5
120	3.702	3.858	3.964	4.044	4.107	4.158	4.202	4.239	4.272	4.301	4.327	4.351	4.372	4.392	4.41	4.426	4.442	4.456	4.469
240	3.672	3.827	3.932	4.011	4.073	4.125	4.168	4.206	4.239	4.268	4.294	4.318	4.339	4.359	4.378	4.394	4.41	4.425	4.439
I nf	3.643	3.796	3.9	3.978	4.04	4.091	4.135	4.172	4.205	4.235	4.261	4.285	4.307	4.327	4.345	4.363	4.379	4.394	4.408

Source: Table B-8 is reproduced with the kind permission by Ms Jessica Waters on behalf of the Wiley-Blackwell Publishers permissions team from the paper by D.B. Duncan, Biometrics, 1955.

Table B-9 Arcsin square root transformation
percentage to angles

%	0	1	2	3	4	5	6	7	8	9
0.0	0.000	0.573	0.810	0.992	1.146	1.281	1.404	1.516	1.621	1.719
0.1	1.812	1.901	1.985	2.066	2.144	2.220	2.292	2.363	2.432	2.498
0.2	2.563	2.626	2.688	2.749	2.808	2.866	2.923	2.978	3.033	3.087
0.3	3.140	3.192	3.243	3.293	3.343	3.392	3.440	3.487	3.534	3.580
0.4	3.626	3.671	3.716	3.760	3.803	3.846	3.889	3.931	3.973	4.014
0.5	4.054	4.095	4.135	4.175	4.214	4.253	4.292	4.330	4.368	4.405
0.6	4.442	4.480	4.516	4.552	4.588	4.624	4.660	4.695	4.730	4.765
0.7	4.799	4.834	4.868	4.901	4.935	4.968	5.001	5.034	5.067	5.099
0.8	5.131	5.164	5.195	5.227	5.259	5.290	5.321	5.352	5.383	5.413
0.9	5.444	5.474	5.504	5.534	5.564	5.593	5.623	5.652	5.681	5.710
1	5.739	6.020	6.289	6.547	6.795	7.035	7.267	7.492	7.710	7.923
2	8.130	8.332	8.530	8.723	8.912	9.097	9.279	9.458	9.633	9.805
3	9.974	10.141	10.305	10.466	10.626	10.782	10.937	11.090	11.241	11.390
4	11.537	11.682	11.826	11.968	12.108	12.247	12.385	12.521	12.656	12.789
5	12.921	13.052	13.181	13.310	13.437	13.563	13.688	13.812	13.936	14.058
6	14.179	14.299	14.418	14.536	14.654	14.770	14.886	15.001	15.116	15.229
7	15.342	15.454	15.565	15.675	15.785	15.894	16.002	16.110	16.218	16.324
8	16.430	16.535	16.640	16.744	16.848	16.951	17.053	17.155	17.256	17.357
9	17.458	17.557	17.657	17.756	17.854	17.952	18.050	18.146	18.243	18.339
10	18.435	18.530	18.625	18.720	18.814	18.907	19.000	19.093	19.186	19.278
11	19.370	19.461	19.552	19.643	19.733	19.823	19.913	20.002	20.091	20.180
12	20.268	20.356	20.444	20.531	20.618	20.705	20.791	20.877	20.963	21.049
13	21.134	21.219	21.304	21.387	21.473	21.557	21.640	21.724	21.807	21.890
14	21.973	22.055	22.137	22.219	22.301	22.382	22.464	22.545	22.626	22.706
15	22.786	22.867	22.946	23.026	23.106	23.185	23.264	23.343	23.421	23.500
16	23.578	23.656	23.734	23.812	23.889	23.966	24.043	24.120	24.197	24.274
17	24.350	24.426	24.502	24.578	24.654	24.729	24.804	24.880	24.955	25.029
18	25.104	25.178	25.253	25.327	25.401	25.475	25.549	25.622	25.696	25.769
19	25.842	25.915	25.988	26.060	26.133	26.205	26.277	26.350	26.421	26.493
20	26.565	26.637	26.708	26.779	26.850	26.921	26.992	27.063	27.134	27.204

(Contd.)

Table B-9 (Contd.)

%	0	1	2	3	4	5	6	7	8	9
21	27.275	27.345	27.415	27.485	27.555	27.625	27.694	27.764	28.834	27.903
22	27.972	28.041	28.110	28.179	28.248	28.316	28.385	28.453	28.522	28.590
23	28.658	28.726	28.794	28.862	28.930	28.997	29.065	29.132	29.200	29.267
24	29.334	29.401	29.468	29.535	29.601	29.668	29.735	29.801	29.868	29.934
25	30.000	30.066	30.132	30.198	30.264	30.330	30.395	30.461	30.526	30.592
26	30.657	30.722	30.788	30.853	30.918	30.983	31.048	31.112	31.177	31.242
27	31.306	31.371	31.435	31.500	31.564	31.628	31.692	31.756	31.820	31.884
28	31.948	32.012	32.076	32.139	32.203	32.266	32.330	32.393	32.456	32.520
29	32.583	32.646	32.709	32.772	32.835	32.898	32.960	33.023	33.086	33.148
30	33.211	33.273	33.336	33.398	33.460	33.523	33.585	33.647	33.709	33.771
31	33.833	33.895	33.957	34.019	34.080	34.142	34.204	34.265	34.327	34.388
32	34.450	34.511	34.573	34.634	34.695	34.756	34.817	34.878	34.940	35.001
33	35.062	35.122	35.183	35.244	35.305	35.366	35.426	35.487	35.547	35.608
34	35.668	35.729	35.789	35.850	35.910	35.970	36.031	36.091	36.151	36.211
35	36.271	36.331	36.391	36.451	36.511	36.571	36.631	36.691	36.750	36.810
36	36.870	36.930	36.989	37.049	37.109	37.168	37.227	37.287	37.346	37.406
37	37.465	37.524	37.584	37.643	37.702	37.761	37.820	37.880	37.939	37.998
38	38.057	38.116	38.175	38.234	38.292	38.351	38.410	38.469	38.528	38.587
39	38.645	38.704	38.763	38.822	38.880	38.939	38.997	39.056	39.114	39.173
40	39.232	39.290	39.348	39.407	39.465	39.524	39.582	39.640	39.698	39.757
41	39.815	39.873	39.932	39.990	40.048	40.106	40.164	40.222	40.280	40.338
42	40.396	40.454	40.513	40.571	40.628	40.686	40.744	40.802	40.860	40.918
43	40.976	41.034	41.092	41.150	41.217	41.265	41.323	41.381	41.438	41.496
44	41.554	41.612	41.669	41.727	41.785	41.842	41.900	41.958	42.015	42.073
45	42.130	42.188	42.246	42.303	42.361	42.418	42.476	42.533	42.591	42.648
46	42.706	42.763	42.821	42.878	42.936	42.993	43.050	43.108	43.165	43.223
47	43.280	43.337	43.395	43.452	43.510	43.567	43.624	43.682	43.739	43.796
48	43.854	43.911	43.968	44.026	44.083	44.140	44.198	44.255	44.312	44.370
49	44.427	44.484	44.542	44.599	44.656	44.714	44.771	44.828	44.885	44.943
50	45.000	45.057	45.114	45.172	45.229	45.286	45.344	45.401	45.458	45.516
51	45.573	45.630	45.688	45.745	45.802	45.860	45.917	45.974	46.032	46.089
52	46.146	46.204	46.261	46.318	46.374	46.433	46.490	46.548	46.605	46.662
53	46.720	46.777	46.835	46.892	46.950	47.007	47.064	47.122	47.179	47.237

(Contd.)

Table B-9 (Contd.)

%	0	1	2	3	4	5	6	7	8	9
54	47.294	47.352	47.409	47.467	47.524	47.582	47.639	47.697	47.754	47.812
55	47.870	47.927	47.985	48.042	48.100	48.158	48.215	48.273	48.331	48.388
56	48.446	48.504	48.562	48.619	48.677	48.735	48.793	48.850	48.908	48.966
57	49.024	49.082	49.140	49.198	49.256	49.313	49.371	49.429	49.487	49.545
58	49.603	49.662	49.720	49.778	49.836	49.894	49.952	50.010	50.068	50.127
59	50.185	50.243	50.301	50.360	50.418	50.476	50.535	50.593	50.652	50.710
60	50.768	50.827	50.885	50.944	51.002	51.061	51.120	51.178	51.237	51.296
61	51.354	51.413	51.472	51.531	51.590	51.648	51.707	51.766	51.825	51.884
62	51.943	52.002	52.061	52.120	52.180	52.239	52.298	52.357	52.416	52.476
63	52.535	52.594	52.654	52.713	52.773	52.832	52.892	52.951	53.011	53.070
64	53.130	53.190	53.250	53.309	53.369	53.429	53.489	53.549	53.609	53.669
65	53.729	53.789	53.849	53.909	53.969	54.030	54.090	54.150	54.210	54.271
66	54.331	54.392	54.453	54.513	54.574	54.634	54.700	54.756	54.817	54.878
67	54.938	54.999	55.060	55.121	55.182	55.244	55.305	55.366	55.427	55.489
68	55.550	55.611	55.673	55.734	55.796	55.858	55.920	55.981	56.043	56.105
69	56.167	56.229	56.291	56.353	56.415	56.477	56.540	56.602	56.664	56.727
70	56.789	56.852	56.914	56.977	57.040	57.102	57.165	57.228	57.291	57.354
71	57.417	57.480	57.544	57.607	57.670	57.734	57.797	57.861	57.924	57.988
72	58.052	58.116	58.180	58.244	58.308	58.372	58.436	58.500	58.565	58.629
73	58.694	58.758	58.823	58.887	58.952	59.017	59.082	59.147	59.212	59.277
74	59.342	59.408	59.473	59.539	59.605	59.670	59.736	59.802	59.868	59.934
75	60.000	60.066	60.132	60.199	60.265	60.332	60.398	60.465	60.532	60.599
76	60.666	60.733	60.800	60.868	60.935	61.003	61.070	61.138	61.206	61.274
77	61.342	61.410	61.478	61.546	61.615	61.684	61.752	61.821	61.890	61.959
78	62.028	62.097	62.166	62.236	62.305	62.375	62.445	62.515	62.585	62.655
79	62.725	62.796	62.866	62.937	63.008	63.078	63.150	63.221	63.292	63.363
80	63.435	63.507	63.578	63.650	63.722	63.795	63.867	63.940	64.012	64.085
81	64.158	64.231	63.304	64.378	64.451	64.525	64.600	64.673	64.747	64.821
82	64.896	64.970	65.045	65.120	65.195	65.271	65.346	65.422	65.498	65.574
83	65.650	65.726	65.803	65.880	65.956	66.034	66.111	66.188	66.266	66.344
84	66.422	66.500	66.578	66.657	66.736	66.815	66.894	66.974	67.053	67.133

(Contd.)

Table B-9 (Contd.)

%	0	1	2	3	4	5	6	7	8	9
85	67.214	67.294	67.374	67.455	67.536	67.617	67.699	67.781	67.863	67.945
86	68.027	68.110	68.193	68.276	68.359	68.443	68.527	68.611	68.696	68.781
87	68.866	68.951	69.037	69.122	69.209	69.295	69.382	69.469	69.556	69.644
88	69.732	69.820	69.909	69.998	70.087	70.177	70.267	70.357	70.448	70.539
89	70.630	70.722	70.814	70.907	70.999	71.093	71.186	71.280	71.375	71.470
90	71.565	71.661	71.757	71.853	71.950	72.048	72.146	72.244	72.343	72.442
91	72.542	72.643	72.744	72.845	72.947	73.049	73.152	73.256	73.360	73.465
92	73.570	73.676	73.782	73.890	73.997	74.106	74.215	74.325	74.435	74.546
93	74.658	74.771	74.884	74.998	75.114	75.229	75.346	75.463	75.582	75.701
94	75.821	75.942	76.064	76.187	76.311	76.437	76.563	76.690	76.819	76.948
95	77.079	77.211	77.344	77.479	77.615	77.753	77.892	78.032	78.174	78.318
96	78.463	78.610	78.759	78.910	79.062	79.217	79.374	79.534	79.695	79.859
97	80.026	80.195	80.367	80.542	80.721	80.902	81.088	81.277	81.470	81.668
98	81.870	82.077	82.290	82.508	82.733	82.965	83.205	83.453	83.711	83.980
99	84.261	84.556	84.868	85.200	85.557	85.945	86.374	86.860	87.437	88.188
100	90.000									

Source: Table worked out on scientific calculator by the author.

Index

491

Reader's Notes